CW00456862

Engineering Applications of Electromagnetic Theory

Engineering Applications of Electromagnetic Theory

Samuel Y. Liao
Professor of Electrical Engineering
California State University, Fresno

WEST PUBLISHING COMPANY
St. Paul New York San Francisco Los Angeles

Design, Editorial, and
Production Services: Quadrata, Inc.
Art: Scientific Illustrators
Composition: Polyglot Pte Ltd.

 By WEST PUBLISHING COMPANY
50 W. Kellogg Boulevard
P.O. Box 64526
St. Paul, MN 55164-1003

Printed in the United States of America

Library of Congress Cataloging-in-Publication Data

Liao, Samuel Y.
Engineering applications of electromagnetic theory/Samuel Y. Liao.
p. cm.
ISBN 0-314-60175-9
1. Microwave transmission lines. 2. Wave guides. 3. Antennas
(Electronics) 4. Electromagnetic theory. I. Title.
TK7876.L4765 1988
621.3—dc 19

For their valuable collective contributions, I dedicate this book to my wife, Lucia Hsiao-Chuang Lee, and my children, Grace in bioengineering, Kathy in electrical engineering, Gary in electronics engineering, and Jeannie in teacher education.

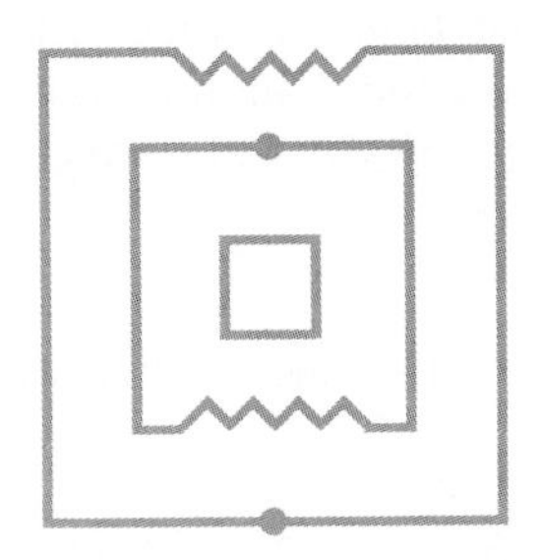

Contents

Preface xiii

Chapter 0 Introduction 1

0-1 Electromagnetic Energy Transmission 1
0-2 Frequency 3
0-3 Electromagnetic Wave Equations 6
0-4 Poynting Theory 10
0-5 Plane-Wave Propagation in Media 14
Problems 32

PART ONE TRANSMISSION LINES 35

Chapter 1 Transmission-Line Equations 39

1-0 Introduction 39
1-1 Types of Transmission-Line Equations 40
1-2 Solutions of Transmission-Line Equations 44

1-3 Characteristic Impedance, Propagation Constant, and Line Impedance 53
1-4 Reflection and Transmission Coefficients 65
1-5 Standing Wave and Standing-Wave Ratio 72
1-6 Coaxial Lines and Coaxial Connectors 82
Problems 85

Chapter 2 Transmission-Line Matching Techniques 89

2-0 Introduction 89
2-1 Smith Chart and Impedance Matching 89
2-2 Single-Stub and Double-Stub Matching 107
2-3 Series-Stub and Other Matching Techniques 118
2-4 N-Junction Matching 124
2-5 VSWR Measurement Techniques 129
References 129
Problems 129

Chapter 3 Striplines 135

3-0 Introduction 135
3-1 Microstrip Lines 135
3-2 Parallel Striplines 149
3-3 Coplanar Striplines 151
3-4 Shielded Striplines 152
References 154
Problems 155

Chapter 4 Digital Transmission Lines 159

4-0 Introduction 159
4-1 Phase Velocity and Group Velocity 159
4-2 Distortion Effects 162
4-3 Skin Effect 166
4-4 Wave Responses 167
4-5 Superconducting Transmission Line 172
4-6 Optical Fiber Transmission Line 176
Reference 176
Problems 176

PART TWO WAVEGUIDES 179

Chapter 5 Rectangular Waveguides 181

5-0 Introduction 181
5-1 Wave Equations in Rectangular Coordinates 182

5-2 TM Modes in Rectangular Waveguides 185
5-3 TE Modes in Rectangular Waveguides 189
5-4 Power Transmission and Power Losses in Rectangular Waveguides 194
5-5 Excitations of Modes and Mode Patterns in Rectangular Waveguides 199
5-6 Characteristics of Standard Rectangular Waveguides 202
5-7 Rectangular-Cavity Resonators and Quality Factor Q 203
Problems 207

Chapter 6 Circular Waveguides 211

6-0 Introduction 211
6-1 Wave Equations in Cylindrical Coordinates 211
6-2 TE Modes in Circular Waveguides 214
6-3 TM Modes in Circular Waveguides 220
6-4 TEM Modes in Circular Waveguides 224
6-5 Power Transmission and Power Losses in Circular Waveguides 226
6-6 Excitation of Modes and Mode Patterns in Circular Waveguides 227
6-7 Characteristics of Standard Circular Waveguides 230
6-8 Circular-Cavity Resonator 231
Problems 233

Chapter 7 Optical-Fiber Waveguides 235

7-0 Introduction 235
7-1 Optical Fibers 235
7-2 Operational Mechanisms of Optical Fibers 241
7-3 Step-Index Fibers 251
7-4 Graded-Index Fiber 254
7-5 Optical-Fiber Communication Systems 258
References 262
Suggested Readings 263
Problems 263

Chapter 8 Dielectric Planar Waveguides 267

8-0 Introduction 267
8-1 Parallel-Plate Waveguides 267
8-2 Dielectric-Slab Waveguides 272
8-3 Coplanar Waveguides 281
8-4 Thin Film-on-Conductor Waveguides 285
8-5 Thin Film-on-Dielectric Waveguides 288
References 290
Problems 290

PART THREE ANTENNAS 293

Chapter 9 Antenna Parameters and Characteristics 295

9-0 Introduction 295
9-1 Field Equations for the Short-Wire Antenna 295
9-2 Near Field and Far Field 298
9-3 Power Pattern, Field Pattern, and Ground Effect 301
9-4 Antenna Beamwidth and Bandwidth 312
9-5 Antenna Gain, Directivity, and Efficiency 313
9-6 Maximum Power Transfer and Effective Aperture 316
9-7 Polarization 321
9-8 Reciprocity Theorem 325
9-9 Antenna Noise Temperature and Signal-to-Noise Ratio 326
9-10 Antenna Impedance and Matching Techniques 329
Problems 336

Chapter 10 Dipole Antennas and Slot Antennas 339

10-0 Introduction 339
10-1 Dipole Antennas 339
10-2 Monopole Antennas 352
10-3 Slot Antennas 353
Problems 362

Chapter 11 Broadband and Array Antennas 365

11-0 Introduction 365
11-1 Log-Periodic Antennas 365
11-2 Phased-Array Antennas 375
11-3 Yagi–Uda Antennas 381
11-4 Antenna Measurement Techniques 383
References 393
Problems 393

Chapter 12 Electromagnetic Energy Transmission System 395

12-0 Introduction 395
12-1 Noise Figure and System Noise Temperature 395
12-2 Energy Transmission Analysis 399
12-3 Electric-Field Measurements 403
12-4 Electric-Field Computations 406
References 411
Problems 411

APPENDIXES

Appendix A Equations for Transmission Lines, Waveguides, and Antennas 415
Appendix B Hyperbolic Functions 417
Appendix C Constants of Materials 421
Appendix D Characteristics of Transmission Lines 423
Appendix E Characteristic Impedances of Common Transmission Lines 425
Appendix F First-Order Bessel-Function Values 427
Appendix G Even-Mode and Odd-Mode Characteristic Impedances for Coupled Microstrip 429
Appendix H Values of Complete Elliptic Integrals of the First Kind 435
Appendix I Television (TV) Channel Frequencies 437
Appendix J Wire Data 441
Appendix K Hankel Functions 443
Appendix L Commercial Lasers and LED Sources 445

Bibliography 447

Index 451

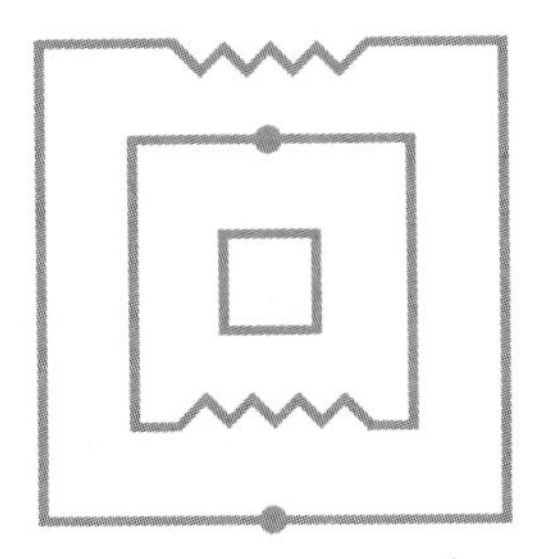

Preface

This book is intended to serve primarily as a text for courses on engineering applications of electromagnetic theory at the senior or beginning graduate level in electrical engineering. The contents of the book grew out of lecture notes that I used in a one-semester course for several years. I assume that students have had previous courses in electrical circuit analysis and electromagnetic theory, including Maxwell's equations. Because the book is largely self-contained, it can also be used as a textbook for physical science students and as a reference book by electronics engineers working in the areas of electromagnetic energy transmission and measurements.

In most universities and four-year colleges, two basic courses in electromagnetic theory and applications are required for all electrical engineering students in the junior or senior year. The first deals with electromagnetic field theory and the second addresses the engineering applications of field theory to transmission lines, waveguides, and antennas. Many good textbooks present either the two courses combined or the theory course alone; only a couple of texts cover the engineering applications course, and they are out of print. This book is intended to serve as a text for the applications course only.

In accordance with the traditional subjects of the second course on electromagnetic theory, this book contains three main parts: transmission lines, waveguides, and antennas. Much new material was incorporated into each part in order to make the book as current as possible.

Part One contains Chapters 1–4 and deals with transmission lines:

Chapter 1 describes transmission lines.

Chapter 2 presents transmission-line matching techniques, such as single-stub matching and double-stub matching.

Chapter 3 covers striplines, such as microstrip lines, parallel striplines, and coplanar striplines.

Chapter 4 discusses digital transmission lines, such as pulse digital lines, superconducting lines, and optical-fiber lines.

Part Two contains Chapters 5–8 and deals with waveguides:

Chapter 5 discusses TE and TM modes in rectangular waveguides and waveguide characteristics.

Chapter 6 analyzes TE and TM modes in circular waveguides and waveguide characteristics.

Chapter 7 investigates optical-fiber waveguides, such as step-index fibers, graded-index fiber, and resonators.

Chapter 8 describes dielectric planar waveguides, such as parallel-plate, dielectric-slab, and coplanar waveguides.

Part Three contains Chapters 9–12 and deals with antennas:

Chapter 9 discusses antenna parameters and characteristics, such as antenna gain, bandwidth, impedance matching, and so on.

Chapter 10 presents dipole antennas and slot antennas.

Chapter 11 covers broadband antennas and array antennas, such as log-periodic antennas, phased-array antennas, Yagi–Uda antennas, and antenna-measurement techniques.

Chapter 12 analyzes an electromagnetic energy transmission system, including antenna temperature, electric-field measurements, and computations.

Instructors have choices in the selection or order of topics to fit either a one-semester or a one-quarter course. Most example problems and field patterns are solved both by conventional calculations and computer methods. The computer solutions are intended to help the student to write computer programs for solving transmission-line problems with complex quantities and hyperbolic functions, determining the modes in waveguides, and plotting electric-field patterns or radiation-power patterns of various antennas. Problems at the end of each chapter are intended to further the student's understanding of the subjects discussed in that chapter. Instructors may obtain a solutions manual from the publisher.

I would like to thank the following reviewers for their many helpful suggestions:

Dr. Ken James, *California State University, Long Beach*; Dr. J. B. Knorr, *Naval Postgraduate School*; Dr. Paul E. Mayes, *University of Illinois at Urbana–Champaign*; Dr. Harold Mott, *University of Alabama*; Dr. Keith D. Paulsen, *University of Arizona*; Dr. Farhad Radpour, *University of Oklahoma*; Dr. Sembian Rengarajan, *California State University, North Ridge*; Dr. David Stephenson, *Iowa State University*; and Dr. V. K. Tripathi, *Oregon State University*.

I would also like to express my appreciation to Michael Slaughter, acquiring editor at West Publishing Company, for his constant encouragement, and Geri Davis and Martha Morong, of Quadrata, Inc., for their skillful design and editorial work.

S. Y. L.

Engineering Applications of Electromagnetic Theory

Electromagnetic energy transmission systems have been used increasingly in such diverse applications as power transmission, television distribution, long-distance telephone transmission, computer links, radio astronomy, space navigation, radar systems, medical equipment, missile systems, satellite communications, and military command and control.

0-2 FREQUENCY

The frequency limitations of transmission lines, waveguides, and antennas are different. We can analyze lines using distributed circuit theory. However, we must use electromagnetic field theory in order to analyze waveguides and antennas.

0-2-1 Frequency Ranges

It seems appropriate to list here the frequency bands, since we refer to frequency ranges throughout this book. In the electronics industry and academic institutions, the Institute of Electrical and Electronics Engineers, Inc. (IEEE) frequency bands, as shown in Table 0-2-1, are commonly used.

Table 0-2-1 IEEE frequency bands

Band Number	Designation	Frequency	Wavelength
2	ELF (Extreme low frequency)	30–300 Hz	10–1 Mm
3	VF (Voice frequency)	300–3000 Hz	1–0.1 Mm
4	VLF (Very low frequency)	3–30 kHz	100–10 km
5	LF (Low frequency)	30–300 kHz	10–1 km
6	MF (Medium frequency)	300–3000 kHz	1–0.1 km
7	HF (High frequency)	3–30 MHz	100–10 m
8	VHF (Very high frequency)	30–300 MHz	10–1 m
9	UHF (Ultrahigh frequency)	300–3000 MHz	100–10 cm
10	SHF (Superhigh frequency)	3–30 GHz	10–1 cm
11	EHF (Extreme high frequency)	30–300 GHz	1–0.1 cm
12	Decimillimeter	300–3000 GHz	1–0.1 mm
	P band	0.23–1 GHz	130–30 cm
	L band	1–2 GHz	30–15 cm
	S band	2–4 GHz	15–7.5 cm
	C band	4–8 GHz	7.5–3.75 cm
	X band	8–12.5 GHz	3.75–2.4 cm
	Ku band	12.5–18 GHz	2.4–1.67 cm
	K band	18–26.5 GHz	1.67–1.13 cm
	Ka band	26.5–40 GHz	1.13–0.75 cm
	Millimeter wave	40–300 GHz	7.5–1 mm
	Submillimeter wave	300–3000 GHz	1–0.1 mm

The frequency designations based on World War II radar security considerations had never been officially sanctioned by any industrial, professional, or government organization until 1969. In August 1969, the U.S. Department of Defense (DoD), Office of the Joint Chiefs of Staff directed all the armed services to use the frequency bands shown in Table 0-2-2. On May 24, 1970, the DoD adopted yet another frequency-band designation, as shown in Table 0-2-3. These three band designations are compared in Table 0-2-4.

Table 0-2-2 Old DoD frequency bands

Designation	Frequency Range, GHz
P-band	0.225– 0.390
L-band	0.390– 1.550
S-band	1.550– 3.900
C-band	3.900– 6.200
X-band	6.200– 10.900
K-band	10.900– 36.000
Q-band	36.000– 46.000
V-band	46.000– 56.000
W-band	56.000–100.000

Table 0-2-3 DoD ECM frequency bands

Designation	Frequency Range, GHz
A-band	0.100– 0.250
B-band	0.250– 0.500
C-band	0.500– 1.000
D-band	1.000– 2.000
E-band	2.000– 3.000
F-band	3.000– 4.000
G-band	4.000– 6.000
H-band	6.000– 8.000
I-band	8.000– 10.000
J-band	10.000– 20.000
K-band	20.000– 40.000
L-band	40.000– 60.000
M-band	60.000–100.000

Table 0-2-4 Comparison of IEEE bands, old DoD bands, and new DoD ECM bands

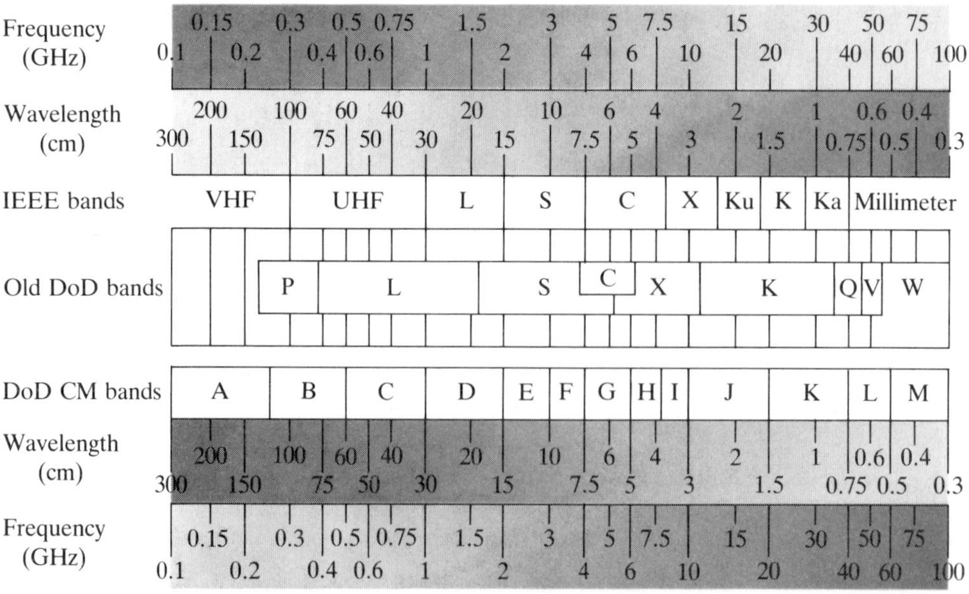

Frequency (GHz)	0.1 0.15 0.2 0.3 0.4 0.5 0.6 0.75 1 1.5 2 3 4 5 6 7.5 10 15 20 30 40 50 60 75 100
Wavelength (cm)	300 200 150 100 75 60 50 40 30 20 15 10 7.5 6 5 4 3 2 1.5 1 0.75 0.6 0.5 0.4 0.3
IEEE bands	VHF UHF L S C X Ku K Ka Millimeter
Old DoD bands	P L S C X K Q V W
DoD CM bands	A B C D E F G H I J K L M
Wavelength (cm)	300 200 150 100 75 60 50 40 30 20 15 10 7.5 6 5 4 3 2 1.5 1 0.75 0.6 0.5 0.4 0.3
Frequency (GHz)	0.1 0.15 0.2 0.3 0.4 0.5 0.6 0.75 1 1.5 2 3 4 5 6 7.5 10 15 20 30 40 50 60 75 100

0-2-2 MKS Units and Physical Constants

We use the rationalized meter–kilogram–second (MKS) system of units—the International System of Units—throughout this book unless otherwise indicated. The most commonly used MKS units are listed in Table 0-2-5. The physical constants commonly used in this book are listed in Table 0-2-6.

Table 0-2-5 MKS units

Quantity	Unit	Symbol
Angstrom	10^{-10} meter	Å
Capacitance	Farad = Coulombs per volt	F
Charge	Coulomb: Ampere-seconds	Q
Conductance	Mho or Siemen	℧
Current	Ampere = $\frac{\text{Coulombs}}{\text{Second}}$	A
Frequency	Cycles per second	Hz
Energy	Joule = Watt-second	J
Field (electric)	Volts per meter	E
Field (magnetic)	Amperes per meter	H
Flux linkage	Weber = volt-seconds	ψ
Inductance	Henry = $\frac{\text{V-s}}{\text{A}}$	H
Length	Meter	m
Micron	10^{-6} meter	μm
Power	Watt = Joules per second	W
Resistance	Ohm	Ω
Time	Second	s
Velocity	Meters per second	v
Voltage	Volt	V

Table 0-2-6 Physical constants

Constant	Symbol	Value
Boltzman constant	k	1.381×10^{-23} J/°K
Electron volt	eV	1.602×10^{-19} J
Electronic charge	q	1.602×10^{-19} C
Electronic mass	m	9.109×10^{-31} Kg
Ratio of charge to mass of an electron	e/m	1.759×10^{11} C/Kg
Permeability of free space	μ_0	$4\pi \times 10^{-7}$ H/m
Permittivity of free space	ε_0	8.854×10^{-12} F/m
Planck's constant	h	6.626×10^{-34} J-s
Velocity of light in vacuum	c	2.998×10^{8} m/s

0-3 ELECTROMAGNETIC WAVE EQUATIONS

0-3-1 Introduction

The principles underlying the transmission of electromagnetic energy are based on the relationships between electricity and magnetism. A changing magnetic field will induce an electric field, and a changing electric field will induce a magnetic field. Also, the induced fields are not confined but ordinarily extend outward into space. The electromagnetic wave causes energy to be interchanged between the magnetic and electric fields in the direction of wave propagation.

At some distance from the source, electric and magnetic fields are perpendicular to each other, and both are normal to the direction of wave propagation. This type of wave is known as the transverse electromagnetic (TEM) wave and generally exists in transmission-line systems. If the electric field is transverse to the direction of wave propagation, the wave is called a TE wave. Similarly, if only the magnetic field is transverse to the direction of wave propagation, the wave is said to be a TM wave. Both TE and TM waves are encountered in either a hollow-pipe or open-wire system. The purposes of this section are to: (1) present the wave equations in the time and frequency domains; and (2) show the plane-wave propagation in free space and lossy media.

0-3-2 Wave Equations in the Time Domain

The fundamental laws of the transmission of electromagnetic energy can be derived, basically, from Maxwell's equations. Maxwell's equations in the time domain are

$$\nabla \times \mathbf{E} = -\frac{\partial \mathbf{B}}{\partial t} \tag{0-3-1}$$

$$\nabla \times \mathbf{H} = \mathbf{J} + \frac{\partial \mathbf{D}}{\partial t} \tag{0-3-2}$$

$$\nabla \cdot \mathbf{D} = \rho_v \tag{0-3-3}$$

and

$$\nabla \cdot \mathbf{B} = 0 \tag{0-3-4}$$

The preceding boldface Roman letters indicate either vector quantities or complex quantities. We define these field variables as follows:

∇ is the vector operator in three coordinate systems.

E is called the electric field strength or intensity and is measured in volts per meter (V/m).

H is called the magnetic field strength or intensity and is measured in amperes per meter (A/m).

D is called the electric flux density and is measured in coulombs per square meter (C/m^2).

B is called the magnetic flux density and is measured in webers per square meter (Wb/m^2), or tesla (1 tesla $= 1\ Wb/m^2 = 10^4$ gauss $= 3 \times 10^{-6}$ ESU).

J is called the electric current density and is measured in amperes per square meter (A/m^2).

ρ_v is called the electric charge density and is measured in coulombs per cubic meter (c/m^3).

The electric current density includes two components:

$$\mathbf{J} = \mathbf{J}_c + \mathbf{J}_o \tag{0-3-5}$$

where

$\mathbf{J}_c = \sigma\mathbf{E}$ and is called the conduction current density; and

$\mathbf{J}_o$ is called the impressed current density, which is independent of the field.

The first Maxwell equation, Eq. (0-3-1), is essentially Faraday's law of induction, which states that a changing magnetic flux density induces a field in a direction opposite to the direction of the inducing magnetic field. The second Maxwell equation, Eq. (0-3-2), is essentially Ampere's circuital law, including displacement current, which states that the curl of a magnetic field intensity is equal to the conduction and displacement current density. Actually, Ampere's circuital law states that the line integral of the magnetic field intensity around any closed path is exactly equal to the current enclosed by that path, or using Stokes's theorem:

$$\oint_l \mathbf{H} \cdot d\mathscr{L} = \int_s (\nabla \times \mathbf{H}) \cdot d\mathbf{S} = \int_s \left(\mathbf{J} + \frac{\partial \mathbf{D}}{\partial t}\right) \cdot d\mathbf{S} = I + I_d \tag{0-3-6}$$

where I is the electric current; and I_d is the displacement current.

The third Maxwell equation, Eq. (0-3-3), states that the magnetic flux density does not have a divergence, because the magnetic lines are always closed. The fourth Maxwell equation, Eq. (0-3-4), is essentially Gauss's law, which states that the divergence of an electric flux density is equal to its volume charge density. Actually, Gauss's law states that the electric flux density passing through any closed surface is equal to the total charge enclosed by that surface. We can show this result by applying Gauss's theorem:

$$\oint_s \mathbf{D} \cdot d\mathbf{S} = \int_v (\nabla \cdot \mathbf{D})\, dv = \int_v \rho_v\, dv = Q \tag{0-3-7}$$

In addition to Maxwell's four equations, we need the characteristics of the medium in which the fields exist in order to specify the flux in terms of the fields in a specific medium. These constitutive relationships for linear, isotropic, and homogeneous media are given by

$$\mathbf{D} = \varepsilon\mathbf{E} \tag{0-3-8}$$

$$\mathbf{B} = \mu\mathbf{H} \tag{0-3-9}$$

and

$$\mathbf{J}_c = \sigma\mathbf{E} \tag{0-3-10}$$

where

$$\varepsilon = \varepsilon_r\varepsilon_0 \tag{0-3-11}$$

$$\mu = \mu_r\mu_0 \tag{0-3-12}$$

and

ε = permittivity or capacitivity of the medium, expressed in farads per meter (F/m);

ε_r = relative permittivity, or dielectric constant (dimensionless);

$\varepsilon_0 = 8.845 \times 10^{-12} = 1/(36\pi) \times 10^{-9}$ F/m (permittivity of free space or vacuum);

μ = permeability or inductivity of the medium, expressed in henries per meter (H/m);

μ_r = relative permeability, or magnetic constant (dimensionless);

$\mu_0 = 4\pi \times 10^{-7}$ H/m (permeability of free space or vacuum); and

σ = conductivity of the medium, expressed in Siemens (℧/m).

We can derive the wave equations in the time domain by solving simultaneously the curl equations of the electric and magnetic fields to obtain two explicit equations: one containing only the electric field and the other containing only the magnetic field. For example, substituting Eq. (0-3-9) into Eq. (0-3-1) and Eqs. (0-3-8) and (0-3-10) into Eq. (0-3-2) results in

$$\nabla \times \mathbf{E} = -\mu \frac{\partial \mathbf{H}}{\partial t} \tag{0-3-13}$$

and

$$\nabla \times \mathbf{H} = \sigma \mathbf{E} + \varepsilon \frac{\partial \mathbf{E}}{\partial t} \tag{0-3-14}$$

Taking the curl of Eq. (0-3-13) on both sides yields

$$\nabla \times \nabla \times \mathbf{E} = -\mu \nabla \times \frac{\partial \mathbf{H}}{\partial t} = -\mu \frac{\partial}{\partial t}(\nabla \times \mathbf{H}) \tag{0-3-15}$$

Substituting Eq. (0-3-14) for the right-hand side of Eq. (0-3-15) yields

$$\nabla \times \nabla \times \mathbf{E} = -\mu \frac{\partial}{\partial t}\left(\sigma \mathbf{E} + \varepsilon \frac{\partial \mathbf{E}}{\partial t}\right) \tag{0-3-16}$$

The vector identity for the curl of the curl of a vector quantity **E** is given by

$$\nabla \times \nabla \times \mathbf{E} = -\nabla^2 \mathbf{E} + \nabla(\nabla \cdot \mathbf{E}) \tag{0-3-17}$$

Substituting Eq. (0-3-17) for the left-hand side of Eq. (0-3-15), with zero divergence of the electric field, yields

$$\nabla^2 \mathbf{E} = \mu \left[\sigma \frac{\partial \mathbf{E}}{\partial t} + \varepsilon \frac{\partial^2 \mathbf{E}}{\partial t^2}\right] \tag{0-3-18}$$

In free space, the space-charge density is zero, and in a perfect conductor, time-varying or static fields do not exist. Hence,

$$\nabla \cdot \mathbf{E} = 0 \tag{0-3-18a}$$

Equation (0-3-18) is the electric-wave equation in the time domain. Similarly, the magnetic wave in the time domain is given by

$$\nabla^2 \mathbf{H} = \mu \left[\sigma \frac{\partial \mathbf{H}}{\partial t} + \varepsilon \frac{\partial^2 \mathbf{H}}{\partial t^2}\right] \tag{0-3-19}$$

The electric and magnetic wave equations in the time domain, Eqs. (0-3-18) and (0-3-19), are the equivalent of the transmission-line equations in the time domain, Eqs. (1-2-9) and (1-2-10), presented in Chapter 1.

Note that the "double del" or "del squared" is a scalar operator. That is,

$$\nabla \cdot \nabla = \nabla^2 \tag{0-3-20}$$

which is a second-order partial differential operator in three different coordinate systems. In rectangular (cartesian) coordinates

$$\nabla^2 = \frac{\partial^2}{\partial x^2} + \frac{\partial^2}{\partial y^2} + \frac{\partial^2}{\partial z^2} \tag{0-3-21}$$

in cylindrical coordinates

$$\nabla^2 = \frac{1}{r}\frac{\partial}{\partial r}\left(r\frac{\partial}{\partial r}\right) + \frac{1}{r^2}\frac{\partial^2}{\partial \phi^2} + \frac{\partial^2}{\partial z^2} \tag{0-3-21a}$$

and in spherical coordinates

$$\nabla^2 = \frac{1}{r^2}\frac{\partial}{\partial r}\left(r^2\frac{\partial}{\partial r}\right) + \frac{1}{r^2 \sin\theta}\frac{\partial}{\partial \theta}\left(\sin\theta\frac{\partial}{\partial \theta}\right) + \frac{1}{r^2 \sin^2\theta}\frac{\partial^2}{\partial \phi^2} \tag{0-3-21b}$$

In free space, the two wave equations in the time domain are, respectively, reduced to

$$\nabla^2 \mathbf{E} = \mu_0 \varepsilon_0 \frac{\partial^2 \mathbf{E}}{\partial t^2} \tag{0-3-22a}$$

and

$$\nabla^2 \mathbf{H} = \mu_0 \varepsilon_0 \frac{\partial^2 \mathbf{H}}{\partial t^2} \tag{0-3-22b}$$

In a lossless, nonconducting dielectric, the two wave equations are

$$\nabla^2 \mathbf{E} = \mu \varepsilon \frac{\partial^2 \mathbf{E}}{\partial t^2} \tag{0-3-22c}$$

and

$$\nabla^2 \mathbf{H} = \mu \varepsilon \frac{\partial^2 \mathbf{H}}{\partial t^2} \tag{0-3-22d}$$

EXAMPLE 0-3-1 Time-Domain Wave Equations

If the electric field equation is given by

$$\mathbf{E}_x = E_0 \cos(\omega t - \beta z)\,\mathbf{x}$$

determine the (a) electric and (b) magnetic wave equations in the time domain in free space.

Solutions:

a) The electric-wave equation is

$$\frac{\partial^2 \mathbf{E}_x}{\partial t^2} = -\omega^2 E_0 \cos(\omega t - \beta z)\,\mathbf{x}$$

$$\frac{\partial^2 \mathbf{E}_x}{\partial z^2} = -\beta^2 E_0 \cos(\omega t - \beta z)\,\mathbf{x}$$

$$\nabla^2 \mathbf{E} = -\omega^2 \mu_0 \varepsilon_0 E_0 \cos(\omega t - \beta z)\,\mathbf{x}$$

b) The magnetic-wave equation is

$$\nabla^2 \mathbf{H} = -\frac{\omega^2}{377}\mu_0 \varepsilon_0 E_0 \cos(\omega t - \beta z)\,\mathbf{y}$$

0-3-3 Wave Equations in the Frequency Domain

Mathematically, the differentiation ∂/dt of a sinusoidal time function in the form $e^{j\omega t}$ can be replaced by $j\omega$. Consider a sinusoidal time function to be

$$E = E_0 e^{j\omega t}$$

Then the result of the differentiation ∂/dt of the function is

$$\frac{\partial E}{\partial t} = j\omega E_0 e^{j\omega t} = j\omega E$$

The Maxwell equations in the frequency domain are expressed as

$$\nabla \times \mathbf{E} = -j\omega\mu\mathbf{H} \tag{0-3-23}$$

$$\nabla \times \mathbf{H} = (\sigma + j\omega\varepsilon)\mathbf{E} \tag{0-3-24}$$

$$\nabla \cdot \mathbf{D} = \rho_v \tag{0-3-25}$$

and

$$\nabla \cdot \mathbf{B} = 0 \tag{0-3-26}$$

When Eqs. (0-3-23) and (0-3-24) are solved simultaneously, they yield the electric- and magnetic-wave equations in the frequency domain:

$$\nabla^2\mathbf{E} = \gamma^2\mathbf{E} \tag{0-3-27}$$

and

$$\nabla^2\mathbf{H} = \gamma^2\mathbf{H} \tag{0-3-28}$$

where

$\gamma = \sqrt{j\omega\mu(\sigma + j\omega\varepsilon)} = \alpha + j\beta$ and is called the intrinsic propagation constant of a medium;

α = attenuation constant with units of Nepers per meter (Np/m);

$\beta = 2\pi/\lambda$ is the phase constant with units of radians per meter (rad/m); and

λ = wavelength in meters (m).

The electric- and magnetic-wave equations in the frequency domain, Eqs. (0-3-27) and (0-3-28), are the equivalent of the transmission line equations in the frequency domain, Eqs. (1-2-6) and (1-2-7) presented in Chapter 1.

0-4 POYNTING THEORY

We are interested in the rate at which electromagnetic energy is transmitted through free space or any medium, stored in electric and magnetic fields, and dissipated as heat. In theory, the time average of any two complex field vectors is defined as the product of those two vectors. However, in actual computation, the time-average power is equal to the real part of the product of one complex vector multiplied by the complex conjugate of the other vector. Hence, the time average of an instantaneous Poynting vector in steady state is given by

$$\langle \mathbf{p} \rangle = \langle \mathbf{E} \times \mathbf{H} \rangle = \tfrac{1}{2}\mathrm{Re}(\mathbf{E} \times \mathbf{H}^*) \tag{0-4-1}$$

where the notation $\langle\ \rangle$ stands for time average, and the factor of $\frac{1}{2}$ appears in the equation of complex power when peak values are used for the complex quantities **E** and **H**. The Re represents the real part of the complex power, and the star (*) indicates complex conjugate. Thus a complex Poynting vector is defined as

$$\mathbf{P} = \tfrac{1}{2}(\mathbf{E} \times \mathbf{H}^*) \tag{0-4-2}$$

0-4-1 Poynting Theorem in the Frequency Domain

Maxwell's equations in the frequency domain for the electric and magnetic fields are

$$\nabla \times \mathbf{E} = -j\omega\mu\mathbf{H} \tag{0-4-3}$$

and

$$\nabla \times \mathbf{H} = \mathbf{J} + j\omega\varepsilon\mathbf{E} \tag{0-4-4}$$

Dot multiplication of Eq. (0-4-3) by **H** and of the conjugate of Eq. (0-4-4) by **E** yields

$$(\nabla \times \mathbf{E}) \cdot \mathbf{H}^* = -j\omega\mu\mathbf{H} \cdot \mathbf{H}^* \tag{0-4-5}$$

and

$$(\nabla \times \mathbf{H}^*) \cdot \mathbf{E} = (\mathbf{J}^* - j\omega\varepsilon\mathbf{E}^*) \cdot \mathbf{E} \tag{0-4-6}$$

Subtracting Eq. (0-4-5) from Eq. (0-4-6) gives

$$\mathbf{E} \cdot (\nabla \times \mathbf{H}^*) - \mathbf{H}^* \cdot (\nabla \times \mathbf{E}) = \mathbf{E} \cdot \mathbf{J}^* - j\omega(\varepsilon|E|^2 - \mu|H|^2) \tag{0-4-7}$$

where $\mathbf{E} \cdot \mathbf{E}^*$ is replaced by $|E|^2$; and $\mathbf{H} \cdot \mathbf{H}^*$ is replaced by $|H|^2$.

The left-hand side of Eq. (0-4-7) equals $-\nabla \cdot (\mathbf{E} \times \mathbf{H}^*)$ by the vector identity. Hence,

$$\nabla \cdot (\mathbf{E} \times \mathbf{H}^*) = -\mathbf{E} \cdot \mathbf{J}^* + j\omega(\varepsilon|E|^2 - \mu|H|^2) \tag{0-4-8}$$

Substituting Eqs. (0-3-5) and (0-4-2) into Eq. (0-4-8) yields

$$-\tfrac{1}{2}\mathbf{E} \cdot \mathbf{J}_0^* = \tfrac{1}{2}\sigma\mathbf{E} \cdot \mathbf{E}^* + j\omega(\tfrac{1}{2}\mu\mathbf{H} \cdot \mathbf{H}^* - \tfrac{1}{2}\varepsilon\mathbf{E} \cdot \mathbf{E}^*) + \nabla \cdot \mathbf{p} \tag{0-4-9}$$

Integrating Eq. (0-4-9) over a volume and applying Gauss's theorem to the last term on the right-hand side of Eq. (0-4-9) give

$$-\int_v \tfrac{1}{2}(\mathbf{E} \cdot \mathbf{J}_0^*)\,dv = \int_v \tfrac{1}{2}\sigma|E|^2\,dv + j2\omega\int_v (w_m - w_e)\,dv + \oint_s \mathbf{p} \cdot d\mathbf{s} \tag{0-4-10}$$

where

$\frac{1}{2}\sigma|E|^2 = \sigma\langle|E|^2\rangle$ and is the time-average dissipated power;

$\frac{1}{4}\mu\mathbf{H} \cdot \mathbf{H}^* = \frac{1}{2}\mu\langle|H|^2\rangle = w_m$ and is the time-average magnetic stored energy;

$\frac{1}{4}\varepsilon\mathbf{E} \cdot \mathbf{E}^* = \frac{1}{2}\varepsilon\langle|E|^2\rangle = w_e$ and is the time-average electric stored energy; and

$-\frac{1}{2}\mathbf{E} \cdot \mathbf{J}_0^*$ is the complex power impressed by the source J_0 into the field.

Equation (0-4-10) is known as the complex Poynting theorem, or the Poynting theorem in the frequency domain. If we let

$P_{in} = -\int_v \frac{1}{2}(\mathbf{E} \cdot \mathbf{J}_0^*)\,dv$ be the total complex power supplied by a source within a region,

$\langle P_d \rangle = \int_v \frac{1}{2}\sigma|E|^2\, dv$ be the time-average power dissipated as heat inside the region,

$\langle W_m - W_e \rangle = \int_v [w_m - w_e]\, dv$ be the difference between time-average magnetic and electric energies stored within the region, and

$P_{tr} = \oint_s \mathbf{p} \cdot ds$ be the complex power transmitted from the region,

we can now simplify the complex Poynting theorem shown in Eq. (0-4-10) to

$$P_{in} = \langle P_d \rangle + j2\omega[\langle W_m - W_e \rangle] + P_{tr} \tag{0-4-11}$$

Equation (0-4-11) states that the total complex power fed into a volume equals the algebraic sum of (1) the active power dissipated as heat; (2) the reactive power proportional to the difference between the time-average magnetic and electric energies stored in the volume; and (3) the complex power transmitted across the surface enclosed by the volume.

0-4-2 Poynting Theorem in the Time Domain

Maxwell's equations in the time domain for the electric and magnetic fields are

$$\nabla \times \mathbf{E} = -\mu \frac{\partial \mathbf{H}}{\partial t} \tag{0-4-12}$$

and

$$\nabla \times \mathbf{H} = \sigma \mathbf{E} + \varepsilon \frac{\partial \mathbf{E}}{\partial t} \tag{0-4-13}$$

Dot multiplication of Eq. (0-4-12) by **H** and of Eq. (0-4-13) by **E** yields

$$\mathbf{H} \cdot (\nabla \times \mathbf{E}) = -\mu \mathbf{H} \cdot \frac{\partial \mathbf{H}}{\partial t} \tag{0-4-14}$$

and

$$\mathbf{E} \cdot (\nabla \times \mathbf{H}) = \mathbf{E} \cdot \left(\sigma \mathbf{E} + \varepsilon \frac{\partial \mathbf{E}}{\partial t} \right) \tag{0-4-15}$$

Subtracting Eq. (0-4-15) from Eq. (0-4-14) gives

$$\mathbf{H} \cdot (\nabla \times \mathbf{E}) - \mathbf{E} \cdot (\nabla \times \mathbf{H}) = -\mu \mathbf{H} \cdot \frac{\partial \mathbf{H}}{\partial t} - \mathbf{E} \cdot \left(\sigma \mathbf{E} + \varepsilon \frac{\partial \mathbf{E}}{\partial t} \right) \tag{0-4-16}$$

The left-hand side of Eq. (0-4-16) equals $\nabla \cdot (\mathbf{E} \times \mathbf{H})$ by the vector identity. Hence,

$$\nabla \cdot (\mathbf{E} \times \mathbf{H}) = -\mu \mathbf{H} \cdot \frac{\partial \mathbf{H}}{\partial t} - \mathbf{E} \cdot \left(\sigma \mathbf{E} + \varepsilon \frac{\partial \mathbf{E}}{\partial t} \right) \tag{0-4-17}$$

since

$$\mu \mathbf{H} \cdot \frac{\partial \mathbf{H}}{\partial t} = \frac{1}{2}\mu \frac{\partial (\mathbf{H} \cdot \mathbf{H})}{\partial t} = \frac{\partial}{\partial t}\left(\frac{1}{2}\mu|H|^2\right) = \frac{\partial}{\partial t}(w_m) \tag{0-4-18}$$

and

$$\varepsilon \mathbf{E} \cdot \frac{\partial \mathbf{E}}{\partial t} = \frac{1}{2}\varepsilon \frac{\partial (\mathbf{E} \cdot \mathbf{E})}{\partial t} = \frac{\partial}{\partial t}\left(\frac{1}{2}\varepsilon|E|^2\right) = \frac{\partial}{\partial t}(w_e) \tag{0-4-19}$$

where

$w_m = \frac{1}{2}\mu|H|^2$ is the magnetic stored-energy density, expressed in J/m^3; and
$w_e = \frac{1}{2}\varepsilon|E|^2$ is the electric stored-energy density, expressed in J/m^3.

Substituting Eqs. (0-4-18) and (0-4-19) into Eq. (0-4-17) then yields

$$-\nabla \cdot (\mathbf{E} \times \mathbf{H}) = \sigma|E|^2 + \frac{\partial}{\partial t}(w_e + w_m) \qquad (0\text{-}4\text{-}20)$$

where $\sigma|E|^2$ = power loss as heat to the medium, expressed in joules per unit volume.

If we let the power source be impressed into the field by $\mathbf{J}_0 \cdot E$, then from the principle of conservation of energy, Eq. (0-4-20) becomes

$$-\mathbf{J}_0 \cdot \mathbf{E} = \sigma|E|^2 + \frac{\partial}{\partial t}(w_e + w_m) + \nabla \cdot (\mathbf{E} \times \mathbf{H}) \qquad (0\text{-}4\text{-}21)$$

Applying Gauss's theorem to the last term on the right-hand side of Eq. (0-4-21) and integrating both sides of the equation over a volume, we obtain

$$-\int_v (\mathbf{J}_0 \cdot \mathbf{E})\,dv = \int_v \sigma|E|^2\,dv + \frac{\partial}{\partial t}\int_v (w_e + w_m)\,dv + \oint_s (\mathbf{E} \times \mathbf{H}) \cdot d\mathbf{s} \qquad (0\text{-}4\text{-}22)$$

Equation (0-4-22) is the Poynting theorem in the time domain, which is very useful later. Now let

$P = -\int_v (\mathbf{J}_0 \cdot \mathbf{E})\,dv$ be the total power source fed into the field;
$P_d = \int \sigma E^2\,dv$ be the total power loss as heat dissipation;
$W_e = \int w_e\,dv$ be the total electric stored energy;
$W_m = \int w_m\,dv$ be the total magnetic stored energy; and
$P_s = \oint \mathbf{P} \cdot d\mathbf{s} = \frac{1}{2}\oint (\mathbf{E} \times \mathbf{H}) \cdot d\mathbf{s}$ be the total power flow.

The Poynting vector $\mathbf{P}$ describes the electromagnetic energy transmission across a surface. The direction of the Poynting vector is the direction of power flow at that point; its magnitude is the value of the power flow per unit area across a surface normal to the direction of the vector. Finally, we can simplify Eq. (0-4-22) to

$$P = p_d + \frac{\partial}{\partial t}[W_e + W_m] + P_s \qquad (0\text{-}4\text{-}23)$$

Equation (0-4-23) is the Poynting theorem in the time domain and says that the source power P fed into the field in a volume by a current source $\mathbf{J}_0$ equals the algebraic sum of (1) the power loss due to the imperfection conductivity σ; (2) the time rate of change of energy stored in the electric and magnetic fields; and (3) the power flow across the surface enclosed by that volume.

EXAMPLE 0-4-1 Power Flow in Free Space

An electric wave propagates into free space and is expressed as

$$E_x = 10\cos(\omega t - \beta z) \qquad \text{V/m} \qquad \text{at } f = 100 \text{ MHz}$$

Determine:

a) the magnetic-wave equation;
b) the phase constant; and
c) the power density flow.

Solutions:

a) The magnetic-wave component is

$$H_y = 10/377 \cos(\omega t - \beta z) \qquad \text{A/m}$$

b) The phase constant is

$$\beta = 2\pi/\lambda = 2\pi/(3 \times 10^8/10^8)$$
$$= 2.09 \text{ rad/m}$$

c) The power flow is

$$P = \tfrac{1}{2}[10e^{j(\omega t - \beta z)} \times 10/377e^{-j(\omega t - \beta z)}]$$
$$= 0.13 \text{ W/m}$$

0-5 PLANE-WAVE PROPAGATION IN MEDIA

0-5-1 Uniform Plane Waves

A plane wave is a wave whose phase is constant over a set of planes. A uniform plane wave is a wave whose magnitude and phase are both constant. A spherical wave is a wave whose equiphase surfaces form a family of concentric spheres.

Electromagnetic waves in free space are typical uniform plane waves. The electric and magnetic fields are perpendicular to each other and to the direction of propagation of the waves. The phases of the two fields are always in time phase, and their magnitudes are always constant. The stored energies are equally divided between the two fields, and the energy flow is transmitted by the two fields in the direction of propagation. Hence, the uniform plane wave is a transverse electromagnetic (TEM) wave.

A nonuniform plane wave is a wave whose amplitude (not phase) may vary within a plane normal to the direction of propagation. Consequently, the electric and magnetic fields are no longer in time phase.

Since a uniform plane wave of electric or magnetic field does not vary in a plane normal to the direction of propagation of the wave,

$$\frac{\partial \mathbf{E}}{\partial x} = \frac{\partial \mathbf{E}}{\partial y} = 0 \quad \text{or} \quad \frac{\partial \mathbf{H}}{\partial x} = \frac{\partial \mathbf{H}}{\partial y} = 0$$

if the direction of propagation is assumed to be in the positive z direction. Based on this assumption and that of a lossless dielectric (that is, $\sigma = 0$), the wave equations, Eqs. (0-3-18) and (0-3-19), in the time domain for the electric and magnetic intensities in free space for rectangular coordinates reduce to

$$\frac{\partial^2 E_x}{\partial z^2} = \mu_0 \epsilon_0 \frac{\partial^2 E_x}{\partial t^2} \qquad \text{(Time domain)} \qquad (0\text{-}5\text{-}1)$$

and

$$\frac{\partial^2 H_y}{\partial z^2} = \mu_0 \varepsilon_0 \frac{\partial^2 H_y}{\partial t^2} \qquad \text{(Time domain)} \tag{0-5-2}$$

We arbitrarily choose the electric intensity in the x direction and the magnetic intensity in the y direction. With no loss in generality, we can assume that the electric intensity for sinusoidal variation is

$$E_x = E_0 e^{j(\omega t - \beta z)} = E_0 e^{j\omega(t - \beta z/\omega)} \tag{0-5-3}$$

We obtain the magnetic intensity in the frequency domain by inserting Eq. (0-5-3) into the curl equation, or

$$\nabla \times \mathbf{E} = -j\omega\mu_0 \mathbf{H} \qquad \text{(Frequency domain)} \tag{0-5-4}$$

For the assumed conditions, the curl equation reduces to

$$\frac{\partial E_x}{\partial z} = -j\omega\mu_0 H_y \tag{0-5-5}$$

Differentiating Eq. (0-5-3) with respect to z and substituting the result into Eq. (0-5-5) yield

$$H_y = \sqrt{\frac{\varepsilon_0}{\mu_0}} E_x \tag{0-5-6}$$

where

$\beta/\omega = \sqrt{\mu_0 \varepsilon_0} = 1/v_p$ is accounted for in the derivation;
$v_p = 1/\sqrt{\mu_0 \varepsilon_0} = 3 \times 10^8$ m/s, which is the phase velocity, or c, the velocity of light in free space;
$\alpha = 0$; and
$\beta = \omega\sqrt{\mu_0 \varepsilon_0}$, which is the phase constant.

The ratio of electric to magnetic intensities is given by

$$\frac{E_x}{H_y} = \eta_0 = \sqrt{\frac{\mu_0}{\varepsilon_0}} = 377\ \Omega \tag{0-5-7}$$

This quantity is called the intrinsic impedance of free space. When electric and magnetic fields are a long distance from their sources, the two fields are perpendicular to each other.

Figure 0-5-1 shows uniform electric and magnetic plane waves in rectangular coordinates. In general, for a uniform plane wave propagating in a lossless dielectric medium ($\sigma = 0$), the characteristics of the wave propagation become

$$\alpha = 0$$

$$\beta = \omega\sqrt{\mu_0 \varepsilon}$$

$$\eta = \sqrt{\frac{\mu_0}{\varepsilon}} = \frac{\eta_0}{\sqrt{\varepsilon_r}}$$

and

$$v_p = \frac{1}{\sqrt{\mu_0 \varepsilon}} = \frac{c}{\sqrt{\varepsilon_r}}$$

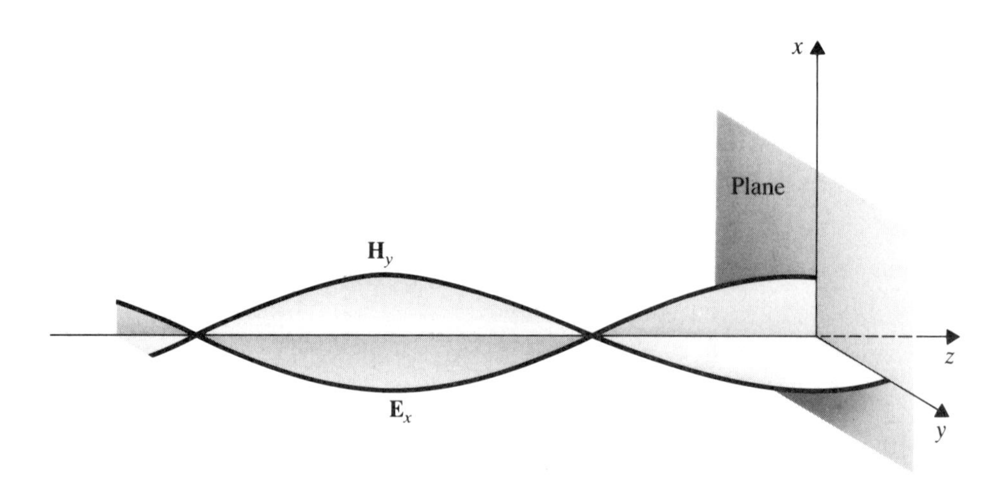

Figure 0-5-1 Uniform plane waves traveling in the z direction.

Note that all dielectrics have approximately the same permeability as free space. (See Appendix C.)

0-5-2 Boundary Conditions

Since Maxwell's equations are in the form of differential equations rather than algebraic equations, we have to apply boundary conditions to a problem if we want to obtain a specific solution. The four basic rules for boundary conditions at the surface between two different materials are:

1. The tangential components of the electric field strength are continuous across the boundary.
2. The normal components of electric flux density (or displacement density) are continuous at the boundary by an amount equal to the surface-charge density on the boundary.
3. The tangential components of magnetic field strength are discontinuous at the boundary by an amount equal to the surface-current density on the boundary.
4. The normal components of magnetic flux density are continuous across the boundary.

We can prove these four statements by applying Faraday's law, Gauss's law, Ampere's law, and $\nabla \cdot \mathbf{B} = 0$ to the boundaries of the diagrams in Fig. 0-5-2.

You can see from the diagrams that

$$\oint_{\ell} \mathbf{E} \cdot d\ell = E_{t1}\,\Delta\ell - E_t\,\Delta\ell = 0$$

$$\oint_{s} \mathbf{D} \cdot d\mathbf{s} = D_{n1}\,\Delta s - D_{n2}\,\Delta s = \rho_s\,\Delta s$$

$$\oint_{\ell} \mathbf{H} \cdot d\ell = H_{t1}\,\Delta\ell - H_{t2}\,\Delta\ell = J_s\,\Delta\ell$$

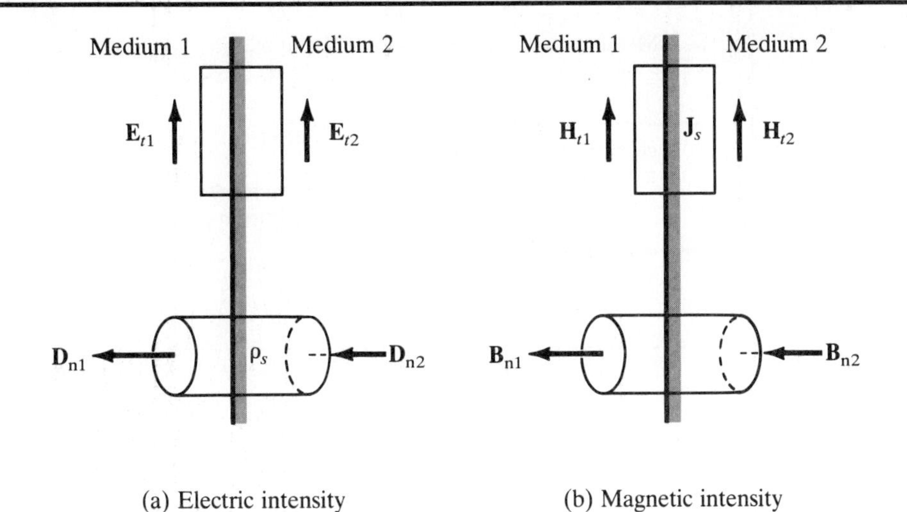

Figure 0-5-2 Boundary conditions.

and

$$\oint_s \mathbf{B} \cdot d\mathbf{s} = B_{n1}\,\Delta s - B_{n2}\,\Delta s = 0$$

where ρ_s is the surface-charge density, in units of coulombs per square meter and J_s is the surface-current density, in units of amperes per meter. Hence the boundary equations are

$$E_{t1} = E_{t2}$$

$$D_{n1} = D_{n2} + \rho_s$$

$$H_{t1} = H_{t2} + J_s$$

and

$$B_{n1} = B_{n2}$$

If medium 2 is a perfect conductor ($\sigma = \infty, \varepsilon_r = 1$, and $\mu_r = 1$) and medium 1 is a perfect dielectric (vacuum or free space, $\sigma = 0$, ε_0, and μ_0),

$$E_{t1} = \frac{D_{t1}}{\varepsilon_0} = 0 \quad \text{and} \quad H_{t1} = 0 \quad \text{if } J_s = 0$$

$$D_n = \varepsilon_0 E_n = \rho_s \quad \text{and} \quad B_n = 0$$

Note that the tangential electric and magnetic fields are vector components.

0-5-3 Uniform Plane-Wave Reflection

1. *Normal-Incidence Reflection*

The simplest reflection condition is that of a uniform plane wave normally incident upon a plane boundary between two different media with no surface-charge density or no surface-current density. This condition is shown in Fig. 0-5-3.

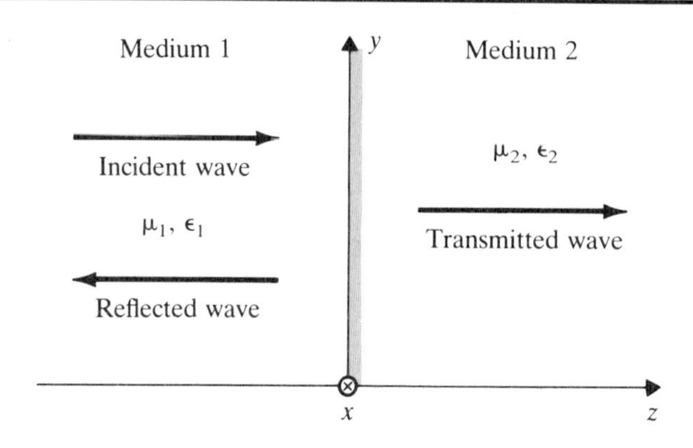

Figure 0-5-3 Uniform plane-wave reflection.

In medium 1 the fields are the sum of an incident wave and a reflected wave, respectively, or

$$E_x^{(1)} = E_0(e^{-j\beta_1 z} + \Gamma e^{j\beta_1 z}) \tag{0-5-8}$$

and

$$H_y^{(1)} = \frac{E_0}{\eta_1}(e^{-j\beta_1 z} - \Gamma e^{j\beta_1 z}) \tag{0-5-9}$$

where

$\beta_1 = \omega\sqrt{\mu_1 \varepsilon_1}$;
$\eta_1 = \sqrt{\mu_1/\varepsilon_1} = \eta_0/\sqrt{\varepsilon_{r1}}$, the intrinsic wave impedance of medium 1; and
Γ = reflection coefficient.

Medium 2 contains transmitted waves, which are

$$E_x^{(2)} = E_0 T e^{-j\beta_2 z} \tag{0-5-10}$$

and

$$H_y^{(2)} = \frac{E_0}{\eta_2} T e^{-j\beta_2 z} \tag{0-5-11}$$

where

$\beta_2 = \omega\sqrt{\mu_2 \varepsilon_2}$;
$\eta_2 = \sqrt{\mu_2/\varepsilon_2} = \eta_0/\sqrt{\varepsilon_{r2}}$, the intrinsic wave impedance of medium 2; and
T = transmission coefficient

From the continuity of tangential components of electric field E and magnetic field H at the boundary, the wave impedances at the boundary must equal

$$Z_z = \left.\frac{E_x^{(1)}}{H_y^{(1)}}\right|_{z=0} = \eta_1 \frac{1+\Gamma}{1-\Gamma} = \left.\frac{E_x^{(2)}}{H_y^{(2)}}\right|_{z=0} = \eta_2 \tag{0-5-12}$$

Hence, the reflection coefficient is

$$\Gamma = \frac{\eta_2 - \eta_1}{\eta_2 + \eta_1} \tag{0-5-13}$$

From the boundary condition, the tangential components of electric field intensity are continuous across the surface. Then

$$E_x^{(1)}\Big|_{z=0} = E_0(1+\Gamma) = E_x^{(2)}\Big|_{z=0} = E_0 T \tag{0-5-14}$$

Hence, the transmission coefficient is expressed as

$$T = 1 + \Gamma = \frac{2\eta_2}{\eta_2 + \eta_1} \tag{0-5-15}$$

If medium 1 is a lossless dielectric (that is, $\sigma = 0$), we define the standing-wave ratio as

$$\text{SWR} = \rho = \frac{E_{\text{max}}^{(1)}}{E_{\text{min}}^{(1)}} = \frac{1+|\Gamma|}{1-|\Gamma|} \tag{0-5-16}$$

The power density transmitted across the boundary for a lossless medium is

$$p_{\text{tr}} = \tfrac{1}{2}(\mathbf{E} \times \mathbf{H}^*)\Big|_{z=0} \cdot \mathbf{u}_z = \frac{E_0^2}{2\eta_1}(1-|\Gamma|^2) \tag{0-5-17}$$

and

$$p_{\text{tr}} = p_{\text{inc}}(1-|\Gamma|^2) \tag{0-5-18}$$

where p_{inc} is the incident power density. The incident power density minus the transmitted power density yields the reflected power density:

$$P_{\text{ref}} = P_{\text{inc}}|\Gamma|^2 \tag{0-5-19}$$

EXAMPLE 0-5-1 Normal-Incident Wave Propagation

A normal-incident wave is propagating from air into a nonmagnetic medium with a dielectric constant of 2.56. The wave equation is

$$E_x = 20\cos(\omega t - \beta z) \qquad \text{V/m}$$

Calculate:

a) the intrinsic impedance η_2 of the medium;
b) the reflection coefficient Γ; and
c) the transmitted power density across the boundary.

Solutions:

a) The intrinsic impedance of medium 2 is

$$\eta_2 = \frac{377}{\sqrt{2.56}} = 235.63\ \Omega$$

b) The reflection coefficient is

$$\Gamma = \frac{235.63 - 377}{235.63 + 377} = -0.23 = 0.23\,\underline{/180^\circ}$$

c) The power density transmitted across the boundary is

$$P_{\text{tr}} = \frac{(20)^2}{2 \times 377}(1 - |0.23|^2)$$

$$= 0.50 \text{ W/m}^2$$

---

2. *Oblique-Incidence Reflection*

a) **E** *is in the Plane of Incidence.*

The direction of propagation and the line normal to the boundary define the plane of incidence of waves. Two linearly polarized uniform plane waves—**E** lying in and **H** normal to the plane of incidence—impinge obliquely on a boundary between two lossless dielectric materials, as shown in Fig. 0-5-4.

There are two special cases for the oblique-incidence reflection. In the first case the electric vector is parallel to the plane of incidence and the magnetic vector is normal to the plane. This case is usually called *vertical polarization.* In the second case the electric vector is normal to the plane of incidence and the magnetic vector is parallel to the plane. This case is usually called *horizontal polarization.*

As shown in Fig. 0-5-4, for a lossless dielectric medium, the phase constants (or propagation constants) of the two media in the x direction on the interface are equal, as

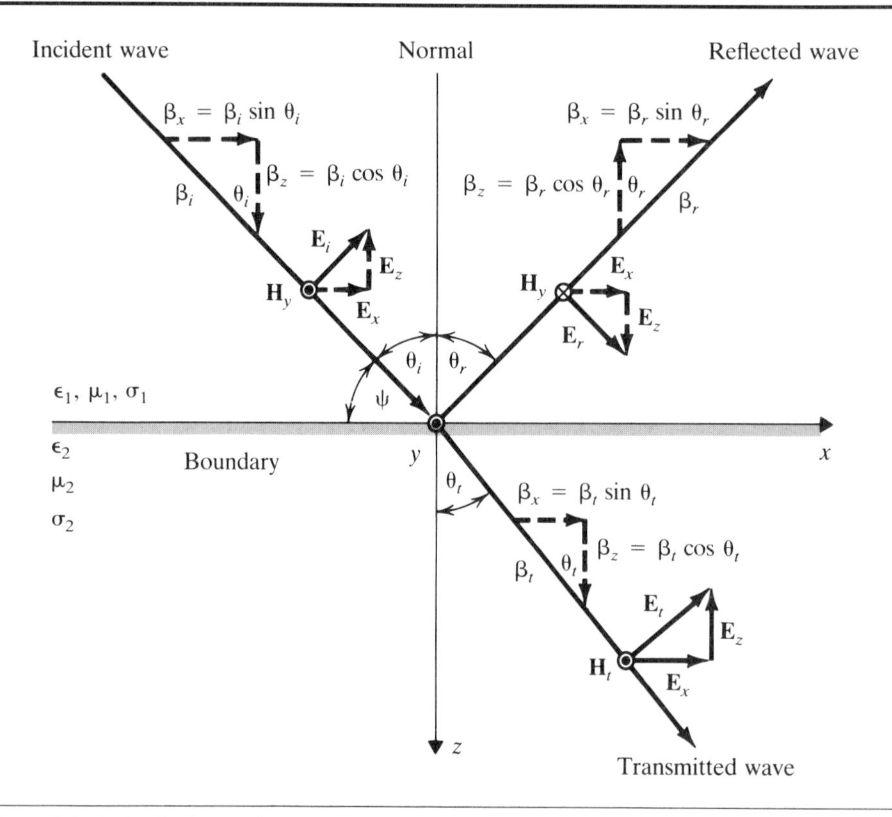

Figure 0-5-4 Reflection and transmission of oblique incidence.

required by the continuity of tangential **E** and **H** on the boundary. Thus

$$\beta_i \sin\theta_i = \beta_r \sin\theta_r \tag{0-5-20}$$

and

$$\beta_i \sin\theta_i = \beta_t \sin\theta_t \tag{0-5-21}$$

From Eq. (0-5-20)—and since $\beta_i = \beta_r = \beta_1$—the angle of reflection equals the angle of incidence. That is,

$$\theta_i = \theta_r \tag{0-5-22}$$

Rearranging Eq. (0-5-21) yields

$$\frac{\sin\theta_t}{\sin\theta_i} = \frac{\beta_i}{\beta_t} = \frac{v_2}{v_1} = \sqrt{\frac{\mu_1\varepsilon_1}{\mu_2\varepsilon_2}} \tag{0-5-23}$$

where v is the phase velocity. Equation (0-5-23) is known as Snell's law. In general, all low-loss dielectrics have equal permeability; that is, $\mu_1 = \mu_2 = \mu_0$. If medium 2 is free space and medium 1 is a nonmagnetic dielectric, the right-hand side of Eq. (0-5-23) becomes $\sqrt{\varepsilon_r}$, which is called the refraction index of the dielectric.

The components of electric intensity **E** are

$$E_x = E_0 \cos\theta_i\, e^{-j\beta_1(x\sin\theta_i + z\cos\theta_i)} \tag{0-5-24}$$

$$E_y = 0 \tag{0-5-25}$$

and

$$E_z = -E_0 \sin\theta_i\, e^{-j\beta_1(x\sin\theta_i + z\cos\theta_i)} \tag{0-5-26}$$

The components of magnetic intensity **H** are

$$H_x = 0 \tag{0-5-27}$$

$$H_y = \frac{E_0}{\eta_1} e^{-j\beta_1(x\sin\theta_i + z\cos\theta_i)} \tag{0-5-28}$$

and

$$H_z = 0 \tag{0-5-29}$$

The wave impedance in the z direction is

$$Z_z = \frac{E_x}{H_y} = \eta\cos\theta \tag{0-5-30}$$

In Eq. (0-5-30) the subscripts of η and θ have been dropped, because the wave impedances of the two regions in the z direction are the same.

We can express the wave impedance in terms of the reflection coefficient of the tangential field components at the boundary $z = 0$. In medium 1,

$$E_x^{(1)} = E_0 \cos\theta_i\,[e^{-j\beta_1 z\cos\theta_i} + \Gamma e^{j\beta_1 z\cos\theta_r}] \tag{0-5-31}$$

$$H_y^{(1)} = \frac{E_0}{\eta_1}[e^{-j\beta_1 z\cos\theta_i} + \Gamma e^{j\beta_1 z\cos\theta_r}] \tag{0-5-32}$$

and

$$Z_z = \left.\frac{E_x^{(1)}}{H_y^{(1)}}\right|_{z=0} = \eta_1 \cos\theta_i \frac{1+\Gamma}{1-\Gamma} \tag{0-5-33}$$

The impedance must equal the z-directed wave impedance in region 2 at the boundary. Substituting $Z_z = \eta_t \cos \theta_t$ into Eq. (0-5-33) yields the reflection coefficient:

$$\Gamma = \frac{\eta_2 \cos \theta_t - \eta_1 \cos \theta_i}{\eta_2 \cos \theta_t + \eta_1 \cos \theta_i} \tag{0-5-34}$$

Then the transmission coefficient is

$$T = \frac{2\eta_2 \cos \theta_t}{\eta_2 \cos \theta_t + \eta_1 \cos \theta_i} \tag{0-5-35}$$

Equations (0-5-34) and (0-5-35) are known as Fresnel's formulas for **E** in the plane of incidence.

b) **H** *is in the Plane of Incidence.*
If **H** is in the plane of incidence, the components of **H** are

$$H_x = H_0 \cos \theta_i \, e^{-j\beta_1(x \sin \theta_i + z \cos \theta_i)} \tag{0-5-36}$$

$$H_y = 0 \tag{0-5-37}$$

and

$$H_z = -H_0 \sin \theta_i \, e^{-j\beta_1(x \sin \theta_i + z \cos \theta_i)} \tag{0-5-38}$$

The components of the electric intensity **E** normal to the plane of incidence are

$$E_x = 0 \tag{0-5-39}$$

$$E_y = -\eta_1 H_0 e^{-j\beta_1(x \sin \theta_i + z \cos \theta_i)} \tag{0-5-40}$$

and

$$E_z = 0 \tag{0-5-41}$$

The wave impedance in the z direction is

$$Z_z = \frac{-E_y}{H_x} = \frac{\eta}{\cos \theta} = \eta \sec \theta \tag{0-5-42}$$

In Eq. (0-5-42) the subscripts of η and θ have been dropped for the same reason they were dropped from Eq. (0-5-30).

Fresnel's formulas for **H** in the plane of incidence are

$$\Gamma = \frac{\eta_2 \sec \theta_t - \eta_1 \sec \theta_i}{\eta_2 \sec \theta_t + \eta_1 \sec \theta_i} \tag{0-5-43}$$

and

$$T = \frac{2\eta_2 \sec \theta_t}{\eta_2 \sec \theta_t + \eta_1 \sec \theta_i} \tag{0-5-44}$$

EXAMPLE 0-5-2 Wave Propagation in the E Plane

An electric wave is propagating from air into a nonmagnetic medium with a dielectric constant of 2.56 at an angle of 30°. The wave equation is

$$E_x = 100 \cos (30^\circ) \, e^{-j\beta_1[x \sin (30^\circ) + z \cos (30^\circ)]} \quad \text{V/m}$$

Calculate:

a) the intrinsic impedances η_1 and η_2;
b) the magnetic field component;
c) the transmission angle θ_t;
d) the reflection coefficient Γ;
e) the transmission coefficient T; and
f) the wave impedances in air and in the medium.

Solutions:

a) The intrinsic impedances are

$$\eta_1 = 377\ \Omega \quad \text{and} \quad \eta_2 = 377/\sqrt{2.56} = 235.63\ \Omega$$

b) The magnetic field component is

$$H_y = 100/377 e^{-j\beta_1[x(0.5)+z(0.87)]}$$
$$= 0.27 e^{-j\beta_1[0.5x+0.87z]}$$

c) The transmission angle is

$$\theta_t = \arcsin\left[\sqrt{\frac{\mu_1\varepsilon_1}{\mu_2\varepsilon_2}}\sin\theta_i\right] = \sin^{-1}\left[\frac{1}{\sqrt{2.56}}\sin(30^\circ)\right]$$
$$= \sin^{-1}(0.31)$$
$$= 18.21^\circ$$

d) The reflection coefficient is

$$\Gamma = \frac{235.63\sec(18.21^\circ) - 377\sec(30^\circ)}{235.63\sec(18.21^\circ) + 377\sec(30^\circ)}$$
$$= 0.28\,\underline{/180^\circ}$$

e) The transmission coefficient is

$$T = \frac{2 \times 235.63\sec(18.21^\circ)}{235.63\sec(18.21^\circ) + 377\sec(30^\circ)}$$
$$= 0.73$$

f) The wave impedance in air is

$$Z_z = 377\cos(30^\circ)$$
$$= 326.49\ \Omega$$

The wave impedance in the medium is

$$Z_z = 325.63\cos(18.21^\circ)$$
$$= 309.32\ \Omega$$

0-5-4 Plane-Wave Propagation at the Earth's Surface

An electromagnetic wave being propagated at the earth's surface is divided into two parts: (1) the ground wave; and (2) the sky, or ionospheric wave. The ground wave is

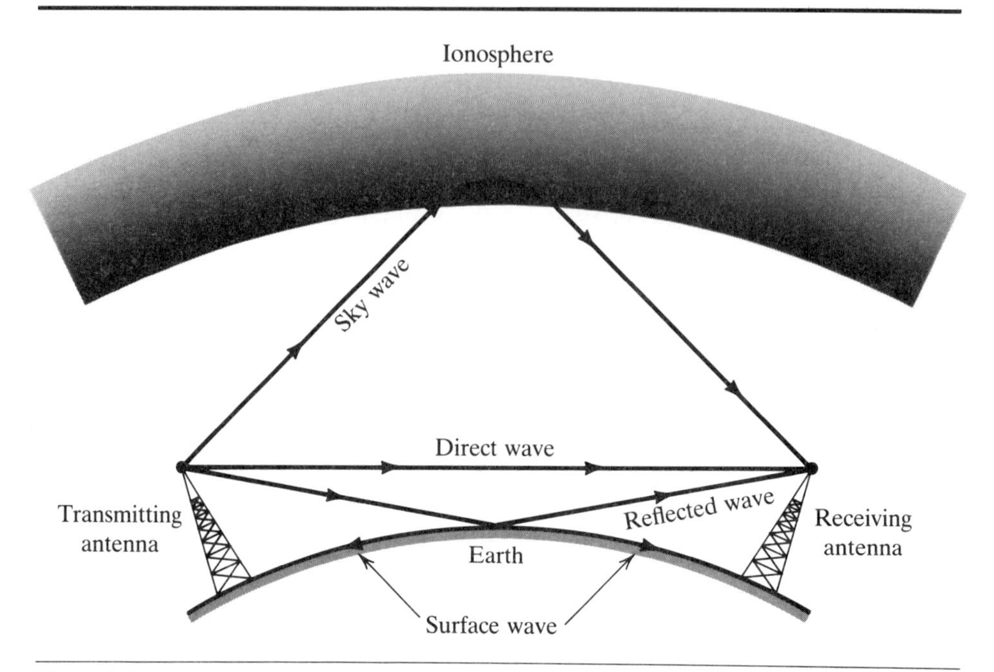

Figure 0-5-5 Wave components near the surface of the earth.

further divided into a direct wave, an earth-reflected wave, and a surface wave. Figure 0-5-5 shows the wave components of an electromagnetic wave from a nondirectional antenna to a receiving station.

The sky wave reaches the receiving station after being reflected by the ionosphere. Although important in many communication systems, the sky wave need not be considered in most microwave applications, because a wavelength shorter than about 4 m will not return to earth from the ionosphere. Energy radiated from the nondirectional transmitting antenna shown in Fig. 0-5-5 strikes the earth at all points between the base of the antenna and the horizon, but only that earth-reflected wave which leaves the antenna in the direction shown reaches the receiver. The surface wave is a wave diffracted around the surface of the earth, or guided around it, by the ground–air interface. This component is important at broadcast frequencies; at the microwave frequencies, however, the surface wave is rapidly attenuated, and at a distance of 2 km from the antenna it has an amplitude of only a fraction of a percent of the direct wave. This component is considered in blind-landing system applications, in which ranges of less than 2 km are important.

The direct wave travels a nearly straight path from the antenna to the receiving station. This component is the only wave that we consider in this book. The term *free space* denotes a vacuum or any other media having essentially the same characteristics as a vacuum, such as open air and an anechoic chamber. When power radiates from the antenna, the power density carried by the spherical wave decreases with distance as the energy in the wave spreads over an ever-increasing surface area. The expression for the power density is

$$p_d = \frac{p_t g_t}{4\pi R^2} \qquad \text{W/m}^2 \tag{0-5-45}$$

where

p_t = transmitting power, in W;

g_t = transmitting antenna gain (numerical); and

R = range between the antenna and the field point, in m.

The power received by the receiving antenna is

$$p_r = p_d A_e = \left(\frac{p_t g_t}{4\pi R^2}\right)\left(\frac{\lambda^2}{4\pi} g_r\right) \qquad \text{W} \tag{0-5-46}$$

where

$A_e = \frac{\lambda^2}{4\pi} g_r$ is the effective antenna area or aperture, expressed in m^2;

$\frac{\lambda^2}{4\pi} = A_a$ is the isotropic antenna area, expressed in m^2; and

g_r = receiving antenna gain (numerical).

Figure 0-5-6 shows the power and power-gain relationships of electromagnetic energy transmission in free space between two antennas. If the received power is expressed in dBW, Eq. (0-5-46) becomes

$$P_r = P_t + G_t + G_r - 20 \log\left(\frac{4\pi R}{\lambda}\right) \qquad \text{dBW} \tag{0-5-47}$$

where P_t is measured in dBW and G_t and G_r are measured in dB.

The term $20 \log (4\pi R/\lambda)$ is known as *free-space attenuation* and is expressed in dB. For example, if the wavelength of a signal is 0.30 m and the range is 20 m, the free-space attenuation is about 79 dB.

Free-space attenuation is entirely different from the dissipative attenuation of a medium such as atmosphere, which absorbs energy from a wave. The factor $4\pi R^2$ in Eq. (0-5-47) simply accounts for the fact that the power density is decreasing inversely and proportionally with the square of the distance when the energy spreads over free space. The factor $\lambda^2/4\pi$ represents the aperture of an isotropic antenna. This factor does not imply that a higher frequency wave decreases in magnitude more rapidly than a lower frequency wave. Rather, it simply indicates that for a given gain the

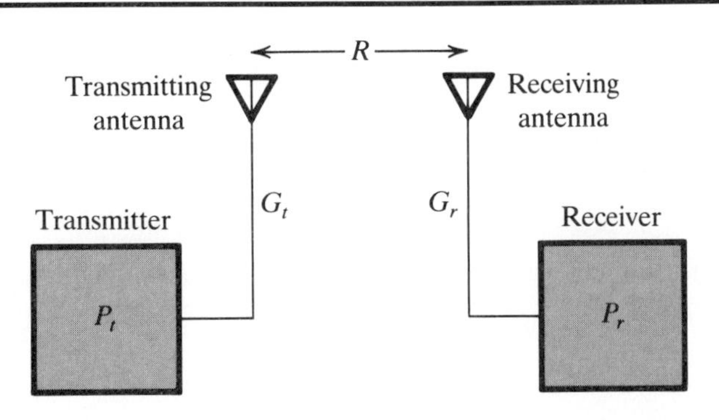

Figure 0-5-6 Electromagnetic energy transmission between two antennas.

aperture of a higher frequency antenna is smaller than that of a lower frequency antenna and thus intercepts a smaller amount of power from a wave.

0-5-5 Plane-Wave Propagation in a Lossless Dielectric

A lossless dielectric, often called *good* or *perfect* dielectric, is characterized by $\sigma = 0$. Hence, the intrinsic impedance for a lossless dielectric can be expressed in terms of air. That is,

$$\eta = \sqrt{\frac{\mu}{\varepsilon}} = \sqrt{\frac{\mu_0}{\varepsilon_r \varepsilon_0}} = \frac{377}{\sqrt{\varepsilon_r}} \quad \Omega \tag{0-5-48}$$

The attenuation constant α is 0, the phase constant β is

$$\beta = \omega\sqrt{\mu\varepsilon} \tag{0-5-49}$$

and the phase velocity v_p is

$$v_p = \frac{1}{\sqrt{\mu\varepsilon}} \tag{0-5-50}$$

0-5-6 Plane-Wave Propagation in Lossy Media

The lossy media are characterized by $\sigma \neq 0$. We discuss three types of lossy media—good conductor, poor conductor, and lossy dielectric—in this section. The presence of loss in the medium by conductivity introduces the wave dispersion. Dispersion makes a general solution in the time domain impossible, except by Fourier expansion methods. Hence, only solutions for the frequency domain (or steady state) are presented.

The electric- and magnetic-wave equations in the frequency domain, Eqs. (0-3-27) and (0-3-28), are repeated here:

$$\nabla^2 \mathbf{E} = j\omega\mu(\sigma + j\omega\varepsilon)\mathbf{E} \tag{0-5-51}$$

and

$$\nabla^2 \mathbf{H} = j\omega\mu(\sigma + j\omega\varepsilon)\mathbf{H} \tag{0-5-52}$$

For one dimension—the positive z direction—in the cartesian coordinate system, Eqs. (0-3-27) and (0-3-28) become

$$\frac{\partial^2 E_x}{\partial z^2} = j\omega\mu(\sigma + j\omega\varepsilon)E_x \tag{0-5-53}$$

and

$$\frac{\partial^2 H_y}{\partial x^2} = j\omega\mu(\sigma + j\omega\varepsilon)H_y \tag{0-5-54}$$

The solutions in the frequency domain for low loss are given by

$$E_x = E_0 e^{-\alpha z} \cos(\omega t - \beta z) \tag{0-5-55}$$

and

$$H_y = \frac{E_0}{\eta} e^{-\alpha z} \cos(\omega t - \beta z) \tag{0-5-56}$$

where

$$\gamma = \sqrt{j\omega\mu(\sigma + j\omega\varepsilon)} = \alpha + j\beta; \text{ and}$$

$$\eta = \sqrt{\frac{j\omega\mu}{\sigma + j\omega\varepsilon}}.$$

0-5-7 Plane Waves in a Good Conductor

A good conductor is characterized by very high conductivity, and consequently, the conduction current is much larger than the displacement current. Mathematically, a good conductor requires that

$$\sigma \gg \omega\varepsilon \tag{0-5-57}$$

The propagation constant γ is expressed as

$$\begin{aligned}
\gamma &= \sqrt{j\omega\mu(\sigma + j\omega\varepsilon)} = j\omega\sqrt{\mu\varepsilon}\sqrt{1 - j\frac{\sigma}{\omega\varepsilon}} \\
&\simeq j\omega\sqrt{\mu\varepsilon}\sqrt{-j\frac{\sigma}{\omega\varepsilon}} \qquad \text{for } \frac{\sigma}{\omega\varepsilon} \gg 1 \\
&= j\sqrt{\omega\mu\sigma}\sqrt{-j} = j\sqrt{\omega\mu\sigma}\left(\frac{1}{\sqrt{2}} - j\frac{1}{\sqrt{2}}\right) \\
&= (1 + j)\sqrt{\pi f \mu\sigma}
\end{aligned} \tag{0-5-58}$$

Hence, the attenuation constant and the phase constant are

$$\alpha = \beta = \sqrt{\pi f \mu\sigma} \tag{0-5-59}$$

The exponential factor $e^{-\alpha z}$ of the traveling wave becomes $e^{-1} = 0.368$ when

$$z = \frac{1}{\sqrt{\pi f \mu\sigma}} \tag{0-5-60}$$

We call this distance the skin depth and denote it by

$$\delta = \frac{1}{\sqrt{\pi f \mu\sigma}} = \frac{1}{\alpha} = \frac{1}{\beta} \tag{0-5-61}$$

At microwave frequencies the skin depth is extremely shallow, and a piece of glass with an evaporated silver coating 5.40 μm thick is an excellent conductor at those frequencies. Table 0-5-1 shows the conductivities of some materials.

The intrinsic impedance of a good conductor is

$$\begin{aligned}
\eta &= \sqrt{\frac{j\omega\mu}{\sigma + j\omega\varepsilon}} = \sqrt{\frac{j\omega\mu}{\sigma}} \qquad \text{for } \sigma \gg \omega\varepsilon \\
&= \sqrt{\frac{\omega\mu}{\sigma}}\,\underline{/45^\circ} = (1 + j)\sqrt{\frac{\omega\mu}{2\sigma}} \\
&= (1 + j)\frac{1}{\sigma\delta} = (1 + j)R_s
\end{aligned} \tag{0-5-62}$$

Table 0-5-1 Table of conductivities

Substance	Type	Conductivity, σ (mho/m)
Quartz, fused	Insulator	10^{-17} (approx.)
Ceresin wax	Insulator	10^{-17} (approx.)
Sulfur	Insulator	10^{-15} (approx.)
Mica	Insulator	10^{-15} (approx.)
Paraffin	Insulator	10^{-15} (approx.)
Rubber, hard	Insulator	10^{-15} (approx.)
Glass	Insulator	10^{-12} (approx.)
Bakelite	Insulator	10^{-9} (approx.)
Distilled water	Insulator	10^{-4} (approx.)
Sea water	Conductor	4 (approx.)
Tellurium	Conductor	5×10^{2} (approx.)
Carbon	Conductor	3×10^{4} (approx.)
Graphite	Conductor	10^{5} (approx.)
Cast iron	Conductor	10^{6} (approx.)
Mercury	Conductor	10^{6}
Nichrome	Conductor	10^{6}
Constantin	Conductor	2×10^{6}
Silicon steel	Conductor	2×10^{6}
German silver	Conductor	3×10^{6}
Lead	Conductor	5×10^{6}
Tin	Conductor	9×10^{6}
Phospher bronze	Conductor	10^{7}
Brass	Conductor	1.1×10^{7}
Zinc	Conductor	1.7×10^{7}
Tungsten	Conductor	1.8×10^{7}
Duralumin	Conductor	3×10^{7}
Aluminum, hard-drawn	Conductor	3.5×10^{7}
Gold	Conductor	4.1×10^{7}
Copper	Conductor	5.8×10^{7}
Silver	Conductor	6.1×10^{7}

in which R_s represents the surface resistance, or the skin effect of a conductor, in ohms per square. This is because the resistance is defined as

$$R = \frac{\text{Specific resistivity} \times \text{length}}{\text{Thickness} \times \text{width}}$$

$$= \frac{\rho\ell}{\delta w} = \frac{\ell}{\sigma\delta w} = \left.\frac{1}{\sigma\delta}\right|_{l=w} \tag{0-5-63}$$

When we choose equal-length units for length ℓ and width w, the resistance is in ohms per square

$$P = \tfrac{1}{2}|H|^2 R_s \tag{0-5-64}$$

where H is the tangential component. The phase velocity within a good conductor is

$$v = \frac{\omega}{\beta} = \omega\delta \tag{0-5-65}$$

EXAMPLE 0-5-3 Wave Propagation in a Good Conductor

An electric wave is normally propagating from air into a copper material at a frequency of 1 GHz. The wave equation is

$$E_x = 100 \cos(\omega t - \beta z) \qquad \text{V/m}$$

Calculate:

a) the propagation constant γ;
b) the skin depth δ;
c) the copper intrinsic impedance η_c; and
d) the surface resistance of copper.

Solutions:

a) The propagation constant is

$$\gamma = (1 + j)\sqrt{\pi \times 10^9 \times 4\pi \times 10^{-7} \times 5.8 \times 10^7}$$
$$= (1 + j)47.85 \times 10^4$$

Thus

$$\alpha = 47.85 \times 10^4 \text{ N/m} \quad \text{and} \quad \beta = 47.85 \times 10^4 \text{ rad/m}$$

b) The skin depth is

$$\delta = \frac{1}{\alpha} = \frac{1}{47.85 \times 10^4}$$
$$= 2.09\ \mu\text{m}$$

c) The copper intrinsic impedance is

$$\eta_c = (1 + j)\sqrt{\frac{2\pi \times 10^9 \times 4\pi \times 10^{-7}}{2 \times 5.8 \times 10^7}}$$
$$= (1 + j)4.65 \times 10^{-6}\ \Omega$$

d) The surface resistance is

$$R_s = 4.65 \times 10^{-6}\ \Omega/\text{square}$$

0-5-8 Plane Waves in a Poor Conductor

Some materials with low conductivity normally are not considered to be either good conductors or good dielectrics. Seawater is a good example. It has a conductivity of 4 ℧/m and a dielectric constant of 20. At some low frequencies the conduction current is greater than the displacement current, whereas at some high frequencies the reverse is true.

In general, the propagation constant and intrinsic impedance for a poor conductor are

$$\gamma = \sqrt{j\omega\mu(\sigma + j\omega\varepsilon)} = j\omega\sqrt{\mu\varepsilon\left(1 + \frac{\sigma}{j\omega\varepsilon}\right)} \qquad \text{(0-5-66)}$$

and

$$\eta = \sqrt{\frac{j\omega\mu}{\sigma + j\omega\varepsilon}} \tag{0-5-67}$$

0-5-9 Plane Waves in a Lossy Dielectric

All dielectric materials have some conductivity, although very small ($\sigma \ll \omega\varepsilon$). When the conductivity cannot be neglected, the electric and magnetic fields in the dielectric are no longer in time phase. This fact can be seen from the intrinsic impedance of the dielectric:

$$\eta = \sqrt{\frac{j\omega\mu}{\sigma + j\omega\varepsilon}} = \sqrt{\frac{\mu}{\varepsilon}}\left[1 - j\frac{\sigma}{\omega\varepsilon}\right]^{-1/2} \tag{0-5-68}$$

We refer to the term $\sigma/\omega\varepsilon$ as the loss tangent and define it as follows:

$$\tan\theta = \frac{\sigma}{\omega\varepsilon} \tag{0-5-69}$$

This relationship is the result of the displacement current density leading the conduction current density by 90°. It is just like the current flowing through a capacitor, which leads the current through a resistor in parallel with it by 90° in an ordinary electric circuit. This phase relationship is shown in Fig. 0-5-7.

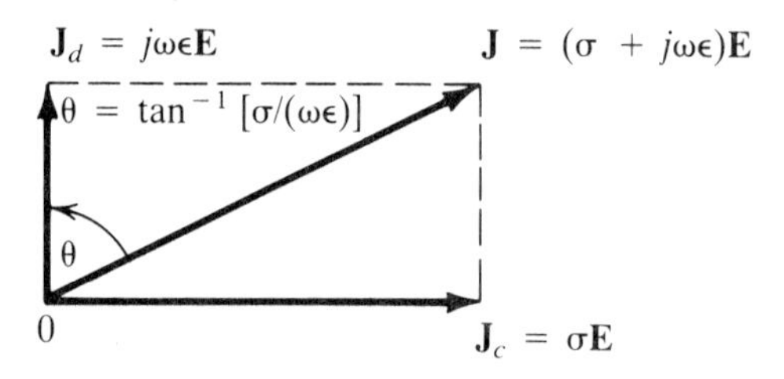

Figure 0-5-7 Loss tangent for a lossy dielectric.

If the loss tangent is very small ($\sigma/\omega\varepsilon \ll 1$), we can calculate the approximate propagation constant and intrinsic impedance using binomial expansion. Since

$$\gamma = j\omega\sqrt{\mu\varepsilon}\sqrt{1 - j\frac{\sigma}{\omega\varepsilon}} \tag{0-5-70}$$

then

$$\gamma = j\omega\sqrt{\mu\varepsilon}\left[1 - j\frac{\sigma}{2\omega\varepsilon} + \frac{1}{8}\left(\frac{\sigma}{\omega\varepsilon}\right)^2 + \cdots\right] \tag{0-5-71}$$

Hence,

$$\alpha \doteq j\omega\sqrt{\mu\varepsilon}\left(-j\frac{\sigma}{2\omega\varepsilon}\right) = \frac{\sigma}{2}\sqrt{\frac{\mu}{\varepsilon}} \tag{0-5-72}$$

and

$$\beta \doteq \omega\sqrt{\mu\varepsilon}\left[1 + \frac{1}{8}\left(\frac{\sigma}{\omega\varepsilon}\right)^2\right] = \omega\sqrt{\mu\varepsilon} \tag{0-5-73}$$

Similarly,

$$\eta = \sqrt{\frac{\mu}{\varepsilon}}\left[1 + j\frac{\sigma}{2\omega\varepsilon} - \frac{3}{8}\left(\frac{\sigma}{\omega\varepsilon}\right)^2 + \cdots\right] \tag{0-5-74}$$

or

$$\eta \doteq \sqrt{\frac{\mu}{\varepsilon}}\left(1 + j\frac{\sigma}{2\omega\varepsilon}\right) \tag{0-5-75}$$

As indicated in Fig. 0-5-4, we assumed that medium 1 has constant ε_1 and μ_1 and that medium 2 has constant ε_2 and μ_2. Then

$$\eta_1 = \sqrt{\frac{\mu_1}{\varepsilon_1}} \qquad \frac{\sin\theta_t}{\sin\theta_i} = \frac{\sqrt{\mu_1\varepsilon_1}}{\sqrt{\mu_2\varepsilon_2}} \qquad \cos\theta_t = \sqrt{1 - \sin^2\theta_t} \tag{0-5-76}$$

and

$$\eta_2 = \sqrt{\frac{\mu_2}{\varepsilon_2}} \qquad \mu_1 = \mu_2 \qquad \cos\theta_t = \sqrt{1 - \frac{\varepsilon_1}{\varepsilon_2}\sin^2\theta_i} \tag{0-5-77}$$

We can now simplify the vertical reflectivity of a lossy dielectric for the tangential components of electric fields, as shown in Eq. (0-5-34):

$$\Gamma_v = \frac{\sqrt{\varepsilon_2/\varepsilon_1 - \sin^2\theta_i} - \varepsilon_2/\varepsilon_1\cos\theta_i}{\sqrt{\varepsilon_2/\varepsilon_1 - \sin^2\theta_i} + \varepsilon_2/\varepsilon_1\cos\theta_i} \tag{0-5-78}$$

If $\Gamma_v = 0$ for total transmission, then

$$\theta_i = \arctan\sqrt{\frac{\varepsilon_2}{\varepsilon_1}} \tag{0-5-79}$$

We call this angle the Brewster, or polarizing, angle. Furthermore, if

$$\varepsilon_2 = \varepsilon\left(1 - j\frac{\sigma}{\omega\varepsilon}\right) \qquad \text{for } \frac{\sigma}{\omega\varepsilon} \ll 1$$

and

$$\varepsilon_1 = \varepsilon_0 \qquad \text{for air}$$

Eq. (0-5-78) becomes

$$\Gamma_v = \frac{\sqrt{(\varepsilon_2 - jx) - \cos^2\psi} - (\varepsilon_2 - jx)\sin\psi}{\sqrt{(\varepsilon_2 - jx) - \cos^2\psi} + (\varepsilon_2 - jx)\sin\psi} \tag{0-5-80}$$

where

$\varepsilon_r = \dfrac{\varepsilon}{\varepsilon_0}$;

$x = \dfrac{\sigma}{\omega\varepsilon_0} = \dfrac{18\sigma}{f_{\text{GHz}}}$; and

$\psi = 90^\circ - \theta_i$, which is called the pseudo-Brewster angle.

Equation (0-5-80) is applicable only to the tangential components of incident and reflected fields that are in the same directions, as shown in Fig. 0-5-4. For total reflection $\Gamma_v = 1$, we set $\theta_t = 90°$ in Eq. (0-5-80), and the incident angle is

$$\theta_i = \theta_c = \arcsin \sqrt{\frac{\mu_2 \varepsilon_2}{\mu_1 \varepsilon_1}} \tag{0-5-81}$$

The angle specified by Eq. (0-5-81) is called the *critical incident angle for total reflection.* A wave incident upon the boundary at an angle equal to or greater than the critical angle will be totally reflected. There is a real critical angle only if $\mu_1 \varepsilon_1 > \mu_2 \varepsilon_2$, or in nonmagnetic material, if $\varepsilon_1 > \varepsilon_2$. Hence, the total reflection occurs only if the wave propagates from a dense medium into a less dense medium, because the value of $\sin \theta_c$ must be less than or equal to unity.

Problems

0-1 A power density of 0.1 W/in^2 is the maximum safe power level that can be applied to the absorber of an anechoic chamber without generating enough heat for combustion. A short-wire antenna with a gain of 2 dB has an effective aperture of $3\lambda^2/(8\pi)$ at 1 GHz. The performance of the absorber is 30 dB at 1 GHz. The shortest distance between the antenna and the absorber is 2 m. Determine the maximum power level from the transmitter that will not damage the absorber.

0-2 A transmitter has an output power of 10 W at a frequency of 3 GHz. The coaxial cable connected to the transmitter has a loss of 3 dB, and the antenna connected to the cable has a gain of 5 dB. The test location in the quiet zone of an anechoic chamber is 3 m from the transmitting antenna. Determine the

a) electric intensity in V/m and power density in W/m^2 at the test site; and
b) electric intensity and power density in dB.

0-3 Seawater has a conductivity of 4 ℧/m and a relative dielectric constant of 20 at a frequency of 4 GHz. Calculate the

a) intrinsic impedance;
b) propagation constant; and
c) phase velocity

0-4 Repeat Problem 0-3 for dry sand ($\sigma = 2 \times 10^{-4}$ ℧/m and $\varepsilon_r = 4$) and copper ($\sigma = 5.8 \times 10^7$ ℧/m) at a frequency of 4 GHz.

0-5 A uniform plane wave is normally incident upon the surface of seawater. The electric intensity of the incident wave is 100×10^{-3} V at a frequency of 4 GHz in the vertical polarization. Calculate the

a) electric field of the reflected wave; and
b) electric field of the transmitted wave.

0-6 Repeat Problem 0-5 for an angle of incidence of 30°.

0-7 Dry ground has a conductivity of 5×10^{-4} ℧/m and a relative dielectric constant of 10 at a frequency of 500 MHz. Calculate the

a) intrinsic impedance;

b) propagation constant; and
c) phase velocity.

0-8 A radar transmitter has an average output power of 1 kW. Calculate the power density in dBW/m^2 at a range of 3000 m and the free-space attenuation in dB. The antenna has a power gain of 10 dB, and the signal frequency is 10 GHz.

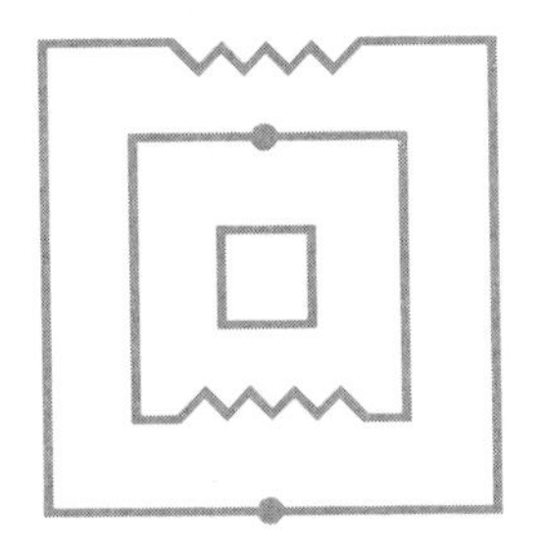

PART ONE

TRANSMISSION LINES

The transmission line is a commonly used energy carrier. In this part we present the transmission-line equations and their analytic and graphic solutions. However, before discussing the transmission line itself, we describe the general types of lines and the ratio of wavelength to physical dimensions.

In ordinary circuit theory we assume that all circuit elements are lumped constants. We cannot make this assumption for a transmission line operating at a high frequency or even for a long line operating at a low frequency. When operating frequencies are high, the inductances of short lengths of conductor and the capacitances betweeen short conductors and their surroundings cannot be neglected. These inductances and capacitances are distributed along the length of a conductor, and their effects intermingle at each point in the conductor. When the wavelength is very short in comparison with the physical dimensions of the line, we cannot represent distributed parameters accurately by means of lumped-parameter equivalent circuits. However, transmission lines form a convenient link between low-frequency circuit concepts and high-frequency field theories, because we can analyze them either in terms of voltage and current or in terms of electric and magnetic fields.

The primary use of transmission lines is to transmit electric power or signals between two points that are separated by a long distance as compared with a quarter wavelength of transmitting frequency. Operating at short wavelengths transmission lines also have many important uses in reactive circuit elements, resonant circuits, impedance transformers, and many other applications.

Basic transmission-line theory is derived from distributed circuit concepts, based on traveling-wave theory, so we can apply transmission-line theory to all types of lines. Ordinarily, transmission lines are classified according to their (1) physical characteristics and (2) functional properties.

The category based on physical characteristics contains four types of line, as shown in Fig. I-0-1:

Type 1—Two-wire transmission line. The two-wire transmission line consists of two parallel conducting wires separated by a uniform distance, as shown in Fig. I-0-1(a). Examples of this type of line are power lines and telephone lines. When the wires are close to each other, the proximity effect of one current on the other causes the current to be confined to the adjacent surfaces. The fields extend far beyond the wires and the radiation losses become excessive at higher frequencies. Hence, two-wire lines are not often used at frequencies above a few hundred MHz.

Type 2—Coaxial transmission line. The coaxial line consists of an inner conductor and a concentric outer conducting sheath separated by a dielectric medium, as shown in Fig. I-0-1(b). The electric and magnetic fields are entirely confined within the dielectric region, and radiation loss is very low. Examples of this type of line are telephone lines, TV cables, and coaxial lines used for precision measurements in laboratories.

Type 3—Stripline. The stripline consists of two parallel or two coplanar conducting strips separated by a dielectric slab of uniform thickness. A stripline may be a parallel stripline, as shown in Fig. I-0-1(c), a coplanar stripline, or a microstrip line. Striplines are commonly used in microwave integrated circuits (MICs).

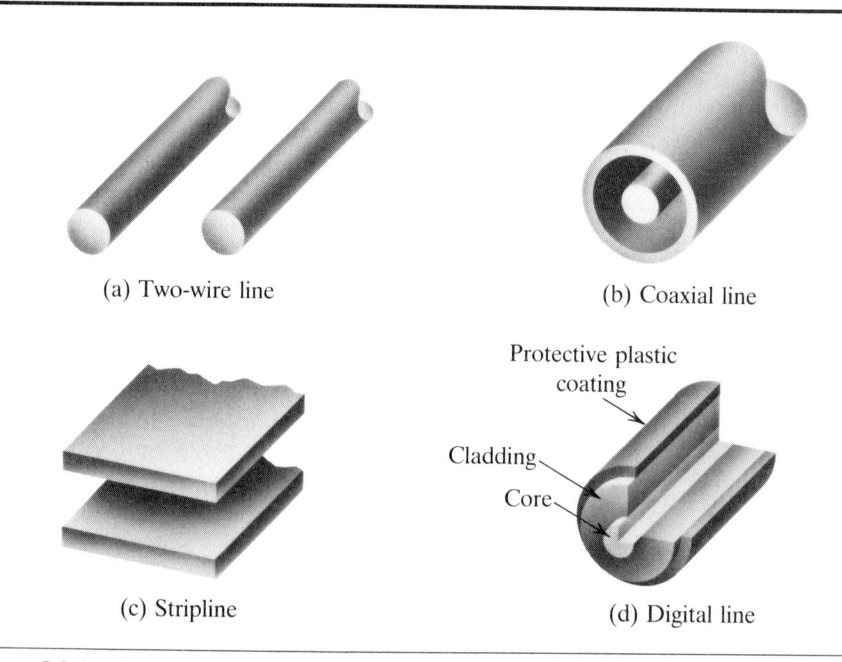

Figure I-0-1 Common types of transmission lines.

Type 4—Digital transmission line. The digital transmission line consists of an inner core of higher refractive index surrounded by an outer cladding of lower refractive index, as shown in Fig. I-0-1(d). Radiation loss is extremely low and major applications include computer links, industrial automation, medical instruments, telecommunications, and military command systems. The optic fiber is a commonly used link in digital communications systems.

The category of lines based on functional properties contains five types:

Type 1—Uniform line. All parameters of the line are considered to be uniformly distributed.

Type 2—Linear line. All parameters of the line are assumed to be independent of signal level.

Type 3—Lossless line. The series resistance and shunt conductance of the line are taken to be zero.

Type 4—Lossy line. The series resistance and shunt conductance of the line are not zero.

Type 5—Distortionless line. The ratio of the series resistance of the line to its series inductance equals the ratio of the shunt conductance of the line to its shunt capacitance.

Solving a transmission-line problem generally consists of three steps:

Step 1. Set up the partial differential equations of line voltage and line current according to Kirchhoff's laws.

Step 2. Apply the initial conditions to the traveling-wave differential equations set up in Step 1.

Step 3. Solve the resultant equations by using the circuit method or the classical differential equation method in terms of line voltage, line current, and transmitted power.

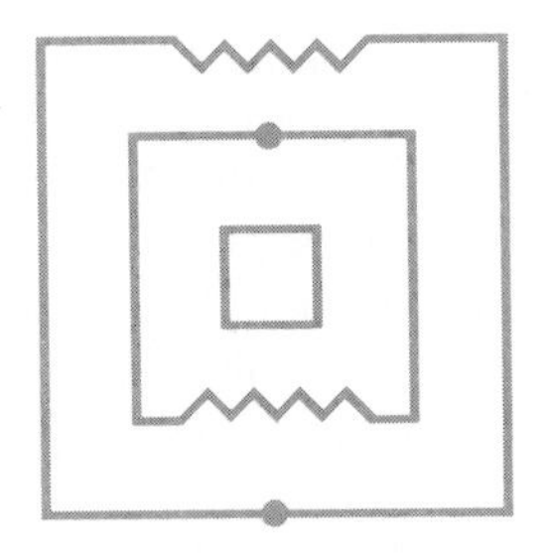

Chapter 1

Transmission-Line Equations

1-0 INTRODUCTION

We can analyze a transmission line either by solving Maxwell's field equations or by using the methods of distributed circuit theory. Solving Maxwell's equations involves three space variables in addition to the time variable. However, the distributed circuit methods involve only one space variable in addition to the time variable. In this chapter we use the distributed circuit method to analyze a transmission line in terms of the voltage, current, and impedance along the line in both the time and frequency domains.

1-0-1 Basic Parameters of Transmission-Line Equations

Each incremental length of a transmission line has its own series resistance, series inductance, shunt conductance and shunt capacitance. Hence, each incremental length of the line is like a small network wherein the effect of a change in the impressed voltage at its sending end of a certain incremental length will take a certain time to reach its receiving end. Similarly, the effect of a change in the impressed voltage at the sending end of the entire transmission line will travel down the line to the receiving end at a speed dependent upon the values of the line parameters and the medium surrounding the line. The resultant waves are called *traveling waves.*

The basic parameters of a transmission line are:

R—series resistance, in ohms per unit length (Ω/ℓ);

L—series inductance, in henries per unit length (H/ℓ);

G—shunt conductance, in mhos per unit length (℧/ℓ); and

C—shunt capacitance, in farads per unit length (F/ℓ).

For simplicity in solving transmission-line problems, we consider these parameters to be constant, even though this assumption is not always true. As the skin effect and the other parameters, such as the conductivity σ, permeability μ, and permittivity ε, are functions of frequency, one or more of R, L, G, and C will be frequency-dependent in some transmission lines. However, these parameters are determined basically by the physical configuration of the line. The parameter equations for commonly used transmission lines are presented in Table 1-0-1.

Table 1-0-1 Distributed parameters of commonly used transmission lines

Parameter	Two-Wire Line	Coaxial Line	Parallel Stripline	Unit
R	$\frac{R_s}{\pi a}$	$\frac{R_s}{2\pi}\left(\frac{1}{a}+\frac{1}{b}\right)$	$\frac{2R_s}{w}$	Ω/m
L	$\frac{\mu}{\pi}\cosh^{-1}\left(\frac{D}{2a}\right)$	$\frac{\mu}{2\pi}\ln\left(\frac{b}{a}\right)$	$\mu\frac{d}{w}$	H/m
G	$\frac{\pi\sigma}{\cosh^{-1}\left(\frac{D}{2a}\right)}$	$\frac{2\pi\sigma}{\ln(b/a)}$	$\sigma\frac{w}{d}$	℧/m
C	$\frac{\pi\varepsilon}{\cosh^{-1}\left(\frac{D}{2a}\right)}$	$\frac{2\pi\varepsilon}{\ln(b/a)}$	$\varepsilon\frac{w}{d}$	F/m
Z_0	$\sqrt{\frac{R+j\omega L}{G+j\omega C}}$	$\frac{60}{\sqrt{\varepsilon r}}\ln\left(\frac{b}{a}\right)$	$\frac{d}{w}\sqrt{\frac{\mu}{\varepsilon}}$	Ω

Note:

$R_s = \sqrt{\frac{\pi f \mu_c}{\sigma_c}}$ and is the surface resistance of a conducting strip in Ω/square.

μ_c is the permeability of a conductor in H/m.

σ_c is the conductivity of a conductor in ℧/m.

D is the distance between the centers of two wires in m.

a is the radius of the wire or center conductor in m.

b is the radius of the outer hollow conductor in m.

w is the width of a stripline in m.

d is the separation distance of a stripline in m.

1-1 TYPES OF TRANSMISSION-LINE EQUATIONS

1-1-1 Transmission-Line Equations in the Time Domain

The electrical equivalent of a physical two-wire transmission line with uniformly distributed constant parameters R, L, G, and C is shown in Fig. 1-1-1.

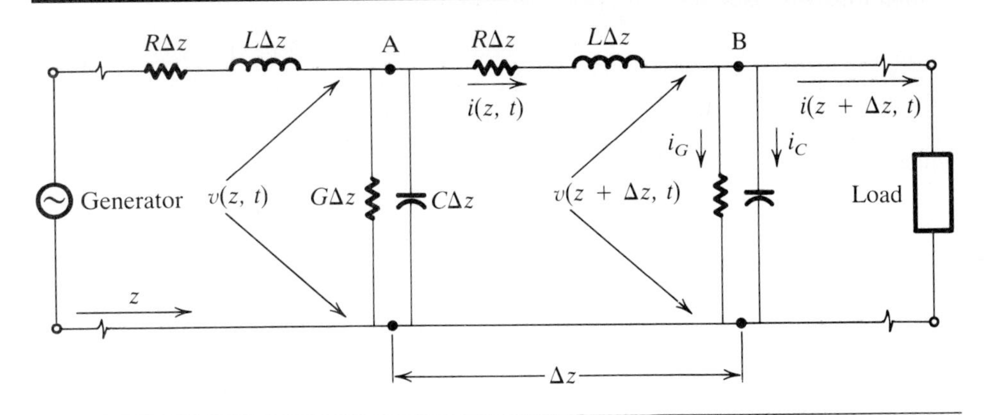

Figure 1-1-1 Equivalent elementary section of a transmission line.

We express the parameters in their respective per-unit length, and we assume the wave propagation to be in the positive z direction. In the following derivation of the transmission-line equations in the time domain, the lowercase letters v and i represent the instantaneous voltage and current along the line; these are scalar functions of position z and time t. The voltage and current are

$$v(z, t) = \mathbf{V}(z)v(t) \tag{1-1-1}$$

and

$$i(z, t) = \mathbf{I}(z)i(t) \tag{1-1-2}$$

where the boldface roman letters **V** and **I** are complex quantities of the sinusoids as functions of position z on the line. These complex quantities are called *phasors*. The phasors express the magnitudes and phases of the sinusoids at each position z. Thus we can express the line voltage and the line current as

$$\mathbf{V}(z) = \mathbf{V}e^{-\gamma z} \tag{1-1-3}$$

and

$$\mathbf{I}(z) = \mathbf{I}e^{-\gamma z} \tag{1-1-4}$$

where

$\mathbf{V}$ = complex voltage; and
$\mathbf{I}$ = complex current.

In addition

$$\gamma = \alpha + j\beta$$

where

γ = propagation constant;
α = attenuation constant in nepers per unit length; and
β = phase constant in radians per unit length.

By Kirchhoff's voltage law, the summation of the voltage drops around the central loop in Fig. 1-1-1 is

$$v(z, t) = i(z, t)R\,\Delta z + L\,\Delta z\frac{\partial i(z, t)}{\partial t} + v(z, t) + \frac{\partial v(z, t)}{\partial z}\Delta z \tag{1-1-5}$$

By rearranging Eq. (1-1-5), dividing it by Δz, and then omitting the argument (z, t), which is understood, we obtain

$$-\frac{\partial v}{\partial z} = Ri + L\frac{\partial i}{\partial t} \tag{1-1-6}$$

Using Kirchhoff's current law, we can express the summation of the currents at point B in Fig. 1-1-1 as

$$\begin{aligned} i(z, t) &= v(z+\Delta z, t)G\,\Delta z + C\,\Delta z\frac{\partial v(z+\Delta z, t)}{\partial t} + i(z+\Delta z, t) \\ &= \left[v(z, t) + \frac{\partial v(z, t)}{\partial z}\Delta z\right]G\,\Delta z + C\,\Delta z\frac{\partial}{\partial t}\left[v(z, t) + \frac{\partial v(z, t)}{\partial z}\Delta z\right] + i(z, t) + \frac{\partial i(z, t)}{\partial z}\Delta z \end{aligned} \tag{1-1-7}$$

By rearranging Eq. (1-1-7), dividing it by Δz, omitting (z, t), and letting $\Delta z = 0$, we have

$$-\frac{\partial i}{\partial z} = Gv + C\frac{\partial v}{\partial t} \tag{1-1-8}$$

Then by differentiating Eq. (1-1-6) with respect to z and Eq. (1-1-8) with respect to t and combining the results, we obtain the final transmission-line equation in voltage form:

$$\frac{\partial^2 v}{\partial z^2} = RGv + (RC + LG)\frac{\partial v}{\partial t} + LC\frac{\partial^2 v}{\partial t^2} \tag{1-1-9}$$

Also by differentiating Eq. (1-1-6) with respect to t and Eq. (1-1-8) with respect to z and combining the results, we obtain the final transmission-line equation in current form:

$$\frac{\partial^2 i}{\partial z^2} = RGi + (RC + LG)\frac{\partial i}{\partial t} + LC\frac{\partial^2 i}{\partial t^2} \tag{1-1-10}$$

Equations (1-1-9) and (1-1-10) are the general wave equations of the voltage and current on a uniform lossy line. For historical reasons, they are known as "the telegrapher's equations."

The important lossless case occurs when $R = G = 0$. Then the traveling-voltage and current-wave equations reduce to

$$\frac{\partial^2 v}{\partial z^2} = LC\frac{\partial^2 v}{\partial t^2} \tag{1-1-11}$$

and

$$\frac{\partial^2 i}{\partial z^2} = LC\frac{\partial^2 i}{\partial t^2} \tag{1-1-12}$$

These are simple one-dimensional wave equations. When you have found one of these equations, you may find the other from the first by applying either Eq. (1-1-6) or Eq. (1-1-7), respectively.

1-1-2 Transmission-Line Equations in the Frequency Domain

Euler's formula states that

$$e^{j\omega t} = \cos(\omega t) + j\sin(\omega t) \tag{1-1-13}$$

which is presented graphically in Fig. 1-1-2. The notation usually used is

$$E\cos(\omega t) = \text{Re}[Ee^{j\omega t}] \tag{1-1-14}$$

where the symbol Re is read *the real part of*. The imaginary part of $e^{j\omega t}$, which is $\sin(\omega t)$, would do just as well in the analysis, but algebraically, it seems more natural to take the real part. So, if we let a time function of e be

$$e = \text{Re}[Ee^{j\omega t}] \tag{1-1-15}$$

then

$$\begin{aligned} \frac{\partial e}{\partial t} &= \text{Re}\frac{\partial}{\partial t}[Ee^{j\omega t}] \\ &= -\omega E\sin(\omega t) \\ &= \frac{\partial}{\partial t}[E\cos(\omega t)] \end{aligned} \tag{1-1-16}$$

It is a common practice to omit the symbol Re and simply to write

$$e = Ee^{j\omega t} \tag{1-1-17}$$

then

$$\frac{\partial e}{\partial t} = Ej\omega e^{j\omega t} = j\omega e \tag{1-1-18}$$

This means that $\frac{\partial}{\partial t}$ is equivalent to $j\omega$ for a function of $e^{j\omega t}$. Similarly, $\frac{\partial}{\partial z}$ is equivalent to $j\beta$ for a function of $e^{j\beta z}$. By substituting $j\omega$ for $\frac{\partial}{\partial t}$ in Eq. (1-1-6) and in Eqs. (1-1-8)–

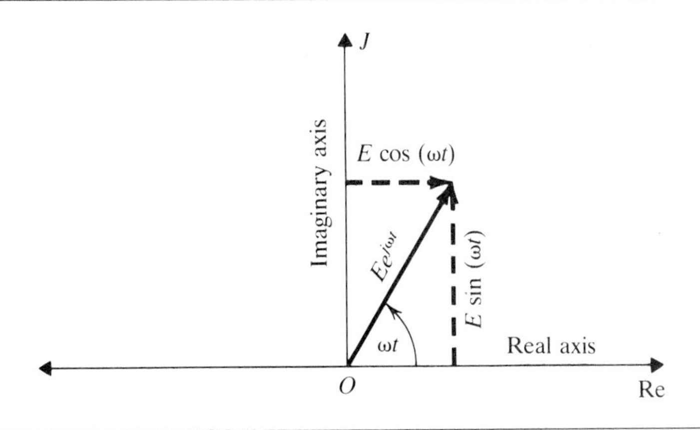

Figure 1-1-2 $Ee^{j\omega t}$ on the complex plane.

(1-1-10) and dividing each equation by $e^{j\omega t}$, we obtain the transmission-line equations in phasor form of the frequency-domain:

$$\frac{d\mathbf{V}}{dz} = -\mathbf{ZI} \tag{1-1-19}$$

$$\frac{d\mathbf{I}}{dz} = -\mathbf{YV} \tag{1-1-20}$$

$$\frac{d^2\mathbf{V}}{dz^2} = \gamma^2\mathbf{V} \tag{1-1-21}$$

and

$$\frac{d^2\mathbf{I}}{dz^2} = \gamma^2\mathbf{I} \tag{1-1-22}$$

in which the following substitutions were made:

$\mathbf{Z} = R + j\omega L$ Ω per unit length;
$\mathbf{Y} = G + j\omega C$ ℧ per unit length; and
$\gamma = \sqrt{\mathbf{ZY}} = \alpha + j\beta$ and is the propagation constant.

For a lossless line $R = G = 0$, and the transmission-line equations in the frequency domain are

$$\frac{d\mathbf{V}}{dz} = -j\omega L\mathbf{I} \tag{1-1-23}$$

$$\frac{d\mathbf{I}}{dz} = -j\omega C\mathbf{V} \tag{1-1-24}$$

$$\frac{d^2\mathbf{V}}{dz^2} = -\omega^2 LC\mathbf{V} \tag{1-1-25}$$

and

$$\frac{d^2\mathbf{I}}{dz^2} = -\omega^2 LC\mathbf{I} \tag{1-1-26}$$

Note that Eqs. (1-1-21) and (1-1-22) for a transmission line are similar to Eqs. (0-3-27) and (0-3-28) for electric and magnetic waves, respectively. The only difference is that the transmission-line equations are one-dimensional.

1-2 SOLUTIONS OF TRANSMISSION-LINE EQUATIONS

In Section 1-1 the transmission-line equations in the time domain and in frequency domain were derived by using Kirchhoff's voltage and current laws. The line equations are one-dimensional, second-order partial differential wave equations with constant coefficients. In this section we describe the solution of wave equations for line voltage and current in the frequency and time domains.

1-2-1 Frequency-Domain Solution of Lossless-Line Equations

The frequency-domain solution is also called the steady-state solution. In the frequency domain, the differential equations for a lossless line, Eqs. (1-1-25) and (1-1-26) are repeated here as

$$\frac{d^2\mathbf{V}}{dz^2} = -\omega^2 LC\mathbf{V} \tag{1-2-1}$$

and

$$\frac{d^2\mathbf{I}}{dz^2} = -\omega^2 LC\mathbf{I} \tag{1-2-2}$$

The *general* solution for Eq. (1-2-1) is

$$\mathbf{V}(z) = \mathbf{V}_+ e^{-j\beta z} + \mathbf{V}_- e^{+j\beta z} \tag{1-2-3}$$

where

$\mathbf{V}_+$ = complex voltage phasor in the positive z direction;
$\mathbf{V}_-$ = complex voltage phasor in the negative z direction;
$e^{-j\beta z}$ = traveling wave in the positive z direction;
$e^{+j\beta z}$ = traveling wave in the negative z direction;
$\beta = \omega\sqrt{LC}$ and is the phase constant in radians per unit length; and
βz = electrical length or distance in radians.

According to Eq. (1-1-23), the current $\mathbf{I}$ is now determined by the relation

$$\begin{aligned}\mathbf{I} &= -\frac{1}{j\omega L}\frac{d\mathbf{V}}{dz}\\ &= -\frac{1}{j\omega L}\frac{d}{dz}(\mathbf{V}_+ e^{-j\beta z} + \mathbf{V}_- e^{+j\beta z})\\ &= \sqrt{\frac{C}{L}}(\mathbf{V}_+ e^{-j\beta z} - \mathbf{V}_- e^{+j\beta z})\\ &= \mathbf{Y}_0\mathbf{V}_+ e^{-j\beta z} - \mathbf{Y}_0\mathbf{V}_- e^{+j\beta z}\\ &= \mathbf{I}_+ e^{-j\beta z} - \mathbf{I}_- e^{+j\beta z}\end{aligned} \tag{1-2-4}$$

in which we define the characteristic impedance of the line as

$$\mathbf{Z}_0 = \frac{1}{\mathbf{Y}_0} \equiv \sqrt{\frac{L}{C}} \tag{1-2-5}$$

The factors $\mathbf{I}_+$ and $\mathbf{I}_-$ represent complex currents traveling in the positive z direction and in the negative z direction, respectively. Since $\mathbf{V}_+$ and $\mathbf{V}_-$ (or $\mathbf{I}_+$ and $\mathbf{I}_-$) are independent quantities determined only by the boundary conditions and not by the differential equations, the complete solution for the line voltage and line current is the sum of the two independent solutions. The voltage and current on the line are shown in Fig. 1-2-1.

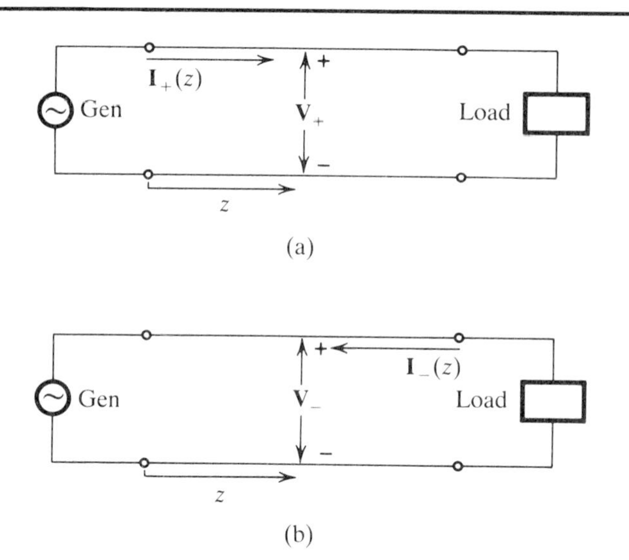

Figure 1-2-1 Line voltages and currents.

1-2-2 Frequency-Domain Solution of Lossy-Line Equations

In terms of a traveling wave, the major power loss along a lossy line is the result of some power dissipation in the line from resistance or conductance. The wave equations for the frequency domain and lossy line, as shown in Eqs. (1-1-21) and (1-1-22), are repeated here as

$$\frac{d^2\mathbf{V}}{dz^2} = \gamma^2\mathbf{V} \tag{1-2-6}$$

and

$$\frac{d^2\mathbf{I}}{dz^2} = \gamma^2\mathbf{I} \tag{1-2-7}$$

1. *Propagation Constants*

In general, the propagation constant γ of a transmission line is a complex quantity. The real part is called the *attenuation constant* α, and the imaginary part is called the *phase constant* β. That is,

$$\gamma = \alpha + j\beta = \sqrt{\mathbf{ZY}} = \sqrt{(R + j\omega L)(G + j\omega C)} \tag{1-2-8}$$

where

α = attenuation constant in Np/m; and

β = phase constant in rad/m.

At high frequencies (or with low losses), when $R \ll \omega L$ and $G \ll \omega C$—and by using the binomial expansion of $(1 + \mathrm{b})^{\pm 1/2} = 1 \pm \mathrm{b}/2$ for $\mathrm{b} \ll 1$—we can express the propagation constant as

$$\begin{aligned}
\gamma &= \pm\sqrt{(R + j\omega L)(G + j\omega C)} \\
&= \sqrt{(j\omega)^2 LC}\sqrt{\left(1 + \frac{R}{j\omega L}\right)\left(1 + \frac{G}{j\omega C}\right)} \\
&\simeq j\omega\sqrt{LC}\left[\left(1 + \frac{1}{2}\frac{R}{j\omega L}\right)\left(1 + \frac{1}{2}\frac{G}{j\omega C}\right)\right] \\
&= j\omega\sqrt{LC}\left[1 + \frac{1}{2}\left(\frac{R}{j\omega L} + \frac{G}{j\omega C}\right)\right] \\
&= \frac{1}{2}\left(R\sqrt{\frac{C}{L}} + G\sqrt{\frac{L}{C}}\right) + j\omega\sqrt{LC}
\end{aligned} \tag{1-2-8a}$$

Therefore the attenuation and phase constants are, respectively,

$$\alpha = \frac{1}{2}\left(R\sqrt{\frac{C}{L}} + G\sqrt{\frac{L}{C}}\right)$$

and

$$\beta = \omega\sqrt{LC} \tag{1-2-8b}$$

As the frequency increases, R becomes negligible compared to ωL, and G becomes negligible compared to ωC. Thus the characteristic impedance of a low-loss line at very high frequency is approximately equal to that of a lossless line.

The general solution for Eq. (1-2-6) is

$$\mathbf{V} = \mathbf{V}_+ e^{-\gamma z} + \mathbf{V}_- e^{\gamma z} \tag{1-2-9}$$

According to Eq. (1-1-19), the current **I** is now determined by the relation

$$\begin{aligned}
\mathbf{I} &= -\frac{1}{\mathbf{Z}}\frac{d\mathbf{V}}{dz} \\
&= -\frac{1}{\mathbf{Z}}\frac{d}{dz}(\mathbf{V}_+ e^{-\gamma z} + \mathbf{V}_- e^{+\gamma z}) \\
&= \frac{\gamma}{\mathbf{Z}}(\mathbf{V}_+ e^{-\gamma z} - \mathbf{V}_- e^{\gamma z}) \\
&= \mathbf{Y}_0(\mathbf{V}_+ e^{-\gamma z} - \mathbf{V}_- e^{\gamma z})
\end{aligned} \tag{1-2-10}$$

where the characteristic impedance is

$$\begin{aligned}
\mathbf{Z}_0 &= \frac{1}{\mathbf{Y}_0} \equiv \sqrt{\frac{\mathbf{Z}}{\mathbf{Y}}} \\
&= \sqrt{\frac{R + j\omega L}{G + j\omega C}} = R_0 \pm jX_0
\end{aligned} \tag{1-2-11}$$

Both the voltage and current waves are attenuated along the line, as shown in Fig. 1-2-2.

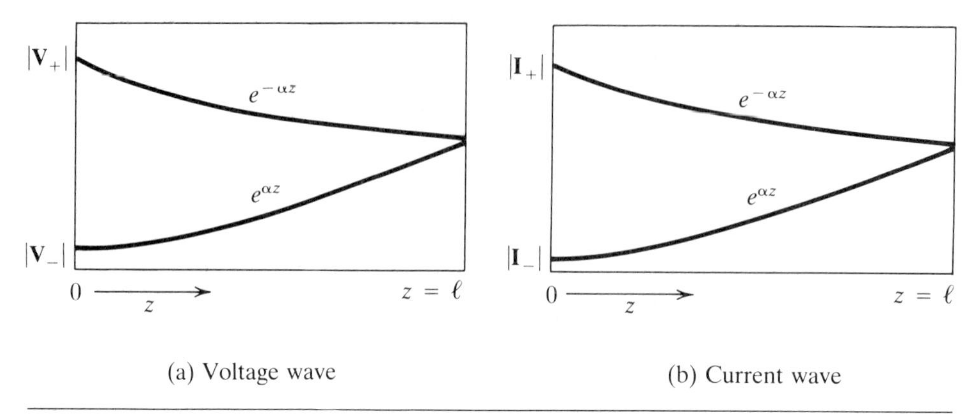

Figure 1-2-2 Magnitude of voltage and current traveling waves on a lossy line.

2. *The Neper and the Decibel*

If the line is lossy, the positive voltage wave along the line is a function of position z, time t, attenuation constant α, and phase constant β. We can express the voltage wave as

$$\begin{aligned} v_+(z, t) &= \text{Re}[\mathbf{V}_+(z)e^{j\omega t}] \\ &= \text{Re}[\mathbf{V}_+ e^{-\gamma z} e^{j\omega t}] \\ &= \text{Re}[\mathbf{V}_+ e^{-\alpha z} e^{j\omega t} e^{-j\beta z}] \\ &= |\mathbf{V}_+| e^{-\alpha z} \cos(\omega t - \beta z + \theta_+) \end{aligned} \tag{1-2-12}$$

where $\mathbf{V}_+ = |\mathbf{V}_+|e^{j\theta_+}$ at some initial position. The voltage wave is moving in the positive z direction, with an amplitude that is decreasing by a factor of $e^{-\alpha z}$. When the voltage wave moves from position 1 to position 2 along the line in the positive z direction for a distance of $\ell = z_2 - z_1$, the ratios of the magnitudes of voltage $\mathbf{V}_+$ and current $\mathbf{I}_+$ at the two points are

$$\left|\frac{\mathbf{V}_2}{\mathbf{V}_1}\right| = \left|\frac{\mathbf{I}_2}{\mathbf{I}_1}\right| = e^{-\alpha\ell} \tag{1-2-13}$$

The quantity $|\alpha\ell|$ indicates the total attenuation of the traveling wave between the two points. Taking the natural logarithm of both sides of Eq. (1-2-13), we obtain an expression for the number of Nepers (Np) of attenuation. That is,

$$\text{Nepers} = \ln\left|\frac{\mathbf{V}_2}{\mathbf{V}_1}\right| = \ln\left|\frac{\mathbf{I}_2}{\mathbf{I}_1}\right| = \ln e^{-\alpha\ell} = |\alpha\ell| \tag{1-2-14}$$

where ln represents the natural logarithm to the base $e = 2.7183$. The decibel (dB) is defined as the logarithm of a power ratio:

$$\text{dB} = 10 \log \frac{P_2}{P_1} \tag{1-2-15}$$

where log represents the logarithm to the base 10. As a result, 1 Neper equals 8.686 dB.

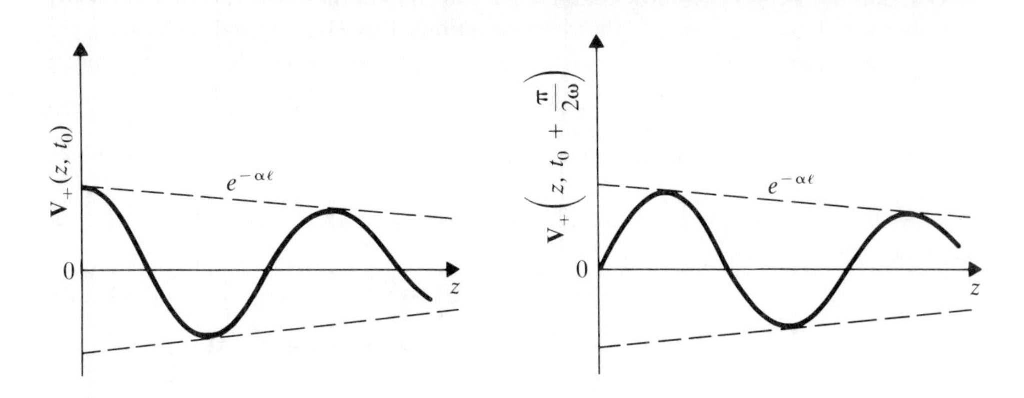

Figure 1-2-3 The positive voltage wave at two instants in time.

Since $P_1 = \mathbf{V}_1^2/R_1$ and $P_2 = \mathbf{V}_2^2/R_2$, the number of decibels is negative for $P_2 < P_1$, when their voltages are measured across the same or equal resistors. That is,

$$\text{dB} = 20 \log \left| \frac{\mathbf{V}_2}{\mathbf{V}_1} \right| \tag{1-2-16}$$

Similarly,

$$\text{dB} = 20 \log \left| \frac{\mathbf{I}_2}{\mathbf{I}_1} \right| \tag{1-2-17}$$

Since the voltage $\mathbf{V}_2$ is smaller than $\mathbf{V}_1$ (or $\mathbf{I}_2 < \mathbf{I}_1$), the number of decibels is negative. That means the traveling wave is attenuated by that number of decibels for a given distance along the lossy line. The curve shown in Fig. 1-2-3, represents a wave moving in the positive z direction, with an amplitude attenuated by a factor of $-\alpha\ell$ as it progresses.

1-2-3 Time-Domain Solution of Lossless-Line Equations

In general, the frequency-domain solutions contain all the information about the wave equations in both time and space for the steady state. However, the time-domain solutions may reveal more important phenomena in the transient state, such as the fault location along the line. So the time-domain solution is sometimes called the transient solution. The wave equations for the time-domain solution on lossless line, Eqs. (1-1-11) and (1-1-12), are repeated here as

$$\frac{\partial^2 v}{\partial z^2} = LC \frac{\partial^2 v}{\partial t^2} \tag{1-2-18}$$

and

$$\frac{\partial^2 i}{\partial z^2} = LC \frac{\partial^2 i}{\partial t^2} \tag{1-2-19}$$

These one-dimensional wave equations represent waves that can travel in either direction at a velocity of $1/\sqrt{LC}$ without change in magnitude or form. In order to verify this condition, we have to show that a wave traveling in the positive z direction with a velocity of $1/\sqrt{LC}$ can satisfy the wave equations, Eqs. (1-2-18) and (1-2-19).

To begin, we let the following voltage function be a solution to the wave equations:

$$v_+ = f_+(t - \sqrt{LC}, z) \tag{1-2-20}$$

Where f_+ represents any single-valued function of argument $(t - \sqrt{LC}, z)$ and phase velocity $v_p = 1/\sqrt{LC}$.

Next, we must apply partial differentiation twice to Eq. (1-2-20). We let the argument of the function be

$$\mathrm{A} = (t - \sqrt{LC}, z) \tag{1-2-21}$$

Then

$$v_+ = f_+(\mathrm{A}) \tag{1-2-22}$$

The partial differentiation theory states that

$$\frac{\partial v_+}{\partial z} = \frac{df_+}{d\mathrm{A}} \frac{\partial \mathrm{A}}{\partial z} \tag{1-2-23}$$

Differentiating Eq. (1-2-21) with respect to z and substituting the result from Eq. (1-2-23) yield

$$\frac{\partial v_+}{\partial z} = -\sqrt{LC}\frac{df_+}{d\mathrm{A}} \tag{1-2-24}$$

Differentiating Eq. (1-2-23) with respect to z once again, we have

$$\frac{\partial^2 v_+}{\partial z^2} = LC\frac{d^2 f_+}{d\mathrm{A}^2} \tag{1-2-25}$$

Similarly,

$$\frac{\partial \mathrm{A}}{\partial t} = 1$$

$$\frac{\partial v_+}{\partial t} = \frac{df_+}{d\mathrm{A}} \frac{\partial \mathrm{A}}{\partial t}$$

$$= \frac{df_+}{d\mathrm{A}}$$

and

$$\frac{\partial^2 v_+}{\partial t^2} = \frac{d^2 f_+}{d\mathrm{A}^2}$$

Substituting the result from Eq. (1-2-25) yields

$$\frac{\partial^2 v_+}{\partial z^2} = LC\frac{\partial^2 v_+}{\partial t^2} \tag{1-2-26}$$

Therefore, the assumed function, $v_+ = f_+(t - \sqrt{LC}, z)$ is one solution of Eq. (1-2-18) with a phase velocity of $v_p = 1/\sqrt{LC}$. Similarly, the wave traveling in the negative z direction, $v_- = f_-(t + \sqrt{LC}, z)$, is also one solution of Eq. (1-2-18). Since Eq. (1-2-18) is linear, the sum of two independent solutions is also a solution. Hence,

$$v = f_+(t - \sqrt{LC}, z) + f_-(t + \sqrt{LC}, z) \tag{1-2-27}$$

is surely a solution. Similarly,

$$i_+ = g_+(t - \sqrt{LC}, z)$$

and

$$i_- = g_-(t + \sqrt{LC}, z)$$

are the solutions of Eq. (1-2-19). Therefore

$$i = g_+(t - \sqrt{LC}, z) + g_-(t + \sqrt{LC}, z) \tag{1-2-28}$$

is surely a solution. For example, we can replace the time-domain solution of Eq. (1-2-20) with the cosine function. If we let

$$v_+ = \mathbf{V}_+ \cos \omega(t - \sqrt{LC}, z) \tag{1-2-28a}$$

then

$$\frac{\partial^2 v_+}{\partial z^2} = -\omega^2 LC\mathbf{V}_+ \cos [\omega(t - \sqrt{LC}, z)] \tag{1-2-28b}$$

and

$$\frac{\partial^2 v_+}{\partial t^2} = -\omega^2 \mathbf{V}_+ \cos [\omega(t - \sqrt{LC}, z)] \tag{1-2-28c}$$

Therefore, v_+ is a solution of Eq. (1-2-18).

1-2-4 Time-Domain Solution of Lossy-Line Equations

Because of losses in series resistance and shunt conductance, the amplitude of a traveling wave is reduced exponentially by a factor of $e^{-\alpha z}$ in the positive z direction and by $e^{+\alpha z}$ in the negative z direction. Even in the absence of a reflected wave, the sinusoidal components of the signal will not, in general, retain either their relative amplitudes or relative phases as they move. That is, the shape of the time function at $(z + \ell)$ may be different from that at z.

The wave equations for the time-domain solution on lossy line, Eqs. (1-1-9) and (1-1-10), are repeated here as

$$\frac{\partial^2 v}{\partial z^2} = RGv + (RC + LG)\frac{\partial v}{\partial t} + LC\frac{\partial^2 v}{\partial t^2} \tag{1-2-29}$$

and

$$\frac{\partial^2 i}{\partial z^2} = RGi + (RC + LG)\frac{\partial i}{\partial t} + LC\frac{\partial^2 i}{\partial t^2} \tag{1-2-30}$$

Let's assume the following solution, in which α and β are unknown coefficients to be determined.

$$v_+ = \mathbf{V}_+ e^{-\alpha z} \cos (\omega t - \beta z) \tag{1-2-31}$$

Then by changing the cosine function into an exponential form for convenience in differentiating, we obtain

$$v_+ = \frac{\mathbf{V}_+}{2}[e^{-(\alpha + j\beta)z}e^{j\omega t} + e^{-(\alpha - j\beta)z}e^{-j\omega t}]$$

$$\frac{\partial v_+}{\partial z} = \frac{\mathbf{V}_+}{2}[-(\alpha + j\beta)e^{-(\alpha + j\beta)z}e^{j\omega t} - (\alpha - j\beta)e^{-(\alpha - j\beta)z}e^{-j\omega t}]$$

$$\frac{\partial^2 v_+}{\partial z^2} = \frac{\mathbf{V}_+}{2}[(\alpha + j\beta)^2 e^{-(\alpha + j\beta)z}e^{j\omega t} + (\alpha - j\beta)^2 e^{-(\alpha - j\beta)z}e^{-j\omega t}]$$

$$\frac{\partial v_+}{\partial t} = \frac{\mathbf{V}_+}{2}[j\omega e^{-(\alpha + j\beta)z}e^{j\omega t} - j\omega e^{-(\alpha - j\beta)z}e^{-j\omega t}]$$

and

$$\frac{\partial^2 v_+}{\partial t^2} = \frac{\mathbf{V}_+}{2}[-\omega^2 e^{-(\alpha + j\beta)z}e^{j\omega t} - \omega^2 e^{-(\alpha - j\beta)z}e^{-j\omega t}]$$

Substituting the results into Eq. (1-2-29), dividing both sides by $(\mathbf{V}_+/2)$, and rearranging terms, we get

$$[(\alpha + j\beta)^2 - RG - (RC + LG)j\omega - LC(j\omega)^2] + [(\alpha + j\beta)^2 - RG - (RC + LG)j\omega - LC(j\omega)^2]^* e^{-j(2\omega t - 2\beta z)} = 0 \quad (1\text{-}2\text{-}32)$$

where (*) represents the complex conjugate of the time domain. Equation (1-2-32) can be satisfied for independently chosen values of t and z if, and only if, the two conjugated factors are zeros. That is,

$$(\alpha + j\beta)^2 - RG - (RC + LG)j\omega - LC(j\omega)^2 = 0 \quad (1\text{-}2\text{-}33)$$

and

$$(\alpha - j\beta)^2 - RG - (RC + LG)(-j\omega) - LC(-j\omega)^2 = 0 \quad (1\text{-}2\text{-}34)$$

Let

$\gamma = \alpha + j\beta$ = propagation constant;
α = attenuation constant in nepers per unit length;
β = phase constant in radians per unit length; and
$\beta\ell$ = electric length in radians or degrees.

From Eq. (1-2-33)

$$\gamma = \sqrt{RG + (RC + LG)j\omega + LC(j\omega)^2} = \sqrt{(R + j\omega L)(G + j\omega C)}$$

Let

$\mathbf{Z} = R + j\omega L$; and
$\mathbf{Y} = R + j\omega C$.

Then

$$\gamma = \sqrt{\mathbf{ZY}} \quad (1\text{-}2\text{-}35)$$

as defined in Eq. (1-2-8). Therefore, Eq. (1-2-31) is surely one solution of the traveling-wave equation, Eq. (1-2-29). Similarly,

$$v_- = \mathbf{V}_- e^{\alpha z} \cos(\omega t + \beta z) \tag{1-2-36}$$

is also a solution. Since Eqs. (1-2-31) and (1-2-36) are independent solutions, the sum of the two solutions, or

$$v = \mathbf{V}_+ e^{-\alpha z} \cos(\omega t - \beta z) + \mathbf{V}_- e^{\alpha z} \cos(\omega t + \beta z) \tag{1-2-37}$$

is also a solution. Similarly,

$$i_+ = \mathbf{I}_+ e^{-\alpha z} \cos(\omega t - \beta z) \tag{1-2-38}$$

$$i_- = \mathbf{I}_- e^{\alpha z} \cos(\omega t - \beta z) \tag{1-2-39}$$

and

$$i = \mathbf{I}_+ e^{-\alpha z} \cos(\omega t - \beta z) + \mathbf{I}_- e^{\alpha z} \cos(\omega t + \beta z) \tag{1-2-40}$$

are the solutions of the traveling-wave equation (1-2-30).

1-3 CHARACTERISTIC IMPEDANCE, PROPAGATION CONSTANT, AND LINE IMPEDANCE

1-3-1 Characteristic Impedance

We define the characteristic impedance of a transmission line as

$$\mathbf{Z}_0 = \sqrt{\frac{\mathbf{Z}}{\mathbf{Y}}} = \sqrt{\frac{R + j\omega L}{G + j\omega C}} = R_0 \pm jX_0 \tag{1-3-1}$$

The characteristic impedance of a line has the following properties:

1. It is independent of the length of the line.
2. It is independent of the termination of the line.
3. It is not the impedance that a line itself possesses.
4. It is determined only by the parameters of the line per unit length.

At high frequencies or with low losses—when $R \ll \omega L$ and $G \ll \omega C$—we can approximate the characteristic impedance from the binomial expansion:

$$\begin{aligned}
\mathbf{Z}_0 &= \sqrt{\frac{\mathbf{Z}}{\mathbf{Y}}} \\
&= \sqrt{\frac{R + j\omega L}{G + j\omega C}} \\
&= \sqrt{\frac{L}{C}}\left(1 + \frac{R}{j\omega L}\right)^{1/2}\left(1 + \frac{G}{j\omega C}\right)^{-1/2} \\
&\simeq \sqrt{\frac{L}{C}}\left(1 + \frac{1}{2}\frac{R}{j\omega L}\right)\left(1 - \frac{1}{2}\frac{G}{j\omega C}\right) \\
&= \sqrt{\frac{L}{C}}\left[1 + \frac{1}{2}\left(\frac{R}{j\omega L} - \frac{G}{j\omega C}\right)\right]
\end{aligned} \tag{1-3-2}$$

but

$$\mathbf{Z}_0 = \sqrt{\frac{L}{C}} \qquad \text{for very high frequency}$$

1. *Admittance*

For a branched transmission line, it is better to solve the line equations for the line voltage, current, and transmitted power in terms of admittance instead of impedance. We define the characteristic admittance and generalized admittance as

$$\mathbf{Y}_0 \equiv \frac{1}{\mathbf{Z}_0} = G_0 \pm jB_0 \tag{1-3-3}$$

and

$$\mathbf{Y} \equiv \frac{1}{\mathbf{Z}} = G \pm jB \tag{1-3-4}$$

EXAMPLE 1-3-1 Characteristic Impedance and Propagation Constant

A coaxial line has the following parameters:

$R = 5\ \Omega/\text{mi}$

$L = 37 \times 10^{-4}\ \text{H/mi}$

$G = 6.2 \times 10^{-3}\ \mho/\text{mi}$

$C = 0.0081 \times 10^{-6}\ \text{F/mi}$

The line operates at a frequency of 100 kHz. Determine its (a) characteristic impedance and (b) propagation constant.

Solutions:

a) $$\mathbf{Z}_0 = \sqrt{\frac{\mathbf{Z}}{\mathbf{Y}}} = \sqrt{\frac{R + j\omega L}{G + j\omega C}}$$

$$= \sqrt{\frac{5 + j2323}{(0.62 + j0.51) \times 10^{-2}}}$$

$$= \sqrt{\frac{2324\angle 90^\circ}{0.0080\angle 39.4^\circ}}$$

$$= 539\angle 25.3^\circ$$

$$= 487 + j230\ \Omega$$

b) $$\gamma = \sqrt{\mathbf{ZY}}$$

$$= \sqrt{(R + j\omega L)(G + j\omega C)}$$

$$= \sqrt{(2324\angle 90^\circ)(0.0080\angle 39.4^\circ)} = 4.31\angle 64.7^\circ$$

$$= 1.85 + j3.90;$$

$\alpha = 1.85$ Np/mi and

$\beta = 3.90$ rad/mi

EXAMPLE 1-3-2 Computer Solution for Characteristic Impedance and Propagation Constant of a Line

```
10C      PROGRAM TO COMPUTE THE CHARACTERISTIC IMPEDANCE ZO AND
11C+     PROPAGATION CONSTANT GAMMA
20C      SLCOM6
30       COMPLEX Z,Y,ZO,GAMMA
40       REAL I,L
50     5 READ 65, R,L,G,C,F
60    65 FORMAT(5F10.5)
65       IF(R ,LE, 0,0) STOP
70       PRINT 85,P,L,G,C,F
79    85 FORMAT(1H ,15HRESISTANCE  R= ,F10.5,
80+      1X,13HOHMS PER MILE/
81+      1X,15HINDUCTANCE   L= ,F10.5,1X,14HHENRY PER MILE/
2+      1X,15HCONDUCTANCE G= ,F10.5,1X,12HMHO PER MILE/
83+      1X,15HCAPACITANCE O= ,F10.5, 1X,19HMICROFARAD PER MILE/
84+      1X,15HFREQUENCY    F= ,F10.5,1X,3HKHZ//)
90       F=F*1000.0
100      C=C*0.000001
110      WL=2.*3.14159*F*L
120      Z=CMPLX(R,WL)
130      WC=2.*3.14159*F*C
140      Y=CMPLX(G,WC)
150      ZO=CSQRT(Z/Y)
160      GAMMA=CSQRT(Z*Y)
170      PRINT 95, ZO
180 95   FORMAT(1X,18HREAL OF ZO       = ,F10.5,1X,4HOHMS/
181+     1X,18HIMAGINARY OF ZO = ,F10.5,1X,4HOHMS/)
210      PRINT 98, GAMMA
220 98   FORMAT(1X,21HREAL OF GAMMA       = ,F10.5,
221+     1X,15HNEPERS PER MILE/
222+     1X,21HIMAGINARY OF GAMMA = ,F10.5,
223+     1X,16HRADIANS PER MILE//)
230      GO TO 5
240      STOP
250      END

READY
RUN

PROGRAM   SLCOM6

?     5.00000    0.00370    0.00620    0.00810 100.00000
 RESISTANCE  R=     5.00000 OHMS PER MILE
 INDUCTANCE  L=      .00370 HENRY PER MILE
 CONDUCTANCE G=      .00620 MHO PER MILE
 CAPACITANCE C=      .00810 MICROFARAD PER MILE
 FREQUENCY   F=   100.00000 KHZ

 REAL OF ZO       =  486.92637 OHMS
 IMAGINARY OF ZO  =  229.62511 OHMS

 REAL OF GAMMA       =     1.85030 NEPERS PER MILE
 IMAGINARY OF GAMMA  =     3.90183 RADIANS PER MILE

?    0.0

SRU       0.828 UNTS.

RUN COMPLETE.
BYE

F228001    LOG OFF     08.09.49.
F228001    SRU       1.000 UNTS.
```

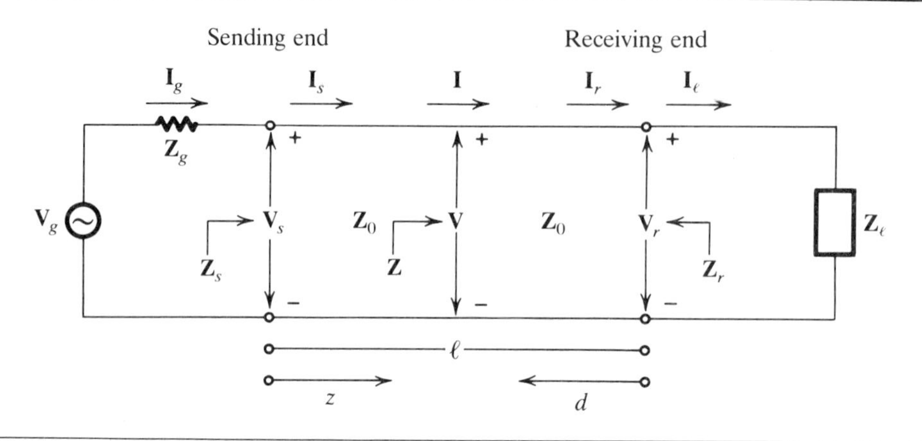

Figure 1-3-1 Diagram of transmission line showing notations.

1-3-2 Line Impedance

The line impedance of a transmission line is the complex ratio of the voltage phasor at any point to the current phasor at that point. We define it as

$$\mathbf{Z} \equiv \frac{\mathbf{V}(z)}{\mathbf{I}(z)} \tag{1-3-5}$$

Figure 1-3-1 shows a diagram for a transmission line.

In general, the voltage (or current) along a line is the sum of the incident wave and the reflected wave. That is,

$$\mathbf{V} = \mathbf{V}_{\text{inc}} + \mathbf{V}_{\text{ref}} = \mathbf{V}_+ e^{-\gamma z} + \mathbf{V}_- e^{\gamma z} \tag{1-3-6}$$

and

$$\mathbf{I} = \mathbf{I}_{\text{inc}} + \mathbf{I}_{\text{ref}} = (\mathbf{V}_+ e^{-\gamma z} - \mathbf{V}_- e^{\gamma z}) Y_0 \tag{1-3-7}$$

where

$\gamma = \alpha + j\beta$; and

$\mathbf{Y}_0 = 1/\mathbf{Z}_0$.

1. ***Impedance Computed from the Sending End (or Input Port)***

Sometimes we find it advantageous to determine the impedance down the line from the sending end. At the sending end, $z = 0$, the voltage and current are

$$\mathbf{V}_s = \mathbf{I}_s \mathbf{Z}_s \quad \text{and} \quad \mathbf{I}_s = \mathbf{I}_s$$

Thus

$$\mathbf{I}_s \mathbf{Z}_s = \mathbf{V}_+ + \mathbf{V}_- \quad \text{and} \quad \mathbf{I}_s \mathbf{Z}_0 = \mathbf{V}_+ - \mathbf{V}_-$$

Solving these two equations for $\mathbf{V}_+$ and $\mathbf{V}_-$, we have

$$\mathbf{V}_+ = \frac{\mathbf{I}_s}{2}(\mathbf{Z}_s + \mathbf{Z}_0) \tag{1-3-8}$$

and

$$\mathbf{V}_- = \frac{\mathbf{I}_s}{2}(\mathbf{Z}_s - \mathbf{Z}_0) \tag{1-3-9}$$

Substituting $\mathbf{V}_+$ and $\mathbf{V}_-$ into Eqs. (1-3-6) and (1-3-7) yields

$$\mathbf{V} = \frac{\mathbf{I}_s}{2}[(\mathbf{Z}_s + \mathbf{Z}_0)e^{-\gamma z} + (\mathbf{Z}_s - \mathbf{Z}_0)e^{\gamma z}] \tag{1-3-10}$$

and

$$\mathbf{I} = \frac{\mathbf{I}_s}{2\mathbf{Z}_0}[(\mathbf{Z}_s + \mathbf{Z}_0)e^{-\gamma z} - (\mathbf{Z}_s - \mathbf{Z}_0)e^{\gamma z}] \tag{1-3-11}$$

Then the line impedance at a point any distance from the sending end, in terms of $\mathbf{Z}_s$ and $\mathbf{Z}_0$, is

$$\mathbf{Z} = \mathbf{Z}_0 \frac{(\mathbf{Z}_s + \mathbf{Z}_0)e^{-\gamma z} + (\mathbf{Z}_s - \mathbf{Z}_0)e^{\gamma z}}{(\mathbf{Z}_s + \mathbf{Z}_0)e^{-\gamma z} - (\mathbf{Z}_s - \mathbf{Z}_0)e^{\gamma z}} \tag{1-3-12}$$

For $z = \ell$, we express the line impedance at the receiving end, in terms of $\mathbf{Z}_s$ and $\mathbf{Z}_0$, as

$$\mathbf{Z}_r = \mathbf{Z}_0 \frac{(\mathbf{Z}_s + \mathbf{Z}_0)e^{-\gamma \ell} + (\mathbf{Z}_s - \mathbf{Z}_0)e^{\gamma \ell}}{(\mathbf{Z}_s + \mathbf{Z}_0)e^{-\gamma \ell} - (\mathbf{Z}_s - \mathbf{Z}_0)e^{\gamma \ell}} \tag{1-3-13}$$

Note that:

1. If the sending-end impedance of a transmission line matches the characteristic impedance $\mathbf{Z}_0$ of the line, or $\mathbf{Z}_s = \mathbf{Z}_0$, the impedance at the receiving end will equal the characteristic impedance, or $\mathbf{Z}_r = \mathbf{Z}_0$.
2. If the sending-end impedance $\mathbf{Z}_s$ does not match the line, that is, $\mathbf{Z}_s \neq \mathbf{Z}_0$, the receiving-end impedance will not equal the characteristic impedance, that is, $\mathbf{Z}_r \neq \mathbf{Z}_0$, and it can be found from Eq. (1-3-13).

We can find the input impedance and current at the sending end from Fig. 1-3-2. The sending-end current and impedance are

$$\mathbf{I}_s = \mathbf{I}_g \tag{1-3-14}$$

and

$$\mathbf{Z}_s = \frac{\mathbf{V}_g}{\mathbf{I}_g} = \mathbf{Z}_g \tag{1-3-15}$$

Substituting Eqs. (1-3-14) and (1-3-15) into Eq. (1-3-12) yields

$$\mathbf{Z} = \mathbf{Z}_0 \frac{\left[\left(\frac{\mathbf{V}_g}{\mathbf{I}_g} + \mathbf{Z}_0 - \mathbf{Z}_g\right)e^{-\gamma z} + \left(\frac{\mathbf{V}_g}{\mathbf{I}_g} - \mathbf{Z}_0 - \mathbf{Z}_g\right)e^{\gamma z}\right]}{\left[\left(\frac{\mathbf{V}_g}{\mathbf{I}_g} + \mathbf{Z}_0 - \mathbf{Z}_g\right)e^{-\gamma z} - \left(\frac{\mathbf{V}_g}{\mathbf{I}_g} - \mathbf{Z}_0 - \mathbf{Z}_g\right)e^{\gamma z}\right]} \tag{1-3-16}$$

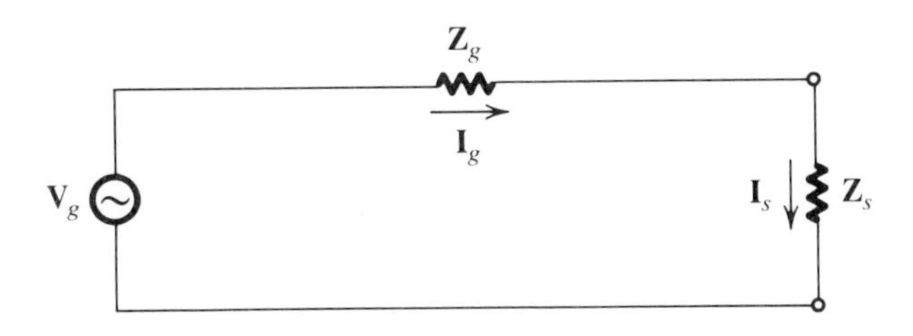

Figure 1-3-2 A simplified circuit at the sending end.

This solution shows that we can find the line impedance at any point from the source voltage $\mathbf{V}_g$, current $\mathbf{I}_g$, and impedance $\mathbf{Z}_g$.

If the internal impedance $\mathbf{Z}_g$ of the generator matches the characteristic impedance $\mathbf{Z}_0$ of the line,

$$\mathbf{Z}_g = \mathbf{Z}_0$$

and we can simplify Eq. (1-3-16) to

$$\mathbf{Z} = \mathbf{Z}_0 \frac{\left[\frac{\mathbf{V}_g}{\mathbf{I}_g} e^{-\gamma z} + \left(\frac{\mathbf{V}_g}{\mathbf{I}_g} - 2\mathbf{Z}_0\right) e^{\gamma z}\right]}{\left[\frac{\mathbf{V}_g}{\mathbf{I}_g} e^{-\gamma z} - \left(\frac{\mathbf{V}_g}{\mathbf{I}_g} - 2\mathbf{Z}_0\right) e^{\gamma z}\right]} \tag{1-3-17}$$

Furthermore, if the sending-end impedance $\mathbf{Z}_s$ of the line also matches its characteristic impedance, $\mathbf{Z}_s = \mathbf{Z}_0$, the line impedance equals the characteristic impedance:

$$\mathbf{Z} = \mathbf{Z}_0 \tag{1-3-18}$$

Thus if both the load impedance and the internal impedance of the generator match the characteristic impedance, the line impedance remains constant everywhere as $\mathbf{Z}_0$.

2. *Impedance Computed from the Receiving End (or Output Port)*

Sometimes we find it desirable to determine the impedance from the receiving end, looking backward along the line. At the receiving end, $z = \ell$, and $\mathbf{V}_R = \mathbf{I}_\ell \mathbf{Z}_\ell$; we can express the line impedance in terms of $\mathbf{Z}_\ell$ and $\mathbf{Z}_0$:

$$\mathbf{I}_\ell \mathbf{Z}_\ell = \mathbf{V}_+ e^{-\gamma\ell} + \mathbf{V}_- e^{\gamma\ell} \tag{1-3-19}$$

and

$$\mathbf{I}_\ell \mathbf{Z}_0 = \mathbf{V}_+ e^{-\gamma\ell} - \mathbf{V}_- e^{\gamma\ell} \tag{1-3-20}$$

Solving Eqs. (1-3-19) and (1-3-20) for $\mathbf{V}_+$ and $\mathbf{V}_-$, we obtain

$$\mathbf{V}_+ = \frac{\mathbf{I}_\ell}{2}(\mathbf{Z}_\ell + \mathbf{Z}_0) e^{\gamma\ell} \tag{1-3-21}$$

and

$$\mathbf{V}_- = \frac{\mathbf{I}_\ell}{2}(\mathbf{Z}_\ell - \mathbf{Z}_0) e^{-\gamma\ell} \tag{1-3-22}$$

Then substituting these results into Eqs. (1-3-6) and (1-3-7), and letting $\ell - z = d$, we have

$$\mathbf{V} = \frac{\mathbf{I}_\ell}{2}[(\mathbf{Z}_\ell + \mathbf{Z}_0) e^{\gamma d} + (\mathbf{Z}_\ell - \mathbf{Z}_0) e^{-\gamma d}] \tag{1-3-23}$$

and

$$\mathbf{I} = \frac{\mathbf{I}_\ell}{2\mathbf{Z}_0}[(\mathbf{Z}_\ell + \mathbf{Z}_0) e^{\gamma d} - (\mathbf{Z}_\ell - \mathbf{Z}_0) e^{-\gamma d}] \tag{1-3-24}$$

Next, we find the line impedance at any point from the receiving end in terms of $\mathbf{Z}_\ell$ and $\mathbf{Z}_0$:

$$\mathbf{Z} = \mathbf{Z}_0 \frac{(\mathbf{Z}_\ell + \mathbf{Z}_0)e^{\gamma d} + (\mathbf{Z}_\ell - \mathbf{Z}_0)e^{-\gamma d}}{(\mathbf{Z}_\ell + \mathbf{Z}_0)e^{\gamma d} - (\mathbf{Z}_\ell - \mathbf{Z}_0)e^{-\gamma d}} \tag{1-3-25}$$

Finally, we get the line impedance at the sending end from Eq. (1-3-25) by setting $d = \ell$, so that

$$\mathbf{Z}_s = \mathbf{Z}_0 \frac{(\mathbf{Z}_\ell + \mathbf{Z}_0)e^{\gamma \ell} + (\mathbf{Z}_\ell - \mathbf{Z}_0)e^{-\gamma \ell}}{(\mathbf{Z}_\ell + \mathbf{Z}_0)e^{\gamma \ell} - (\mathbf{Z}_\ell - \mathbf{Z}_0)e^{-\gamma \ell}} \tag{1-3-26}$$

Note that:

1. If the load impedance matches the characteristic impedance $\mathbf{Z}_0$ of the transmission line, or $\mathbf{Z}_\ell = \mathbf{Z}_0$, the line impedance at the sending end will equal the characteristic impedance, or $\mathbf{Z}_s = \mathbf{Z}_0$.
2. If the load impedance does not match the line, that is, $\mathbf{Z}_\ell \neq \mathbf{Z}_0$, the sending-end impedance will not equal the characteristic impedance, that is, $\mathbf{Z}_s \neq \mathbf{Z}_0$, and it can be found from Eq. (1-3-26).

3. *Transfer Impedance*

We define the transfer impedance of a transmission line as the ratio of the sending-end voltage to the receiving-end current. By setting $d = \ell$ in Eq. (1-3-23) and changing $\mathbf{V}$ to $\mathbf{V}_s$ and $\mathbf{I}_\ell$ to $\mathbf{I}_R$, we get

$$\mathbf{Z}_{\text{tr}} = \frac{\mathbf{V}_s}{\mathbf{I}_R} = \frac{1}{2}[(\mathbf{Z}_\ell + \mathbf{Z}_0)e^{\gamma \ell} + (\mathbf{Z}_\ell - \mathbf{Z}_0)e^{-\gamma \ell}] \tag{1-3-27}$$

Equation (1-3-27) is useful for finding the sending-end voltage from the known quantities at the receiving end.

Similarly, we can find the ratio of the sending-end current to the receiving-end current from Eq. (1-3-24), or

$$\frac{\mathbf{I}_s}{\mathbf{I}_R} = \frac{(\mathbf{Z}_\ell + \mathbf{Z}_0)e^{\gamma \ell} - (\mathbf{Z}_\ell - \mathbf{Z}_0)e^{-\gamma \ell}}{2\mathbf{Z}_0} \tag{1-3-28}$$

4. *Impedance in Terms of Hyperbolic or Circular Functions*

The hyperbolic functions are

$$e^{\pm \gamma z} = \cosh(\gamma z) \pm \sinh(\gamma z) \tag{1-3-29}$$

Substituting the hyperbolic functions into Eq. (1-3-12) yields the line impedance at any point from the sending end in terms of the hyperbolic functions:

$$\mathbf{Z} = \frac{\mathbf{Z}_s \cosh(\gamma z) - \mathbf{Z}_0 \sinh(\gamma z)}{\mathbf{Z}_0 \cosh(\gamma z) - \mathbf{Z}_s \sinh(\gamma z)}$$

$$= \mathbf{Z}_0 \frac{\mathbf{Z}_s - \mathbf{Z}_0 \tanh(\gamma z)}{\mathbf{Z}_0 - \mathbf{Z}_s \tanh(\gamma z)} \tag{1-3-30}$$

Similarly, when we substitute the hyperbolic functions into Eq. (1-3-25), we obtain the line impedance from the receiving end in terms of the hyperbolic functions:

$$\mathbf{Z} = \mathbf{Z}_0 \frac{\mathbf{Z}_\ell \cosh(\gamma d) + \mathbf{Z}_0 \sinh(\gamma d)}{\mathbf{Z}_0 \cosh(\gamma d) + \mathbf{Z}_\ell \sinh(\gamma d)}$$

$$= \mathbf{Z}_0 \frac{\mathbf{Z}_\ell + \mathbf{Z}_0 \tanh(\gamma d)}{\mathbf{Z}_0 + \mathbf{Z}_\ell \tanh(\gamma d)} \tag{1-3-31}$$

For a lossless line, $\gamma = j\beta$. By using the following relationships between hyperbolic and circular functions,

$$\sinh(j\beta z) = j \sin(\beta z) \tag{1-3-32}$$

and

$$\cosh(j\beta z) = \cos(\beta z) \tag{1-3-33}$$

we can express the impedance of a lossless transmission line ($\mathbf{Z}_0 = R_0$) in terms of the circular functions from the sending end:

$$\mathbf{Z} = R_0 \frac{\mathbf{Z}_s \cos(\beta z) - jR_0 \sin(\beta z)}{R_0 \cos(\beta z) - j\mathbf{Z}_s \sin(\beta z)}$$

$$= R_0 \frac{\mathbf{Z}_s - jR_0 \tan(\beta z)}{R_0 - j\mathbf{Z}_s \tan(\beta z)} \tag{1-3-34}$$

Similarly, we can express the impedance of a lossless line in terms of the circular functions from the receiving end:

$$\mathbf{Z} = R_0 \frac{\mathbf{Z}_\ell \cos(\beta d) + jR_0 \sin(\beta d)}{R_0 \cos(\beta d) + j\mathbf{Z}_\ell \sin(\beta d)}$$

$$= R_0 \frac{\mathbf{Z}_\ell + jR_0 \tan(\beta d)}{R_0 + j\mathbf{Z}_\ell \tan(\beta d)} \tag{1-3-35}$$

Two special cases, which we discuss later in detail, are very important because of the use of shorted lines as tuning stubs:

1. *Short circuit*. If the load is zero, that is, $\mathbf{Z}_\ell = 0$, the input impedance becomes

$$\mathbf{Z}_{in} = -jR_0 \tan(\beta d) \qquad \text{(inductor)} \tag{1-3-35a}$$

2. *Open circuit*. If the load is open, that is, $\mathbf{Z}_\ell = \infty$, the impedance becomes

$$\mathbf{Z}_{in} = -jR_0 \cot(\beta d) \qquad \text{(capacitor)} \tag{1-3-35b}$$

5. *Measuring Characteristic Impedance*

A common procedure for determining the characteristic impedance and propagation constant of a given transmission line is to make two measurements. First, we measure the sending-end impedance with the receiving end short-circuited, that is, by letting $d = \ell$ and $\mathbf{Z}_\ell = 0$ in Eq. (1-3-31). Then

$$\mathbf{Z}_{sc} = \mathbf{Z}_0 \tanh(\gamma\ell) \tag{1-3-35c}$$

Second, we measure the sending-end impedance with the receiving end open circuited, that is, by letting $d = \ell$ and $\mathbf{Z}_\ell = \infty$ in Eq. (1-3-31). Then

$$\mathbf{Z}_{oc} = \mathbf{Z}_0 \coth(\gamma\ell) = \frac{\mathbf{Z}_0}{\tanh(\gamma\ell)} \tag{1-3-35d}$$

Now, we can calculate the characteristic impedance from

$$\mathbf{Z}_0 = \sqrt{\mathbf{Z}_{oc}\mathbf{Z}_{sc}} \tag{1-3-35e}$$

and the propagation constant from

$$\gamma\ell = \alpha\ell + j\beta\ell = \tanh^{-1}\sqrt{\frac{\mathbf{Z}_{sc}}{\mathbf{Z}_{oc}}} \tag{1-3-35f}$$

For a lossless or low-loss line, the line impedance or admittance repeats at intervals of $\lambda/2$. You can see this condition in Eq. (1-3-35). Because the line impedance repeats itself every half-wavelength, we cannot place the bead insulators in a coaxial line at half-wavelengths apart; to do so is equivalent to connecting them all in parallel. If we insert the beads a quarter-wavelength apart, the reflections from the beads will cancel each other, except for the losses from the beads themselves.

a) Half-Wavelength Lines

A section of half-wavelength line can serve as an impedance transformer if we let $d = n\lambda/2$ in Eq. (1-3-35), where $n = 1, 2, 3, \ldots$. Then

$$\mathbf{Z}_s = \mathbf{Z}_\ell$$

Hence, the half-wave transmission line behaves like a one-to-one transformer in so far as the impedance is concerned. The voltage and current are a one-to-one ratio, but their phases are 180° different.

b) Quarter-Wavelength Lines

In the analysis of line impedance in Section 1-3, we showed that a quarter-wavelength section of a lossless or low-loss line just acts as an *impedance inverter*. We can also show this result by letting $d = \lambda/4$ in Eq. (1-3-35). Then

$$\mathbf{Z}_s = \frac{\mathbf{Z}_0^2}{\mathbf{Z}_\ell} \tag{1-3-36}$$

We can also express Eq. (1-3-36) in terms of normalized impedances with respect to the characteristic impedance $\mathbf{Z}_0$:

$$z_s = \frac{Z_0}{Z_\ell} = \frac{1}{Z_\ell/Z_0} = \frac{1}{z_\ell} \tag{1-3-37}$$

This process ensures that the impedance is inverted for every odd number quarter-wavelength and is sometimes referred to as *quarter-wave transformation*. If a given resistive load $\mathbf{Z}_\ell$ cannot match the main transmission line, it can feed through a quarter-wavelength line whose characteristic impedance $\mathbf{Z}_{0q}$ has a sending-end impedance just equal to the characteristic impedance $\mathbf{Z}_0$ of the main line. The characteristic impedance of the quarter-wavelength line is

$$\mathbf{Z}_{0q} = \sqrt{\mathbf{Z}_\ell \mathbf{Z}_0} \quad \Omega \tag{1-3-38}$$

Note that $\mathbf{Z}_{0q}$, $\mathbf{Z}_\ell$, and $\mathbf{Z}_0$ are pure resistances.

EXAMPLE 1-3-3 Quarter-Wave Transformer

A half-wave, center-fed antenna has a driving-point impedance of 73 Ω, but the transmission line connected to the antenna has a characteristic impedance of 50 Ω.

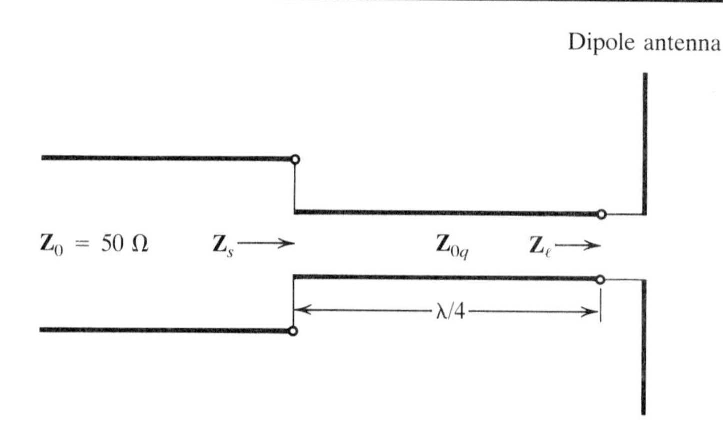

Figure 1-3-3 Quarter-wave transformer for Example 1-3-3.

Determine the characteristic impedance of a quarter-wavelength line to be used for matching the dipole antenna to the line. (See Fig. 1-3-3.)

For a proper match, $\mathbf{Z}_0 = \mathbf{Z}_s = 50\ \Omega$; then from Eq. (1-3-38), we get

$$\mathbf{Z}_{0q} = \sqrt{73 \times 50} = 60\ \Omega$$

EXAMPLE 1-3-4 Determination of Line Impedance at a Distance from Load

A certain open-wire telephone line has the following parameters at a frequency of 2 kHz:

$R = 6.75\ \Omega/\text{mi.}$

$L = 0.00340\ \text{H/mi.}$

$G = 0.400\ \mu\mho/\text{mi.}$

$C = 0.00862\ \mu\text{F/mi.}$

$D = 100\ \text{mi.}$

$\mathbf{Z}_\ell = 200 - j200\ \Omega.$

Calculate the (a) characteristic impedance $\mathbf{Z}_0$, (b) propagation constant γ, and (c) line impedance $\mathbf{Z}_{\text{line}}$ at a distance D from the receiving load $\mathbf{Z}_\ell$.

Solutions:

a) The characteristic impedance is

$$\mathbf{Z} = R + j\omega L = 6.75 + j6.28 \times 2 \times 10^3 \times 3.4 \times 10^{-3}$$
$$= 6.75 + j42.7$$
$$\mathbf{Y} = G + j\omega C = 0.40 \times 10^{-6} + j6.28 \times 2 \times 10^3 \times 8.62 \times 10^{-9}$$
$$= (0.4 + j108.26) \times 10^{-6}$$
$$\mathbf{Z}_0 = \sqrt{\frac{\mathbf{Z}}{\mathbf{Y}}} = \sqrt{\frac{6.75 + j42.7}{(0.40 + j108.26) \times 10^{-6}}} = 632\underline{/-4.4^\circ}$$
$$= 630 - j48.3 \quad \Omega$$

b) The propagation constant is

$$\gamma = \sqrt{\mathbf{ZY}} = (6.75 + j42.7)(0.40 + j108.26) \times 10^{-6} = 0.0685\,\underline{/85.4^\circ}$$

$$= 0.00545 + j0.068$$

$$\alpha = 0.00545 \text{ Np/mi} \quad \text{and}$$

$$\beta = 0.0680 \text{ rad/mi}$$

c) From Eq. (1-3-31), the line impedance is

$$\mathbf{Z}_{\text{line}} = \mathbf{Z}_0 \frac{\mathbf{Z}_\ell \cosh(\gamma d) + \mathbf{Z}_0 \sinh(\gamma d)}{\mathbf{Z}_0 \cosh(\gamma d) + \mathbf{Z}_\ell \sinh(\gamma d)}$$

So

$$\text{Cosh}(\gamma d) = \text{Cosh}(\alpha + j\beta)d = \cosh(\alpha d)\cos(\beta d) + j\sinh(\alpha d)\sin(\beta d)$$

$$= \cosh(0.548)\cos(6.83) + j\sinh(0.548)\sin(6.83)$$

$$= 1.154 \times 0.857 + j0.576 \times 0.514$$

$$= 0.990 + j0.296$$

$$\text{Sinh}(\gamma d) = \sinh(\alpha d + j\beta d) = \sinh(\alpha d)\cos(\beta d) + j\cosh(\alpha d)\sin(\beta d)$$

$$= \sinh(0.548)\cos(6.83) + j\cosh(0.548)\sin(6.83)$$

$$= 0.576 \times 0.857 + j1.154 \times 0.576$$

$$= 0.494 + j0.593$$

and

$$\mathbf{Z}_{\text{line}} = (630 - j48.3)$$

$$\times \frac{(200 - j200)(0.990 + j0.296) + (630 - j483)(0.494 + j0.593)}{(630 - j48.3)(0.990 + j0.296) + (200 - j200)(0.494 + j0.593)}$$

$$= 460.1\,\underline{/4.5^\circ}$$

$$= 458.6 + j36.80 \quad \Omega$$

EXAMPLE 1-3-5 Computer Solution for Impedance on the Line from the Receiving Load

```
PROGRAM   LINEZ

010C      PROGRAM TO COMPUTE THE IMPEDANCE ON A LINE
020C+     FROM THE LOAD
030C      LINEZ
040       COMPLEX Z,Y,Z0,ZL,ZLINE,GAMMA,A,B
045       REAL L
050     1 READ 4, R,L,G,C,F
060       IF(R ,LE, 0.0) STOP
065       READ 5, D,ZL
070       PRINT 6,R,L,G,C,F,D,ZL
080       F=F*1000.0
090       C=C*0.000001
```

(continued)

EXAMPLE 1-3-5 *(continued)*

```
100       G=G*0.000001
110       PI=3.141593
120       WL=2.*PI*F*L
130       Z=CMPLX(R,WL)
140       WC=2.*PI*F*C
150       Y=CMPLX(G,WC)
160       ZO=CSQRT(Z/Y)
165       ZOMAG=CABS(ZO)
167       ZOANG=(ATAN(AIMAG(ZO)/REAL(ZO)))*180.0/3.141593
170       GAMMA=CSQRT(Z*Y)
172       GAMMAM=CABS(GAMMA)
174       GAMMAA=(ATAN(AIMAG(GAMMA)/REAL(GAMMA)))*180./3.141593
175       IF((REAL(GAMMA).LT.0.0).AND.(AIMAG(GAMMA).LT.0.0))
176+      GAMMAA=GAMMAA+180.0
177       IF((REAL(GAMMA).LT.0.0).AND.(AIMAG(GAMMA).GT.0.0))
178+      GAMMAA=GAMMAA-180.0
180       PRINT 8, ZO, ZOMAG,ZOANG,GAMMA,GAMMAM,GAMMAA
190C      SINH(GAMMA*D)=SINH(ALPHA)*COS(BETA)+
200C+                   JCOSH(ALPHA)*SIN(BETA)
210C      COSH(GAMMA*D)=COSH(ALPHA)*COS(BETA)+
220C+                   JSINH(ALPHA)*SIN(BETA)
230       ALPHA=REAL(GAMMA)
240       BETA=AIMAG(GAMMA)
250       SH=SINH(ALPHA*D)
260       CS=COS(BETA*D)
270       CH=COSH(ALPHA*D)
280       SN=SIN(BETA*D)
290       AR=SH*CS
300       AI=CH*SN
310       BR=CH*CS
320       BI=SH*SN
325       PRINT 9, SH,CS,CH,SN,AR,AI,BR,BI
330       A=CMPLX(AR,AI)
340       B=CMPLX(BR,BI)
350       ZLINE=ZO*(ZL*B+ZO*A)/(ZO*B+ZL*A)
352       ZLINEM=CABS(ZLINE)
354       ZLINEA=(ATAN(AIMAG(ZLINE)/REAL(ZLINE)))*180./3.141593
355       IF((REAL(ZLINE).LT.0.0).AND.(AIMAG(ZLINE).LT.0.0))
356+      ZLINEA=ZLINEA+180.0
357       IF((REAL(ZLINE).LT.0.0).AND.(AIMAG(ZLINE).GT.0.0))
358+      ZLINEA=ZLINEA-180.0
360       PRINT 10, ZLINE,ZLINEM,ZLINEA
370     4 FORMAT(5F10.5)
375     5 FORMAT(3F10.5)
380     6 FORMAT(/1H ,15HRESISTANCE  R= ,F10.5,
390+      1X,13HOHMS PER MILE/
400+      1X,15HINDUCTANCE  L= ,F10.5,1X,14HHENRY PER MILE/
410+      1X,15HCONDUCTANCE G= ,F10.5,1X,17HMICROMHO PER MILE/
420+      1X,15HCAPACITANCE C= ,F10.5,1X,19HMICROFARAD PER MILE/
430+      1X,15HFREQUENCY   F= ,F10.5,1X,3HKHZ/
440+      1X,15HDISTANCE    D= ,F10.5,1X,5HMILES/
445+      1X,15HREAL OF ZL   = ,F10.5,1X,4HOHMS/
447+      1X,15HIMAG OF ZL   = ,F10.5,1X,4HOHMS/)
450     8 FORMAT(1X,21HREAL OF ZO           = ,F10.5,1X,4HOHMS/
460+      1X,21HIMAGINARY OF ZO     = ,F10.5,1X,4HOHMS/
462+      1X,21HMAGNITUDE OF ZO     = ,F10.5,1X,4HOHMS/
464+      1X,21HANGLE OF ZO         = ,F10.5,1X,7HDEGREES/
```

EXAMPLE 1-3-5 *(continued)*

```
470+     1X,21HREAL OF GAMMA        = ,F10.5,1X,8HNEPER/MI/
480+     1X,21HIMAGINARY OF GAMMA = ,F10.5,1X,6HRAD/MI/
482+     1X,21HMAGNITUDE OF GAMMA = ,F10.5/
484+     1X,21HANGLE OF GAMMA      = ,F10.5,1X,7HDEGREES/)
485    9 FORMAT(8F7.4)
490   10 FORMAT(/1H ,21HREAL OF ZLINE       = ,F10.5,1X,4HOHMS/
492+     1X,21HIMAGINARY OF ZLINE = ,F10.5,1X,4HOHMS/
494+     1X,21HMAGNITUDE OF ZLINE = ,F10.5,1X,4HOHMS/
496+     1X,21HANGLE OF ZLINE      = ,F10.5,1X,7HDEGREES/)
500      GO TO 1
510      STOP
520      END
READY.
RUN

PROGRAM   LINEZ

?    6.75000    0.00340    0.40000    0.00862    2.00000
?  100.00000 200.00000-200.00000

 RESISTANCE  R=     6.75000 OHMS PER MILE
 INDUCTANCE  L=      .00340 HENRY PER MILE
 CONDUCTANCE G=      .40000 MICROMHO PER MILE
 CAPACITANCE C=      .00862 MICROFARAD PER MILE
 FREQUENCY   F=     2.00000 KHZ
 DISTANCE    D=   100.00000 MILES
 REAL OF ZL   =   200.00000 OHMS
 IMAG OF ZL   =  -200.00000 OHMS

 REAL OF Z0            =  630.07028 OHMS
 IMAGINARY OF Z0       =  -48.29366 OHMS
 MAGNITUDE OF Z0       =  631.91838 OHMS
 ANGLE OF Z0           =   -4.38304 DEGREES
 REAL OF GAMMA         =     .00548 NEPER/MI
 IMAGINARY OF GAMMA    =     .06823 RAD/MI
 MAGNITUDE OF GAMMA    =     .06845
 ANGLE OF GAMMA        =   85.40538 DEGREES

  .5762  .8577 1.1541  .5141  .4943  .5933  .9900  .2962

 REAL OF ZLINE         =  458.62372 OHMS
 IMAGINARY OF ZLINE    =   36.79190 OHMS
 MAGNITUDE OF ZLINE    =  460.09712 OHMS
 ANGLE OF ZLINE        =    4.58658 DEGREES

?    0.0
SRU       1.173 UNTS.

RUN COMPLETE.
```

1-4 REFLECTION AND TRANSMISSION COEFFICIENTS

1-4-1 Reflection Coefficient

In the analysis of the solutions of the transmission-line equation in Section 1.2, the traveling wave along the line contains two components: one traveling in the positive z direction, and the other traveling in the negative z direction. If the load impedance

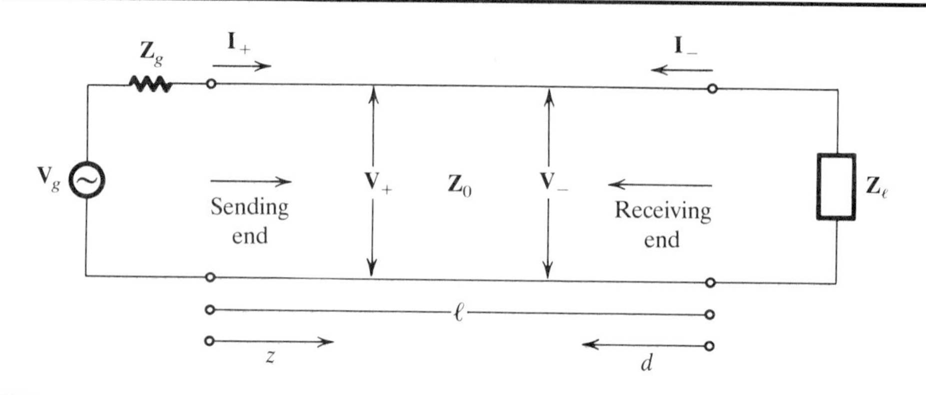

Figure 1-4-1 Transmission line terminated in an impedance $\mathbf{Z}_\ell$.

equals the line characteristic impedance, however, there is no reflected traveling wave. Figure 1-4-1 shows a transmission line terminated in a load impedance $\mathbf{Z}_\ell$.

We usually find it more convenient to start solving a transmission-line problem from the receiving end rather than from the sending end. The reason is that the voltage–current relationship at the load point is fixed by the load impedance. The instantaneous incident voltage and current waves traveling in the transmission line in frequency domain are

$$\mathbf{V} = \mathbf{V}_+ e^{-\gamma z} + \mathbf{V}_- e^{+\gamma z} \tag{1-4-1}$$

and

$$\mathbf{I} = \mathbf{I}_+ e^{-\gamma z} + \mathbf{I}_- e^{+\gamma z} \tag{1-4-2}$$

in which we can express the current wave in terms of the voltage:

$$\mathbf{I} = \frac{\mathbf{V}_+}{\mathbf{Z}_0} e^{-\gamma z} - \frac{\mathbf{V}_-}{\mathbf{Z}_0} e^{+\gamma z} \tag{1-4-3}$$

If the line is of length ℓ, the voltage and current at the receiving end become

$$\mathbf{V} = \mathbf{V}_+ e^{-\gamma\ell} + \mathbf{V}_- e^{\gamma\ell} \tag{1-4-4}$$

and

$$\mathbf{I} = \frac{1}{\mathbf{Z}_0}(\mathbf{V}_+ e^{-\gamma\ell} - \mathbf{V}_- e^{\gamma\ell}) \tag{1-4-5}$$

The ratio of the voltage to the current at the receiving end is the load impedance. That is,

$$\mathbf{Z}_\ell = \frac{\mathbf{V}_\ell}{\mathbf{I}_\ell} = \mathbf{Z}_0 \frac{\mathbf{V}_+ e^{-\gamma\ell} + \mathbf{V}_- e^{\gamma\ell}}{\mathbf{V}_+ e^{-\gamma\ell} - \mathbf{V}_- e^{\gamma\ell}} \tag{1-4-6}$$

We define the reflection coefficient, designated by $\mathbf{\Gamma}$, as

$$\text{Reflection coefficient} \equiv \frac{\text{Reflected voltage or negative reflected current}}{\text{Incident voltage or current}}$$

or

$$\Gamma \equiv \frac{\mathbf{V}_{\text{ref}}}{\mathbf{V}_{\text{inc}}} = \frac{-\mathbf{I}_{\text{ref}}}{\mathbf{I}_{\text{inc}}} \tag{1-4-7}$$

at any point on the line.

If we solve Eq. (1-4-6) for the ratio of the reflected voltage at the receiving end, $\mathbf{V}_- e^{\gamma \ell}$, to the incident voltage at the same receiving end, $\mathbf{V}_+ e^{-\gamma \ell}$, the result is the reflection coefficient at the receiving end:

$$\Gamma_\ell = \frac{\mathbf{V}_- e^{\gamma \ell}}{\mathbf{V}_+ e^{-\gamma \ell}} = \frac{\mathbf{Z}_l - \mathbf{Z}_0}{\mathbf{Z}_l + \mathbf{Z}_0} \tag{1-4-8}$$

If the load impedance and/or the characteristic impedance are complex quantities, as is usually the case, the reflection coefficient is generally a complex quantity, which we can express as

$$\Gamma_\ell = |\Gamma_\ell| e^{j\theta_\ell} \tag{1-4-9}$$

where $|\Gamma_\ell|$ is the magnitude and never greater than 1 for a positive load. Note that θ_ℓ is the phase in radians between the incident voltage and the reflected voltage at the receiving end. It is usually called the phase angle of the reflection coefficient.

The general solution of the reflection coefficient at any point on the line, then, corresponds to the incident and reflected waves at that point. Each wave is attenuated in the direction of its own progress along the line. The generalized reflection coefficient is

$$\Gamma \equiv \frac{\mathbf{V}_- e^{+\gamma z}}{\mathbf{V}_+ e^{-\gamma z}} \tag{1-4-10}$$

If we let $z = \ell - d$ (see Fig. 1-4-1), the reflection coefficient at some point located distance d from the receiving end is

$$\begin{aligned}\Gamma_d &= \frac{\mathbf{V}_- e^{\gamma(\ell - d)}}{\mathbf{V}_+ e^{-\gamma(\ell - d)}} = \frac{\mathbf{V}_- e^{\gamma \ell}}{\mathbf{V}_+ e^{-\gamma \ell}} e^{-2\gamma d} \\ &= \Gamma_\ell e^{-2\gamma d}\end{aligned} \tag{1-4-11}$$

Next, we can express the reflection coefficient at that point in terms of Γ_ℓ:

$$\begin{aligned}\Gamma_d &= \Gamma_\ell e^{-2\alpha d} e^{-j2\beta d} \\ &= |\Gamma_\ell| e^{-2\alpha d} e^{j(\theta_\ell - 2\beta d)}\end{aligned} \tag{1-4-12}$$

This equation is very useful for determining the reflection coefficient at any point along the line. For a lossy line, both the magnitude and phase of the reflection coefficient are changing in an inward-spiral way as shown in Fig. 1-4-2.

For a lossless line, $\alpha = 0$, the magnitude of the reflection coefficient remains constant. Only the phase of Γ is changing circularly toward the generator with $-2\beta d$ as shown in Fig. 1-4-3.

It is evident that Γ_ℓ will be zero and that there will be no reflection from the receiving end when the terminating impedance is equal to the characteristic impedance of the line. Thus a terminating impedance that differs from the characteristic impedance $\mathbf{Z}_0$ will create a reflected wave traveling toward the source from the termination. The reflection, upon reaching the sending end, will itself be reflected if the source impedance is different from the characteristic impedance at the sending end, $\mathbf{Z}_s$.

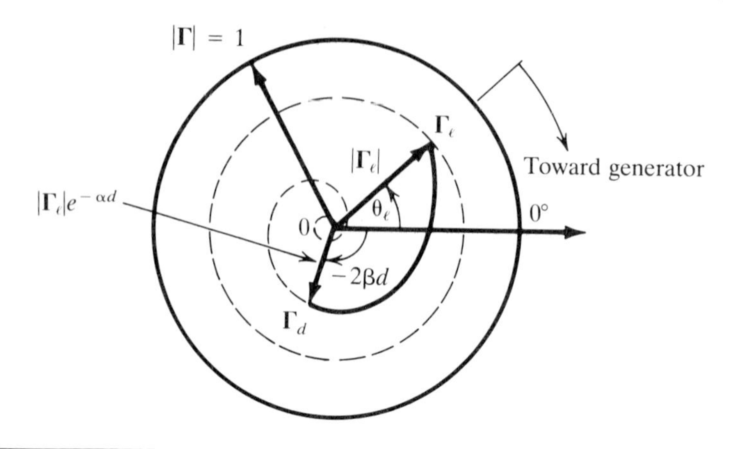

Figure 1-4-2 Reflection coefficient for lossy line.

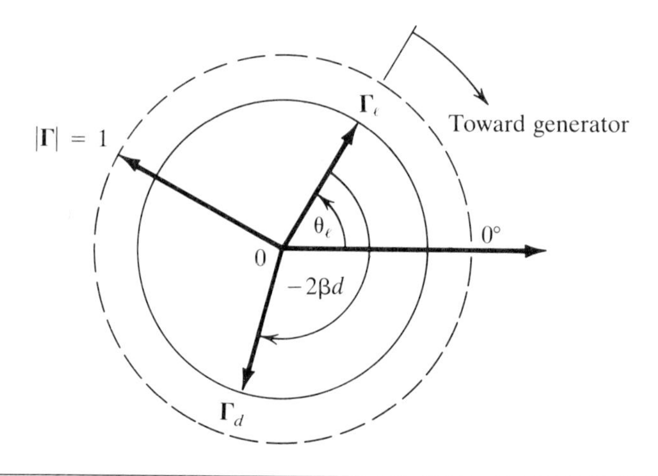

Figure 1-4-3 Reflection coefficient for lossless line.

EXAMPLE 1-4-1 Determination of Reflection Coefficient $\mathbf{\Gamma}$ at Load

A certain transmission line has a characteristic impedance of $50 - j0.01\ \Omega$ and terminates in an impedance of $73 + j42.50\ \Omega$. Determine $\mathbf{\Gamma}$ at the load.

Solution:

$$\mathbf{\Gamma} = \frac{\mathbf{Z}_\ell - \mathbf{Z}_0}{\mathbf{Z}_\ell + \mathbf{Z}_0} = \frac{73 + j42.50 - (50 - j0.01)}{73 + j42.50 + (50 - j0.01)}$$

$$= 0.377\,\underline{/42.7^\circ}$$

$$= 0.2735 + j0.2525$$

EXAMPLE 1-4-2 Computer Solution for Reflection Coefficient Γ at Load

```
010C        PROGRAM TO COMPUTE REFLECTION COEFFICIENT
020C+       GAMMA AT LOAD
030C        GAMMA
040         COMPLEX ZO,ZL,BGAML
050       1 READ 2, ZO,ZL
060         IF(ZO ,LE, 0.0) STOP
070         PRINT 6, ZO,ZL
080         BGAML=(ZL-ZO)/(ZL+ZO)
090         BGAMLM=CABS(BGAML)
100         BGAMLA=(ATAN(AIMAG(BGAML)/REAL(BGAML)))*180./3.141593
110         IF((REAL(BGAML).LT.0.0).AND.(AIMAG(BGAML).LT.0.0))
120+        BGAMLA=BGAMLA + 180.0
130         IF((REAL(BGAML).LT.0.0).AND.(AIMAG(BGAML).GT.0.0))
140+        BGAMLA=BGAMLA - 180.0
150         PRINT 8, BGAML,BGAMLM,BGAMLA
160       2 FORMAT(4F10.5)
170       6 FORMAT(//1H ,31HCHARACTERISTIC IMPEDANCE   ZO=
180+        1X,F10.5,2H+J,F10.5,1X,4HOHMS/
190+        1X,31HLOAD IMPEDANCE                  ZL=
200+        1X,F10.5,2H+J,F10.5,1X,4HOHMS/)
210       3 FORMAT(1H ,33HREFLECTION COEFFICIENT AT LOAD IS/
220+        1X,21HREAL OF BGAML      = ,1X,F10.5/
230+        1X,21HIMAGINARY OF BGAML = ,1X,F10.5/
240+        1X,21HMAGNITUDE OF BGAML = ,1X,F10.5/
250+        1X,21HANGLE OF BGAML     = ,1X,F10.5,1X,7HDEGREES/)
260         GO TO 1
270         STOP
280         END
READY.
RUN

PROGRAM   GAMMA

?   50.00000  -0.01000  73.00000  42.50000

 CHARACTERISTIC IMPEDANCE   ZO=    50.00000+J    -.01000 OHMS
 LOAD IMPEDANCE             ZL=    73.00000+J   12.50000 OHMS

 REFLECTION COEFFICIENT AT LOAD IS
 REAL OF BGAML      =        .27372
 IMAGINARY OF BGAML =        .25105
 MAGNITUDE OF BGAML =        .37142
 ANGLE OF BGAML     =     42.52710 DEGREES

? 0.0

SRU       0.811 UNTS.

RUN COMPLETE.
```

1-4-2 Transmission Coefficient

A transmission line terminated in its characteristic impedance $\mathbf{Z}_0$ is called a properly terminated, or nonresonant, line. Otherwise, it is called an improperly terminated, or resonant, line. As we described in Section 1.4.1, a reflection coefficient Γ exists at any point along an improperly terminated line. From the principle of conservation of energy, the incident power minus the reflected power must equal the power transmitted to the load, or

$$1 - \mathbf{\Gamma}^2 = \mathbf{T}^2 \frac{\mathbf{Z}_0}{\mathbf{Z}_\ell} \tag{1-4-13}$$

We will verify Eq. (1-4-13) later. The **T** represents the transmission coefficient, which we define as

$$\mathbf{T} \equiv \frac{\text{Transmitted voltage}}{\text{Incident voltage or current}}$$

or

$$\mathbf{T} = \frac{\mathbf{V}_{\text{tr}}}{\mathbf{V}_{\text{inc}}} \tag{1-4-14}$$

Figure 1-4-4 shows the power transmission along a transmission line, where P_{inc} is the incident power, P_{ref} is the reflected power, and P_{tr} is the transmitted power.

Let the traveling waves at the receiving end be

$$\mathbf{V}_+ \mathbf{e}^{-\gamma z} + \mathbf{V}_- e^{\gamma z} = \mathbf{V}_{\text{tr}} e^{-\gamma z} \tag{1-4-15}$$

and

$$\frac{\mathbf{V}_+}{\mathbf{Z}_0} e^{-\gamma z} - \frac{\mathbf{V}_-}{\mathbf{Z}_0} e^{\gamma z} = \frac{\mathbf{V}_{\text{tr}}}{\mathbf{Z}_\ell} e^{-\gamma z} \tag{1-4-16}$$

When we multiply Eq. (1-4-16) by $\mathbf{Z}_\ell$ and substitute the result into Eq. (1-4-15), we get

$$\mathbf{\Gamma}_\ell \equiv \frac{\mathbf{V}_- e^{\gamma z}}{\mathbf{V}_+ e^{-\gamma z}} = \frac{\mathbf{Z}_\ell - \mathbf{Z}_0}{\mathbf{Z}_\ell + \mathbf{Z}_0} \tag{1-4-17}$$

which, upon substitution back into Eq. (1-4-15), results in

$$\mathbf{T} \equiv \frac{\mathbf{V}_{\text{tr}}}{\mathbf{V}_+} = \frac{2\mathbf{Z}_\ell}{\mathbf{Z}_\ell + \mathbf{Z}_0} \tag{1-4-18}$$

The net average power carried by the incident and reflected waves is

$$\langle P_{\text{ir}} \rangle = \frac{|\mathbf{V}_+ e^{-\alpha z}|^2}{2\mathbf{Z}_0} - \frac{|\mathbf{V}_- e^{+\alpha z}|^2}{2\mathbf{Z}_0} \tag{1-4-19}$$

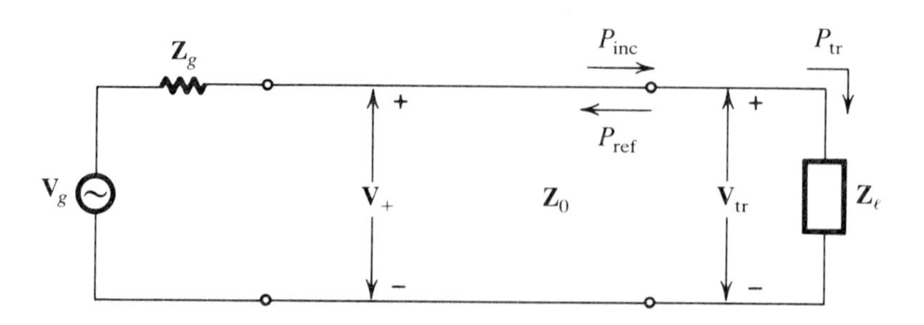

Figure 1-4-4 Power transmission along a line.

and the average power carried to the load by the transmitted wave is

$$\langle P_{tr} \rangle = \frac{|\mathbf{V}_{tr} e^{-\alpha z}|^2}{2\mathbf{Z}_\ell} \tag{1-4-20}$$

Setting $\langle P_{ir} \rangle = \langle P_{tr} \rangle$ and using Eqs. (1-4-17) and (1-4-18), we obtain

$$\mathbf{T}^2 = \frac{\mathbf{Z}_\ell}{\mathbf{Z}_0}(1 - \Gamma^2) \tag{1-4-21}$$

This relation verifies our previous statement that the transmitted power equals the algebraic sum of the incident and reflected power, as shown in Eq. (1-4-13).

EXAMPLE 1-4-3 Reflection Coefficient and Transmission Coefficient

A certain transmission line has a characteristic impedance of $75 + j0.01\ \Omega$ and is terminated in a load impedance of $70 + j50\ \Omega$.

Calculate:

(a) the reflection coefficient and (b) the transmission coefficient.

Verify:

(c) the relationship shown in Eq. (1-4-21) and (d) that the transmission coefficient equals the algebraic sum of 1 plus the reflection coefficient, as shown in Eq. (0-5-15).

Solutions:

a) From Eq. (1-4-8) the reflection coefficient is

$$\begin{aligned}\mathbf{\Gamma} &= \frac{\mathbf{Z}_\ell - \mathbf{Z}_0}{\mathbf{Z}_\ell + \mathbf{Z}_0} = \frac{70 + j50 - (75 + j0.01)}{70 + j50 + (75 + j0.01)} \\ &= \frac{50.24\,\underline{/95.71^\circ}}{153.38\,\underline{/19.03^\circ}} = 0.33\,\underline{/76.68^\circ} \\ &= 0.08 + j0.32\end{aligned}$$

b) From Eq. (1-4-18) the transmission coefficient is

$$\begin{aligned}\mathbf{T} &= \frac{2\mathbf{Z}_\ell}{\mathbf{Z}_\ell + \mathbf{Z}_0} = \frac{2(70 + j50)}{70 + j50 + (75 + j0.01)} \\ &= \frac{172.05\,\underline{/35.54^\circ}}{153.38\,\underline{/19.03^\circ}} = 1.12\,\underline{/16.51^\circ} \\ &= 1.08 + j0.32\end{aligned}$$

c)

$$\mathbf{T}^2 = (1.12\,\underline{/16.51^\circ})^2 = 1.25\,\underline{/33.02^\circ}$$

$$\begin{aligned}\frac{\mathbf{Z}_\ell}{\mathbf{Z}_0}(1 - \mathbf{\Gamma}^2) &= \frac{70 + j50}{75 + j0.01}[1 - (0.33\,\underline{/76.68^\circ})^2] \\ &= \frac{86\,\underline{/35.54^\circ}}{75\,\underline{/0^\circ}} \times 1.10\,\underline{/-2.6^\circ} \\ &= 1.25\,\underline{/33^\circ}\end{aligned}$$

Thus we have verified Eq. (1-4-21).

d) From Eq. (0-5-15) we obtain

$$\mathbf{T} = 1.08 + j0.32 = 1 + 0.08 + j0.32$$
$$= 1 + \mathbf{\Gamma}$$

1-5 STANDING WAVE AND STANDING-WAVE RATIO

1-5-1 Standing Wave

The general solutions of the transmission-line equation involve the analysis of two waves traveling in opposite directions with unequal amplitudes. The line voltage and current equations are:

$$\mathbf{V} = \mathbf{V}_+ e^{-\gamma z} + \mathbf{V}_- e^{\gamma z} \tag{1-5-1}$$

and

$$\mathbf{I} = Y_0(\mathbf{V}_+ e^{-\gamma z} - \mathbf{V}_- e^{\gamma z}) \tag{1-5-2}$$

where $\gamma = \alpha + j\beta$.

We can rewrite Eq. (1-5-1) as

$$\begin{aligned}\mathbf{V} &= \mathbf{V}_+ e^{-\alpha z} e^{-j\beta z} + \mathbf{V}_- e^{\alpha z} e^{j\beta z}\\ &= \mathbf{V}_+ e^{-\alpha z}[\cos(\beta z) - j\sin(\beta z)] + \mathbf{V}_- e^{\alpha z}[\cos(\beta z) + j\sin(\beta z)]\\ &= (\mathbf{V}_+ e^{-\alpha z} + \mathbf{V}_- e^{\alpha z})\cos(\beta z) - j(\mathbf{V}_+ e^{-\alpha z} - \mathbf{V}_- e^{\alpha z})\sin(\beta z)\end{aligned} \tag{1-5-3}$$

With no loss in generality, we can assume that $\mathbf{V}_+ e^{-\alpha z}$ and $\mathbf{V}_- e^{+\alpha z}$ are real; we can then express the voltage standing-wave equation as

$$\mathbf{V}_s = \mathbf{V}_0 e^{-j\phi_0} \tag{1-5-4}$$

Equation (1-5-4) is called the equation of the voltage standing wave, in which

$$\mathbf{V}_0 = [(\mathbf{V}_+ e^{-\alpha z} + \mathbf{V}_- e^{\alpha z})^2 \cos^2(\beta z) + (\mathbf{V}_+ e^{-\alpha z} - \mathbf{V}_- e^{\alpha z})^2 \sin^2(\beta z)]^{1/2} \tag{1-5-5}$$

and is called the standing-wave pattern of the voltage wave, or the amplitude of the standing wave; and

$$\phi_0 = \arctan\left[\frac{\mathbf{V}_+ e^{-\alpha z} - \mathbf{V}_- e^{\alpha z}}{\mathbf{V}_+ e^{-\alpha z} + \mathbf{V}_- e^{\alpha z}} \tan(\beta z)\right] \tag{1-5-6}$$

which is called the phase pattern of the standing wave.

We can find the maximum and minimum values of Eq. (1-5-5) in the usual manner by differentiating the equation with respect to βz and equating the resultant to zero. When we do so—and substitute the proper values of βz into Eq. (1-5-5)—we find that:

1. The maximum amplitude is

$$\mathbf{V}_{\text{max}} = \mathbf{V}_+ e^{-\alpha z} + \mathbf{V}_- e^{\alpha z} \tag{1-5-7}$$

which occurs at $\beta z = n\pi$, where $n = 0, \pm 1, \pm 2, \ldots$

2. The minimum amplitude is

$$\mathbf{V}_{\text{min}} = \mathbf{V}_+ e^{-\alpha z} - \mathbf{V}_- e^{\alpha z} \tag{1-5-8}$$

which occurs at $\beta z = (2n - 1)(\pi/2)$, where $n = 0, \pm 1, \pm 2, \ldots$

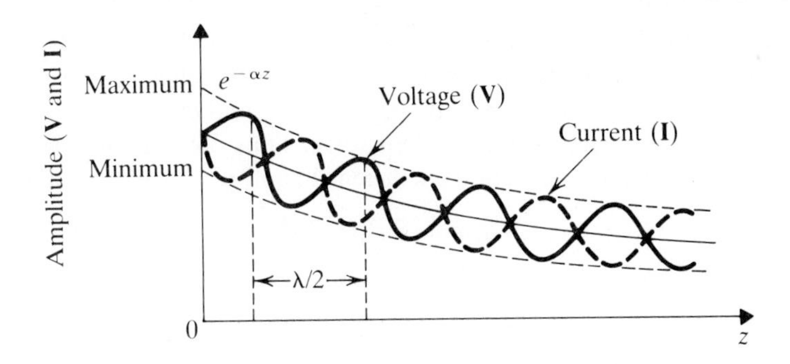

Figure 1-5-1 Standing-wave pattern in a lossy line.

3. The distance between any two successive maxima or minima is one-half wavelength, since $\beta z = n\pi$, $n = 0, \pm 1, \pm 2, \ldots$, so

$$z = \frac{n\pi}{\beta} = \frac{n\pi}{(2\pi)/\lambda} = n\frac{\lambda}{2}$$

Then

$$z_1 = \frac{\lambda}{2} \qquad \text{for } n = 1$$

It is evident that there are no 0's in the minimum. Figure 1-5-1 shows the standing-wave pattern of two oppositely traveling waves with unequal amplitude in a lossy line.

Similarly, the maximum and minimum currents are

$$\mathbf{I}_{\text{max}} = \mathbf{I}_{+}e^{-\alpha z} + \mathbf{I}_{-}e^{\alpha z} \tag{1-5-9}$$

and

$$\mathbf{I}_{\text{min}} = \mathbf{I}_{+}e^{-\alpha z} - \mathbf{I}_{-}e^{\alpha z} \tag{1-5-10}$$

Figure 1-5-2 shows the standing-wave pattern of two oppositely traveling waves with unequal amplitude in a lossless line.

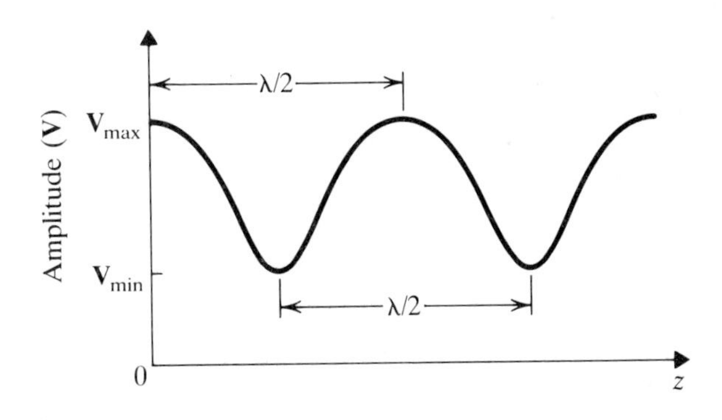

Figure 1-5-2 Standing-wave pattern with unequal amplitude in a lossless line.

1. *Lossy Line*

If the transmission line is lossy, that is, $\alpha \neq 0$, the incident wave traveling in the positive z direction undergoes an attenuation of $e^{-\alpha z}$. The reflected wave is attenuated exponentially as it travels in the negative z direction, and eventually its amplitude is negligible compared to that of the incident wave. Thus the maxima and minima are functions of position z and reflection coefficient $\mathbf{\Gamma}$, and we can express them as

$$\mathbf{V}_{\text{max}} = \mathbf{V}_{+} e^{-\alpha z}(1 + |\mathbf{\Gamma}|) \tag{1-5-11}$$

and

$$\mathbf{V}_{\text{min}} = \mathbf{V}_{+} e^{-\alpha z}(1 - |\mathbf{\Gamma}|) \tag{1-5-12}$$

Similarly,

$$\mathbf{I}_{\text{max}} = \mathbf{I}_{+} e^{-\alpha z}(1 + |\mathbf{\Gamma}|) \tag{1-5-13}$$

and

$$\mathbf{I}_{\text{min}} = \mathbf{I}_{-} e^{-\alpha z}(1 - |\mathbf{\Gamma}|) \tag{1-5-14}$$

2. *Lossless Line*

If the transmission line is lossless, that is, $\alpha = 0$, maximum and minimum amplitudes remain constant, and we can express them as

$$\mathbf{V}_{\text{max}} = \mathbf{V}_{+}(1 + |\mathbf{\Gamma}|) \tag{1-5-15}$$

and

$$\mathbf{V}_{\text{min}} = \mathbf{V}_{+}(1 - |\mathbf{\Gamma}|) \tag{1-5-16}$$

Similarly,

$$\mathbf{I}_{\text{max}} = \mathbf{I}_{+}(1 + |\mathbf{\Gamma}|) \tag{1-5-17}$$

and

$$\mathbf{I}_{\text{min}} = \mathbf{I}_{-}(1 - |\mathbf{\Gamma}|) \tag{1-5-18}$$

Further consideration of Eq. (1-5-5) reveals that when $\mathbf{V}_{+} \neq 0$ and $\mathbf{V}_{-} = 0$, the standing-wave pattern becomes

$$\mathbf{V}_{0} = \mathbf{V}_{+} e^{-\alpha z} \tag{1-5-19}$$

and the standing wave is

$$\mathbf{V}_{s} = \mathbf{V}_{+} e^{-\gamma z} \tag{1-5-20}$$

which is a pure traveling wave.

When the positive wave and the negative wave have equal amplitudes, that is, $|\mathbf{V}_{+} e^{-\alpha z}| = |\mathbf{V}_{-} e^{\alpha z}|$, or the magnitude of the reflection coefficient equals 1, the standing-wave pattern with a zero phase is

$$\mathbf{V}_{s} = 2\mathbf{V}_{+} e^{-\alpha z} \cos(\beta z) \tag{1-5-21}$$

which is called a pure standing wave.

Similarly, for the current the equation of a pure standing wave is

$$\mathbf{I}_{s} = -j2\mathbf{Y}_{0}\mathbf{V}_{+} e^{-\alpha z} \sin(\beta z) \tag{1-5-22}$$

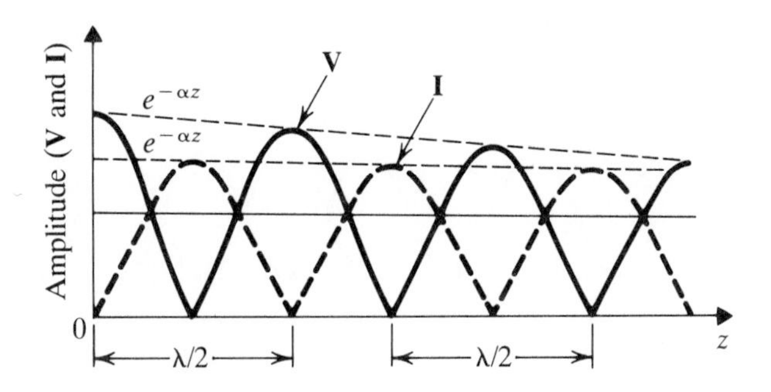

Figure 1-5-3 Pure standing wave of voltage and current.

Equations (1-5-21) and (1-5-22) show that the voltage and current are 90° out of phase along the line. The points of zero voltage are called the voltage nodes, and those of zero current are called the current nodes. The voltage nodes and the current nodes are interlaced a quarter-wavelength apart.

We can express the voltage and current as real functions of time and space:

$$v_s(z, t) = \text{Re}[\mathbf{V}_s(z)e^{j\omega t}] = 2\mathbf{V}_+ e^{-\alpha z} \cos(\beta z) \cos(\omega t) \tag{1-5-23}$$

and

$$i_s(z, t) = \text{Re}[\mathbf{I}_s(z)e^{j\omega t}] = 2\mathbf{Y}_0\mathbf{V}_+ e^{-\alpha z} \sin(\beta z) \sin(\omega t) \tag{1-5-24}$$

The amplitudes of Eqs. (1-5-23) and (1-5-24) vary sinusoidally with time, so the voltage is at maximum at the instant when the current is zero—and vice versa. Figure 1-5-3 shows the pure standing-wave patterns of the phasor of Eqs. (1-5-21) and (1-5-22) for an open-terminated line.

From the complex Poynting theorem, we can determine the time average of the electric and magnetic energy densities for a pure standing wave from Eqs. (1-5-21) and (1-5-22):

$$\begin{aligned} \langle w_{es} \rangle &= \tfrac{1}{4} C |\mathbf{V}_s|^2 \\ &= \tfrac{1}{4} C |2\mathbf{V}_+ e^{-\alpha z}|^2 \cos^2(\beta z) \end{aligned} \tag{1-5-25}$$

and

$$\begin{aligned} \langle w_{ms} \rangle &= \tfrac{1}{4} L |\mathbf{I}_s|^2 \\ &= \tfrac{1}{4} L |2 Y_0 \mathbf{V}_+ e^{-\alpha z}|^2 \sin^2(\beta z) \end{aligned} \tag{1-5-26}$$

For a low-loss or lossless line, that is, $\alpha z \ll 1$ or $\alpha z = 0$, we can simplify Eqs. (1-5-25) and (1-5-26) to

$$\langle W_{es} \rangle = C |\mathbf{V}_+|^2 \cos^2(\beta z) \tag{1-5-27}$$

and

$$\langle W_{ms} \rangle = C |\mathbf{V}_+|^2 \sin^2(\beta z) \tag{1-5-28}$$

in which we replaced $\mathbf{Y}_0^2$ by C/L.

Figure 1-5-4 shows the time average of the electric and magnetic energy densities in a pure standing wave.

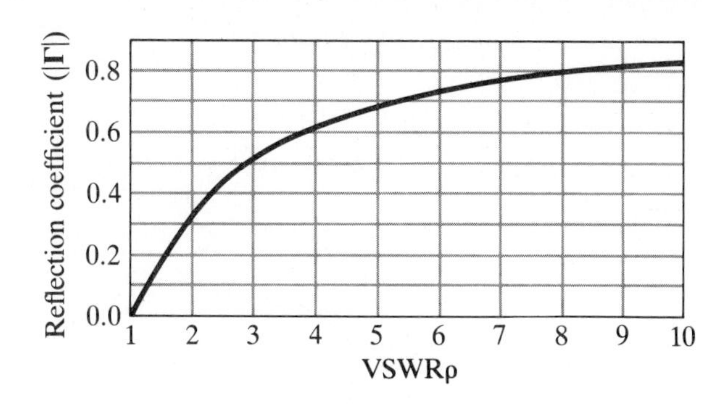

Figure 1-5-5 VSWR versus reflection coefficient.

flat line. The standing-wave ratio cannot be defined on a lossy line because the standing-wave pattern changes markedly from one position to another. On a low-loss line the ratio remains fairly constant, and it may be defined for some region. On a lossless line, the ratio stays the same throughout the line.

Since the reflected wave is defined as the product of an incident wave and its reflection coefficient, the standing-wave ratio ρ is related to the reflection coefficient $\mathbf{\Gamma}$ as follows:

$$\rho = \frac{1 + |\mathbf{\Gamma}|}{1 - |\mathbf{\Gamma}|} \tag{1-5-33}$$

and vice versa:

$$|\mathbf{\Gamma}| = \frac{\rho - 1}{\rho + 1} \tag{1-5-34}$$

This relationship between ρ and $\mathbf{\Gamma}$ is useful for determining the reflection coefficient from the standing-wave ratio, which you can usually find from the Smith chart (see p. 93). The curve in Fig. 1-5-5 shows the relationship between the reflection coefficient $|\mathbf{\Gamma}|$ and the standing-wave ratio ρ.

Since $|\mathbf{\Gamma}| \leq 1$, you can see from Eq. (1-5-33) that the standing-wave ratio is a positive real number and is never less than 1, that is, $\rho \geq 1$. Similarly, you can see from Eq. (1-5-34) that the magnitude of the reflection coefficient is never greater than 1.

1. *The Crank, or Bicycle Diagram*

For a lossless or low-loss line ($\alpha\ell = 0$ or $\alpha\ell \ll 1$), we can easily determine the standing-wave patterns of voltage and current along the line by means of the crank diagram shown in Fig. 1-5-6.

The steps in constructing the crank diagram are:

1. Lay off a horizontal line of arbitrary length to represent the incident complex voltage $\mathbf{V}_+$ at the sending end.
2. Calculate the reflected voltage at the receiving end from

$$\mathbf{V}_{r-} = \mathbf{\Gamma}_\ell \mathbf{V}_{r+} \tag{1-5-35}$$

where

$$\mathbf{\Gamma}_\ell = \frac{\mathbf{Z}_\ell - \mathbf{Z}_0}{\mathbf{Z}_\ell + \mathbf{Z}_0} = |\mathbf{\Gamma}_\ell| e^{\cdots} \tag{1-5-36}$$

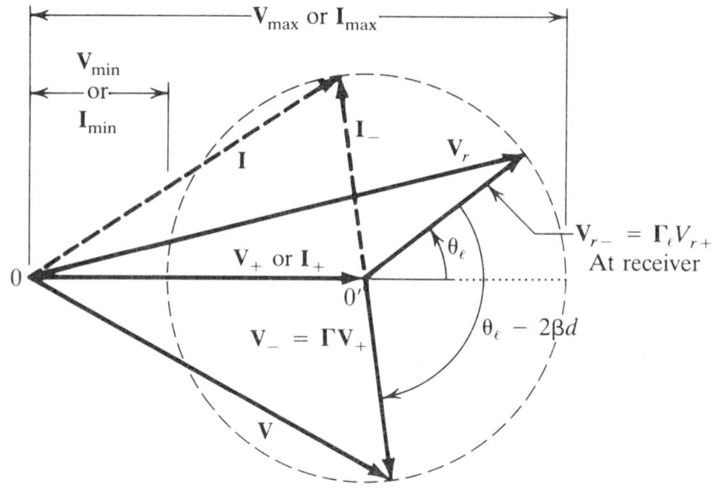

Figure 1-5-6 Crank diagram for a low-loss and a lossless line.

Since $\alpha\ell \ll 1$, the incident voltage at the receiving end equals the incident voltage at the sending end. Then $\mathbf{V}_{r+} = \mathbf{V}_+$ and $\mathbf{V}_{r-} = \mathbf{V}_-$, so

$$\mathbf{V}_- = \mathbf{\Gamma V}_+ \tag{1-5-37}$$

3. Lay off a line representing $\mathbf{V}_{r-}$ from the point 0′. Then the sum of $\mathbf{V}_{r-}$ and $\mathbf{V}_+$ represents the voltage at the receiving point.
4. Calculate the reflection coefficient $\mathbf{\Gamma}$ at a distance d from the receiving point by using Eq. (1-5-12) for $\alpha d \ll 1$. Then

$$\mathbf{\Gamma} = |\mathbf{\Gamma}_\ell| e^{j(\theta_\ell - 2\beta d)} \tag{1-5-38}$$

Since $\alpha\ell \ll 1$, the incident voltage remains constant, and the reflected voltage at a distance d from the receiving point is $\mathbf{V}_- = \mathbf{\Gamma V}_+$.

5. Finally, draw a line from 0′ to represent $\mathbf{V}$:

$$\mathbf{V} = \mathbf{V}_+ + \mathbf{\Gamma V}_+ = \mathbf{V}_+(1 + \mathbf{\Gamma}) \tag{1-5-39}$$

It is evident from the crank diagram that

$$|\mathbf{V}_{max}| = |\mathbf{V}_+| + |\mathbf{V}_-| = |\mathbf{V}_+|(1 + |\mathbf{\Gamma}|) \tag{1-5-40}$$

and

$$|\mathbf{V}_{min}| = |\mathbf{V}_+| - |\mathbf{V}_-| = |\mathbf{V}_+|(1 - |\mathbf{\Gamma}|) \tag{1-5-41}$$

Therefore, we can express the standing-wave ratio ρ as

$$\rho = \frac{|\mathbf{V}_{max}|}{|\mathbf{V}_{min}|} = \frac{1 + |\mathbf{\Gamma}|}{1 - |\mathbf{\Gamma}|} \tag{1-5-42}$$

Similarly, we can also find the standing-wave ratio from the current quantities in the crank diagram:

$$|\mathbf{I}_{max}| = |\mathbf{I}_+| + |\mathbf{I}_-| \tag{1-5-43}$$

and

$$|\mathbf{I}_{min}| = |\mathbf{I}_+| - |\mathbf{I}_-| \tag{1-5-44}$$

Since $|\mathbf{V}_+| = |\mathbf{I}_+|\mathbf{Z}_0$, and $|\mathbf{V}_-| = |\mathbf{I}_-|\mathbf{Z}_0$, we can express the maximum and minimum voltage of a line having a characteristic impedance $\mathbf{Z}_0$ as

$$|\mathbf{V}_{max}| = |\mathbf{I}_{max}|\mathbf{Z}_0 \tag{1-5-45}$$

and

$$|\mathbf{V}_{min}| = |\mathbf{I}_{min}|\mathbf{Z}_0 \tag{1-5-46}$$

In view of the standing-wave pattern when the voltage is maximum and the current is minimum, the impedance at that point must be maximum and purely resistive. Therefore,

$$\mathbf{Z}_{max} = \frac{|\mathbf{V}_{max}|}{|\mathbf{I}_{min}|} \tag{1-5-47}$$

and

$$\mathbf{Z}_{min} = \frac{|\mathbf{V}_{min}|}{|\mathbf{I}_{max}|} \tag{1-5-48}$$

The power transmitted by a line is defined as

$$\mathbf{P} = |\mathbf{V}_{max}||\mathbf{I}_{min}| = |\mathbf{V}_{min}||\mathbf{I}_{max}| \tag{1-5-49}$$

so we can express the power in terms of the line voltage as

$$\begin{aligned} \mathbf{P} &= \frac{|\mathbf{V}_{max}||\mathbf{V}_{min}|}{\mathbf{Z}_0} \\ &= \frac{|\mathbf{V}_{max}|^2}{\mathbf{Z}_{max}} \\ &= \frac{|\mathbf{V}_{min}|^2}{\mathbf{Z}_{min}} \end{aligned} \tag{1-5-50}$$

Similarly, we can express the transmitted power in terms of the line current:

$$\begin{aligned} \mathbf{P} &= |\mathbf{I}_{max}||\mathbf{I}_{min}|\mathbf{Z}_0 \\ &= |\mathbf{I}_{max}|^2\mathbf{Z}_{min} \\ &= |\mathbf{I}_{min}|^2\mathbf{Z}_{max} \end{aligned} \tag{1-5-51}$$

Substituting Eqs. (1-5-40) and (1-5-41) into Eq. (1-5-50) yields

$$\mathbf{P} = \frac{|\mathbf{V}_+|^2}{\mathbf{Z}_0} - \frac{|\mathbf{V}_-|^2}{\mathbf{Z}_0} \tag{1-5-52}$$

Also, substituting Eqs. (1-5-43) and (1-5-44) into Eq. (1-5-51) yields

$$\mathbf{P} = |\mathbf{I}_+|^2\mathbf{Z}_0 - |\mathbf{I}_-|^2\mathbf{Z}_0 \tag{1-5-53}$$

From Eqs. (1-5-52) and (1-5-53), it follows that the first term is the power associated with the incident wave, and the second term is the power associated with the reflected wave. This relation can be true only when the characteristic impedance is a pure resistance.

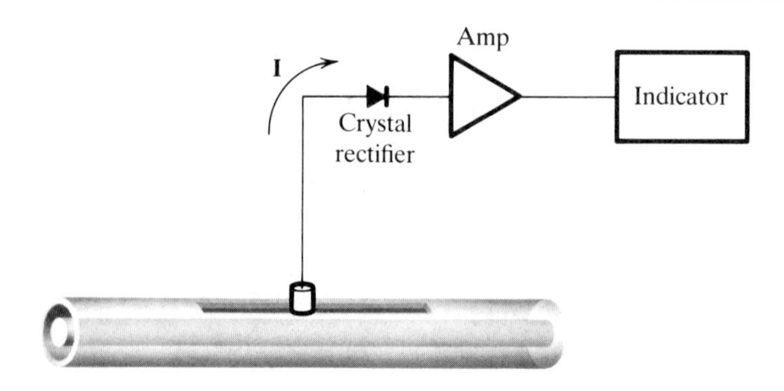

Figure 1-5-7 Slotted line with a probe and devices.

The instrument ordinarily used to measure the standing-wave ratio at higher frequencies is a probe, which is connected to a crystal device. The probe is moved along the transmission line to sample the relative amplitude of the electric field intensity. The output of the probe is rectified and/or amplified and connected directly to an indicator such as microammeter, electronic voltmeter, or oscilloscope. Figure 1-5-7 shows the probe on a slotted line.

From the equation of the voltage standing wave, Eq. (1-5-5), the magnitude of a voltage wave is maximum at some point of z, and there is a minimum voltage at another point of z. The current shown by the indicator is directly proportional to the induced probe voltage according to the following relationships:

$$\mathbf{I}_{\max} = k\mathbf{V}^n_{\max} \tag{1-5-54}$$

and

$$\mathbf{I}_{\min} = k\mathbf{V}^n_{\min} \tag{1-5-55}$$

where

k = constant; and

n = index, or law, of the detector.

EXAMPLE 1-5-1 Standing-Wave Ratio ρ for an Unmatched Line

A certain transmission line has an impedance of $50 + j0.01\ \Omega$ and is terminated in an impedance of $73 - j42.50\ \Omega$. Calculate the ρ of the line.

Solution:

$$\Gamma = \frac{\mathbf{Z}_\ell - \mathbf{Z}_0}{\mathbf{Z}_\ell + \mathbf{Z}_0} = \frac{73 - j42.50 - (50 + j0.01)}{73 - j42.50 + (50 + j0.01)}$$

$$= 0.377 \underline{/-42.7^\circ}$$

So

$$\rho = \frac{1 + |\Gamma|}{1 - |\Gamma|} = \frac{1 + 0.377}{1 - 0.377}$$

$$= 2.21$$

EXAMPLE 1-5-2 Computer Solution for Standing-Wave Ratio ρ for an Unmatched Line

```
PROGRAM   VSWR
010C      PROGRAM TO COMPUTE STANDING-WAVE RATIO
020C+     RHO FOR AN UNMATCHED LINE
030C      VSWR
040       COMPLEX Z0,ZL,BGAML
050     1 READ 2, Z0,ZL
060       IF(Z0 .LE. 0.0) STOP
070       PRINT 4, Z0,ZL
080       BGAML=(ZL-Z0)/(ZL+Z0)
090       BGAMLM=CABS(BGAML)
100       RHO=(1.0+BGAMLM)/(1.0-BGAMLM)
110       PRINT 6, BGAMLM,RHO
120     2 FORMAT(4F10.5)
130     4 FORMAT(//1H ,29HCHARACTERISTIC IMPEDANCE Z0=
140+      1X,F10.5,1X,2H+J,F10.5,1X,4HOHMS/
150+      1X,29HLOAD IMPEDANCE             ZL=
160+      1X,F10.5,1X,2H+J,F10.5,1X,4HOHMS/)
170     6 FORMAT(1H ,27HMAGNITUDE OF GAMMA BGAMLM=
180+      1X,F10.5/
190+      1X,27HSTANDING-WAVE RATIO    RHO= ,F10.5/)
200       GO TO 1
210       STOP
220       END
READY.
RUN

PROGRAM   VSWR
?   50.00000   0.01000  73.00000 -42.50000

 CHARACTERISTIC IMPEDANCE Z0=    50.00000 +J    .01000 OHMS
 LOAD IMPEDANCE           ZL=    73.00000 +J -42.50000 OHMS

 MAGNITUDE OF GAMMA BGAMLM=       .37142
 STANDING-WAVE RATIO    RHO=    2.18176

? 0.0
SRU       0.750 UNTS.
RUN COMPLETE.
```

EXAMPLE 1-5-3

A standing-wave detector is equipped with a crystal having an index of $n = 1.8$. The meter reads 50 μA at a maximum and 10 μA at a minimum, as the probe is moved along the line. What is the standing-wave ratio ρ on the line?

Solution:

$$\rho = \frac{|\mathbf{V}_{\text{max}}|}{|\mathbf{V}_{\text{min}}|} = \left[\frac{\mathbf{I}_{\text{max}}}{\mathbf{I}_{\text{min}}}\right]^{1/n} = \left[\frac{50 \times 10^{-6}}{10 \times 10^{-6}}\right]^{1/1.8}$$

$$= 2.4$$

1-6 COAXIAL LINES AND COAXIAL CONNECTORS

Coaxial lines and coaxial connectors are widely used in electronic circuit systems. The reasons are that the coaxial lines operate in the TEM mode and that their characteristic impedances are constant. However, when the frequency is up to the microwave range, the connectors cause a serious problem with regard to the connections.

1-6-1 Coaxial Lines

A coaxial line is formed by placing one conductor within and on the center axis of another hollow conductor. The space between the two conductors is filled with a frequency independent dielectric. The shape of a coaxial line is either round or square. For any uniform, lossless, reciprocal, and TEM-mode round coaxial line, the element parameters are

$$R = \frac{R_s}{2\pi}\left(\frac{1}{a} + \frac{1}{b}\right) \tag{1-6-1}$$

$$L = \frac{\mu_c}{2\pi} \ln\left(\frac{b}{a}\right) \tag{1-6-2}$$

$$G = \frac{2\pi\sigma_c}{\ln (b/a)} \tag{1-6-3}$$

and

$$C = \frac{2\pi\varepsilon}{\ln (b/a)} \tag{1-6-4}$$

where

$R_s = \sqrt{\frac{\pi f \mu_c}{\sigma_c}}$ is the surface resistance of the conductor, in Ω/square;

σ_c is the conductivity of the conductor, in ℧/m;

μ_c is the permeability of the conductor, in H/m;

a is the radius of the center conductor, in m;

b is the radius of the outer hollow conductor, in m. and

ε is the permittivity, in F/m.

The characteristic impedance of a round coaxial line is

$$\mathbf{Z}_0 = 60\sqrt{\frac{\mu_r}{\varepsilon_r}} \ln\left(\frac{b}{a}\right) \tag{1-6-5}$$

or

$$\mathbf{Z}_0 = 138\sqrt{\frac{\mu_r}{\varepsilon_r}} \log\left(\frac{b}{a}\right) \tag{1-6-6}$$

where

ε_r = the relative permittivity of the insulator between the two conductors; and

μ_r = the relative permeability of the insulator between the two conductors.

For a square coaxial line the characteristic impedance is

$$\mathbf{Z}_0 = 138\sqrt{\frac{\mu_r}{\varepsilon_r}}\log\left(1.0787\frac{b}{a}\right) \tag{1-6-7}$$

1-6-2 Coaxial Connectors

For high-frequency operation the average circumference of a coaxial line must be limited to about one wavelength in order to reduce multimodal wave propagation and eliminate erratic reflection coefficients, power losses, and signal distortion. The standardization of coaxial connectors during World War II was mandatory for microwave operation to maintain a low reflection coefficient or low voltage standing-wave ratio. Since then many modifications and new designs for microwave connectors have been developed. Figure 1-6-1 shows several commonly used coaxial connectors.

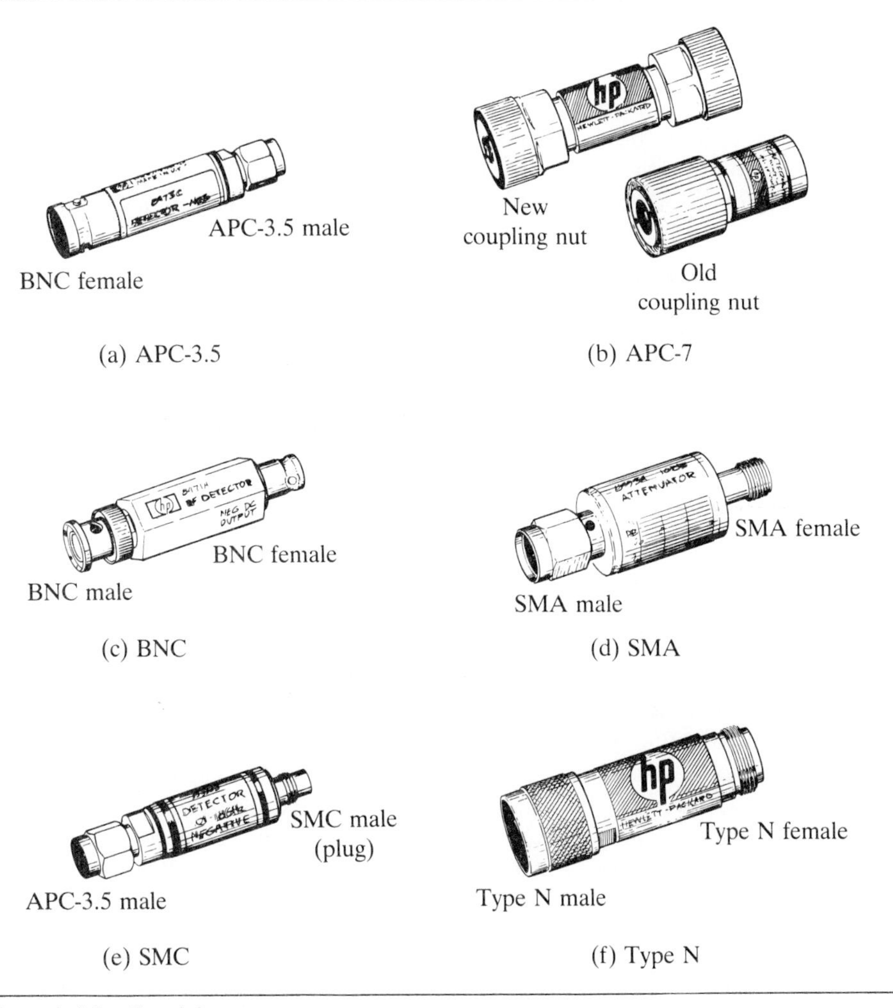

Figure 1-6-1 Coaxial connectors.

APC-3.5 The APC-3.5 (amphenol precision connector—3.5 mm) was originally developed by Hewlett-Packard but is now manufactured by Amphenol. The connector provides repeatable connections and has a very low VSWR. Either the plug (positive, or male) or jack (negative, or female) end of this 50-ohm connector can mate with the opposite type of SMA connector. The APC-3.5 connector can work at frequencies of up to 34 GHz.

APC-7 The APC-7 (amphenol precision connector—7 mm) was also developed by Hewlett-Packard in the mid-1960s but was later improved and is now manufactured by Amphenol. The connector provides a coupling mechanism without male or female distinction and is the most repeatable connecting device used for very accurate 50-ohm measurement applications. Its VSWR is extremely low: in the range of 1.02 for the frequency up to 18 GHz. Maury Microwave also produces an MPC series.

BNC The BNC (bayonet navy connector) was originally designed for military applications during World War II. The connector operates very well at frequencies of up to about 4 GHz; beyond that it tends to radiate electromagnetic energy. The BNC can accept flexible cables with a diameter of up to 6.35 mm (0.25 in.) and a characteristic impedance of 50 or 75 ohms. It is now the most commonly used connector for frequencies of less than 1 GHz.

SMA The SMA (sub-miniature A) connector was originally designed by Bendix Scintilla Corporation, but it has been manufactured by Omni-Spectra, Inc., as the OSM connector, and by many other electronic companies. The main application of SMA connectors is on components of microwave systems. The connector is seldom used for frequencies of more than 24 GHz because of higher order modes.

SMC The SMC (sub-miniature C) is a 50-ohm connector that is smaller than the SMA and is manufactured by Sealectro Corporation. The connector can accept flexible cables with diameters of up to 3.17 mm (0.125 in.) for a frequency range of up to 7 GHz.

TNC The TNC (threaded navy connector) is merely a threaded BNC. The function of the thread is to stop radiation at higher frequencies, so that the connector can work at frequencies of up to 12 GHz.

Type N The type N (navy) connector was originally designed for military systems during World War II and is the most popular measurement connector for the frequency range of 1–18 GHz. It is a 50- or 75-ohm connector and its VSWR is extremely low: less than 1.02.

In microwave measurements the test fixtures are usually connected to test equipment, such as the Hp (Hewlett-Packard) automatic network analyzer (8409B ANA), through some microwave connectors and/or adaptors. Difficulties often occur at the transition section because the connection is not well-enough matched. For example, the small round pin of an APC-7 connector has a very small contact area with the flat surface of a 50-ohm microstrip line. Positioning the center pin of an APC-7 connector on the microstrip line without causing unforseeable gap is very difficult and also creates problems. In order to improve the transition section and minimize its VSWR, the shape of the APC-7 center pin should be modified to a square type and the position of the center pin should be stabilized with a teflon fiberglass.

Problems

1-1 An open-wire line has a characteristic impedance of 300 Ω and is terminated in a load of $300 - j300$ Ω. The propagation constant of the line is $(0.054 + j3.53)$/m. Determine the

a) reflection coefficient at the load;
b) transmission coefficient at the load; and
c) reflection coefficient at a point 2 m from the load.

1-2 A lossless coaxial line has a characteristic impedance of 50 Ω and is terminated in a load of 100 Ω. The magnitude of a voltage wave incident to the line is 20 V(rms). Determine the

a) VSWR on the line;
b) maximum voltage $\mathbf{V}_{max}$ and minimum voltage $\mathbf{V}_{min}$ on the line;
c) maximum current $\mathbf{I}_{max}$ and minimum current $\mathbf{I}_{min}$ on the line; and
d) power transmitted by the line.

1-3 Repeat Problem 1-2 for 0 and ∞ loads.

1-4 A telephone line has $R = 6.70$ Ω/mi, $L = 3.55 \times 10^{-3}$ H/mi, $G = 0.45 \times 10^{-6}$ ℧/mi, and $C = 8.42 \times 10^{-9}$ F/mi. The signal frequency is 1 kHz. Determine the

a) quantities $\mathbf{Z}_0$, α, and β;
b) phase velocity v; and
c) wavelength λ.

1-5 A telephone transmission line has a characteristic impedance of $650 - j83$ Ω and propagation constant of $(0.00540 + j0.0353)$/mi at a frequency of 1 kHz. The line is energized by a generator with an output voltage of 10 V(rms) and output impedance of $\mathbf{Z}_0$. The line terminates in an open circuit 100 mi from the sending end. Compute the magnitude and the phase of the voltage across the receiving end.

1-6 A coaxial cable with a solid polyethylene dielectric ($\epsilon_r = 2.25$) is to be used at a frequency of 0.3 GHz. Its characteristic impedance $\mathbf{Z}_0$ is 50 Ω and its attenuation constant α is 0.0156 Np/m. The velocity constant, which is defined as the ratio of phase velocity to velocity of light in free space, is 0.660. The line is 100 m long and terminates in its characteristic impedance. A generator having an open-circuit voltage of 50 V(rms) and an internal impedance of 50 Ω is connected to the sending end of the line. The frequency is tuned at 0.3 GHz. Calculate the

a) magnitude of the sending-end and receiving-end voltages;
b) magnitude of the sending-end and receiving-end power; and
c) wavelengths of the line.

1-7 An open-wire transmission line has $R = 5$ Ω/m, $L = 5.36 \times 10^{-8}$ H/m, $G = 6.2 \times 10^{-3}$ ℧/m, and $C = 2.07 \times 10^{-11}$ F/m. The signal frequency is 0.1 GHz. Calculate the

a) characteristic impedance of the line in both rectangular and polar form;
b) propagation constant of the line;
c) normalized impedance of a load $100 + j100$;
d) reflection coefficient at the load; and
e) sending-end impedance, assuming that the line is a quarter-wavelength long.

1-8 A telephone line is 100 mi long and has a characteristic impedance of $683 - j138\ \Omega$. Its propagation constant is $0.0074 + j0.0356$ at a frequency of 1 kHz. The line terminates in its characteristic impedance. The generator has an open-circuit voltage output of 20 V(rms) and an internal impedance of $500\ \Omega$.

a) Find the sending-end current, voltage, and power.
b) Determine the receiving-end voltage, current, and power.

1-9 A coaxial line has $R = 6\ \Omega/\text{m}$, $L = 5.2 \times 10^{-8}$ H/m, $G = 6 \times 10^{-3}$ ℧/m, and $C = 2.136 \times 10^{-10}$ F/m. The line is operated at a frequency of 100 kHz and terminated by a load of $100 + j100\ \Omega$. Calculate the

a) characteristic impedance and the propagation constant of the line;
b) reflection and transmission coefficients at the load;
c) reflection and transmission coefficients at a point 5 m from the load.

1-10 A coaxial line has $\mathbf{Z}_0 = 49.4 + j7.6$, $\mathbf{Z}_\ell = 100 + j100$, $\alpha = 0.207$ Np/m, and $\beta = 0.69$ rad/m. The line is 1 m long and operates at a frequency of 1 GHz.

a) Determine the reflection coefficients at the load and at the point 1 m from the load.
b) Find the standing-wave ratios at the load and at the point 1 m from the load.

1-11 A lossless short-circuited stub with a characteristic impedance of $300\ \Omega$ is to have an input reactance of $400\ \Omega$. The signal frequency is 2 GHz. Determine the

a) length of the stub if the stub is to be inductive; and
b) length of the stub if the stub is to be capacitive.

1-12 A coaxial line has a characteristic impedance of $75\ \Omega$ and terminates in $100 + j100\ \Omega$. The line is 2 wavelengths long, and has $\alpha\ell = 0.40$ Np. Find the sending-end impedance of the line.

1-13 A 100-mi long telephone line has a characteristic impedance of $100 - j100\ \Omega$. Its propagation constant is $(0.0052 + j0.035)$/mi at a frequency of 1 kHz. The receiving end is open-circuited, and the sending-end voltage is 15 V(rms).

a) Determine the sending-end impedance.
b) Find the magnitude and phase angle of the receiving-end voltage.

1-14 A coaxial line has a characteristic impedance of $50\ \Omega$. The relative dielectric constant of the polyethylene that separates the center and outer conductors is 2.25. The attenuation constant is 0.061 Np/m at a frequency of 200 MHz. A section of this 2-wavelength–long line terminates in an open circuit. The sending-end voltage is 50 V(rms).

a) Determine the sending-end impedance.
b) Find the magnitude of the receiving-end voltage.

1-15 Repeat Problem 1-14 for the loads of 0 and $\mathbf{Z}_0$.

1-16 A lossless line has a characteristic impedance of $75\ \Omega$ and terminates in a load of $300\ \Omega$. The line is energized by a generator having an open-circuit output voltage of 20 V(rms) and output impedance of $75\ \Omega$. The line is assumed to be $2\frac{1}{4}$ wavelengths long.

a) Find the sending-end impedance.

b) Determine the magnitude of the receiving-end voltage.
c) Calculate the receiving-end power at the load.

1-17 A lossless coaxial line has a characteristic impedance of 100 Ω and terminates in a load of 75 Ω. The line is 0.75 wavelength long. Determine the

a) sending-end impedance; and
b) reactance, which if connected across the sending end of the line, will make the input impedance a pure resistance.

1-18 A quarter-wave lossless line has a characteristic impedance of 50 Ω and terminates in a load of 100 Ω. The line is energized by a generator of 20 V(rms) with an internal resistance of 50 Ω. Calculate the

a) sending-end impedance; and
b) magnitude of the receiving-end voltage; and
c) power delivered to the load.

1-19 Repeat Problem 1-18 for a termination of open circuit.

1-20 Repeat Problem 1-18 for a termination of the same characteristic impedance.

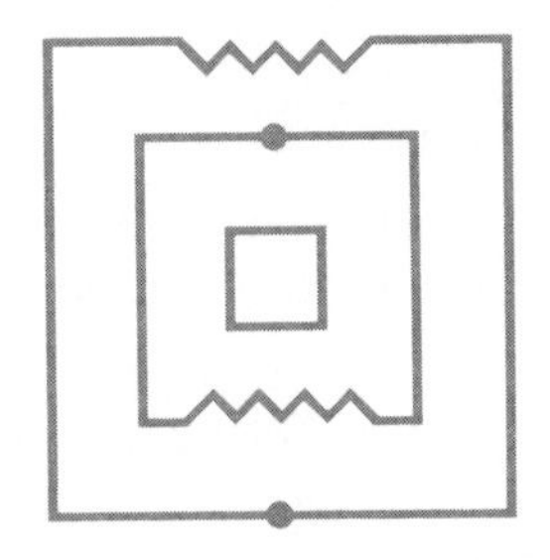

Chapter 2

Transmission-Line Matching Techniques

2-0 INTRODUCTION

Many of the calculations required to solve transmission-line problems involve the use of rather complicated equations. Solving such problems is tedious and difficult because accurate manipulation of numerous equations is necessary. In order to simplify our work, we need a graphic method to help us arrive at an answer quickly.

A number of impedance charts have been designed to facilitate the graphic solution of transmission-line problems. Basically, all these charts are derived from the fundamental relationships expressed in the transmission equations. The most popular chart is that developed by Phillip H. Smith. [1] In this chapter we present the graphic solution of transmission-line problems using the Smith chart.

2-1 SMITH CHART AND IMPEDANCE MATCHING

2-1-1 Smith Chart

Basically, the Smith chart consists of a plot of the normalized impedance or admittance, with the angle and magnitude of a generalized complex reflection coefficient in a unit circle. The chart is applicable to the analysis of both lossless and lossy lines. By simply rotating the chart, we can determine the effect of the position on the line. To see how a Smith chart works, let's consider the equation of the reflection coefficient at the load for a transmission line, as shown in Eq. (1-4-8):

$$\boldsymbol{\Gamma}_\ell = \frac{\mathbf{Z}_\ell - \mathbf{Z}_0}{\mathbf{Z}_\ell + \mathbf{Z}_0} = |\boldsymbol{\Gamma}_\ell| e^{j\theta_\ell} = \Gamma_r + j\Gamma_i \tag{2-1-1}$$

Since $|\boldsymbol{\Gamma}_\ell| \leq 1$, the value of $\boldsymbol{\Gamma}_\ell$ must lie on or within the unit circle that has a radius of 1. The reflection coefficient at any other location along a line is expressed in terms of $\boldsymbol{\Gamma}_\ell$ in Eq. (1-4-12). That is,

$$\boldsymbol{\Gamma}_d = |\boldsymbol{\Gamma}_\ell| e^{-2\alpha d} e^{j(\theta_\ell - 2\beta d)} = |\boldsymbol{\Gamma}_d| e^{j(\theta_\ell - 2\beta d)} \tag{2-1-2}$$

which is also within the unit circle.

Figure 2-1-1 shows circles for a constant reflection coefficient $\boldsymbol{\Gamma}$ and constant electric-length radials βd.

As we described in Eq. (1-3-37), the normalized impedance along a line is

$$z = \frac{\mathbf{Z}}{\mathbf{Z}_0} = \frac{1 + \boldsymbol{\Gamma}_\ell e^{-2\gamma d}}{1 - \boldsymbol{\Gamma}_\ell e^{-2\gamma d}} \tag{2-1-3}$$

With no loss in generality, we can assume that $d = 0$; then

$$z = \frac{1 + \boldsymbol{\Gamma}_\ell}{1 - \boldsymbol{\Gamma}_\ell} = \frac{\mathbf{Z}_\ell}{\mathbf{Z}_0} = \frac{R_\ell + jX_\ell}{\mathbf{Z}_0} = r + jx \tag{2-1-4}$$

and

$$\boldsymbol{\Gamma}_\ell = \frac{z - 1}{z + 1} = \Gamma_r + j\Gamma_i \tag{2-1-5}$$

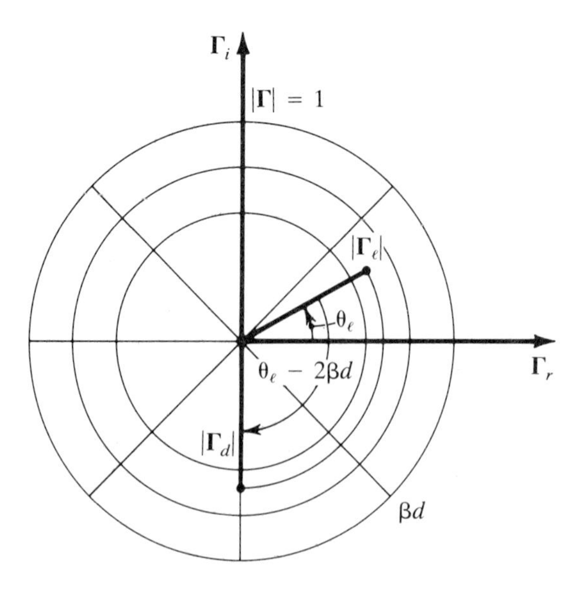

Figure 2-1-1 Constant $\boldsymbol{\Gamma}$ circles and electric-length radials βd.

Substituting Eq. (2-1-5) into Eq. (2-1-4) yields

$$r = \frac{1 - \Gamma_r^2 - \Gamma_i^2}{(1 - \Gamma_r)^2 + \Gamma_i^2} \tag{2-1-6}$$

and

$$x = \frac{2\Gamma_i}{(1 - \Gamma_r)^2 + \Gamma_i^2} \tag{2-1-7}$$

We can rearrange Eqs. (2-1-6) and (2-1-7) as

$$\left(\Gamma_r - \frac{r}{1+r}\right)^2 + \Gamma_i^2 = \left(\frac{1}{1+r}\right)^2 \tag{2-1-8}$$

and

$$(\Gamma_r - 1)^2 + \left(\Gamma_i - \frac{1}{x}\right)^2 = \left(\frac{1}{x}\right)^2 \tag{2-1-9}$$

Equation (2-1-8) represents a family of circles, in which each circle has a constant resistance r. The radius of any circle is $1/(1 + r)$, and the center of any circle is at $\Gamma_r = r/(1 + r)$ and $\Gamma_i = 0$ along the real axis of the unit circle, where r varies from 0 to ∞. The constant-resistance circles plotted in Fig. 2-1-2 were obtained from Eq. (2-1-8) for the r values shown.

Equation (2-1-9) also describes a family of circles, but each of these circles

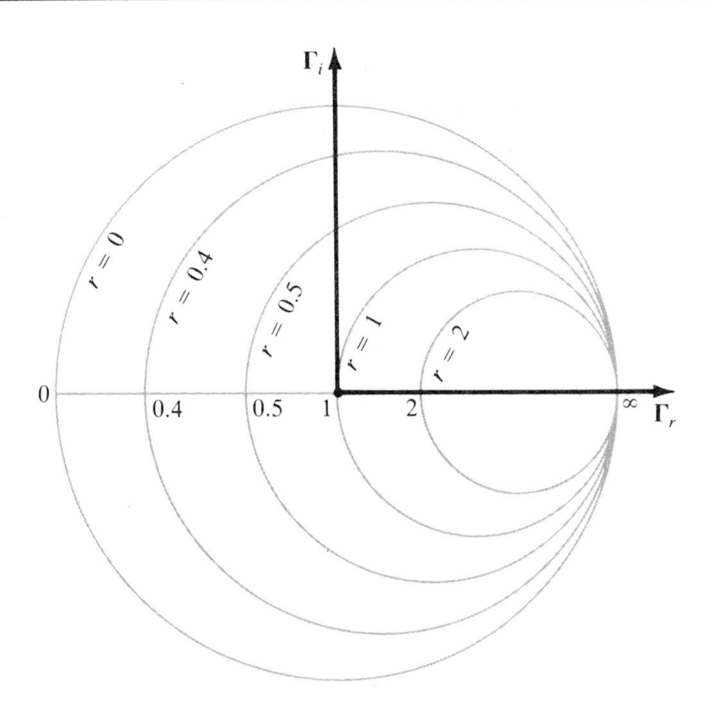

Figure 2-1-2 Constant resistance (r) circles.

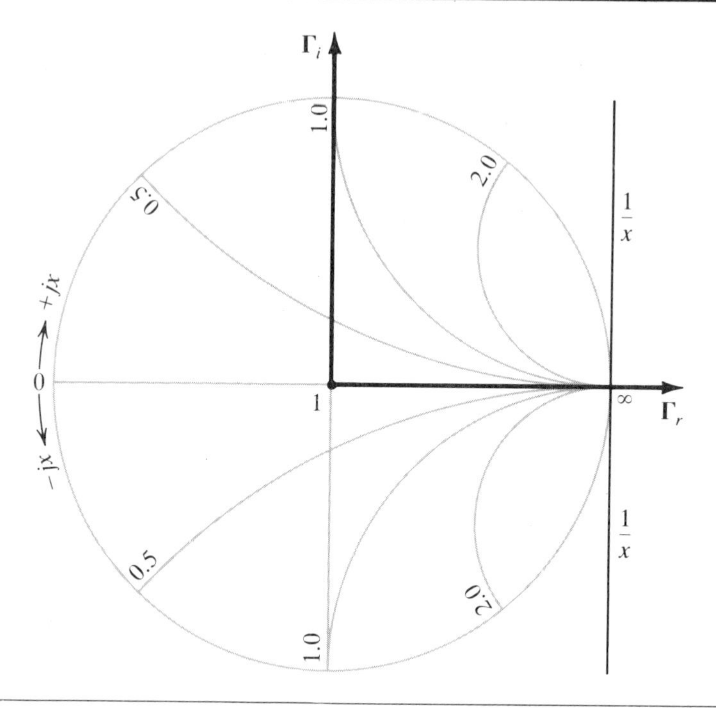

Figure 2-1-3 Constant reactance (x) circles.

represents a constant reactance x. The radius of any circle is $1/x$, and the center of any circle is at $\Gamma_r = 1$ and $\Gamma_i = 1/x$, where $-\infty \leq x \leq \infty$. The constant reactance circles plotted in Fig. 2-1-3 were obtained from Eq. (2-1-9) for the x values shown.

The Smith chart also contains relative distance scales (in wavelength) along the circumference and a phase scale specifying the angle of the reflection coefficient.

When you locate a normalized impedance z on the chart, you can then find the normalized impedance of any other location along the line by using Eq. (2-1-3):

$$z = \frac{1 + \Gamma_\ell e^{-2\gamma d}}{1 - \Gamma_\ell e^{-2\gamma d}} \tag{2-1-10}$$

where

$$\Gamma_\ell e^{-2\gamma d} = |\Gamma_\ell| e^{-2\alpha d} e^{j(\theta_\ell - 2\beta d)}$$

You can also use the Smith chart to determine normalized admittance. Since

$$\mathbf{Y}_0 = \frac{1}{\mathbf{Z}_0} = G_0 + jB_0 \quad \text{and} \quad \mathbf{Y} = \frac{1}{\mathbf{Z}} = G + jB \tag{2-1-11}$$

the normalized admittance is

$$y = \frac{\mathbf{Y}}{\mathbf{Y}_0} = \frac{\mathbf{Z}_0}{\mathbf{Z}} = \frac{1}{z} = g + jb \tag{2-1-12}$$

Figure 2-1-4 shows a Smith chart that combines Figs. 2-1-2 and 2-1-3 into one chart. Figure 2-1-5 shows a complete, commercially available Smith chart. Note that the reflection coefficient is the radial coordinate and that the circles concentric with the

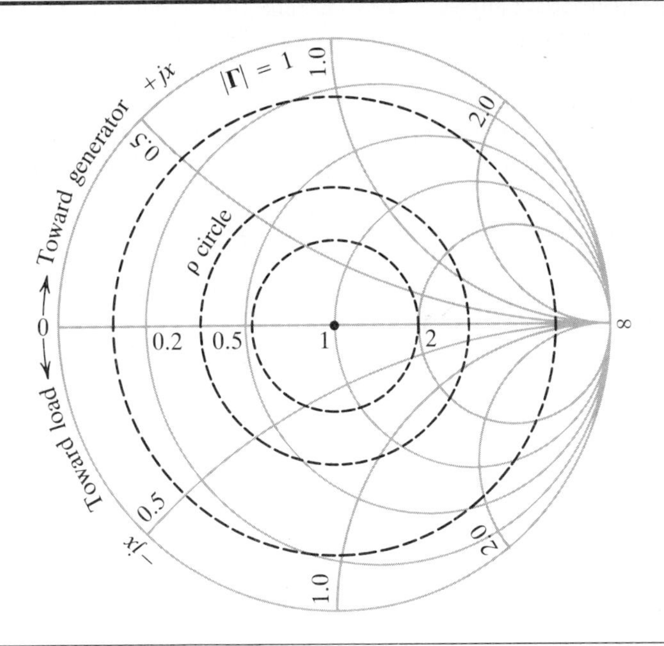

Figure 2-1-4 Smith chart.

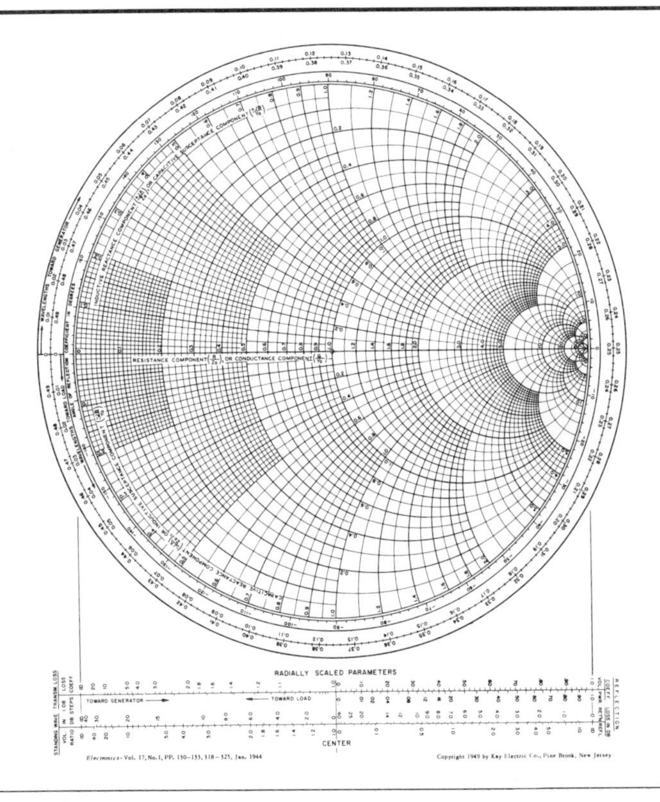

Figure 2-1-5 Smith chart.

center of the unit circle are circles of constant reflection coefficient. Since the standing-wave ratio is determined only by the magnitude of the reflection coefficient, these circles are also contours of constant standing-wave ratio. Since the standing-wave ratio is never less than one, the scale for the standing-wave ratio varies from 1 to ∞ on the real axis. Note also that the distances are given in wavelengths toward both the generator and the load, so that we can easily determine in which direction to advance as position on the line changes.

We can summarize the characteristics of the Smith chart as follows:

1. The constant r and constant x loci form two families of orthogonal circles on the chart.
2. The constant r and constant x circles all pass through the point $(\Gamma_r = 1, \Gamma_i = 0)$.
3. The upper half of the diagram represents $+jx$.
4. The lower half of the diagram represents $-jx$.
5. For admittance, the constant r circles become constant g circles, and the constant x circles become constant susceptance b circles.
6. The distance around the Smith chart once is one-half wavelength, or $\lambda/2$.
7. At a point of $z_{\min} = 1/\rho$, there is a $\mathbf{V}_{\min}$ on the line.
8. At a point of $z_{\max} = \rho$, there is a $\mathbf{V}_{\max}$ on the line.
9. The horizontal radius to the right of the chart's center corresponds to $\mathbf{V}_{\max}$, $\mathbf{I}_{\min}$, $z_{\max}$, and ρ (SWR).
10. The horizontal radius to the left of the chart's center corresponds to $\mathbf{V}_{\min}$, $\mathbf{I}_{\max}$, $z_{\min}$, and $1/\rho$.
11. Since the normalized admittance y is a reciprocal of the normalized impedance z, the corresponding quantities in the admittance chart are 180° out of phase with those in the impedance chart.
12. The normalized impedance or admittance is repeated for every one-half wavelength of distance.

The magnitude of the reflection coefficient is related to the standing-wave ratio by the following expression:

$$|\Gamma| = \frac{\rho - 1}{\rho + 1} \tag{2-1-13}$$

You can use a Smith chart or a slotted line to measure a standing-wave pattern directly and then calculate from it the magnitude of the reflection coefficient, reflected power, transmitted power, and the load impedance. Typical values are shown in Fig. 2-1-6 and at the bottom of the Smith chart (Fig. 2-1-5.) We illustrate the use of the Smith chart in the following examples.

EXAMPLE 2-1-1 Location Determination of Voltage Maximum and Minimum from Load

For a normalized load impedance $z_\ell = 1 - j1$ and an operating wavelength $\lambda = 5$ cm, determine (a) the first $\mathbf{V}_{\max}$; (b) the first $\mathbf{V}_{\min}$ from the load; and (c) the VSWR ρ. (See Fig. 2-1-7.)

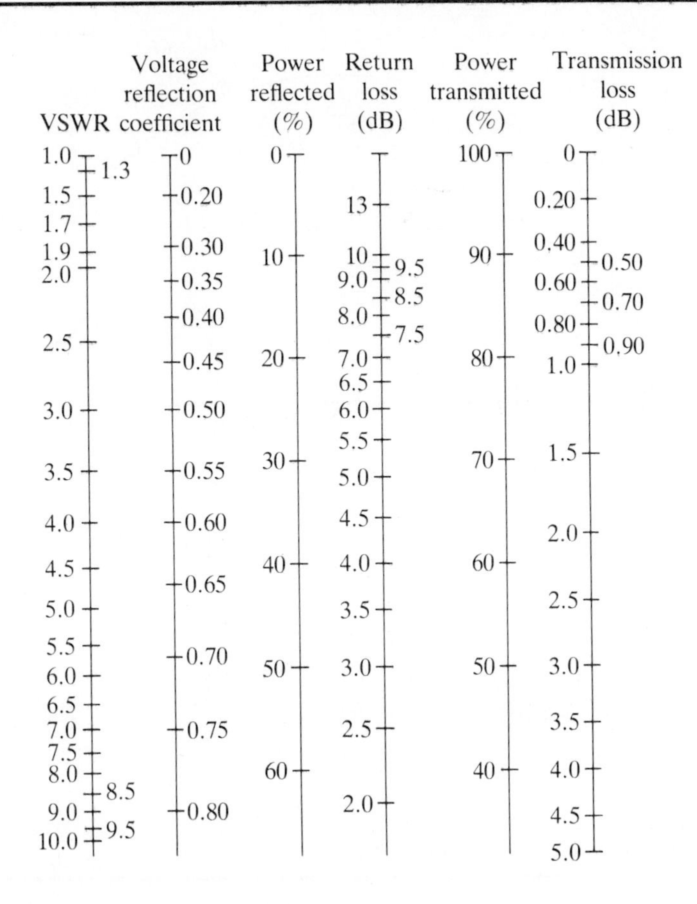

Figure 2-1-6 Nomograph for a transmission line.

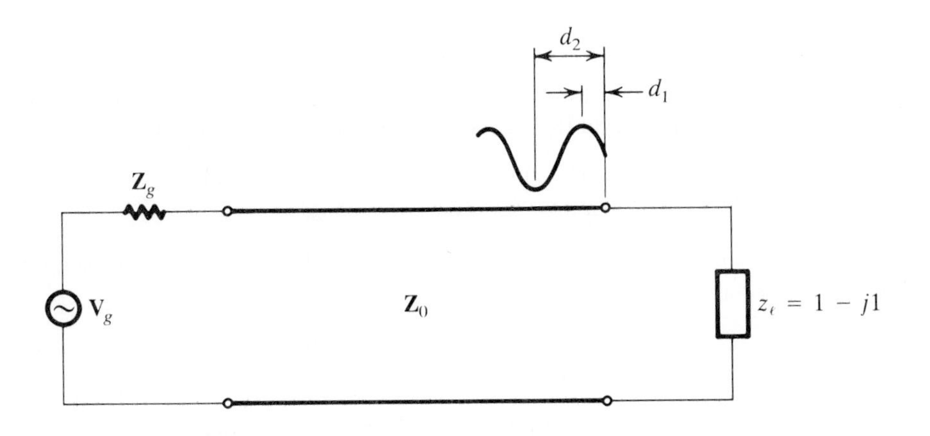

Figure 2-1-7 Circuit diagram for Example 2-1-1.

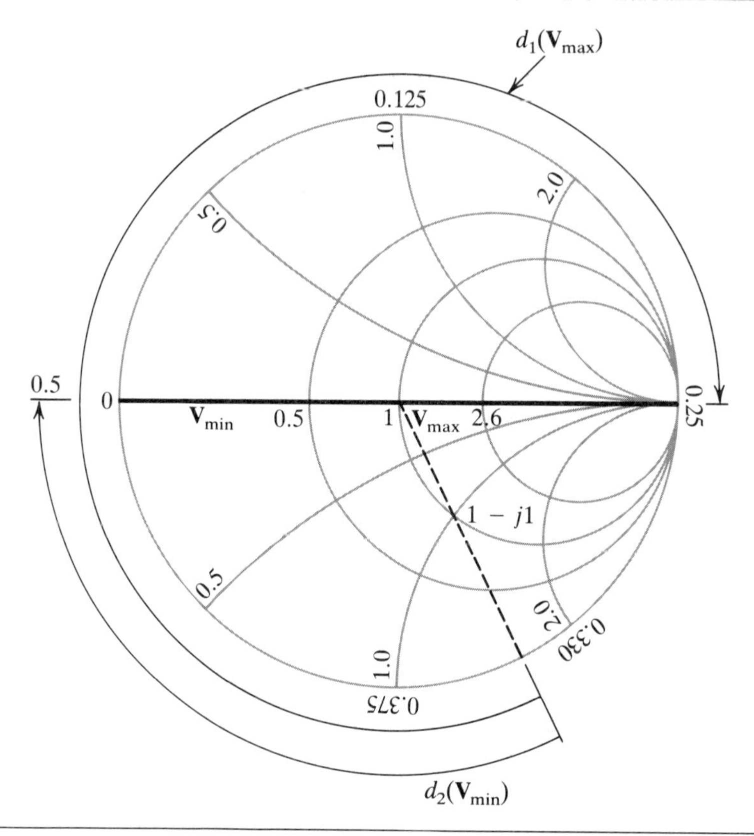

Figure 2-1-8 Graphic solution for Example 2-1-1.

Solutions:

a) Determine the first $\mathbf{V}_{max}$:

1. Enter $z_\ell = 1 - j1$ on the chart, as shown in Fig. 2-1-8.
2. Read 0.338λ on the distance scale by drawing a dashed straight line from the center of the chart through the load point and intersecting the distance scale.
3. Move from the point at 0.338λ toward the generator and stop first at the voltage maximum on the right-hand real axis at 0.25λ. Then

$$d_1(\mathbf{V}_{max}) = (0.25 + 0.162)\lambda = 0.412(5)$$
$$= 2.06 \text{ cm}$$

b) Similarly, move from the point at 0.338λ toward the generator and stop first at the voltage minimum on the left-hand real axis at 0.5λ. Then

$$d_2(\mathbf{V}_{min}) = (0.5 - 0.338)\lambda = 0.162(5)$$
$$= 0.81 \text{ cm}$$

c) Make a standing-wave ratio circle with the center at (1, 0) and pass the circle through the point of $1 - j1$. The intersection of the standing-wave ratio circle and the right portion of the real axis is the SWR, or $\rho = 2.6$.

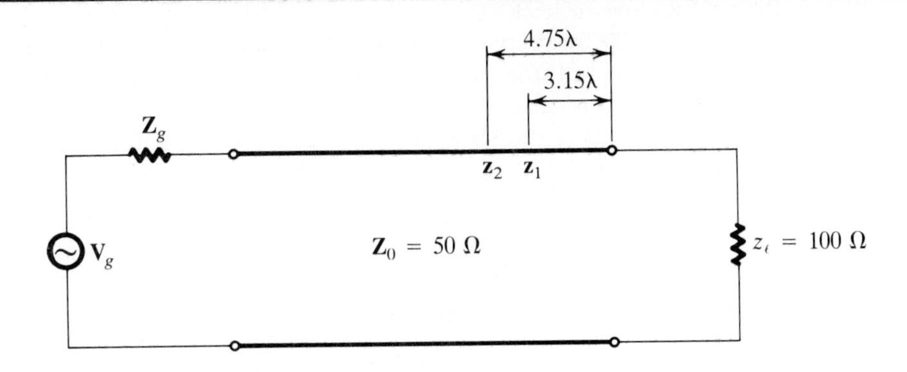

Figure 2-1-9 Load diagram for Example 2-1-2.

EXAMPLE 2-1-2 Line Impedance Determination with Restrictive Load

For a line having a characteristic impedance of $\mathbf{Z}_0 = 50\ \Omega$ and a terminating impedance of $\mathbf{Z}_\ell = 100\ \Omega$, what impedance occurs at points 3.15λ and 4.75λ from the termination? (See Fig. 2-1-9.)

Solutions:

a) Compute the normalized load impedance $z_\ell = 2$ and enter it on the chart, as shown in Fig. 2-1-10.

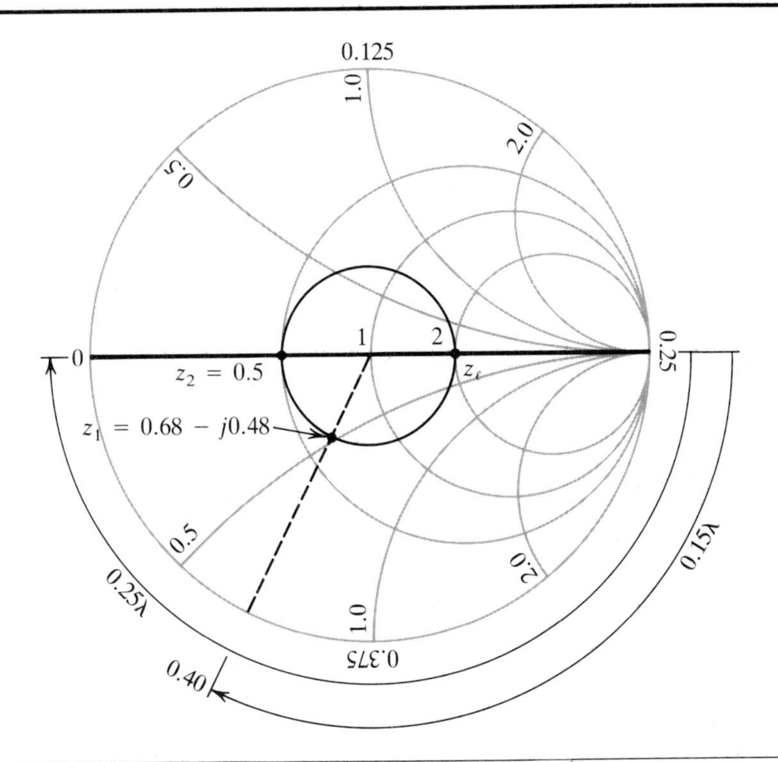

Figure 2-1-10 Graphic solution for Example 2-1-2.

b) Plot the VSWR circle ($\rho = 2$).

c) Move 3.15λ and 4.75λ, respectively, from point 0.25 along the distance scale toward the generator and stop at points 0.40 and 0.50. Note that the distance around the chart once is 0.5λ.

d) The intercepting points of the SWR circle ($\rho = 2$) and the lines linking the center to points 0.40 and 0.50 are

$$z_1 = 0.68 - j0.48$$

$$z_2 = 0.5$$

e) Convert the normalized values to line impedances:

$$\mathbf{Z}_1 = (0.68 - j0.48)(50) = 34 - j24\ \Omega$$

$$\mathbf{Z}_2 = (0.5)(50) = 25\ \Omega$$

EXAMPLE 2-1-3 Line Impedance Determination with a Complex Load Admittance

For a characteristic impedance $\mathbf{Z}_0 = 100\ \Omega$ and a load admittance $\mathbf{Y}_\ell = 0.0025 - j0.0025\ \mho$, determine (a) the SWR ρ at a point 4.15λ from the load; and (b) the impedance at that point. (See Fig. 2-1-11.)

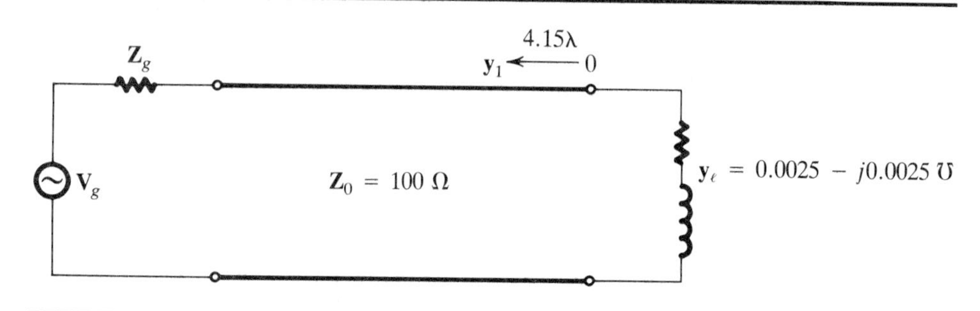

Figure 2-1-11 Load diagram for Example 2-1-3.

Solutions:

a) Find the SWR ρ.

1. Calculate the normalized admittance:

$$y_\ell = \frac{\mathbf{Y}_\ell}{\mathbf{Y}_0} = \mathbf{Y}_\ell \mathbf{Z}_0 = (0.0025 - j0.0025)(100)$$

$$= 0.25 - j0.25$$

2. Enter y_ℓ on the chart, as shown in Fig. 2-1-12.
3. Plot an SWR ρ circle and read $\rho = 4.2$.

b) Find the impedance.

1. Move 4.15λ from the load y_ℓ at 0.458 on the distance scale toward the

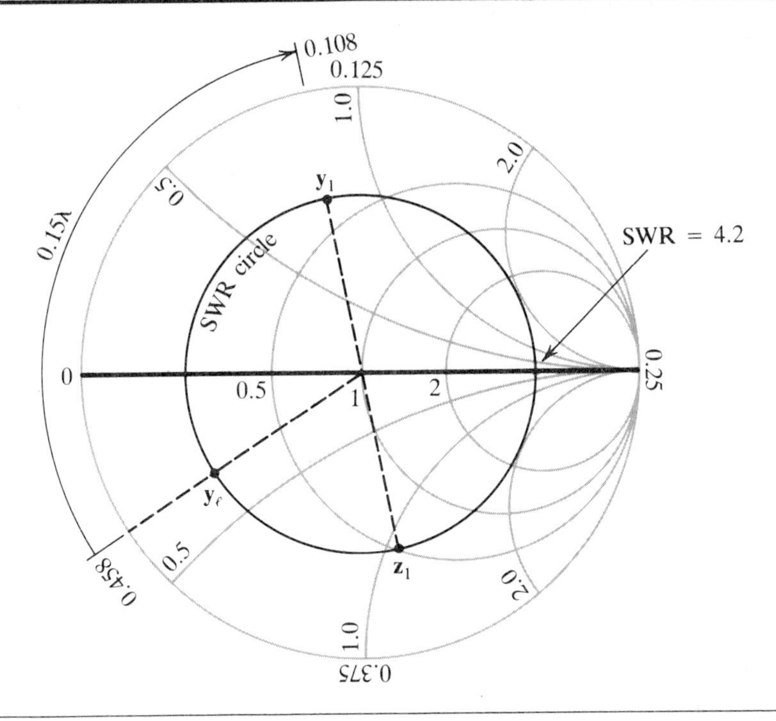

Figure 2-1-12 Graphic solution for Example 2-1-3.

generator and stop at 0.108. Remember that 4λ does not count because the admittance repeats in that interval.

2. Read the normalized admittance on the ρ circle:

$$y_1 = 0.38 + j0.74.$$

3. Read the normalized impedance on the ρ circle 180° out of phase with y_1:

$$z_1 = 0.55 - j1.10.$$

4. The impedance at that point is

$$\begin{aligned}\mathbf{Z}_1 &= (0.55 - j1.10)(100) \\ &= 550 - j110\ \Omega\end{aligned}$$

EXAMPLE 2-1-4 Impedance Determination with Short-Circuit Minima Shift

The location of a minimum instead of a maximum is usually specified, because it can be determined more accurately than that of the maximum. The characteristic impedance of a line $\mathbf{Z}_0 = 100\ \Omega$, and the SWR $\rho = 2$ when the line is loaded. When the load is shorted, the minima shift 1.15λ toward the load. Determine the load impedance. (See Fig. 2-1-13.)

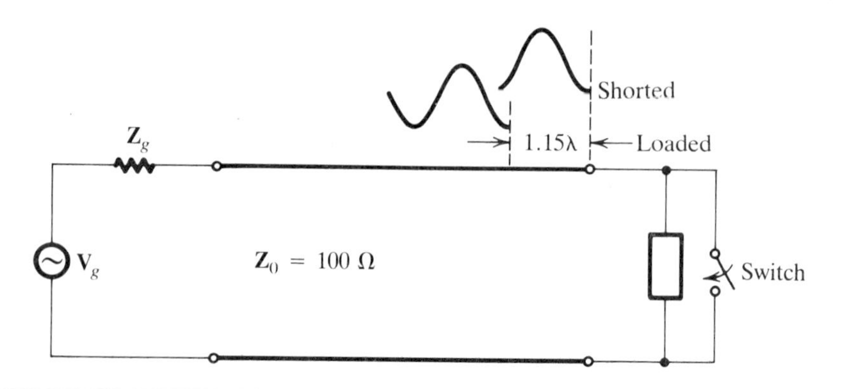

Figure 2-1-13 Load diagram for Example 2-1-4.

Solution:

a) When the line is shorted, the first voltage minimum occurs at the load.
b) When the line is loaded, the first voltage minimum shifts 1.15λ from the load. The distance between two successive minima is one-half wavelength.
c) Plot an SWR circle for $\rho = 2$, as shown in Fig. 2-1-14.
d) Move 1.15λ from the minimum point along the distance scale toward the load and stop at 0.15λ.
e) Draw a line from this point to the center of the chart.

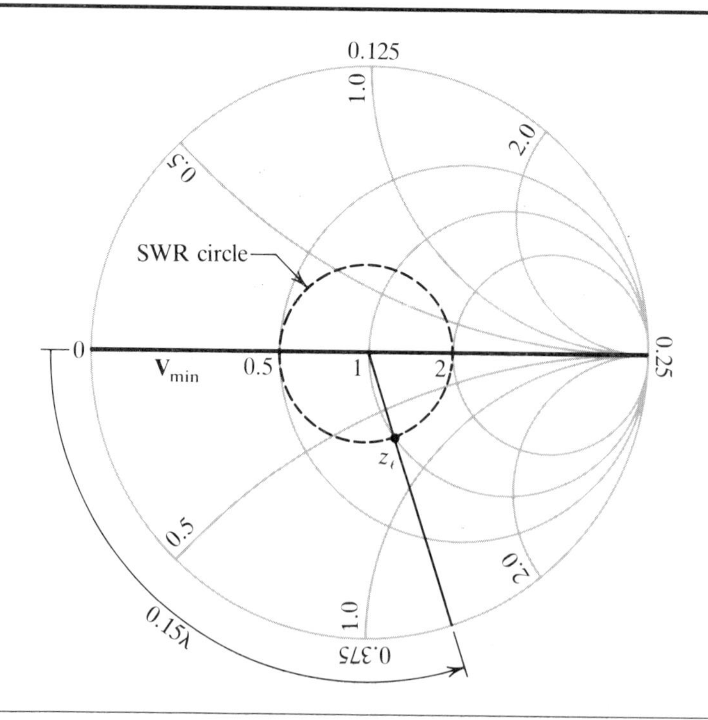

Figure 2-1-14 Graphic solution for Example 2-1-4.

f) The intersection between the line and the SWR circle is

$$z_\ell = 1 - j0.65$$

g) The load impedance is

$$\mathbf{Z}_\ell = (1 - j0.65)(100) = 100 - j65\ \Omega$$

EXAMPLE 2-1-5 Determination of Γ, ρ, Return Loss, and Transmission Loss on a Lossy Line

A coaxial line has the following parameters:

$\mathbf{Z}_0 = 50\ \Omega$;
$\mathbf{Z}_\ell = 114 + j84\ \Omega$;
$\alpha = 0.207$ Np/wavelength;
$\beta = 6.3$ rad/wavelength;
$f = 1$ GHz;
$\ell = 1$ m; and
$\varepsilon_r = 2.25$

a) Determine the reflection coefficient $\mathbf{\Gamma}$ at the load and at 0.1 m from the load.
b) Determine the standing-wave ratio ρ at the load and at 0.1 m from the load.
c) Determine the return loss and transmitted power.
d) Determine the transmission-loss coefficients.

(See Fig. 2-1-15.)

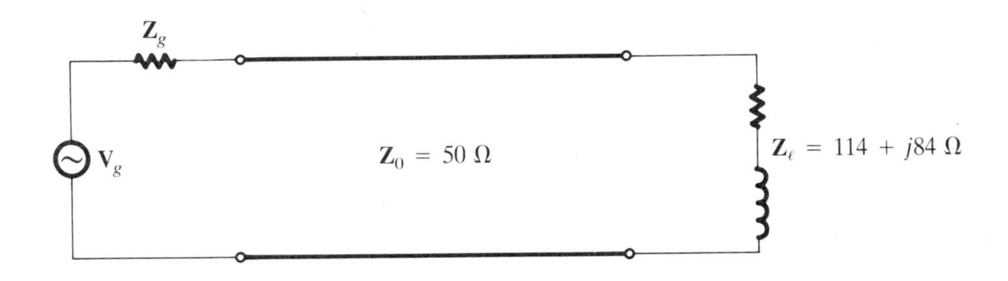

Figure 2-1-15 Circuit diagram for Example 2-1-5.

Solutions:

In Fig. 2-1-16 the voltage reflection coefficient $\mathbf{\Gamma}$ is shown on the upper right-hand scale at the bottom of the chart, and the voltage standing-wave ratio is shown on the lower left-hand scale. The return loss is shown on the lower right-hand scale, indicating the reverse power in dB traveling in a line from a mismatched load below the reference point incident on that mismatched load. The transmission loss is shown on the upper left-hand scale, indicating the power loss in a lossy line.

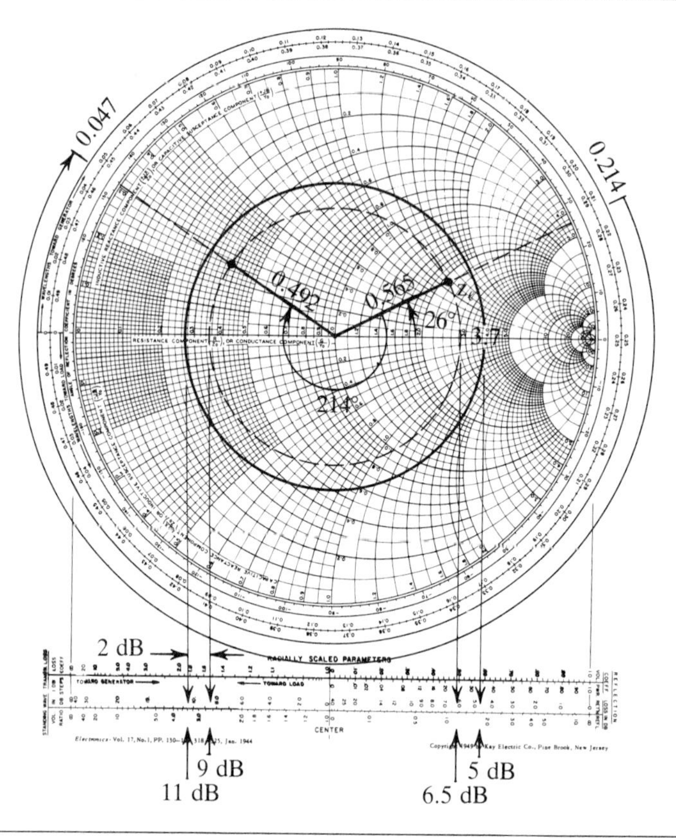

Figure 2-1-16 Smith chart for Example 2-1-5.

a) Find $\mathbf{\Gamma}$.

1. Enter the normalized load $z_\ell = 2.28 + j1.68$ on the chart in Fig. 2-1-16.
2. Draw the SWR circle and read $\rho = 3.7$.
3. Measure the distance from the center to the point z_ℓ and the angle of z_ℓ:

$$\mathbf{\Gamma}_\ell = |\mathbf{\Gamma}_\ell| e^{j\theta_\ell} = 0.565 e^{j26^\circ}$$

4. Read the power reflection coefficient on the lower right-hand scale:

$$|\mathbf{\Gamma}|^2 = 0.32$$

5. Change the unit of 0.1 m to wavelength:

$$\ell = \frac{0.1}{\lambda} = \frac{0.1}{3 \times 10^8/10^9}$$
$$= 0.333\lambda$$

6. The effective reflection coefficient at 0.1 m from the load is

$$|\mathbf{\Gamma}| = |\mathbf{\Gamma}_\ell| e^{-2\alpha\ell} = 0.565 e^{-(2)(0.207)(0.333)}$$
$$= 0.492$$

7. Plot a circle with a radius of 0.492 centered at (1, 0). This circle is the new SWR circle, and the new SWR is 2.9.
8. Move 0.333λ from 0.214 on the distance scale toward the generator and stop at 0.047.
9. Draw a line from the point at 0.047 through the new SWR circle to the center at (1, 0) and read the value of the intersection for the normalized impedance:

$$z = 0.37 + j0.27$$

10. The new reflection coefficient is

$$\Gamma = 0.492e^{j(\theta_\ell - 2\beta\ell)} = 0.942e^{j(26-240)}$$

$$= 0.492e^{-j214^\circ}$$

11. The power reflection coefficient is

$$|\Gamma|^2 = 0.24$$

b) Find ρ.

1. The SWR at the load $\rho_\ell = 3.7$.
2. The SWR at 0.1 m from the load $\rho = 2.9$.
3. The SWRs (in dB) are measured on the standing-wave scale at the lower left-hand edge of the chart in Fig. 2-1-16:

$$\rho_\ell = 11 \text{ dB}$$

$$\rho = 9 \text{ dB}$$

Note that because of the loss on the line, the standing-wave ratio decreases in a spiral way from $\rho_\ell = 3.7$ to $\rho = 2.9$, as indicated on the chart. This condition results from power being transmitted toward the load on a lossy line and power being reflected back toward the generator suffer an attenuation in both directions.

4. The attenuation loss is 2 dB, as shown on the chart.

c) Find the return loss and transmitted power.

1. Draw a line perpendicular to the return-loss scale at the lower right-hand edge of the chart from the points where the SWR circle intersects the real axis on the right, and read

$$\text{Return loss (at load)} = 5.0 \text{ dB}$$

$$\text{Return loss (at 0.1 m from load)} = 6.5 \text{ dB}$$

These values mean that the power reflected from the points is 5 dB (or 6.5 dB) below the power incident upon that point.

2. Read the transmitted power (reflection loss) from the lower side of the scale used in (b-3):

$$\text{Transmitted power (at load)} \doteq 2 \text{ dB}$$

$$\text{Transmitted power (at 1 m from load)} \doteq 1 \text{ dB}$$

These values indicate that the power absorbed by the load is about 2 dB below the power incident upon the load and that the power transmitted at the point 0.1 m from the load is about 1 dB below the power incident on the load.

d) Find the transmission-loss coefficients:

1. Draw a line perpendicular to the transmission-loss scale at the upper left-hand scale of the bottom of the chart in Fig. 2-1-16, from the points at which the SWR circles intersect the real axis on the left, and read

$$\text{Transmission-loss coefficient (at load)} \doteq 2.0$$

$$\text{Transmission-loss coefficient (at 0.1 m from load)} \doteq 1.6$$

2-1-2 Impedance Matching

Impedance matching is very desirable for radio-frequency (RF) transmission lines. Standing waves lead to increased losses and frequently cause the transmitter to malfunction. However, a line terminated in its characteristic impedance has a standing-wave ratio of one and transmits a given power without reflection. Also, transmission efficiency is optimum when there is no reflected power. A "flat" line is nonresonant; that is, its input impedance always remains at the same $\mathbf{Z}_0$ when the frequency changes.

Matching a transmission line has a special meaning, differing from that used in circuit theory to indicate equal impedance as seen in both directions from a given terminal pair for maximum power transfer. In circuit theory, maximum power transfer requires the load impedance to be equal to the complex conjugate of the generator. We sometimes refer to this condition as a *conjugate match*. In transmission-line problems, *matching* simply means terminating the line in its characteristic impedance.

A common application of radio frequency transmission lines is the one in which there is a feeder connection between a transmitter and an antenna. Usually, the input impedance to the antenna itself is not equal to the characteristic impedance of the line. Furthermore, the output impedance of the transmitter may not be equal to the characteristic impedance $\mathbf{Z}_0$ of the line. Matching devices are necessary to flatten the line. A complete matched transmission-line system is shown in Fig. 2-1-17.

For a low-loss or lossless transmission line at radio frequency, the characteristic impedance $\mathbf{Z}_0$ of the line is resistive. At every point the impedances viewed in opposite directions are conjugate. If $\mathbf{Z}_0$ is real, it is its own conjugate. We can try matching first in the load side to flatten the line, then adjust the transmitter side to provide maximum power transfer. For audio frequencies an iron-cored transformer is almost universally used as an impedance-matching device. Occasionally an iron-cored transformer is also used for radio frequencies. In an actual transmission-line system, the transmitter is

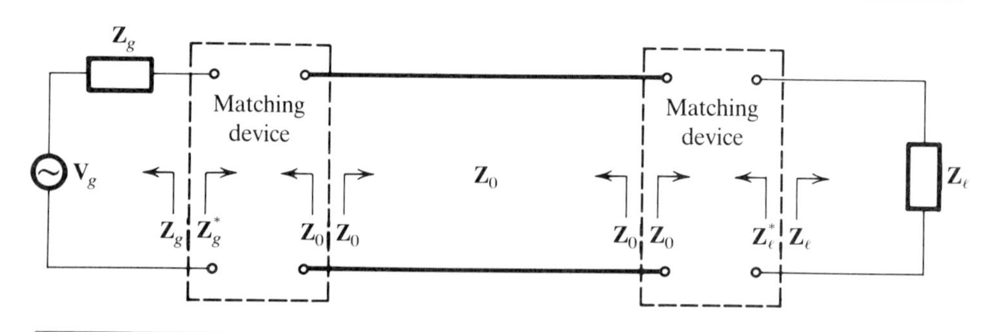

Figure 2-1-17 Matched transmission-line system.

ordinarily matched to the coaxial cable for maximum power transfer. Because of the variable loads, however, an impedance-matching device is often required at the load side.

Matching involves parallel connections on the transmission line, so it is necessary to solve matching problems using admittances rather than impedances. We can use the Smith chart to convert the normalized impedance to admittance by a rotation of 180°, as described earlier in this section.

2-1-3 Double-Susceptance Matching

Sometimes we can accomplish matching by using a double susceptance. Assume that the load is an antenna, which because of design considerations, can be matched only to a normalized impedance of $0.4 - j0.4$, with an SWR $\rho = 2.9$. At some point on the line the conductance equals the characteristic admittance of the line. In order to achieve a characteristic admittance of 1 (that is, $g = 1$) at that point, we have to insert a shunt susceptance to cancel out the susceptive part of the admittance there. In general, the admittance at that point is

$$\mathbf{Y} = G_0 \pm jB_0 \tag{2-1-14}$$

and the normalized admittance for matching purposes is

$$y = 1 \pm jb \tag{2-1-15}$$

EXAMPLE 2-1-6 Double-Susceptance Matching

A transmission line has the following parameters:

$\mathbf{Y}_\ell = 0.004 - j0.004$ ℧;
$\mathbf{Z}_0 = 100\ \Omega$;
$\rho = 2.9$; and
$f = 1$ GHz

Determine (a) the value of a lumped inductance or capacitance that can match the transmission line; and (b) the distance from the load to the point where the tuner should be placed. (See Fig. 2-1-18.)

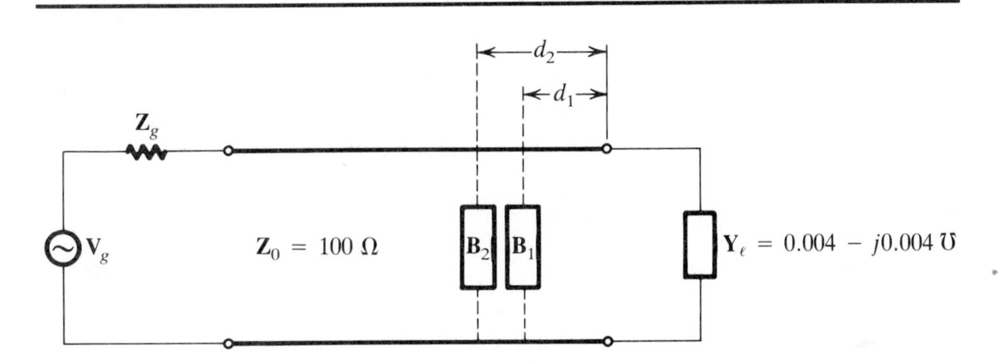

Figure 2-1-18 Circuit diagram for double-susceptance matching for Example 2-1-6.

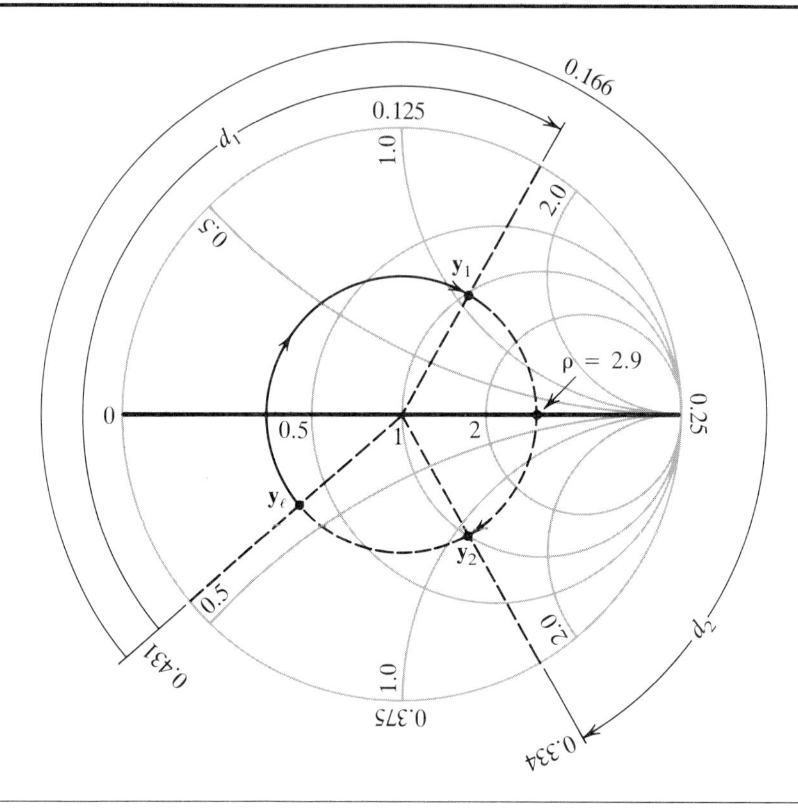

Figure 2-1-19 Graphic solution for Example 2-1-6.

Solutions:

a) Determine matching inductance or capacitance.

1. Calculate the normalized load admittance and enter it on the chart in Fig. 2-1-19.

$$y_\ell = \mathbf{Y}_\ell \mathbf{Z}_0 = (0.004 - j0.004)(100)$$
$$= 0.4 - j0.4$$

2. Plot an SWR ρ circle on the chart.
3. The normalized line admittances to be matched are

$$y_1 = 1 + j1.15$$
$$y_2 = 1 - j1.15$$

4. For an inductive susceptance B_1,

$$\mathbf{Y}_1 = -jb_1\mathbf{Y}_0 = -j(1.15)\left(\frac{1}{100}\right) = -jB_1 \qquad \mho$$

$$B_1 = 0.0115 = \frac{1}{\omega L}$$

and

$$L = \frac{1}{2\pi(10^9)(0.0115)}$$

$$= 0.0138\ \mu\text{H}$$

5. For a capacitive susceptance B_2,

$$\mathbf{Y}_2 = +jB_2 = +jb_2\mathbf{Y}_0 = +j(1.15)\left(\frac{1}{100}\right) = j0.0115\ \mho$$

$$B_2 = 0.0115 = \omega C$$

and

$$C = \frac{B_2}{\omega} = \frac{0.0115}{2\pi(10^9)} = 1.83\ \mu\mu\text{F}$$

b) Compute the distance.

1. The distance to the tuner from the load based on inductive susceptance is

$$d_1 = [0.166 + (0.50 - 0.431)]\lambda = 0.231 \times 30$$

$$= 6.93 \text{ cm}$$

2. The distance to the tuner from the load based on capacitive susceptance is

$$d_2 = [0.334 + (0.50 - 0.431)]\lambda = 0.403 \times 30$$

$$= 12.09 \text{ cm}$$

2-2 SINGLE-STUB AND DOUBLE-STUB MATCHING

2-2-1 Single-Stub Matching

Although single lumped inductors or capacitors can match the transmission line, the susceptive properties of short-circuited sections of transmission lines are more commonly used. Short-circuited sections are preferable to open-circuited ones, because a good short circuit is easier to obtain than a good open circuit.

For a lossless line with $\mathbf{Y}_g = \mathbf{Y}_0$, maximum power transfer requires $\mathbf{Y}_{11} = \mathbf{Y}_0$, where $\mathbf{Y}_{11}$ is the total admittance of the line and stub, looking to the right junction point 1 as shown in Fig. 2-2-1 in Example 2-2-1. The stub must be located at that point on the line where the real part of the admittance, looking toward the generator, is $\mathbf{Y}_0$. In a normalized unit, y_{11} must be in the form

$$y_{11} = y_d \pm y_s = 1 \tag{2-2-1}$$

if the stub has the same characteristic impedance as that of the line. Otherwise, total admittances must be used:

$$\mathbf{Y}_{11} = \mathbf{Y}_d \pm \mathbf{Y}_s = \mathbf{Y}_0 \tag{2-2-2}$$

We then adjust the stub length so that its susceptance just cancels out the susceptance of the line at the junction.

EXAMPLE 2-2-1 Single-Stub Matching

A lossless line of characteristic impedance $R_0 = 50\ \Omega$ is to be matched to a load $\mathbf{Z}_\ell = 5.58 - j10.41\ \Omega$ by means of a lossless short-circuited stub. The characteristic impedance of the stub is $100\ \Omega$. Find the stub position (closest to the load) and length so that match is obtained. (See Fig. 2-2-1.)

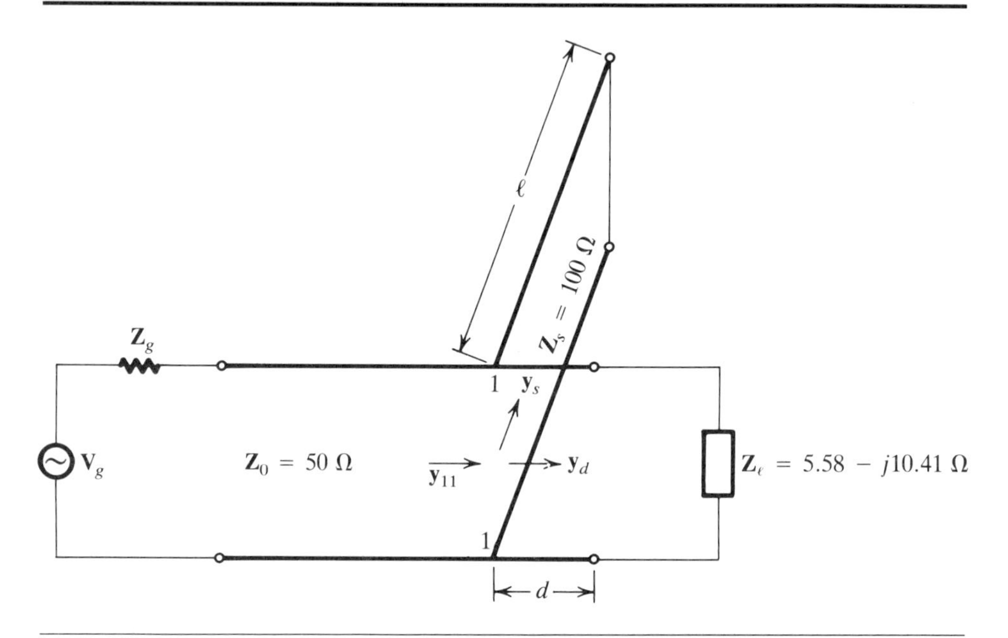

Figure 2-2-1 Single-stub matching circuit for Example 2-2-1.

Solution:

a) Calculate the normalized load admittance and enter it on the Smith chart, as shown in Fig. 2-2-2.

$$y_\ell = \frac{1}{z_\ell} = \frac{\mathbf{Z}_0}{\mathbf{Z}_\ell} = 2 + j3.732$$

b) Draw an SWR circle through the point y_ℓ; the circle intersects the unit circle at y_d; and $y_d = 1 - j2.6$. Note that there is an infinite number of y_d. Take the one that permits the stub to be attached as closely as possible to the load in order to minimize stub-line losses.

c) Since the characteristic impedance of the stub is different from that of the line, the condition for impedance matching at the junction requires that

$$\mathbf{Y}_{11} = \mathbf{Y}_d + \mathbf{Y}_s$$

where $\mathbf{Y}_s$ is the susceptance that the stub will contribute.

It is clear that the stub and the portion of the line from the load to the junction are in parallel, as indicated by the main line extending to the generator. The admittances must be converted to normalized values for matching on the Smith chart. Then Eq. (2-2-2) becomes

$$y_{11}\mathbf{Y}_0 = y_d\mathbf{Y}_0 + y_s\mathbf{Y}_{0s}$$

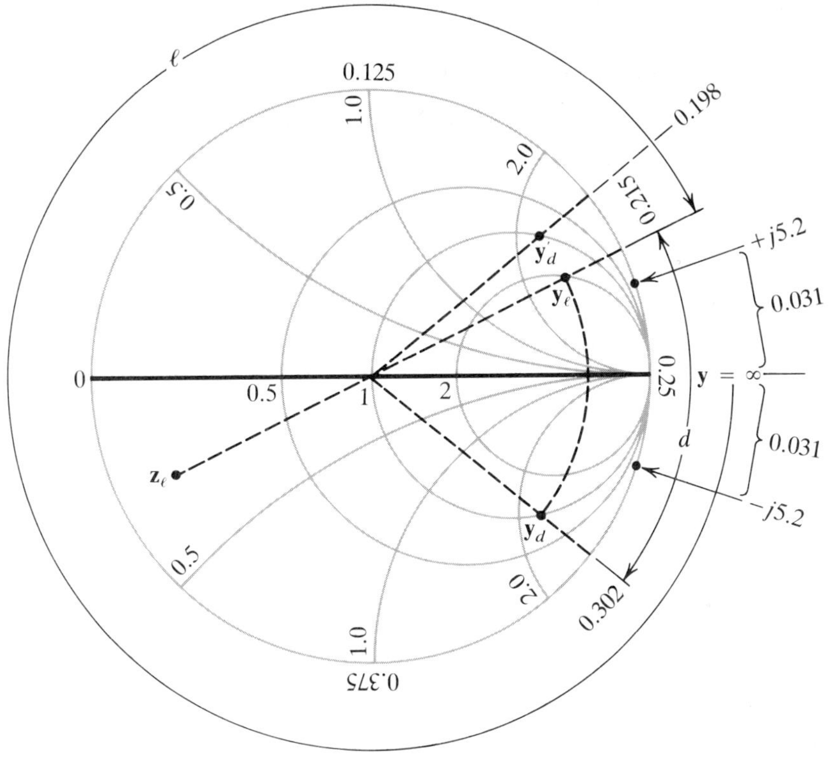

Figure 2-2-2 Graphic solution for Example 2-2-1.

and

$$y_s = (y_{11} - y_d)\left(\frac{\mathbf{Y}_0}{\mathbf{Y}_{0s}}\right) = [1 - (1 - j2.6)]\frac{100}{50}$$

$$= +j5.20$$

d) Calculate the distance between the load and the stub position from the distance scale:

$$d = (0.302 - 0.215)\lambda = 0.087\lambda$$

e) Since the stub contributes a susceptance of $+j5.20$, enter $+j5.20$ on the chart and determine the required distance ℓ from the short-circuited end ($z = 0$, $y = \infty$), which corresponds to the right-hand side of the real axis on the chart, by transversing the chart toward the generator until the point reaching $+j5.20$. Then

$$\ell = (0.50 - 0.031)\lambda = 0.469\lambda$$

When a line is matched at the junction, there will be no standing wave in the line from the stub to the generator.

f) If an inductive stub is required,

$$y'_d = 1 + j2.6$$

and the susceptance of the stub will be

$$y'_s = -j5.2$$

g) The position of the stub from the load is

$$d' = [0.50 - (0.215 - 0.198)]\lambda = 0.483\lambda$$

and the length of the short-circuited stub is

$$\ell' = 0.031\lambda.$$

Recall that for maximum transmission of energy in a line, it is equally necessary to match the internal impedance of the generator and the load impedance to the line. The following example demonstrates the techniques that you can use to match the generator to the line.

EXAMPLE 2-2-2 Generator-Matching Stub

A lossless line with $\mathbf{Z}_0 = 100\ \Omega$ has a length of 5 wavelengths and terminates in $100 + j200\ \Omega$. The generator has an output of 100 V and an internal impedance of $100 + j100$. A short-circuited stub is attached to the line near the generator in order to maximize the power to the load. The wavelength is 100 cm. Find (a) the shortest distance d and the shortest stub ℓ to accomplish the matching; and (b) determine the power delivered to the load under the proper matching condition. (See Fig. 2-2-3.)

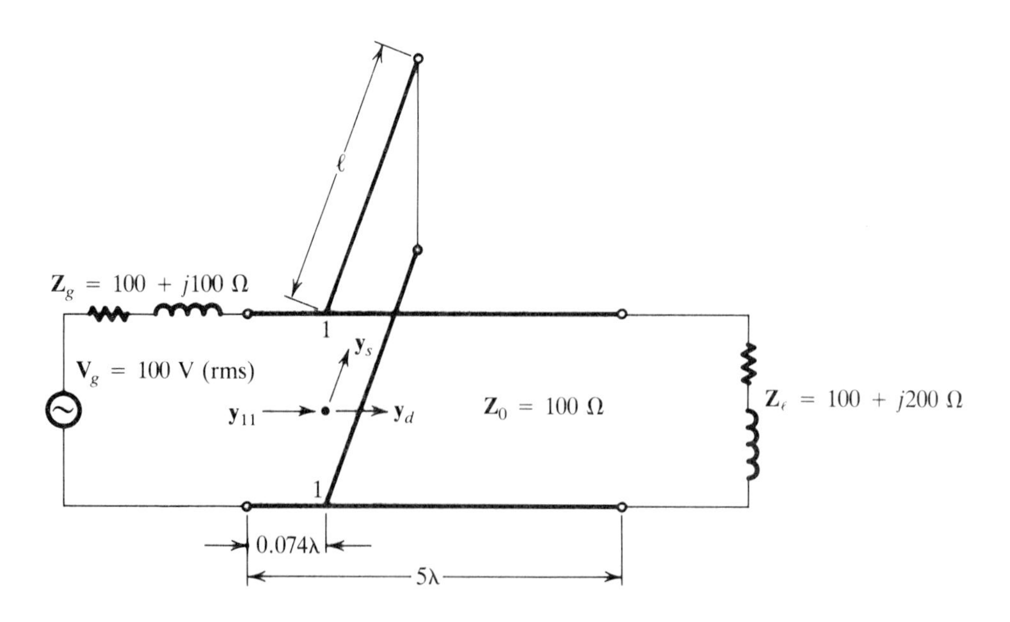

Figure 2-2-3 Generator-matching stub circuit for Example 2-2-2.

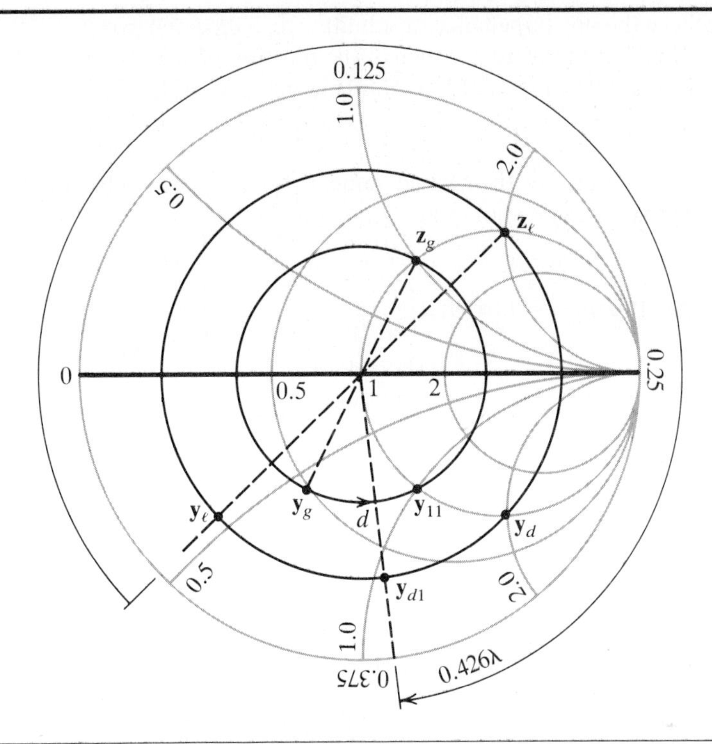

Figure 2-2-4 Graphic solution for Example 2-2-2.

Solutions:

a) Find d and ℓ.

1. Calculate the normalized load impedance z_ℓ and enter it on the chart, as shown in Fig. 2-2-4:

$$z_\ell = \frac{100 + j200}{100} = 1 + j2$$

2. Calculate the normalized internal impedance of the generator and enter it on the chart:

$$z_g = \frac{100 + j100}{100} = 1 + j1$$

3. Read y_ℓ and y_g from the chart:

$$y_\ell = 0.2 - j0.4$$

$$y_g = 0.38 - j0.5$$

4. To match y_g for maximum energy, move y_d toward the load for a distance d and read

$$y_{11} = 1 - j1$$

Thus

$$d = 0.074\lambda = 7.4 \text{ cm}$$

5. Since the line impedance or admittance is repeated every $\lambda/2$, the normalized load admittance at the position of the stub is just 0.426λ from the load ($0.426\lambda = \lambda/2 - 0.074\lambda$) and has the value

$$y_{d1} = 0.4 - j1.08$$

This result is not the desired value. For a proper match, the normalized load admittance should be

$$y_d = 1 - j2.0$$

6. The stub must contribute

$$y_s = y_{11} - y_d = 1 - j1 - (1 - j2.0)$$
$$= +j2$$

and the length of the stub is

$$\ell = (0.25 + 0.176)\lambda = 42.6 \text{ cm}$$

b) Determine the power delivered.

1. For a proper match the input current must be

$$I_s = \frac{100}{100 + 100}$$
$$= 0.5 \text{ A}$$

2. The power delivered to the load is

$$P_\ell = I^2R = (0.5)^2(100)$$
$$= 25 \text{ W}$$

2-2-2 Double-Stub Matching

Single-stub matching is impractical at times because the stub cannot be placed physically in the ideal location, and so double-stub matching is needed. Double-stub devices consist of two short-circuited stubs connected in parallel with a fixed length between them. The length of the fixed section is usually one-eighth, three-eighths, or five-eighths, of a wavelength. We use the stub nearest the load to adjust the susceptance. The stub is located at a fixed wavelength from the constant conductance unit circle (that is, $g = 1$) on an appropriate constant SWR circle. The admittance of the line at the second stub, as shown in Fig. 2-2-5 in Example 2-2-3 is

$$y_{22} = y_{d2} \pm y_{s2} = 1 \tag{2-2-3}$$

or

$$\mathbf{Y}_{22} = \mathbf{Y}_{d2} \pm \mathbf{Y}_{s2} = \mathbf{Y}_0 \tag{2-2-4}$$

Equations (2-2-3) and (2-2-4) are based on the assumption that the stubs and the main line have the same characteristic admittance. If we choose the proper positions and lengths of the stubs, there will be no standing wave on the line to the left of the second stub, as measured from the load.

EXAMPLE 2-2-3 Double-Stub Matching

The terminating impedance $\mathbf{Z}_\ell$ in a line is $200 + j200\ \Omega$. The characteristic impedance $\mathbf{Z}_0$ of the line and stubs is $100\ \Omega$. The first stub is 0.40λ from the load. The spacing between the two stubs is $\frac{3}{8}\lambda$. Determine the length of the short-circuited stubs when matching is achieved. What terminations are forbidden for matching the line by the double-stub device? (See Fig. 2-2-5.)

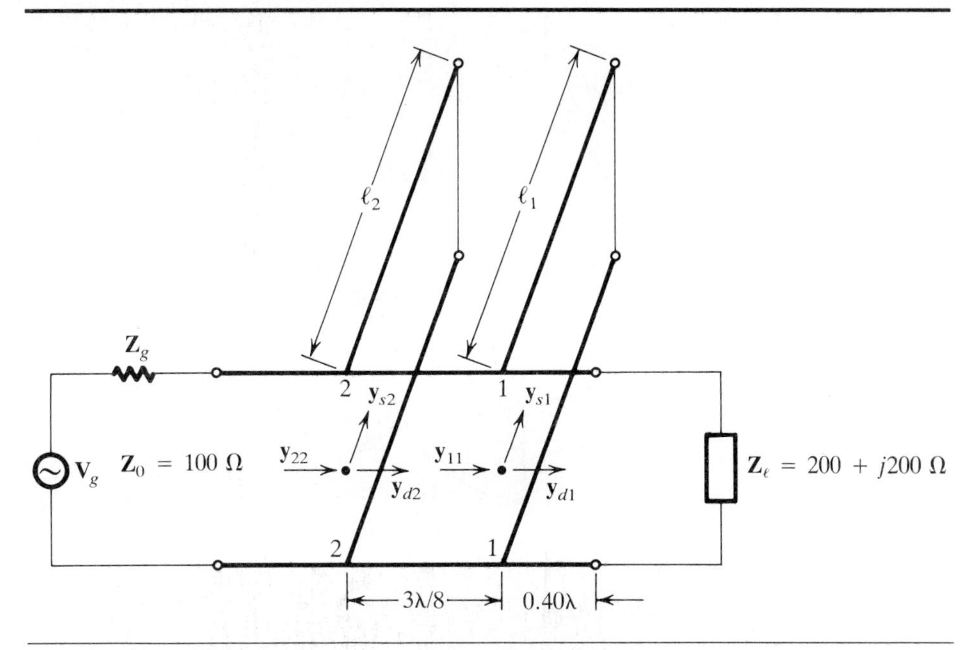

Figure 2-2-5 Double-stub matching circuit for Example 2-2-3.

Solution:

a) Calculate the normalized load impedance z_ℓ and enter it on the chart, as shown in Fig. 2-2-6:

$$z_\ell = \frac{200 + j200}{100} = 2 + j2$$

b) Plot an SWR ρ circle and read the normalized load admittance 180° out of phase with z_ℓ on the SWR circle:

$$y_\ell = 0.25 - j0.25$$

c) Draw the spacing circle of $\frac{3}{8}\lambda$ by rotating the constant conductance unit circle (that is, $g = 1$) through a distance of $2\beta d = 2\beta\frac{3}{8}\lambda = \frac{3}{2}\pi$ toward the load. Point y_{11} must be on this spacing circle, since y_{d2} will be on the $g = 1$ circle (y_{11} and y_{d2} are $\frac{3}{8}\lambda$ apart).

d) Move y_ℓ for a distance of 0.40λ from 0.458 to 0.358 along the SWR ρ circle toward the generator and read y_{d1} from the chart:

$$y_{d1} = 0.55 - j1.08$$

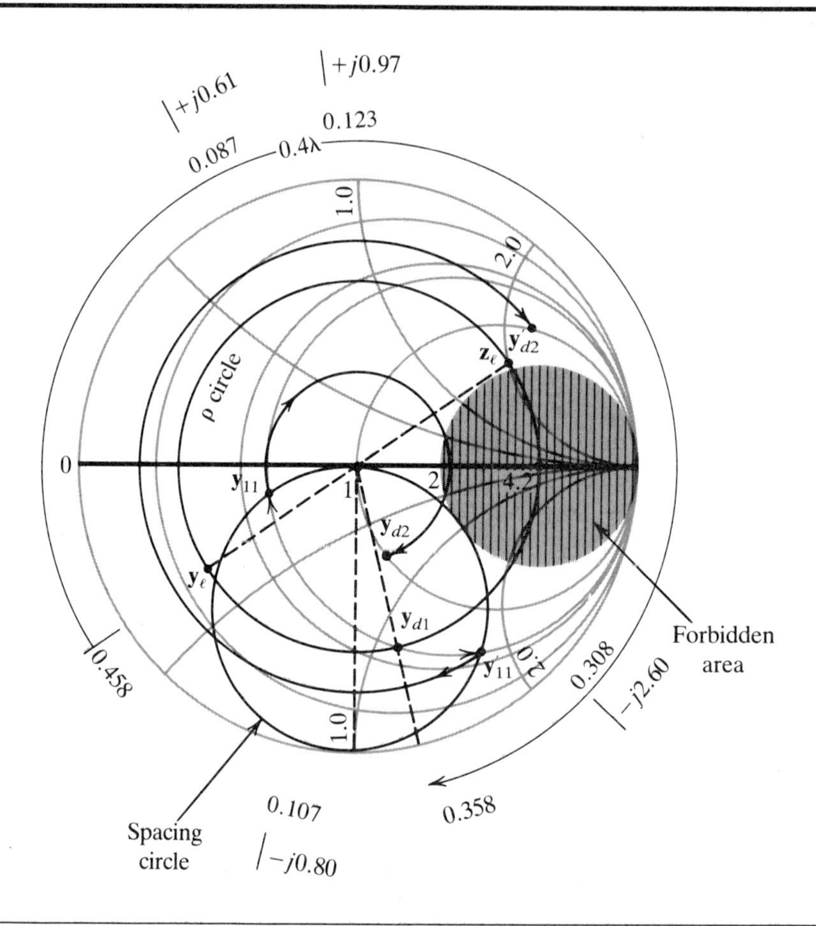

Figure 2-2-6 Graphic solution for Example 2-2-3.

e) There are two possible solutions for y_{11}. They can be found by carrying y_{d1} along the constant conductance $g = 0.55$ circle, which intersects the spacing circle at two points:

$$y_{11} = 0.55 - j0.11$$

$$y'_{11} = 0.55 - j1.88$$

f) At the junction 1-1,

$$y_{11} = y_{d1} + y_{s1}$$

Then

$$y_{s1} = y_{11} - y_{d1} = (0.55 - j0.11) - (0.55 - j1.08)$$
$$= +j0.97$$

Similarly,

$$y'_{s1} = -j0.80$$

g) The lengths of stub 1 are

$$\ell = (0.25 + 0.123)\lambda$$
$$= 0.373\lambda$$
$$\ell'_1 = (0.25 - 0.107)\lambda$$
$$= 0.143\lambda$$

h) The $\frac{3}{8}\lambda$ section of line transforms y_{11} to y_{d2} and y'_{11} to y'_{d2} along their constant standing-wave circles, respectively. That is,

$$y_{d2} = 1 - j0.61$$
$$y'_{d2} = 1 + j2.60$$

i) Stub 2 must contribute

$$y_{s2} = +j0.61$$
$$y'_{s2} = -j2.60$$

j) The lengths of stub 2 are

$$\ell_2 = (0.25 + 0.087)\lambda$$
$$= 0.337\lambda$$
$$\ell'_2 = (0.308 - 0.25)\lambda$$
$$= 0.058\lambda$$

k) Figure 2-2-6 shows that a normalized admittance y_ℓ located inside the shaded area cannot be brought to lie on the locus of y_{11} or y'_{11} for a possible match by the parallel connection of any short-circuited stub, because the spacing circle and the $g = 2$ circle are mutually tangent. Thus the area of a $g = 2$ circle is called the *forbidden region* of the normalized load admittance.

Normally we can solve a double-stub matching problem backward from the load toward the generator, since we know the load and can arbitrarily choose the distance of the first stub away from the load. In quite a few practical matching problems, however, some stubs have a $\mathbf{Z}_0$ different from that of the line, the length of a stub may be fixed, and so on. Thus it is hard to describe a definite procedure for solving double-matching problems. The following example demonstrates an alternative matching method.

EXAMPLE 2-2-4 Double-Stub Matching

The characteristic impedance R_0 of a line is 100 Ω and the load impedance $\mathbf{Z}_\ell$ is $100 - j100$ Ω. Stub 1 is located at a distance d from the load and has a characteristic impedance of 100 Ω. Stub 2 is located $\frac{5}{8}\lambda$ from stub 1 and has an R_0 of 200 Ω and a length of 0.1λ. Both stubs are lossless and terminate in short-circuits.

a) Find distance d and length ℓ_1 of stub 1.
b) With the line and load properly matched, determine the SWR ρ on the section of line between the stubs.

(See Fig. 2-2-7.)

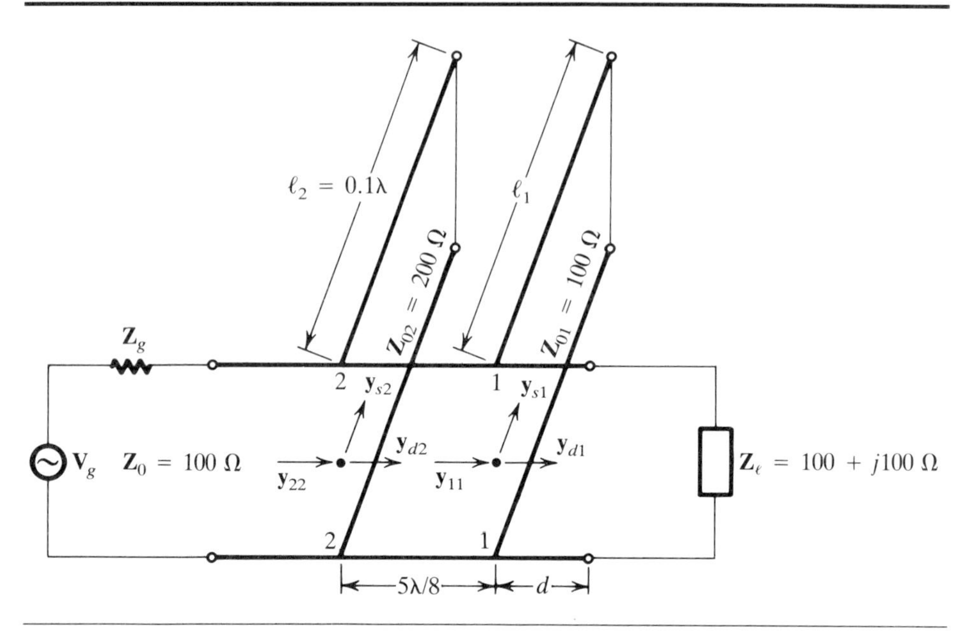

Figure 2-2-7 Double-stub matching circuit for Example 2-2-4.

Solutions:

a) Find d and ℓ_1 for stub 1.

1. Calculate the normalized load impedance and enter it on the chart, as shown in Fig. 2-2-8:

$$z_\ell = \frac{\mathbf{Z}_\ell}{\mathbf{Z}_0} = \frac{100 + j100}{100}$$

$$= 1 + j1$$

2. Draw an SWR ρ circle and read the normalized load admittance:

$$y_\ell = 0.5 - j0.5$$

3. Draw the spacing circle of $\frac{5}{8}\lambda$, as shown on the chart.
4. Since the length of stub 2 is given as $\ell_2 = 0.1\lambda$ for a short-circuited termination, translate a distance of 0.1λ from 0.25λ ($y = \infty$) to 0.35λ at ($-j1.38$). Thus stub 2 must contribute

$$y_{s2} = -j1.38$$

5. At junction 2, stub 2 has a $\mathbf{Z}_0$ different from that of the main line, so

$$\mathbf{Y}_{22} = \mathbf{Y}_{d2} + \mathbf{Y}_{s2} = \mathbf{Y}_0$$

and

$$y_{d2} = 1 - y_{s2}\left(\frac{\mathbf{Z}_0}{\mathbf{Z}_{02}}\right)$$

$$= 1 - (-j1.38)\left(\frac{100}{200}\right)$$

$$= 1 + j0.69$$

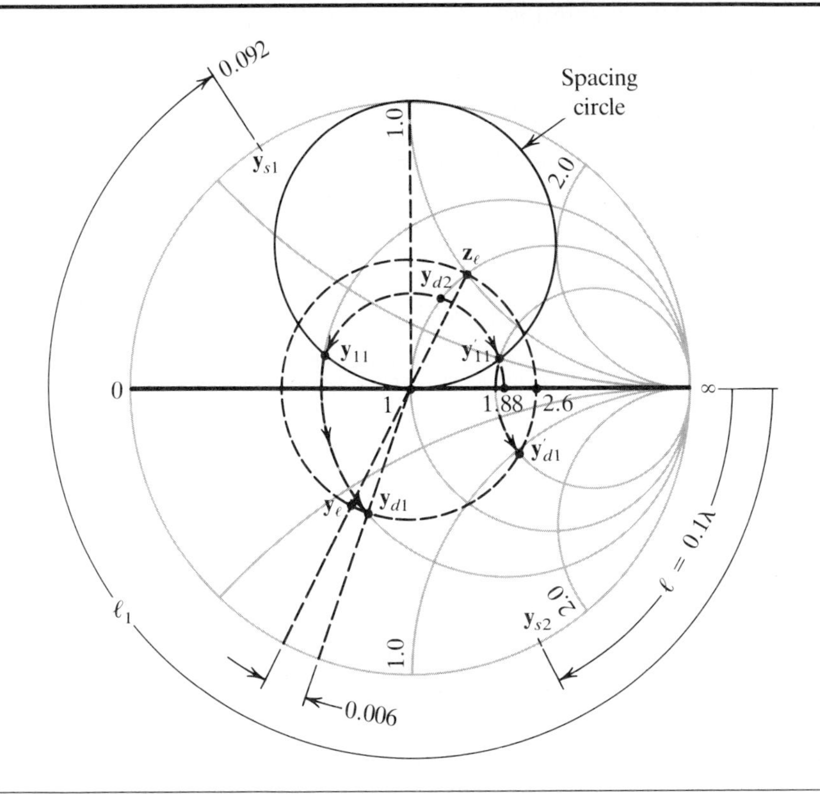

Figure 2-2-8 Graphic solution for Example 2-2-4.

6. Draw an SWR ρ circle through the point y_{d2}; the circle intersects the spacing circle at

$$y_{11} = 0.52 + j0.125$$

Note: Another point $y'_{11} = 1.80 + j0.40$ can also be used to match the admittance.

7. Transform y_{11} along the constant conductance circle $g = 0.52$ to y_{d1} on the SWR circle ($\rho = 2.6$) and read:

$$y_{d1} = 0.52 - j0.53$$

Note: Another point $y'_{d1} = 1.80 - j1.00$ can also be used for matching.

8. Since stub 1 has the same $\mathbf{Z}_0$ as the main line, the normalized admittance is

$$y_{11} = y_{d1} + y_{s1}$$

and

$$y_{s1} = y_{11} - y_{d1} = (0.52 + j0.125) - (0.52 - j0.53)$$
$$= j0.655$$

Thus stub 1 contributes $+j0.655$.

9. The length of stub 1 is

$$\ell_1 = (0.25 + 0.092)\lambda = 0.342\lambda$$

b) Determine the SWR ρ between the stubs.

1. The distance d of stub 1 from the load is

$$d = (0.50 - 0.006)\lambda = 0.494\lambda$$

2. The SWR ρ in the section of line between the stubs ≈ 2.

2-3 SERIES-STUB AND OTHER MATCHING TECHNIQUES

2-3-1 Series-Stub Matching

Sometimes we find it desirable to place a single-stub tuner in series with a line instead of a single-stub or double-stub tuner in parallel with the line. We can solve a series-stub matching problem using impedance in a Smith chart. The impedance, looking into the load at the point where the tuner is connected to the generator side of the line, must equal the sum of the impedance contributed by the stub and the impedance at the point of the stub connected to the load side of the line. That is,

$$\mathbf{Z}_{11} = \mathbf{Z}_s + \mathbf{Z}_d \tag{2-3-1}$$

If the characteristic impedance $\mathbf{Z}_{0s}$ of the stub equals that of the line, Eq. (2-3-1) becomes

$$z_{11} = z_s + z_d \tag{2-3-2}$$

where z_{11}, z_s, and z_d are the normalized impedances, respectively, of $\mathbf{Z}_{11}$, $\mathbf{Z}_s$, and $\mathbf{Z}_d$, as defined in Fig. 2-3-1 in Example 2-3-1.

EXAMPLE 2-3-1 Series-Stub Matching

A single short-circuited stub is to be placed in series with a line, as shown in Fig. 2-3-1, to match the load $\mathbf{Z}_\ell$ of $150 + j100\ \Omega$ to the line. The characteristic impedance R_0 of

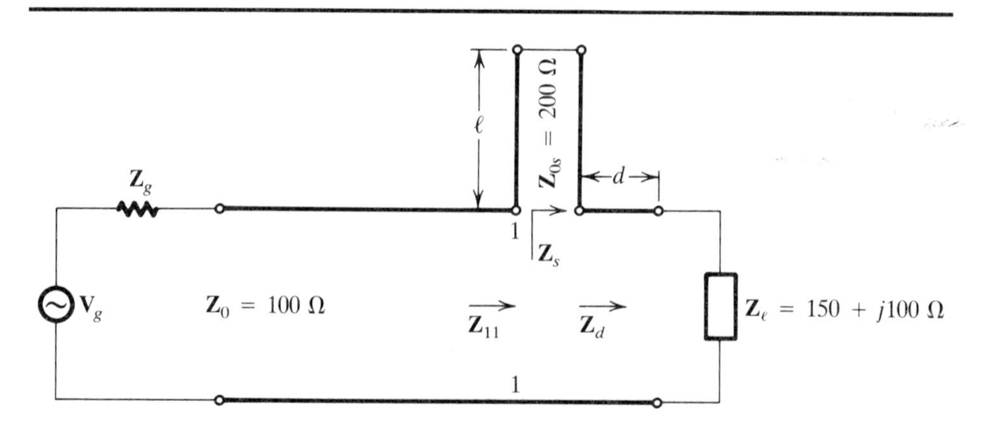

Figure 2-3-1 Series-stub matching circuit for Example 2-3-1.

the line is 100 Ω, but that of the tuner is 200 Ω. Determine the length ℓ of the tuner and its distance d from the load required to match the line.

Solution:

a) Calculate the normalized load impedance z_ℓ and enter it on the chart, as shown in Fig. 2-3-2:

$$z_\ell = \frac{150 + j100}{100} = 1.5 + j1$$

b) Draw an SWR ρ circle through z_ℓ toward the generator; the ρ circle intersects the constant resistance unit circle $r = 1$ at z_d. Read $z_d = 1 - j0.92$; $z'_d = 1 + j0.92$ is also a solution.

c) Measure the distance between z_ℓ and z_d on the distance scale:

$$\begin{aligned} d &= (0.341 - 0.192)\lambda \\ &= 0.14\lambda \end{aligned}$$

which means that the stub can be placed in series with the line 0.14λ from the load.

d) At the junction, since the stub and the line have different characteristic impedances, the impedance $\mathbf{Z}_{11}$ can be calculated from Eq. (2-3-1), or

$$\mathbf{Z}_{11} = \mathbf{Z}_s + \mathbf{Z}_d$$

Then

$$z_{11}\mathbf{Z}_0 = z_s\mathbf{Z}_{0s} + z_d\mathbf{Z}_0$$

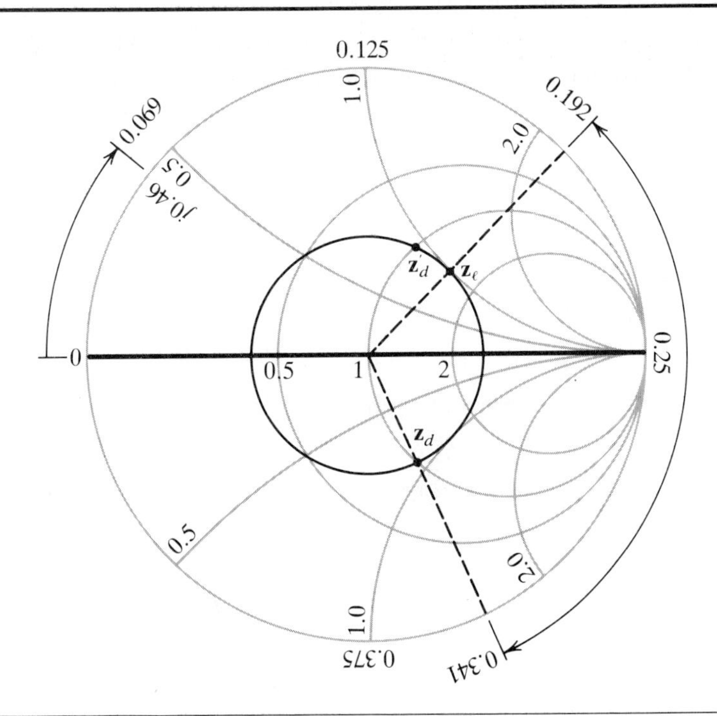

Figure 2-3-2 Graphic solution for Example 2-3-1.

and

$$z_s = (z_{11} - z_d)\frac{\mathbf{Z}_0}{\mathbf{Z}_{0s}} = (1 - 1 + j0.92)\left(\frac{100}{200}\right)$$

$$= +j0.46$$

which indicates that the series stub must contribute $+j0.46$.

e) The length of the stub is

$$\ell = 0.069\lambda$$

Note that a short-circuited series stub has a termination impedance $\mathbf{Z} = 0$.

2-3-2 Dielectric-Bead and Dielectric-Slug Lines

Dielectric beads, slugs, or slabs often are placed at intervals along a transmission line to support one or more conductors and maintain the desired spacing between conductors. Since even a single dielectric bead disturbs a condition of match, a large number of beads distributed along a line might have a serious effect on transmission properties if the beads happen to be so located that their effects are repeated. The following example illustrates the effect of a single bead on a transmission line.

EXAMPLE 2-3-2 Dielectric-Bead Line

The end of a coaxial line is supported by a dielectric bead of dielectric constant $\varepsilon_r = 4$, as shown in Fig. 2-3-3. The characteristic impedance in the dielectric section is the same as in the air-filled section. The bead itself introduces no reflections. A movable single-stub tuner is used to match the load to the line. The operating frequency is 3 GHz.

a) Find the position d (in cm) closest to the bead where a tuner would be located if the tuner represents a parallel admittance.

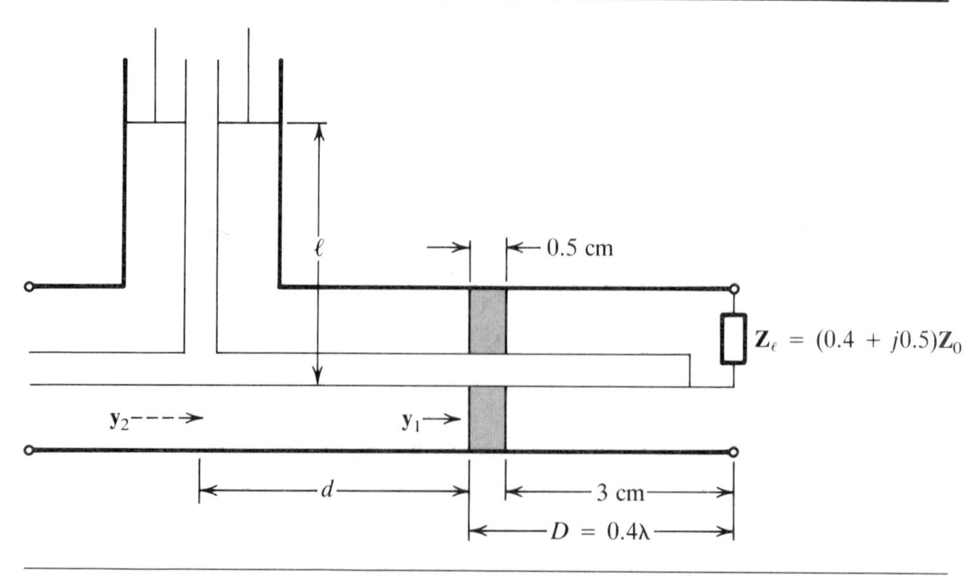

Figure 2-3-3 Dielectric-bead line circuit for Example 2-3-2.

b) Find the necessary length ℓ (in cm) of the short-circuited tuner if the tuner has the same characteristic impedance $\mathbf{Z}_0$ as the line.

Solutions:

a) Find d.

1. Enter $z_\ell = 0.4 + j0.5$ on the chart, as shown in Fig. 2-3-4 and read:

$$y_\ell = 1 - j1.2$$

2. The wavelength λ_0 in air is

$$\lambda_0 = \frac{c}{f} = \frac{3 \times 10^{10}}{3 \times 10^9} = 10 \text{ cm}$$

The wavelength λ_ε in dielectric is

$$\lambda_\varepsilon = \frac{\lambda_0}{\sqrt{\varepsilon_r}} = \frac{10}{\sqrt{4}}$$

$$= 5 \text{ cm}$$

3. The total distance on the chart from the load through the bead is

$$D = \frac{3}{\lambda_0} + \frac{0.5}{\lambda_\varepsilon} = \frac{3}{10} + \frac{0.5}{5}$$

$$= 0.4\ \lambda_0$$

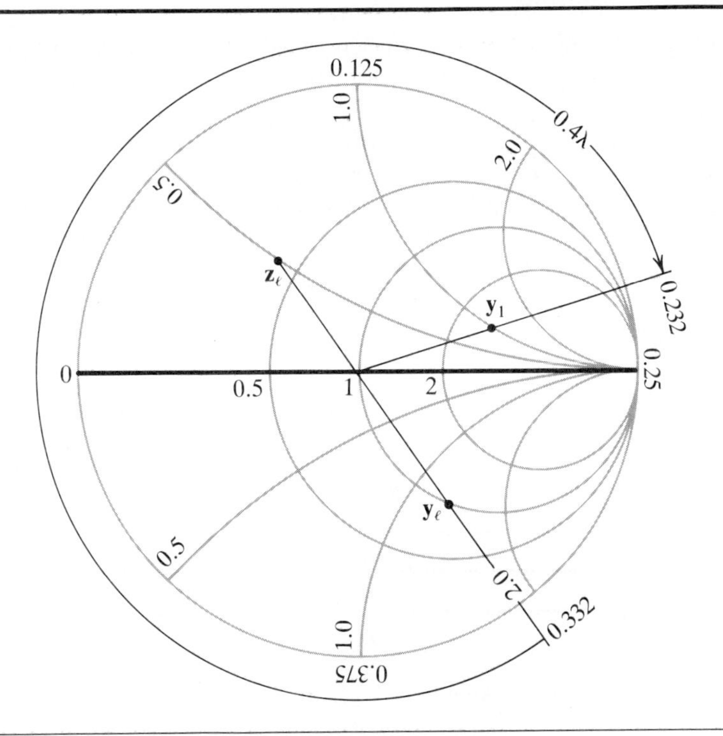

Figure 2-3-4 Graphic solution for Example 2-3-2.

4. Move the normalized load admittance y_ℓ a distance of 0.4 λ_0 toward the generator and read y_1 on the same SWR ρ circle:

$$y_1 = 2.8 + j0.85$$

5. For a proper match, y_2 must be on the unit circle; read y_2 at the intersection of the SWR and unit circles:

$$y_2 = 1 - j1.2$$

6. The short-circuited single-stub tuner must contribute

$$y_s = +j1.2$$

The distance d is

$$d = 0.1\lambda_0 = (0.1)(10)$$
$$= 1 \text{ cm}$$

b) Find ℓ.

1. The length ℓ of the tuner is

$$\ell = (0.25 + 0.139)\lambda_0$$
$$= 3.89 \text{ cm}$$

2-3-3 Balun Transformer

Another type of quarter-wave transformer is called a *balun* transformer (or bazooka transformer) because of its ability to transform from a *bal*anced (or ungrounded) line or load to an *un*balanced (or grounded) line or load without disturbing the equilibrium conditions on either line. A balun transformer can be used as a transition between an unbalanced coaxial line and a balanced antenna or two-wire line. A two-wire line should be balanced with respect to ground so that the conductors carry equal and opposite currents. If a coaxial line is connected to an unbalanced load, a current will be transmitted along the outside of the coaxial cable. This will result in high loss. One type of balun transformer is shown in Fig. 2-3-5 and consists of a short-circuited quarter-

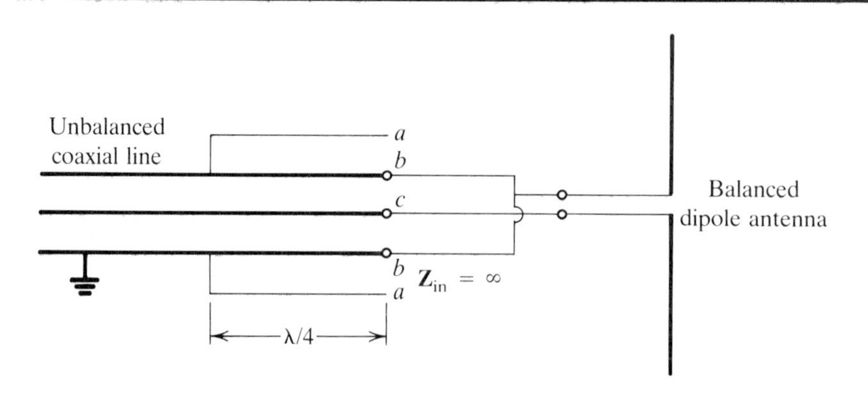

Figure 2-3-5 Matching a coaxial cable to a dipole antenna using a balun transformer.

wave–long sleeve mounted concentrically around the end of a coaxial cable. The outer conductor of the coaxial line is grounded. Since the balun sleeve is a quarter-wave long and short-circuited, the input impedance at the open ends *a-b* is infinite. Hence, conductor *b* is isolated from ground, and the ends *b-c* of the coaxial cable may be connected to a balanced dipole antenna.

2-3-4 Quarter-Wave Coaxial Sleeves

Another widely used impedance matching device is the quarter-wave sleeve. In coaxial cables, conducting sleeves that fit closely to either the inner or outer conductor are commonly used, as shown in Fig. 2-3-6. The characteristic impedance in the sleeved section is smaller than the normal value of $\mathbf{Z}_0$ of the line.

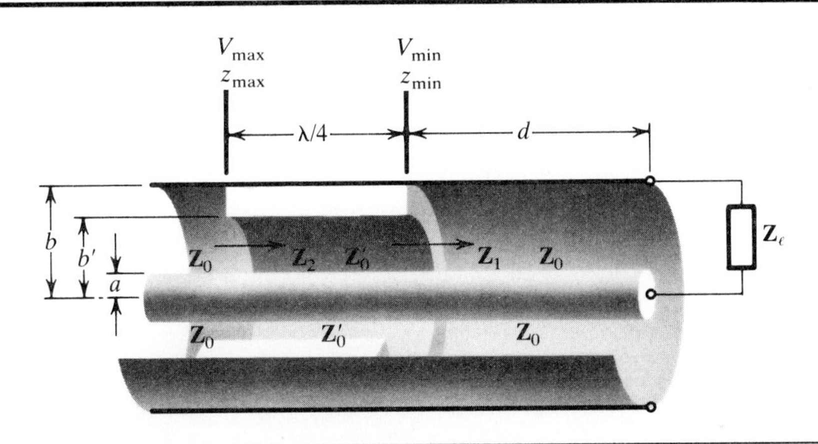

Figure 2-3-6 Quarter-wave coaxial sleeve matching circuit.

The characteristic resistance of the sleeved section,

$$Z_0' = 60\sqrt{\frac{\mu_r}{\varepsilon_r}}\ln\frac{b'}{a} \tag{2-3-3}$$

is smaller than the line characteristic impedance, Z_0. That is, $Z_0/Z_0' > 1$.

The sleeve is ordinarily fitted to the conductors between the points of minimum and maximum voltages from the load, where the line impedance is a pure resistance. To match the load to the line by a quarter-wave sleeve, the following conditions must be met:

$$\frac{Z_2}{Z_0} = 1 \tag{2-3-4}$$

and

$$Z_0' = \sqrt{Z_1 Z_2} = \sqrt{Z_1 Z_0} \tag{2-3-5}$$

Also,

$$\frac{Z_1}{Z_0'} = \frac{Z_1}{Z_0}\frac{Z_0}{Z_0'} > \frac{Z_1}{Z_0} \tag{2-3-6}$$

$$\frac{Z_2}{Z_0} = \frac{Z_2}{Z_0'}\frac{Z_0'}{Z_0} < \frac{Z_2}{Z_0'} \tag{2-3-7}$$

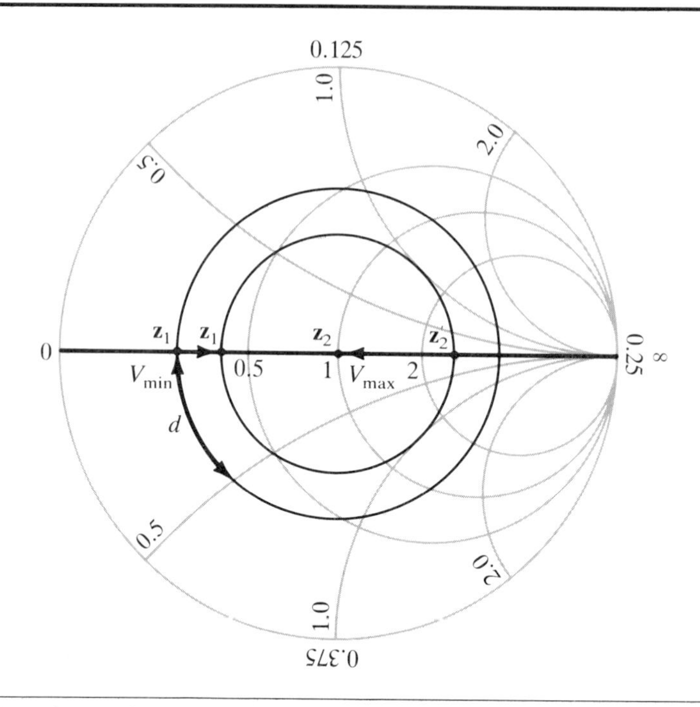

Figure 2-3-7 Diagram for designing a quarter-wave sleeve.

and

$$Z_0' = \frac{Z_0}{\sqrt{\text{SWR}}}$$

From the Smith chart,

$$z_{\text{max}} = \frac{Z_2}{Z_0} = \text{SWR} = \rho$$

$$z_{\text{min}} = \frac{Z_1}{Z_0} = \frac{1}{\rho}$$

We can determine the standing-wave ratio by measurement. When we know Z_1 and Z_0, we can find the required impedance in the sleeved section from Eqs. (2-3-4) and (2-3-5). Finally, we can calculate the thickness b' of the sleeve from Eq. (2-3-3). Figure 2-3-7 shows the values of the required impedances.

2-4 N-JUNCTION MATCHING

Electromagnetic energy can be transmitted efficiently by a transmission line when no reflected wave is present. This condition requires that the load impedance and the characteristic impedance of the line be equal. To obtain energy transmission with maximum efficiency, the line must be completely free from discontinuities and reflection. Discontinuities such as a line branched into two lines, a shunt resistor across two conductors, an N-junction line, and so on, always result in a reflected wave, which in turn, causes a standing wave along the line and decreases transmission efficiency.

If one or more transmission lines have the same or different characteristic impedance and are joined at a junction with or without a lumped resistance network, an incident wave of a line arriving at the junction may create a reflected wave back along the same line and result in transmitted waves on all the other lines. In order to solve this type of transmission-line problem we first have to determine the reflection and transmission coefficients of each line and then find the line voltage and current.

Figure 2-4-1 shows two different transmission lines connected in tandem.

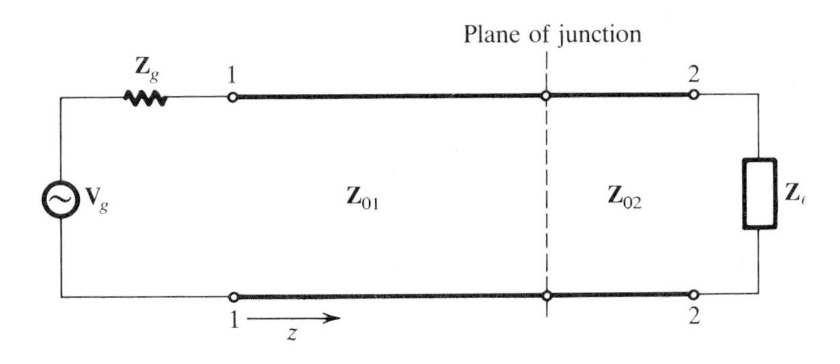

Figure 2-4-1 Tandem-connected line circuit.

From left to right at the plane of junction, the reflection coefficient is

$$\Gamma_{11} = \frac{Z_{02} - Z_{01}}{Z_{02} + Z_{01}} \tag{2-4-1}$$

From right to left at the plane of junction, the reflection coefficient is

$$\Gamma_{22} = \frac{Z_{01} - Z_{02}}{Z_{01} + Z_{02}} \tag{2-4-2}$$

The line voltages are

$$v_1(z) = v_{\text{in}}(z) + \Gamma_{11} v_{\text{in}}(z) \tag{2-4-3}$$

and

$$v_2(z) = T_{21} v_{\text{in}}(z) \tag{2-4-4}$$

where T_{21} is the transmission coefficient on line 1, which indicates that the wave is transmitted into line 2 from line 1.

The voltage at the plane of junction is

$$v_1(z) = v_2(z) \tag{2-4-5}$$

Then

$$T_{21} = 1 + \Gamma_{11} = \frac{2Z_{02}}{Z_{02} + Z_{01}} \tag{2-4-6}$$

Similarly,

$$T_{12} = 1 + \Gamma_{22} = \frac{2Z_{01}}{Z_{01} + Z_{02}} \tag{2-4-7}$$

Thus each transmitted voltage wave is the sum of the incident and reflected voltages on the other line. We customarily represent both the reflection coefficient Γ and transmission coefficient T by a single letter S for scattering. We can now express the combined equations for the line voltage waves as

$$V_1 = S_{11}v_1 + S_{12}v_2 \tag{2-4-8}$$

and

$$V_2 = S_{21}v_1 + S_{22}v_2 \tag{2-4-9}$$

The first of the two subscripts of S, as shown in Eqs. (2-4-8) and (2-4-9), indicates the voltage wave transmitted into that line; the second subscript represents the wave departing from that line at the junction. We can further simplify Eqs. (2-4-8) and (2-4-9) to a matrix form:

$$[\mathbf{V}_d] = [\mathbf{S}][\mathbf{V}_{in}] \tag{2-4-10}$$

where

$\mathbf{V}_d$ = voltage departing from the plane of junction; and
$\mathbf{V}_{in}$ = voltage incident upon the plane of junction.

We can write the coefficients as

$$[\mathbf{S}] = \begin{bmatrix} S_{11} & S_{12} \\ S_{21} & S_{22} \end{bmatrix} \tag{2-4-11}$$

The matrix $[\mathbf{S}]$ is called the scattering matrix, or the S parameters. If the network is symmetrical, $S_{11} = S_{22}$. If the network is reciprocal, $S_{12} = S_{21}$. We can express the reflection coefficient at the sending end in terms of the S parameters and the reflection coefficient Γ_ℓ at the load:

$$\Gamma_s = S_{11} + \frac{S_{12}S_{21}\Gamma_\ell}{1 - S_{22}\Gamma_\ell} \tag{2-4-12}$$

For the N-junction line the scattering matrix becomes

$$[\mathbf{S}_{ij}] = \begin{bmatrix} S_{11} & S_{12} & \cdots & S_{1n} \\ S_{21} & S_{22} & \cdots & S_{2n} \\ S_{n1} & S_{n2} & \cdots & S_{nn} \end{bmatrix} \tag{2-4-13}$$

The diagonal element $\mathbf{S}_{ij}$ of the scattering matrix is the reflection coefficient at the jth port, and the off-diagonal element $\mathbf{S}_{ij}$ is the transmission coefficient at the jth port with the wave transmitted into the ith port.

EXAMPLE 2-4-1 Branched Line

Three semi-infinite, lossless transmission lines are joined at point A, as shown in Fig. 2-4-2.

The wire size and spacing are the same on all three lines; their inductance is

$$L = 0.333 \times 10^{-6}\ \text{H/m}$$

The dielectric material of line 3 has $C_3 = 4C_1$. The characteristic impedance of line 1 is 100 Ω. A 10-V, 1-μs rectangular pulse travels to the right along line 1 and the leading edge strikes point A at time $t = 0$. The rectangular pulse passes point B and its leading

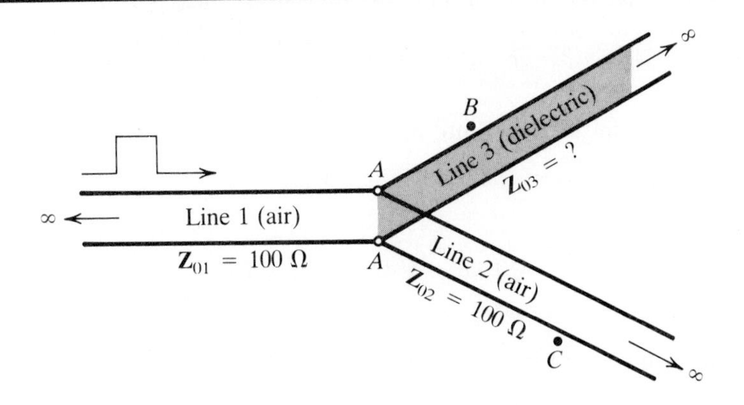

Figure 2-4-2 Diagram for branched line for Example 2-4-1.

edge arrives at $t = 2\ \mu s$.

a) What is the capacitance of line 1?
b) What is the height of the voltage pulse passing point B?
c) What is the distance between points A and B?
d) Repeat (b) for line 2.

Solutions:

a) The capacitance is

$$C_1 = \frac{L}{Z_{01}^2} = \frac{0.333 \times 10^{-6}}{(100)^2} = 3.33 \times 10^{-11}\ \text{F/m}$$

b) An infinite line has the characteristic of a line terminated in its characteristic impedance.

1. The characteristic impedance of line 3 is

$$Z_{03} = \sqrt{\frac{L}{C_3}} = \sqrt{\frac{L}{4C_1}} = \frac{1}{2}\sqrt{\frac{L}{C_1}} = \frac{1}{2}(100)$$
$$= 50\ \Omega$$

2. Since lines 2 and 3 are in parallel, as viewed from line 1, the impedance at point A for lines 2 and 3 is

$$Z_{23} = \frac{100 \times 50}{100 + 50}$$
$$= 33.33\ \Omega$$

3. The reflection coefficient Γ_{11} of line 1 at point A is

$$\Gamma_{11} = \frac{Z_{23} - Z_{01}}{Z_{23} + Z_{01}} = \frac{33.33 - 100}{33.33 + 100}$$
$$= -\frac{1}{2}$$

At point A the reflected pulse $V = -5$ V.

4. The transmission coefficient into line 3 from line 1 is

$$T_{31} = 1 + \Gamma_{11} = 1 - \frac{1}{2} = \frac{1}{2}$$

The transmitted pulse into line 3 is then $V_{tr} = 5$ V.

c) The phase velocity of the wave in line 3 is

$$v_3 = \sqrt{\frac{1}{LC_3}} = \sqrt{\frac{1}{L4C_1}} = \frac{1}{2}\sqrt{\frac{1}{LC_1}} = \frac{3 \times 10^8}{2}$$

$$= 1.5 \times 10^8 \text{ m/s}$$

Thus the distance between A and B on line 3 is

$$D_{AB} = v_3 t = 1.5 \times 10^8 \times 2 \times 10^{-6}$$

$$= 300 \text{ m}$$

d) The distance between A and C on line 2 is

$$D_{AC} = vt = 3 \times 10^8 \times 2 \times 10^{-6}$$

$$= 600 \text{ m}$$

EXAMPLE 2-4-1 Computer Solution for Branched Line

```
010C       PROGRAM TO COMPUTE VOLTAGE ALONG
015C+      A BRANCHED LINE
020C       BRANCH
030        REAL L
040      1 READ 2, Z01,VG,L,TP,TL
050        IF(L .LE. 0.0) STOP
060        VEL=3.000E+08
110        Z03=0.5*Z01
120        Z23=Z01*Z03/(Z01+Z03)
130        G11=(Z23-Z01)/(Z23+Z01)
140        VA=G11*VG
150        T31=1 + G11
160        VTR=T31*VG
170        VEL3=0.5*VEL
180        DAB=VEL3*TL
190        DAC=VEL*TL
200        PRINT 4, Z03,Z23,G11,VA
210        PRINT 6, T31,VTR,VEL3,DAB,DAC
220      2 FORMAT(2F7.2, 3E10.3)
230      4 FORMAT(//1H ,3X,"Z03(OHMS)",2X,"Z23(OHMS)",
240+       5X,"G11",4X,"VA(VOLTS)"/
250+       3X,F8.3,3X,F8.3,1X,F8.3,3X,F8.3//)
260      6 FORMAT(5X,"T31",5X,"VTR(VOLTS)",1X,"VEL3(M/S)",
270+       3X,"DAB(METERS)",2X,"DAC(METERS)"/
280+       1X,E10.3,1X,E10.3,1X,E10.3,2X,E10.3,4X,E10.3/)
290        GO TO 1
300        STOP
310        END
READY.
RUN
```

EXAMPLE 2-4-1 *(continued)*

```
PROGRAM    BRANCH

?  100.00   10.00 0.333E-06 0.100E-05 0.200E-05
    Z03(OHMS)   Z23(OHMS)        G11      VA(VOLTS)
     50.000      33.333       -.500       -5.000

     T31        VTR(VOLTS) VEL3(M/S)    DAB(METERS)   DAC(METERS)
   .500E+00     .500E+01    .150E+09     .300E+03      .600E+03

?  0.0

SRU        0.731 UNTS.

RUN COMPLETE.
```

2-5 VSWR MEASUREMENT TECHNIQUES

From linear network theory, we can interchange voltage standing-wave ratio (VSWR) measurements between the frequency domain and the time domain (or distance domain) by using the Laplace transformation, or convolution, technique. We can analyze actual linear transmission-line systems by using the Hewlett-Packard automatic network analyzer (ANA) in terms of frequencies. However, the VSWR in the frequency domain is merely a cumulative total of the reflections measured in the entire linear transmission-line system. We have no way of telling which connection is bad and how bad it is. If we can display the VSWR in the time or distance domain, the reflectometer can precisely point out the VSWR at a specific location. This method requires the use of a deconvolution technique in order to change frequency-domain data to time-domain data and then display the information on an oscilloscope. Recently, Made-It Microwave Associates developed a measurements and microwave analysis (MAMA) program that works with the computer controller in the H-P automatic network analyzer (ANA) system to display the VSWR in either the frequency domain or the time domain. This technique is very helpful in the design and analysis of systems by electronics engineers, both in industry and research laboratories.

REFERENCES

[1] Smith, Phillip H. "Transmission line calculator," *Electronics*, vol. 12, pp. 29–31, 1939. "An improved transmission line calculator," *Electronics*, vol. 17, pp. 130–133, 318–325, 1944.

Smith Charts—Their Development and Use. A series published at intervals by the Kay Electric Co. No. 1 is dated March 1962; No. 9 is dated December 1966.

Problems

2-1 A half-wave dipole antenna has a driving-point impedance of $73 + j42.5\ \Omega$. The transmission line connected to a TV set has a characteristic impedance of $300\ \Omega$. Design a shorted stub having the same characteristic impedance as the line in order to match the antenna to the line. The stub may be placed close to the antenna. The assumed reception is Channel 24 at a frequency of 531.25 MHz.

2-2 A lossless transmission line has a characteristic impedance of $100\ \Omega$ and is loaded by $100 + j100\ \Omega$. A single shorted stub with the same characteristic impedance is inserted at $\frac{1}{4}\lambda$ from the load to match the line. The measured load current is 2 A. The length of stub is $\frac{1}{8}\lambda$. Determine the

a) magnitude and the phase of the voltage across the stub location; and
b) magnitude and the phase of the current flowing through the end of the stub.

2-3 A double-stub matching line is shown in Fig. P2-3. The characteristic resistances of both the line and the stubs are 50 Ω. The spacing between the two stubs is $\frac{1}{8}\lambda$. The load is $100 + j100$ Ω. One stub is located at the load. Determine the

a) reactances contributed by the stubs; and
b) lengths of the two shorted double-stub tuners.

(*Note:* There are two sets of solutions.)

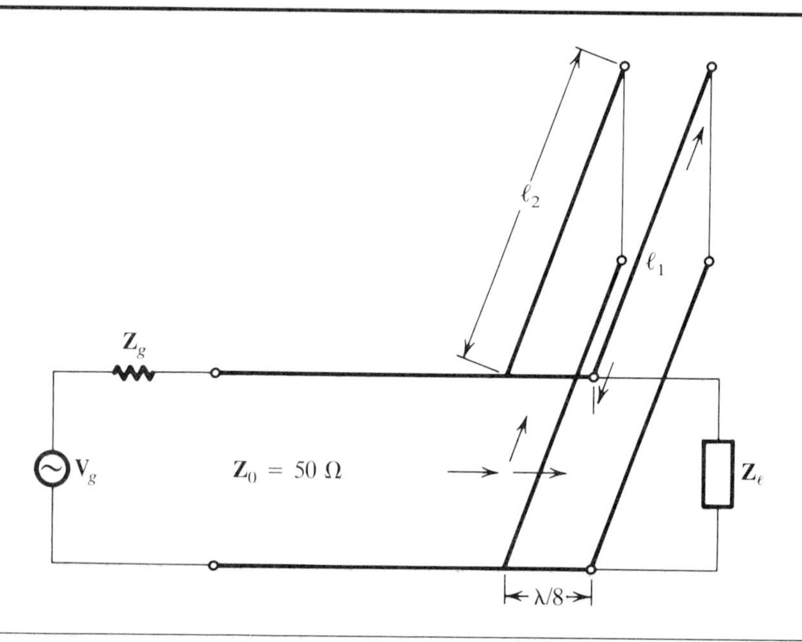

Figure P2-3

2-4 A double-stub tuner is to match a 60-Ω line to a load of 28 Ω. The distance from the load to the first stub is $\frac{1}{6}\lambda$. The spacing between stubs is $\frac{3}{8}\lambda$. Determine the proper lengths for the two shorted stubs, as fractions of the wavelength. (There are two possible sets of solutions.)

2-5 The normalized load at the end of a lossless transmission line is $z_\ell = 1 + j1$. The guide wavelength is 5 cm.

a) Use a Smith chart to find the distances from the load to where the impedances are real. Since there is an infinite number, find the two closest to the load.
b) What are the values of these impedances?
c) What is the magnitude of the voltage standing-wave ratio (VSWR)?

2-6 The characteristic impedance Z_0 of a lossless transmission line is 50 Ω and the load is $60 - j80$ Ω. A double-stub matching device is designed to match the load to the line. One stub is at the load and the second is $\frac{3}{8}\lambda$ from the first.

a) Find the lengths of the shorted stubs needed to achieve a match.
b) Locate and crosshatch the "forbidden region" of the normalized load.

2-7 The normalized input impedance of a lossy transmission line is $0.4 - j0.2$ if the end of the line is shorted. Determine the

a) attenuation loss in dB of the line by using a Smith chart;
b) length of the line in terms of wavelengths;
c) load impedance of the same line if the input impedance of the line is measured as $0.6 - j0.35\ \Omega$ per unit length.

2-8 A lossless transmission line has a characteristic impedance Z_0 of 100 Ω and is loaded by an unknown impedance. Its VSWR is 4 and the first voltage maximum is $\frac{1}{8}\lambda$ from the load.

a) Find the load impedance.
b) To match the load to the line, a quarter-section of different line with a characteristic impedance $Z_{01} < Z_0$ is to be inserted somewhere between (in cascade with) the load and the original line. Determine the minimum distance between the load and matching section and the characteristic impedance Z_{01} in terms of Z_0.

2-9 A lossless transmission line has a characteristic impedance of 100 Ω and is loaded by an unknown impedance. The VSWR along the line is 2. The adjacent minima are located at $z = -10$ cm and $z = -35$ cm from the load, where $z = 0$. Determine the load impedance.

2-10 A matched transmission line is shown in Fig. P2-10.

a) Find the lengths ℓ_1 and ℓ_2 that will provide a proper match.
b) With the line and load properly matched, determine the VSWR on the section of line between the stubs.

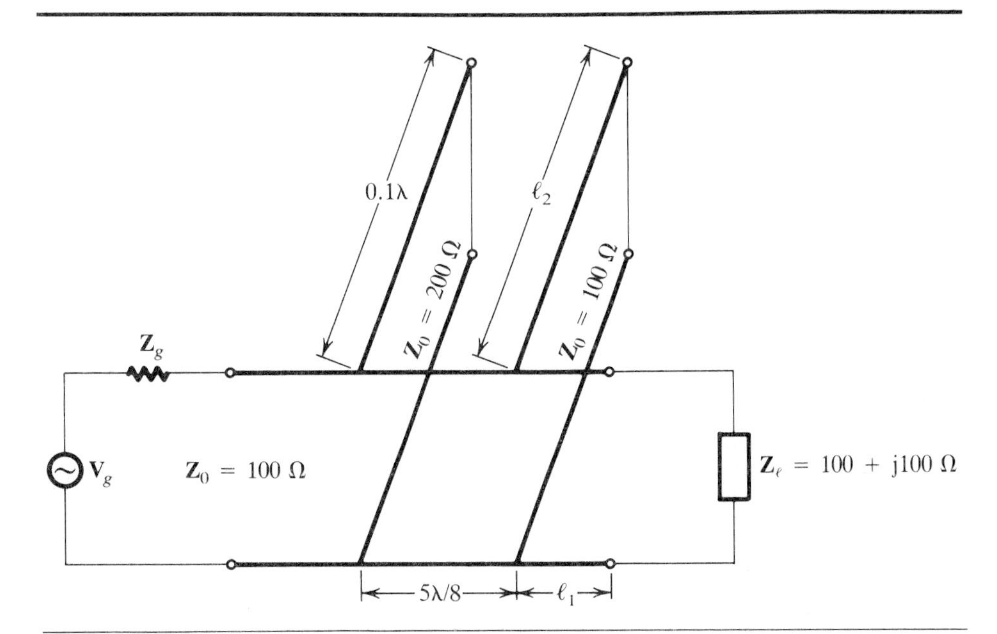

Figure P2-10

2-11 A lossless transmission line to be matched with double-stub tuners is shown in Fig. P2-11. The normalized load admittance is $y_\ell = 0.55 + j0.27$ and the characteristic impedance Z_0 of the line is 50 Ω. The signal frequency is 1 GHz.

a) Determine ℓ_1 and ℓ_2 in cm.
b) Find the susceptances y_{s1} and y_{s2} contributed by the stubs.
c) Determine the VSWR between the tuners.

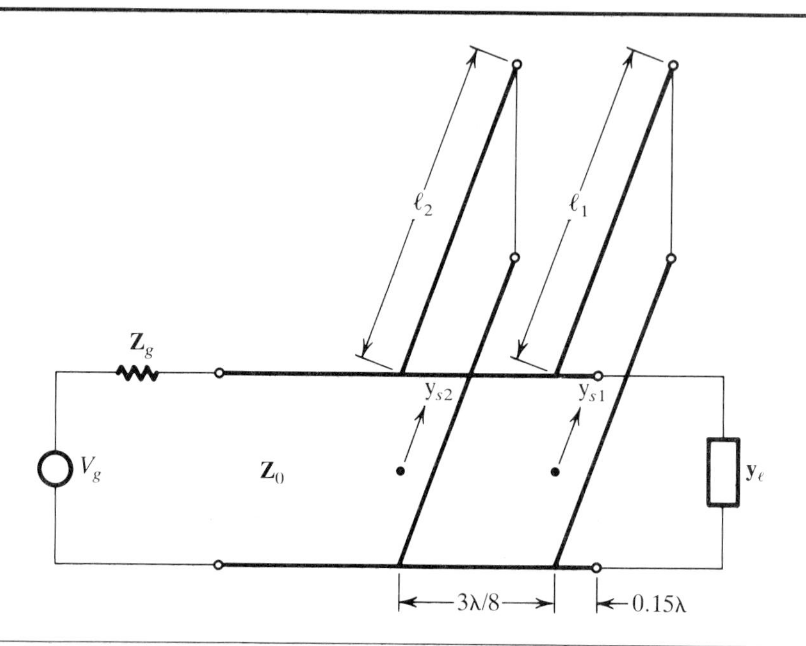

Figure P2-11

2-12 A dipole antenna has a driving-point impedance of 73 Ω, but the lossless transmission line connecting the antenna to a TV set has a characteristic impedance of 300 Ω. Determine the characteristic impedance of a quarter-wave line to be used for matching the dipole antenna to the line.

2-13 a) When the end of a lossy transmission line is open, the normalized input impedance is $1 + j1$. Determine the attenuation loss of the line in dB.

b) Determine the input impedance of the line when the same lossy line is loaded by a normalized load of $2 - j2$. Use a Smith chart to obtain the solution.

c) Find the length of the line in terms of wavelengths.

2-14 A lossless transmission line has a characteristic impedance of 300 Ω and is operated at a frequency of 10 GHz. The load is $z_\ell = 0.4 - j0.95$ and the observed VSWR on the line is 5.0. A short-circuited stub is to be used to reduce the standing wave.

a) Determine the distance in centimeters from a voltage minimum at which the stub should be located. (There are two possible solutions.)
b) Find the length of the stub in centimeters. (There are two possible solutions.)

2-15 A lossless line has a characteristic impedance of 50 Ω and is loaded by $60 - j60$ Ω. One stub is at the load, and the second is $\frac{3}{8}\lambda$ from the first.

a) Determine the required lengths of the stubs in centimeters to achieve a match.
b) Locate and crosshatch the "forbidden region" of the normalized load.

2-16 A lossless transmission line has a characteristic impedance of 300 Ω and terminates in an impedance Z_ℓ. The observed VSWR on the line is 6, and the distance of the first-voltage minimum from the load is 0.166λ. Determine the

a) load impedance Z_ℓ; and
b) lengths (in cm) of two shorted stubs, one at the load and one at $\frac{1}{8}\lambda$ from the load, which are required to match the load to the line.

2-17 A single-stub tuner is to match a line of 400 Ω to a load of $800 + j300$ Ω. The frequency is 200 MHz. Find the

a) distance in meters from the load to the tuning stub; and
b) length in meters of the short-circuited stub.

2-18 A single-stub tuner is to match a line of 300 Ω to a load of $200 + j200$ Ω. The frequency is 500 MHz. Find the

a) shortest distance in meters from the load to the tuning stub; and
b) proper length in meters of the short-circuited stub.

2-19 A double-tuning device is to match a line of 60 Ω to a load of $28 + j28$ Ω. The distance from the load to the first stub is $\frac{1}{18}\lambda$. The spacing between stubs is $\frac{1}{18}\lambda$. The frequency is 1 GHz. Determine the

a) proper lengths (in cm) of the two stubs; and
b) susceptances contributed by the stubs.

2-20 Repeat Problem 2-19 for a distance of $\frac{1}{6}\lambda$ between the load and the first stub.

2-21 A lossless transmission line has a characteristic impedance of 50 Ω and terminates in a load of $5 - j10$ Ω. The characteristic impedance of the stub is 100 Ω. Determine the

a) distance (in wavelength) from the load for a short-circuited stub to match the line to the load; and
b) length (in wavelength) of the stub.

2-22 A lossless transmission line has a characteristic impedance of 100 Ω and terminates in an impedance Z_ℓ. The observed VSWR on the line is 8. The distance between successive minima is 10 cm, and the distance from the load to the first voltage minimum is 8.12 cm. The frequency is 1 GHz. Determine the

a) load impedance Z_ℓ; and
b) distance (in cm) from the load and the length (in cm) of a single stub to be used to match the load to the line.

2-23 A lossless transmission line has a characteristic impedance of 100 Ω and terminates in an impedance Z_ℓ. The observed VSWR on the line is 6.5, and the distance from the load to the first voltage minimum is 0.168λ. Find the

a) load impedance Z_ℓ; and
b) lengths (in wavelength) of two shorted stubs, one at the load and one at $\frac{1}{4}\lambda$ from the load, required to match the load to the line.

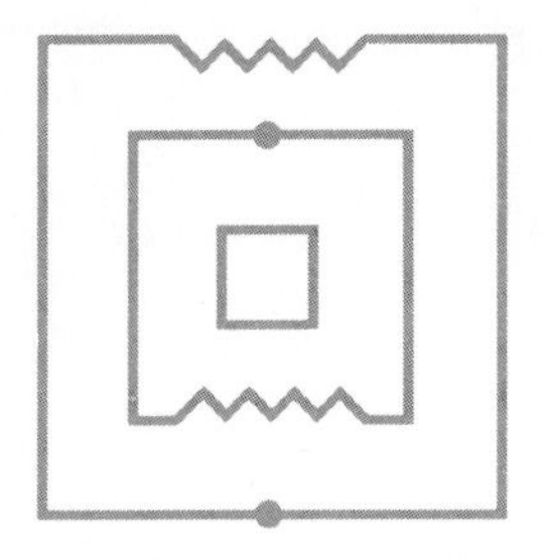

Chapter 3

Striplines

3-0 INTRODUCTION

Prior to 1965 nearly all microwave equipment utilized coaxial, waveguide, or parallel stripline circuits. In recent years—with the introduction of microwave monolithic integrated circuits (MMICs)—microstrip lines and coplanar striplines have been used extensively, because they provide one free and accessible surface on which solid-state devices can be placed. In this chapter we describe parallel, coplanar, and shielded striplines and microstrip lines, which are shown in Figure 3-0-1 on the following page [1].

3-1 MICROSTRIP LINES

In Chapters 1 and 2, we described and discussed conventional transmission lines in detail. All electrical and electronic devices with high-power output commonly use conventional transmission lines, such as coaxial lines or waveguides, for power transmission. However, the microwave solid-state device is usually fabricated as a semiconducting chip with a volume on the order of 0.008–0.08 mm^3. The method of applying signals to the chips and extracting output power from them is entirely different from that used for vacuum-tube devices. Microwave integrated circuits with microstrip lines are commonly used with the chips. The microstrip line is also called "open-strip line." In engineering applications, MKS units have not been universally adopted for use in designing the microstrip line. In this section we use either English units or MKS units, depending on the application, for practical purposes.

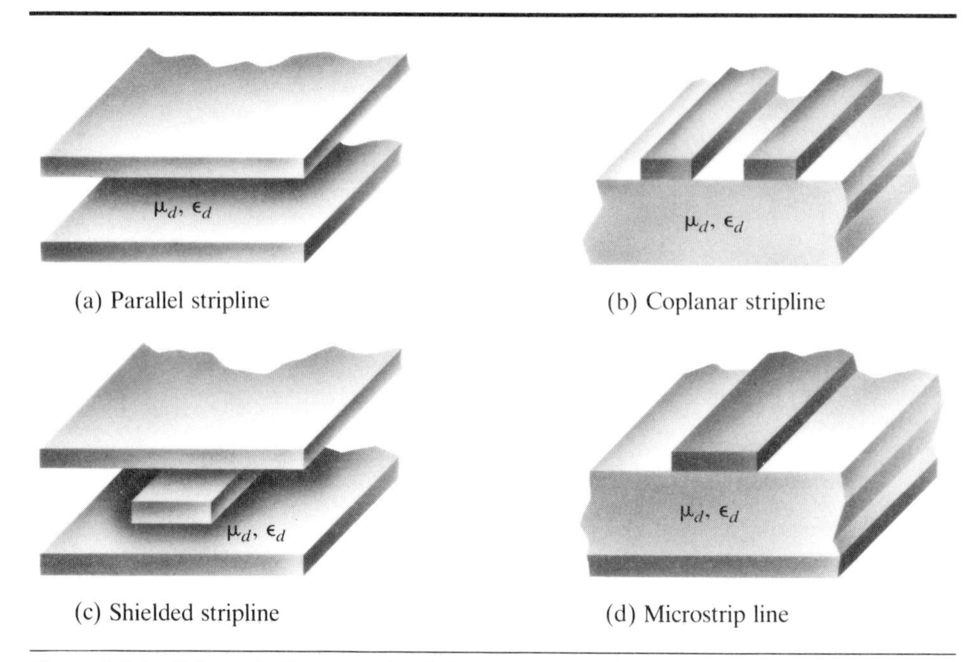

Figure 3-0-1 Schematic diagrams of striplines.

Modes on microstrip line are only quasi-TEM (Transverse Electric and Magnetic). Thus the theory of TEM-coupled lines applies only approximately. Radiation loss in microstrip lines is a problem, particularly at such discontinuities as short-circuit posts, corners, and so on. However, the use of thin, high-dielectric materials reduces considerably the radiation loss of the open strip. Microstrip line has an advantage over balanced-strip line, because the open strip line has better interconnection features and is easier to fabricate. Several researchers have analyzed the circuit of a microstrip line mounted on an infinite dielectric substrate over an infinite ground plane [2–5]. Numerical analysis of microstrip lines, however, requires the use of large digital computers, whereas microstrip-line problems can generally be solved by conformal transformations, without our needing complete numerical calculations.

3-1-1 Characteristic Impedance of Microstrip Lines

Microstrip lines are used extensively to interconnect high-speed logic circuits in digital computers, because they can be fabricated by automated techniques and they provide the required uniform signal paths. Figure 3-1-1 shows cross-sections of a microstrip line and a wire-over-ground line for purposes of comparison.

In Fig. 3-1-1(a) you can see that the characteristic impedance of a microstrip line is a function of the strip-line width, the strip-line thickness, the distance between the line and the ground plane, and the homogeneous dielectric constant of the board material. Several different methods for determining the characteristic impedance of a microstrip line have been developed. The field-equation method was employed by several authors for calculating an accurate value of the characteristic impedance [3–5]. However, it requires the use of a large digital computer and is extremely complicated. Another method is to derive the characteristic-impedance equation of a microstrip line from a well-known equation, making some changes [2]. This method is called a *comparative*,

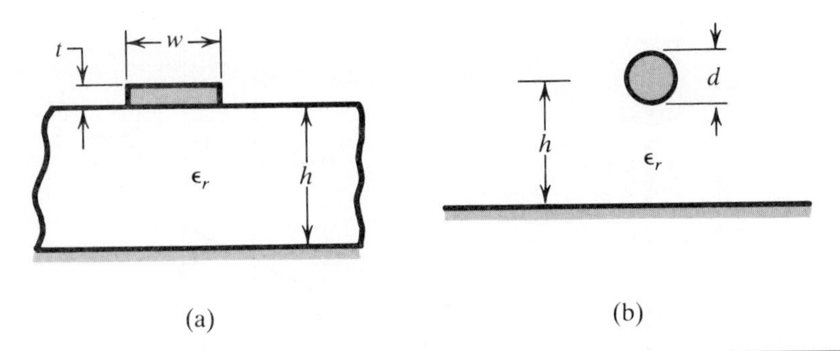

Figure 3-1-1 Cross-sections of (a) a microstrip line; and (b) a wire-over-ground line.

or an *indirect*, method. The well-known equation of the characteristic impedance of a wire-over-ground transmission line, as shown in Fig. 3-1-1(b) is

$$Z_0 = \frac{60}{\sqrt{\varepsilon_r}} \ln \frac{4h}{d} \qquad \text{for } h \gg d \tag{3-1-1}$$

where

ε_r is the dielectric constant of the ambient medium;

h is the height from the center of the wire to the ground plane; and

d is the diameter of the wire.

If we can determine the effective or equivalent values of the relative dielectric constant ε_r of the ambient medium and the diameter d of the wire for the microstrip line, we can calculate the characteristic impedance of the microstrip line.

1. *Effective Dielectric Constant* ε_{re}

For a homogeneous dielectric medium, the propagation-delay time per unit length is

$$T_d = \sqrt{\mu\varepsilon} \tag{3-1-2}$$

where μ is the permeability of the medium; and ε is the permittivity of the medium. In free space, the propagation-delay time is

$$T_{df} = \sqrt{\mu_0 \varepsilon_0} = 3.333 \quad \text{ns/m} \quad \text{or} \quad 1.016 \quad \text{ns/ft} \tag{3-1-3}$$

where

$\mu_0 = 4\pi \times 10^{-7}$ H/m, or 3.83×10^{-7} H/ft; and

$\varepsilon_0 = 8.854 \times 10^{-12}$ F/m, or 2.69×10^{-12} F/ft.

In transmission lines used for interconnections, the relative permeability is 1. Consequently, the propagation-delay time for a line in a nonmagnetic medium is

$$T_d = 1.016\sqrt{\varepsilon_r} \qquad \text{ns/ft} \tag{3-1-4}$$

The effective relative dielectric constant for a microstrip line can be related to the relative dielectric constant of the board material. DiGiacomo and his co-workers discovered an empirical equation for the effective relative dielectric constant of a

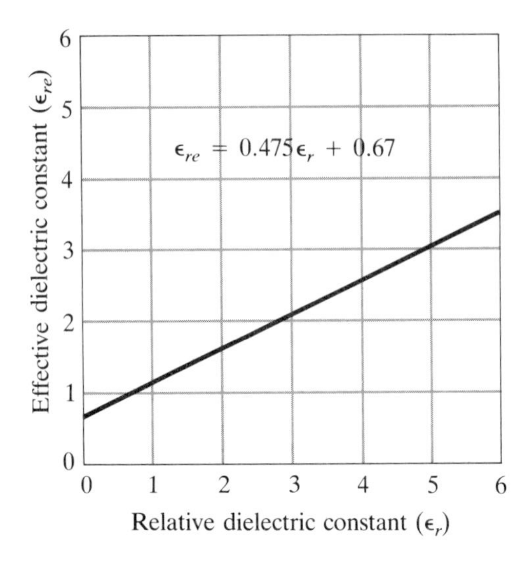

Figure 3-1-2 Effective dielectric constant as a function of relative dielectric constant for a microstrip line.

SOURCE: AFTER H. R. KAUPP [2]; © 1967 IEEE.

microstrip line by measuring the propagation-delay time and the relative dielectric constant of several board materials, such as fiberglass-epoxy and nylon phenolic [6]. His empirical equation, as shown in Fig. 3-1-2, is expressed as

$$\varepsilon_{re} = 0.475\varepsilon_r + 0.67 \tag{3-1-5}$$

where ε_r is the relative dielectric constant of the board material; and ε_{re} is the effective relative dielectric constant for a microstrip line.

2. *Transformation of a Rectangular Conductor into an Equivalent Circular Conductor*

The cross-section of a microstrip line is rectangular, but we need to transform the rectangular conductor into an equivalent circular conductor. Springfield discovered an empirical equation for the transformation [7]. His equation is

$$d = 0.67w\left(0.8 + \frac{t}{w}\right) \tag{3-1-6}$$

where

d is the diameter of the wire over ground;

w is the width of the microstrip line; and

t is the thickness of the microstrip line.

The limitation of the ratio of thickness to width is between 0.1 and 0.8, as indicated in Fig. 3-1-3.

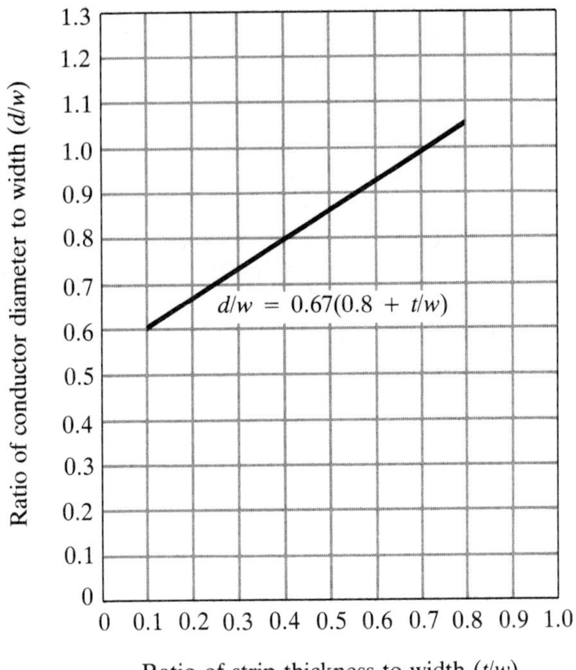

Figure 3-1-3 Relationship between a round conductor and a rectangular conductor far from its ground plane.

SOURCE: AFTER H. R. KAUPP [2]; © 1967 IEEE.

3. *Characteristic Impedance Equation*

Substituting Eq. (3-1-3) for the dielectric constant and Eq. (3-1-6) for the equivalent diameter in Eq. (3-1-1) yields

$$Z_0 = \frac{87}{\sqrt{\varepsilon_r + 1.41}} \ln\left[\frac{5.98h}{0.8w + t}\right] \quad \text{for } (h < 0.8w) \tag{3-1-7}$$

where

ε_r is the relative dielectric constant of the board material;
h is the height from the microstrip line to the ground;
w is the width of the microstrip line; and
t is the thickness of the microstrip line.

Equation (3-1-7) is the equation of characteristic impedance for a narrow microstrip line. The velocity of propagation is

$$v = \frac{c}{\sqrt{\varepsilon_{re}}} = \frac{3 \times 10^8}{\sqrt{\varepsilon_{re}}} \quad \text{m/s} \tag{3-1-8}$$

The characteristic impedance for a wide microstrip line was derived by Assadourian and others [7] and is expressed by

$$Z_0 = \frac{h}{w}\sqrt{\frac{\mu}{\varepsilon}} = \frac{377}{\sqrt{\varepsilon_r}}\frac{h}{w} \qquad \text{for } (w \gg h) \tag{3-1-9}$$

4. ***Limitations of Equation (3-1-7)***

Most microstrip lines are made from boards of copper with a thickness of 1.4 or 2.8 mils (1 or 2 ounces of copper per square foot). The narrowest widths of lines in production are about 0.005–0.010 in. Line widths are usually less than 0.020 in; consequently, ratios of thickness to width of less than 0.1 are uncommon. The straight-line approximation from Eq. (3-1-6) is an accurate value of characteristic impedance, or the ratio of thickness to width between 0.1 and 0.8.

Since the dielectric constant of the materials used does not vary excessively with frequency, the dielectric constant of microstrip line can be considered independent of frequency. The validity of Eq. (3-1-7) is doubtful for values of dielectric thickness h that are greater than 80 percent of the line width w. Typical values for the characteristic impedance of a microstrip line vary from 50 Ω to 150 Ω, if the values of the parameters vary from $\varepsilon_r = 5.23$, $t = 2.8$ mils, $w = 10$ mils, and $h = 8$ mils to $\varepsilon_r = 2.9$, $t = 2.8$ mils, $w = 10$ mils, and $h = 67$ mils [2].

EXAMPLE 3-1-1 Characteristic Impedance of Microstrip Line

A certain microstrip line has the following parameters:

$\varepsilon_r = 5.23$

$h = 8$ mils

$t = 2.8$ mils

$w = 10$ mils

Calculate the characteristic impedance Z_0 of the line.

Solution:

$$\begin{aligned} Z_0 &= \frac{87}{\sqrt{\varepsilon_r + 1.41}} \ln\left[\frac{5.98h}{0.8w + t}\right] \\ &= \frac{87}{\sqrt{5.23 + 1.41}} \ln\left[\frac{5.98 \times 8}{0.8 \times 10 + 2.8}\right] \\ &= 50.50\ \Omega \end{aligned}$$

EXAMPLE 3-1-2 Computer Solution for Characteristic Impedance of Microstrip Lines

```
010C        PROGRAM TO COMPUTE CHARACTERISTIC IMPEDANCE
020C+       OF A MICROSTRIP LINE
030C        STRIPL
040       1 READ 2, ER,H,T,W
050         IF(ER. LE. 0.0) STOP
060         PRINT 4,ER,H,T,W
```

EXAMPLE 3-1-2 *(continued)*

```
070        ZO=87./SQRT(ER+1.41)*ALOG(5.98*H/(.8*W+T))
080        PRINT 6, ZO
090      2 FORMAT(4F10.5)
100      4 FORMAT(//1H ,33HRELATIVE DIELECTRIC CONSTANT ER= ,
110+       1X,F10.5/
120+       1X,33HHEIGHT OF LINE ABOVE GROUND    H= ,
130+       1X,F10.5,1X,4HMILS/
140+       1X,33HTHICKNESS OF MICROSTRIP LINE   T= ,
150+       1X,F10.5,1X,4HMILS/
160+       1X,33HWIDTH OF MICROSTRIP LINE       W= ,
170+       1X,F10.5,1X,4HMILS/)
180      6 FORMAT(1H ,29HCHARACTERISTIC IMPEDANCE ZO= ,
190+       1X,F10.5,1X,4HOHMS/)
200        GO TO 1
210        STOP
220        END
READY.
RUN

PROGRAM   STRIPL

?     5.23000    8.00000    2.80000   10.00000

 RELATIVE DIELECTRIC CONSTANT ER=       5.23000
 HEIGHT OF LINE ABOVE GROUND    H=      8.00000 MILS
 THICKNESS OF MICROSTRIP LINE   T=      2.80000 MILS
 WIDTH OF MICROSTRIP LINE       W=     10.00000 MILS

 CHARACTERISTIC IMPEDANCE ZO=     50.24934 OHMS

? 0.0

SRU        0.721 UNTS.

RUN COMPLETE.
BYE
```

3-1-2 Losses in Microstrip Lines

Microstrip transmission lines consisting of a conductive ribbon attached to a dielectric sheet with conductive backing (as shown in Fig. 3-1-4) are widely used in both microwave and computer technology. Because such lines are easily fabricated by printed-circuit manufacturing techniques, they have economic and technical merit.

We analyzed the characteristic impedance and wave-propagation velocity of a microstrip line in Section 3-1-1. The other characteristic of the microstrip line is its attenuation. The attenuation constant of the dominant microstrip mode depends on geometric factors, electrical properties of the substrate and conductors, and frequency. For a nonmagnetic dielectric substrate, two types of losses occur in the dominant microstrip mode: (1) dielectric loss in the substrate; and (2) ohmic skin loss in the strip conductor and the ground plane. We can express the sum of these two losses as losses per unit length in terms of an attenuation factor α. From ordinary transmission-line theory, the power carried by a wave traveling in the positive z direction is

$$P = \tfrac{1}{2}VI^* = \tfrac{1}{2}(V_+e^{-\alpha z}I_+e^{-\alpha z}) = \tfrac{1}{2}\frac{|V_+|^2}{Z_0}e^{-2\alpha z} = P_0e^{-2\alpha z} \tag{3-1-10}$$

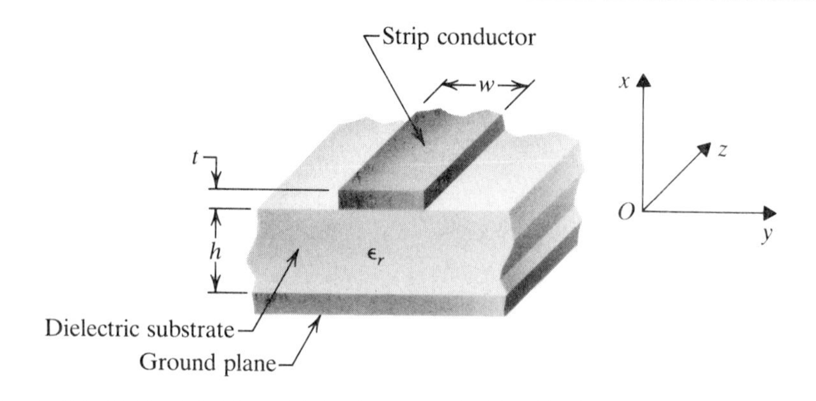

Figure 3-1-4 Schematic diagram of a microstrip line.

where $P_0 = |V_+|^2/(2Z_0)$ is the power at $z = 0$. We can then express the attenuation constant α as

$$\alpha = -\frac{dP/dz}{2P(z)} = \alpha_d + \alpha_c \tag{3-1-11}$$

where α_d is the dielectric attenuation constant, and α_c is the ohmic attenuation constant. We can further express the gradient of power in the z direction in Eq. (3-1-11) in terms of the power loss per unit length dissipated by the resistance and the power loss per unit length in the dielectric. That is

$$\begin{aligned} -\frac{dP(z)}{dz} &= -\frac{d}{dz}\left(\tfrac{1}{2}VI^*\right) \\ &= \tfrac{1}{2}\left(-\frac{dV}{dz}\right)I^* + \tfrac{1}{2}\left(-\frac{dI^*}{dz}\right)V \\ &= \tfrac{1}{2}(RI)I^* + \tfrac{1}{2}\sigma V^*V \\ &= \tfrac{1}{2}|I|^2R + \tfrac{1}{2}|V|^2\sigma = P_c + P_d \end{aligned} \tag{3-1-12}$$

where σ is the conductivity of the dielectric substrate board. Substituting Eq. (3-1-12) into Eq. (3-1-11), we get

$$\alpha_d \simeq \frac{P_d}{2P(z)} \qquad \text{Np/cm} \tag{3-1-13}$$

and

$$\alpha_c \simeq \frac{P_c}{2P(z)} \qquad \text{Np/cm} \tag{3-1-14}$$

1. *Dielectric Losses*

As we stated in Section 0-5-9, when the conductivity of a dielectric cannot be neglected, the electric and magnetic fields in the dielectric are no longer in time phase.

In that case the dielectric attenuation constant, as expressed in Eq. (0-5-72), is

$$\alpha_d = \frac{\sigma}{2}\sqrt{\frac{\mu}{\varepsilon}} \qquad \text{Np/cm} \tag{3-1-15}$$

where σ is the conductivity of the dielectric substrate board in ℧/cm. We can express this dielectric constant in terms of dielectric loss tangent, as expressed in Eq. (0-5-69):

$$\tan\theta = \frac{\sigma}{\omega\varepsilon} \tag{3-1-16}$$

Then the dielectric attenuation constant is

$$\alpha_d = \frac{\omega}{2}\sqrt{\mu\varepsilon}\tan\theta \qquad \text{Np/cm} \tag{3-1-17}$$

Since the microstrip line is a nonmagnetic, mixed dielectric system, the upper dielectric above the microstrip ribbon is air, in which no loss occurs. Welch and Pratt derived an expression for the attenuation constant of a dielectric substrate [8]. Later, Pucel and his co-workers modified Welch's equation [9], resulting in the expression:

$$\begin{aligned}\alpha_d &= 4.34\frac{q\sigma}{\sqrt{\varepsilon_{re}}}\sqrt{\frac{\mu_0}{\varepsilon_0}} \\ &= 1.634 \times 10^3 \frac{q\sigma}{\sqrt{\varepsilon_{re}}} \qquad \text{dB/cm}\end{aligned} \tag{3-1-18}$$

In Eq. (3-1-18) the conversion factor of 1 Np = 8.686 dB is used, ε_{re} is the effective dielectric constant of the substrate, as expressed in Eq. (3-1-5), and q denotes the dielectric filling factor, defined by Wheeler [3] as

$$q = \frac{\varepsilon_{re} - 1}{\varepsilon_r - 1} \tag{3-1-19}$$

We usually express the attenuation constant per wavelength:

$$\alpha_d = 27.3\left(\frac{q\varepsilon_r}{\varepsilon_{re}}\right)\frac{\tan\theta}{\lambda_g} \qquad \text{dB}/\lambda_g \tag{3-1-20}$$

where

$\lambda_g = \dfrac{\lambda_0}{\sqrt{\varepsilon_{re}}}$, and λ_0 is the wavelength in free space; or

$\lambda_g = \dfrac{c}{f\sqrt{\varepsilon_{re}}}$, and c is the velocity of light in vacuum.

If the loss tangent, $\tan\theta$, is independent of frequency, the dielectric attenuation per wavelength is also independent of frequency. Moreover, if the substrate conductivity is independent of frequency, as for a semiconductor, the dielectric attenuation per unit is also independent of frequency. Since q is a function of ε_r and w/h, the filling factor for the loss tangent $q\varepsilon_n/\varepsilon_{re}$ and for the conductivity $q/\sqrt{\varepsilon_{re}}$ are also functions of these quantities. Figure 3-1-5 shows the loss-tangent filling factor against w/h for a range of dielectric constants that are suitable for microwave integrated circuits. For most

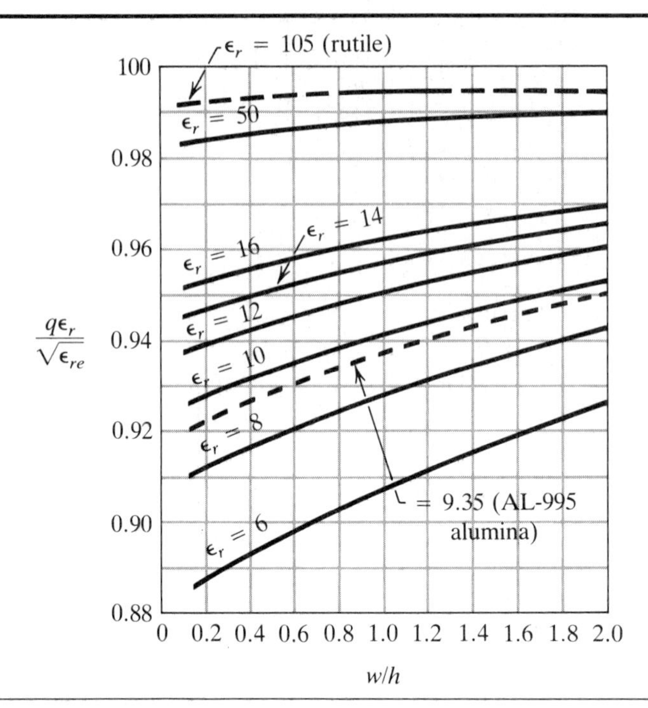

Figure 3-1-5 Filling factor for loss tangent of microstrip substrate as a function of w/h.

SOURCE: AFTER R. A. PUCEL [9]; © 1968 IEEE.

practical purposes, we consider this factor to be 1. Figure 3-1-6 illustrates the product $\alpha_d \rho$ against w/h for two semiconducting substrates that are used for integrated microwave circuits: silicon and gallium arsenide. For design purposes, we can ignore the conductivity filling factor, which exhibits only a mild dependence on w/h.

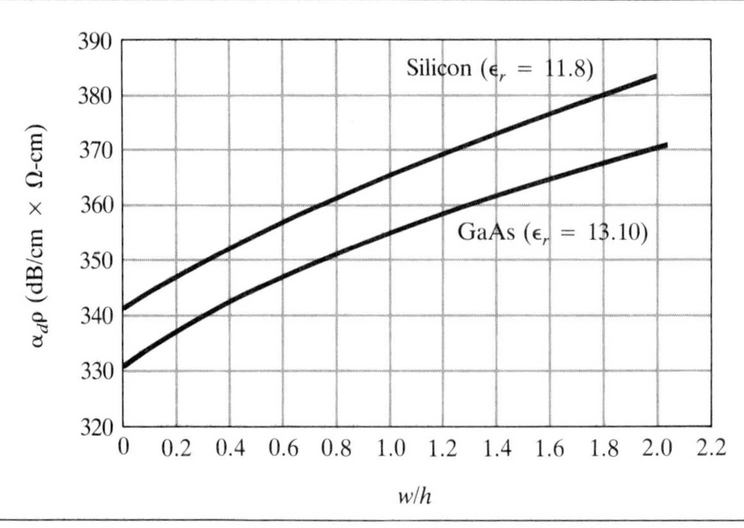

Figure 3-1-6 Dielectric attenuation factor of microstrip as a function of w/h for silicon and gallium arsenide substrates.

SOURCE: AFTER R. A. PUCEL [9]; © 1968 IEEE.

2. *Ohmic Losses*

In a microstrip line over a low-loss dielectric substrate, the predominant sources of losses at microwave frequencies are the nonperfect conductors. The current density in the conductors of a microstrip line is concentrated in a sheet that is approximately a skin depth thick inside the conductor surface and exposed to the electric field. We assume that both the strip conductor thickness and the ground plane thickness are at least three or four skin depths thick. The current density in the strip conductor and the ground conductor is not uniform in the transverse plane. The microstrip conductor contributes the major part of the ohmic loss. A diagram of the current density J for a microstrip line is shown in Fig. 3-1-7.

Because of mathematical complexity, exact expressions for the current density of a microstrip line with nonzero thickness have never been derived [9]. Several researchers [7] have assumed, for simplicity, that the current distribution is uniform and equal to I/w in both conductors and confined to the region $|x| < w/2$. With this assumption, we can express the conducting attenuation constant of a wide microstrip line as

$$\alpha_c \simeq \frac{8.686 R_s}{Z_0 w} \quad \text{dB/cm} \qquad \text{for} \frac{w}{h} > 1 \tag{3-1-21}$$

where

$R_s = \sqrt{\frac{\pi f \mu}{\sigma}}$ and is the surface skin resistance, in Ω/square;

$R_s = \frac{1}{\delta\sigma} \Omega$/square; and

$\delta = \sqrt{\frac{1}{\pi f \mu \sigma}}$ is the skin depth in cm.

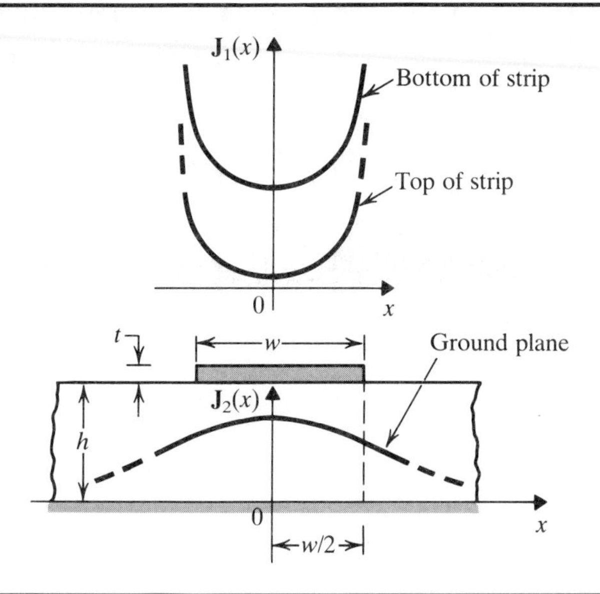

Figure 3-1-7 Current distribution on microstrip conductors.

SOURCE: AFTER R. A. PUCEL [9]; © 1968 IEEE.

For a narrow microstrip line with $w/h < 1$, however, Eq. (3-1-21) is not applicable. The reason is that the current distribution in the conductor is not uniform, as assumed. Purcel and his co-workers [9, 10] derived the following three formulas from the results of Wheeler's work [3]:

$$\frac{\alpha_c Z_0 h}{R_s} = \frac{8.68}{2\pi}\left[1 - \left(\frac{w'}{4h}\right)^2\right]\left[1 + \frac{h}{w'} + \frac{h}{\pi w'}\left(\ln\frac{4\pi w}{t} + \frac{t}{w}\right)\right] \quad \text{for } \frac{w}{h} \le \frac{1}{2\pi} \tag{3-1-22}$$

$$\frac{\alpha_c Z_0 h}{R_s} = \frac{8.68}{2\pi}\left[1 - \left(\frac{w'}{4h}\right)^2\right]\left[1 + \frac{h}{w'} + \frac{h}{w'}\left(\ln\frac{2h}{t} - \frac{t}{h}\right)\right] \quad \text{for } \frac{1}{2\pi} < \frac{w}{h} \le 2 \tag{3-1-23}$$

and

$$\frac{\alpha_c Z_0 h}{R_s} = \frac{8.68}{\left\{\frac{w'}{h} + \frac{2}{\pi}\ln\left[2\pi e\left(\frac{w'}{2h} + 0.94\right)\right]\right\}^2}\left[\frac{w'}{h} + \frac{w'/(\pi h)}{\frac{w'}{2h} + 0.94}\right] \times \left[1 + \frac{h}{w'} + \frac{h}{\pi w'}\left(\ln\frac{2h}{t} - \frac{t}{h}\right)\right] \quad \text{for } 2 \le \frac{w}{h} \tag{3-1-24}$$

where α_c is expressed in dB/cm and

$$e = 2.718$$

$$w' = w + \Delta w \tag{3-1-25}$$

$$\Delta w = \frac{t}{\pi}\left(\ln\frac{4\pi w}{t} + 1\right) \quad \text{for } \frac{2t}{h} < \frac{w}{h} \le \frac{\pi}{2} \tag{3-1-26}$$

$$\Delta w = \frac{t}{\pi}\left(\ln\frac{2h}{t} + 1\right) \quad \text{for } \frac{w}{h} \ge \frac{\pi}{2} \tag{3-1-27}$$

The values of α_c obtained from solving Eqs. (3-1-22) through (3-1-24) are plotted in Fig. 3-1-8. For purposes of comparison, values of α_c based on Assadourian and Rimai's Eq. (3-1-21) are also shown.

3. *Radiation Losses*

In addition to conductor and dielectric losses, microstrip line also has radiation loss. The radiation loss depends upon the substrate's thickness and dielectric constant, as well as its geometry. Lewin [11] has calculated the radiation loss for several discontinuities using the following approximations:

1. TEM transmission.
2. Uniform dielectric in the neighborhood of the strip, equal in magnitude to an effective value.
3. Neglect of radiation from the transverse electric (TE) field component parallel to the strip.
4. Substrate thickness much less than the free-space wavelength.

Lewin's results show that the ratio of radiated power to total dissipated power for an open-circuited microstrip line is

$$\frac{P_{\text{rad}}}{P_t} = 240\pi^2\left(\frac{h}{\lambda_0}\right)^2\frac{F(\varepsilon_{re})}{Z_0} \tag{3-1-28}$$

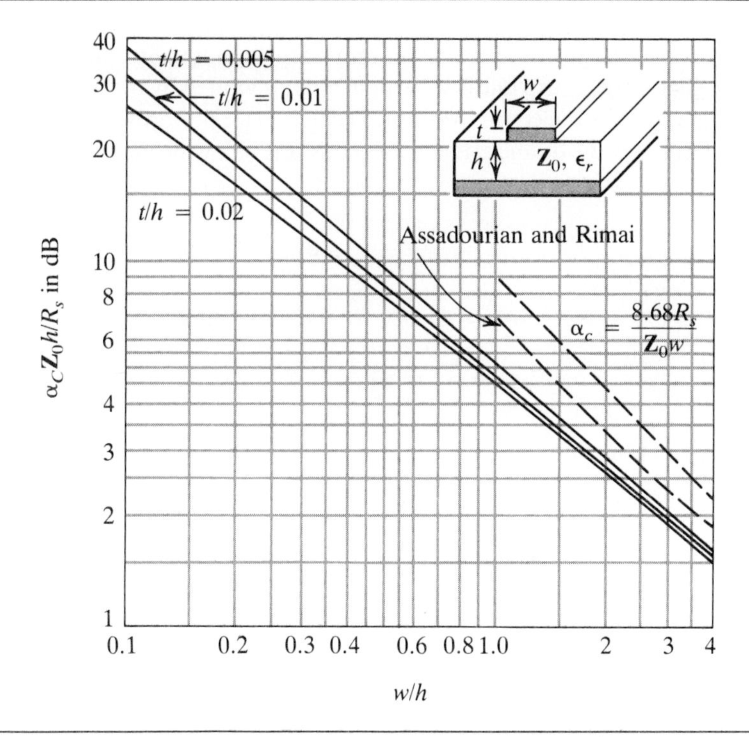

Figure 3-1-8 Theoretical conductor attenuation factor of microstrip as a function of w/h.

SOURCE: AFTER R. A. PUCEL [9]; © 1968 IEEE.

where $F(\varepsilon_{re})$ is a radiation factor:

$$F(\varepsilon_{re}) = \frac{\varepsilon_{re} + 1}{\varepsilon_{re}} - \frac{(\varepsilon_{re} - 1)}{2\varepsilon_{re}\sqrt{\varepsilon_{re}}} \ln \frac{\sqrt{\varepsilon_{re}} + 1}{\sqrt{\varepsilon_{re}} - 1} \tag{3-1-29}$$

ε_{re} is the effective dielectric constant, and $\lambda_0 = c/f$ is the free-space wavelength. The radiation factor decreases with increasing substrate dielectric constant. So, alternatively, we can express Eq. (3-1-28) as

$$\frac{P_{\text{rad}}}{P_t} = \frac{R_r}{Z_0} \tag{3-1-30}$$

where R_r is the radiation resistance of an open-circuited microstrip and it is expressed by

$$R_r = 240\pi^2 \left(\frac{h}{\lambda_0}\right)^2 F(\varepsilon_{re}) \tag{3-1-31}$$

The ratio of the radiation resistance R_r to the real part of the characteristic impedance Z_0 of the microstrip line is equal to a small fraction of the power radiated from a single open-circuit discontinuity. In view of Eq. (3-1-28), the radiation loss decreases when the characteristic impedance increases. For lower dielectric-constant substrates, radiation is significant at higher impedance levels. For higher dielectric-constant substrates, radiation becomes significant until very low impedance levels are reached.

3-1-3 The Quality Factor Q of Microstrip Lines

Many microwave integrated circuits require very high quality factor Q resonant circuits. The Q of a microstrip line is very high, but it is limited by the radiation losses of the substrates and a low dielectric constant. Recall that for uniform current distribution in the microstrip line, the ohmic attenuation constant of a wide microstrip line is given by Eq. (3-1-21), or

$$\alpha_c = \frac{8.686R_s}{Z_0 w} \qquad \text{dB/cm}$$

and that the characteristic impedance of a wide microstrip line, Eq. (3-1-9), is

$$Z_0 = \frac{h}{w}\sqrt{\frac{\mu}{\varepsilon}} = \frac{377}{\sqrt{\varepsilon_r}}\frac{h}{w} \qquad \Omega$$

The wavelength in the microstrip line is

$$\lambda_g = \frac{30}{f\sqrt{\varepsilon_r}} \qquad \text{cm} \tag{3-1-32}$$

where f is the frequency in GHz. Since Q_c is related to the conductor attenuation constant by

$$Q_c = \frac{27.3}{\alpha_c} \tag{3-1-33}$$

where α_c is in dB/λ_g, we can express the Q_c of a wide microstrip line as

$$Q_c = 39.5\left(\frac{h}{R_s}\right) f_{\text{GHz}} \tag{3-1-34}$$

where h is measured in cm, R_s is expressed as

$$R_s = \sqrt{\frac{\pi f \mu}{\sigma}} = 2\pi\sqrt{\frac{f_{\text{GHz}}}{\sigma}} \qquad \Omega/\text{square.} \tag{3-1-35}$$

Finally, the quality factor Q_c of a wide microstrip line is

$$Q_c = 0.63h\sqrt{\sigma f_{\text{GHz}}} \tag{3-1-36}$$

where σ is the conductivity of dielectric substrate board in ℧/m. For a copper strip, $\sigma = 5.8 \times 10^7$ ℧/m and Q_c becomes

$$Q_{\text{Cu}} = 4780h\sqrt{f_{\text{GHz}}} \tag{3-1-37}$$

For 25-mil alumina at 10 GHz, the maximum Q_c achievable from wide microstrip lines is 954 [12].

Similarly, a quality factor Q_d is related to the dielectric attenuation constant:

$$Q_d = \frac{27.3}{\alpha_d} \tag{3-1-38}$$

where α_d is in dB/λ_g. Substituting Eq. (3-1-20) into Eq. (3-1-38) yields

$$Q_d = \frac{\lambda_0}{\sqrt{\varepsilon_{re}}\tan\theta} \simeq \frac{1}{\tan\theta} \tag{3-1-39}$$

where λ_0 is the free-space wavelength in cm. Note that the Q_d for the dielectric attenuation constant of a microstrip line is approximately the reciprocal of the dielectric loss tangent θ and is relatively constant with frequency.

3-2 PARALLEL STRIPLINES

A parallel stripline consists of two perfectly parallel strips separated by a perfect dielectric slab of uniform thickness, as shown in Fig. 3-2-1. The plate width is w, the separation distance is d, and the relative dielectric constant of the slab is ε_{rd}.

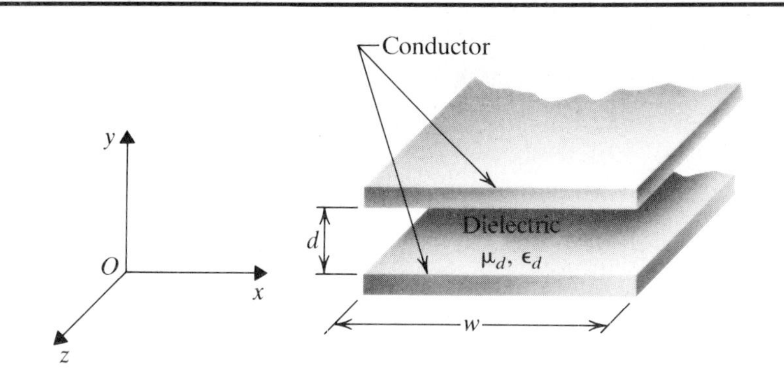

Figure 3-2-1 Schematic diagram of a parallel stripline.

3-2-1 Distributed Parameters

In a microwave integrated circuit a stripline can be easily fabricated on a dielectric substrate by using printed-circuit techniques. A parallel stripline is similar to a two-conductor transmission line, so it can support a quasi-TEM mode. Let's consider a TEM mode propagating in the positive z direction in a lossless stripline ($R = G = 0$). The electric field is in the y direction, and the magnetic field is in the x direction. If the width w is much larger than the separation distance d, the fringing capacitance is negligible. Thus we can write the equation for the inductance along the two conducting strips as

$$L = \frac{\mu_c d}{w} \quad \text{H/m} \tag{3-2-1}$$

where μ_c is the permeability of the conductor. We can also express the capacitance between the two conducting strips as

$$C = \frac{\varepsilon_d w}{d} \quad \text{F/m} \tag{3-2-2}$$

where ε_d is the permittivity of the dielectric slab.

If the two parallel strips have some surface resistance and the dielectric substrate has some shunt conductance, however, the parallel stripline would have some losses. The series resistance for both strips is

$$R = \frac{2R_s}{w} = \frac{2}{w}\sqrt{\frac{\pi f \mu_c}{\sigma_c}} \quad \Omega/\text{m} \tag{3-2-3}$$

where $R_s = \sqrt{(\pi f \mu_c)/\sigma_c}$ is the conductor surface resistance in Ω/square; and σ_c is the conductor conductivity in ℧/m. The shunt conductance of the stripline is

$$G = \frac{\sigma_d w}{d} \qquad \mho/\text{m} \tag{3-2-4}$$

where σ_d is the conductivity of the dielectric substrate.

3-2-2 Characteristic Impedance

The characteristic impedance of a lossless parallel stripline is

$$Z_0 = \sqrt{\frac{L}{C}} = \frac{d}{w}\sqrt{\frac{\mu_d}{\varepsilon_d}} = \frac{377}{\sqrt{\varepsilon_{rd}}}\frac{d}{w} \qquad \text{for } w \gg d \tag{3-2-5}$$

The phase velocity along a parallel stripline is

$$v_p = \frac{\omega}{\beta} = \frac{1}{\sqrt{LC}} = \frac{1}{\sqrt{\mu_d \varepsilon_d}} = \frac{c}{\sqrt{\varepsilon_{rd}}} \qquad \text{m/s} \qquad \text{for } \mu_c = \mu_0 \tag{3-2-6}$$

We can approximate the characteristic impedance of a lossy parallel stripline at microwave frequencies ($R \ll \omega L$ and $G \ll \omega C$) as

$$Z_0 \simeq \sqrt{\frac{L}{C}} = \frac{377}{\sqrt{\varepsilon_{rd}}}\frac{d}{w} \qquad \text{for } w \gg d \tag{3-2-7}$$

3-2-3 Attenuation Losses

We can express the propagation constant of a parallel stripline at microwave frequencies as

$$\begin{aligned} \gamma &= \sqrt{(R + j\omega L)(G + j\omega C)} \qquad \text{for } R \ll \omega L \quad \text{and} \quad G \ll \omega C \\ &\simeq \frac{1}{2}\left(R\sqrt{\frac{C}{L}} + G\sqrt{\frac{L}{C}}\right) + j\omega\sqrt{LC} \end{aligned} \tag{3-2-8}$$

Thus the attenuation and phase constants are

$$\alpha = \frac{1}{2}\left(R\sqrt{\frac{C}{L}} + G\sqrt{\frac{L}{C}}\right) \qquad \text{Np/m} \tag{3-2-9}$$

and

$$\beta = \omega\sqrt{LC} \qquad \text{rad/m} \tag{3-2-10}$$

Substituting the distributed parameters of a parallel stripline into Eq. (3-2-9), we obtain the attenuation constants for the conductor and dielectric losses:

$$\alpha_c = \frac{1}{2}R\sqrt{\frac{C}{L}} = \frac{1}{d}\sqrt{\frac{\pi f \varepsilon_d}{\sigma_c}} \qquad \text{Np/m} \tag{3-2-11}$$

and

$$\alpha_d = \frac{1}{2}G\sqrt{\frac{L}{C}} = \frac{188\sigma_d}{\sqrt{\varepsilon_{rd}}} \qquad \text{Np/m} \tag{3-2-12}$$

EXAMPLE 3-2-1 Characteristics of a Parallel Stripline

A lossless parallel stripline has a conducting strip width w. The substrate dielectric separating the two conducting strips has a relative dielectric constant ε_{rd} of 6 (beryllia or beryllium oxide BeO) and a thickness d of 4 mm.

Calculate:

a) the required width w of the conducting strip in order to have a characteristic impedance of 50 Ω;
b) the stripline capacitance;
c) the stripline inductance; and
d) the phase velocity of the wave in the parallel stripline.

Solutions:

a) Using Eq. (3-2-5), we obtain the width of the conducting strip:

$$w = \frac{377}{\sqrt{\varepsilon_{rd}}} \frac{d}{Z_0} = \frac{377}{\sqrt{6}} \frac{4 \times 10^{-3}}{50}$$

$$= 12.31 \times 10^{-3} \text{ m}$$

b) The stripline capacitance is

$$C = \frac{\varepsilon_d w}{d} = \frac{8.854 \times 10^{-12} \times 6 \times 12.31 \times 10^{-3}}{4 \times 10^{-3}}$$

$$= 163.50 \text{ pF/m}$$

c) The stripline inductance is

$$L = \frac{\mu_c d}{w} = \frac{4\pi \times 10^{-7} \times 4 \times 10^{-3}}{12.31 \times 10^{-3}}$$

$$= 0.41\ \mu\text{H/m}$$

d) The phase velocity is

$$v_p = \frac{c}{\sqrt{\varepsilon_{rd}}} = \frac{3 \times 10^8}{\sqrt{6}}$$

$$= 1.22 \times 10^8 \text{ m/s}$$

3-3 COPLANAR STRIPLINES

A coplanar stripline consists of two conducting strips on one substrate surface with one strip grounded, as shown in Fig. 3-3-1. The coplanar stripline has advantages over the conventional parallel stripline described in Section 3-2, because its two strips are on the same substrate surface for convenient connections. In microwave integrated circuits (MICs) the wire bonds have always presented reliability and reproducibility problems. The coplanar striplines eliminate the difficulties involved in connecting the shunt elements between the hot and ground strips. As a result, reliability is increased, reproducibility is enhanced, and production cost is decreased.

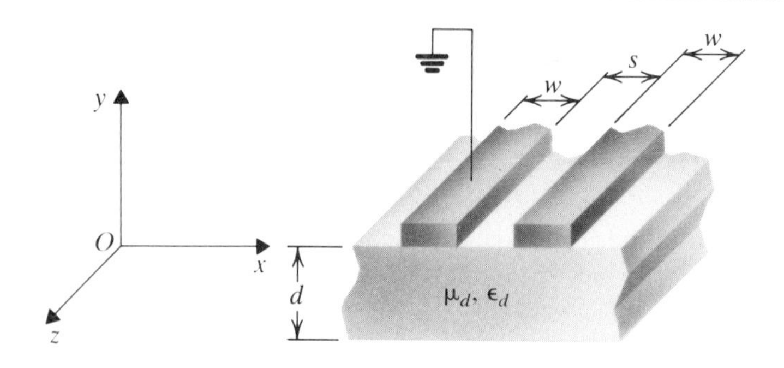

Figure 3-3-1 Schematic diagram of a coplanar stripline.

The characteristic impedance of a coplanar stripline is

$$Z_0 = \frac{2P_{avg}}{I_0^2} \tag{3-3-1}$$

where I_0 is the total peak current in one strip, and P_{avg} is the average power flowing in the positive z direction. We can express the average flowing power as

$$P_{avg} = \tfrac{1}{2}\text{Re} \iint (\mathbf{E} \times \mathbf{H}^*) \cdot \mathbf{u}_z \, dx \, dy \tag{3-3-2}$$

where

$\mathbf{E}_x$ = electric field intensity in the positive x direction;
$\mathbf{H}_y$ = magnetic field intensity in the positive y direction; and
$*$ = conjugate.

EXAMPLE 3-3-1 Characteristic Impedance of a Coplanar Stripline

A coplanar stripline carries an average power of 250 mW and a peak current of 100 mA. Determine the characteristic impedance of the coplanar stripline.

Solution:

From Eq. (3-3-1), the characteristic impedance of the coplanar stripline is

$$Z_0 = \frac{2 \times 250 \times 10^{-3}}{(100 \times 10^{-3})^2} = 50\ \Omega$$

3-4 SHIELDED STRIPLINES

A partially shielded stripline has its stripconductor embedded in a dielectric medium, and its top and bottom ground planes have no connection, as shown in Fig. 3-4-1.

The characteristic impedance for a wide strip ($w/d \geq 0.35$) [13] is

$$Z_0 = \frac{94.15}{\sqrt{\varepsilon_r}} \left(\frac{w}{d} K + \frac{C_f}{8.854\varepsilon_r} \right)^{-1} \tag{3-4-1}$$

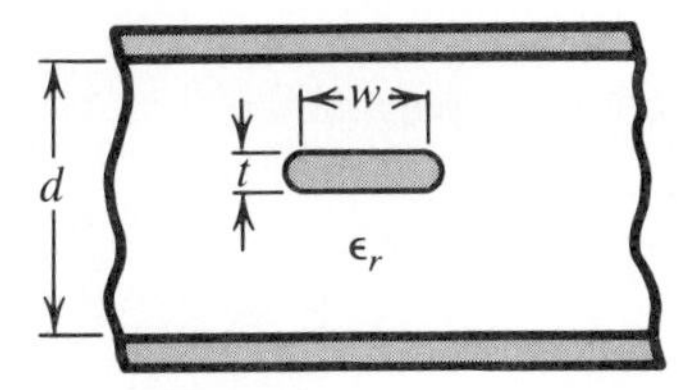

Figure 3-4-1 Partially shielded stripline.

where

$$K = \frac{1}{1 - t/d};$$

t is the strip thickness;

d is the distance between the two ground planes; and

$C_f = \dfrac{8.854\varepsilon_r}{\pi}[2K \ln(K+1) - (K-1)\ln(K^2-1)]$ and is the fringe capacitance in pF/m.

Figure 3-4-2 shows the characteristic impedance $\mathbf{Z}_0$ for a partially shielded stripline, with the t/d ratio as a parameter.

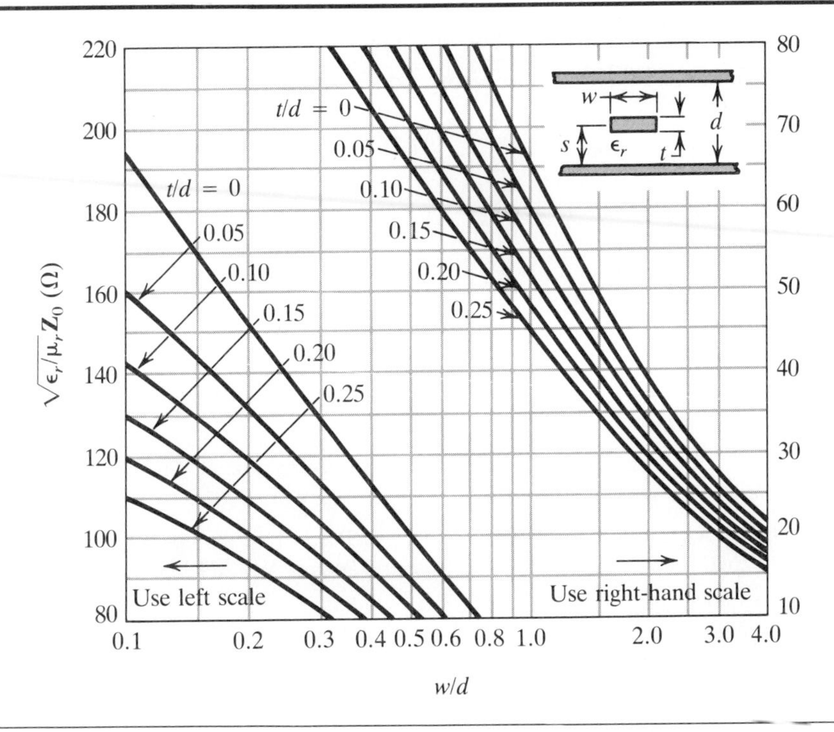

Figure 3-4-2 Characteristic impedance $\mathbf{Z}_0$ of a partially shielded stripline with the t/d ratio as a parameter.

SOURCE: AFTER S. COHN [13]; © 1954 IEEE.

EXAMPLE 3-4-1 Characteristic Impedance of a Shielded Stripline

A shielded stripline has the following parameters:

Dielectric constant of the insulator (polystyrene)	$\varepsilon_r = 2.56$
Strip width	$w = 25$ mils
Strip thickness	$t = 14$ mils
Shield depth	$d = 70$ mils

Calculate:

a) the K factor;
b) the fringe capacitance; and
c) the characteristic impedance of the line.

Solutions:

a) Using Eq. (3-4-1), we obtain the K factor:

$$K = \left(1 - \frac{t}{d}\right)^{-1} = \left(\frac{1 - 14}{70}\right)^{-1} = 1.25$$

b) From Eq. (3-4-1), the fringe capacitance is

$$C_f = \frac{8.854 \times 2.56}{3.1416}[2 \times 1.25 \ln(1.25 + 1) - (1.25 - 1)\ln(1.25^2 - 1)]$$

$$= 15.61 \text{ pF/m.}$$

c) The characteristic impedance from Eq. (3-4-1) is

$$Z_0 = \frac{94.15}{\sqrt{2.56}}\left[\frac{25}{70}(1.25) + \frac{15.61}{8.854 \times 2.56}\right]^{-1}$$

$$= 50.29\ \Omega$$

REFERENCES

[1] Liao, S. Y., *Microwave Devices and Circuits*, 2nd ed., Chapter 13. Englewood Cliffs, N.J.: Prentice-Hall, 1985.

[2] Kaupp, H. R., "Characteristics of Microstrip Transmission Lines," *IEEE Trans., Electronic Computers*. vol. EC-16, no. 2, pp. 185–193, April 1967.

[3] Wheeler, H. A., "Transmission-line Properties of Parallel Strips Separated by a Dielectric Sheet," *IEEE Trans., Microwave Theory and Techniques*, vol. MTT-3, no. 3, pp. 172–185, March 1965.

[4] Bryant, T. G., and Weiss, J. A., "Parameters of Microstrip Transmission Lines and of Coupled Pairs of Microstrip Lines," *IEEE Trans., Microwave Theory and Techniques*, vol. MTT-6, no. 12, pp. 1021–1027, December 1968.

[5] Stinehelfer, H. E., "An Accurate Calculation of Uniform Microstrip Transmission Lines," *IEEE Trans., Microwave Theory and Techniques*, vol. MTT-16, no. 7, pp. 439–443, July 1968.

[6] DiGiacomo, J. J., et al., "Design and Fabrication of Nanosecond Digital Equipment," RCA, March 1965.

[7] Assodourian, F., and Rimol, E., "Simplified Theory of Microwave Transmission Systems," *Proceedings IRE*, vol. 40, pp. 1651–1657, December 1952.

[8] Welch, J. D., and Pratt, H. J., "Losses in Microstrip Transmission Systems for Integrated Microwave Circuits," *NEREM Rec.*, vol. 8, pp. 100–101, 1966.

[9] Pucel, Robert A., Masse, Daniel J., and Hartwig, Curtis P., "Losses in Microstrip," *IEEE Trans., Microwave Theory and Techniques*, vol. MTT-16, no. 6, pp. 342–350, June 1968.

[10] Pucel, Robert A., Masse, Daniel J., and Hartwig, Curtis P., "Correction to 'Losses in Microstrip'," *IEEE Trans., Microwave Theory and Techniques*, vol. MTT-16, no. 12, p. 1064, December 1968.

[11] Lewin, L., "Radiation from Discontinuities in Strip-Line," *IEEE Monograph No. 358E*, February 1960.

[12] Vendeline, George D., "Limitations on Stripline Q," *Microwave Journal*, pp. 63–69, May 1970.

[13] Cohn, S., "Characteristic Impedance of the Shielded-strip Transmission Line," *IRE Trans., Microwave Theory and Techniques*, vol. MTT-2, no. 7, p. 52, July 1954.

Problems

3-1 A microstrip line has the following parameters:

$\varepsilon_r = 5.23$ and is the relative dielectric constant of the fiberglass board material.

$h = 8$ mils.

$t = 2.8$ mils.

$w = 10$ mils.

Write a FORTRAN program to compute the characteristic impedance Z_0 of the line. Use a READ statement to read in the input values, the F10.5 format for numerical outputs, and the Hollerith format for character outputs.

3-2 Since modes on microstrip lines are only quasi-transverse electric and magnetic (TEM), the theory of TEM-coupled lines applies only approximately. From the basic theory of a lossless line, show that the inductance L and capacitance C of a microstrip line are

$$L = \frac{Z_0}{v} = \frac{Z_0\sqrt{\varepsilon_r}}{c}$$

and

$$C = \frac{1}{Z_0 v} = \frac{\sqrt{\varepsilon_r}}{Z_0 c}$$

where

Z_0 is the characteristic impedance of the microstrip line;

v is the wave velocity in the microstrip line;

$c = 3 \times 10^8$ m/s, the velocity of light in vacuum; and

ε_r is the relative dielectric constant of the board material.

3-3 A microstrip line is constructed of a perfect conductor and a lossless dielectric board. The relative dielectric constant of the fiberglass-epoxy board is 5.23, and the line characteristic impedance is 50 Ω. Calculate the line inductance and the line capacitance.

3-4 A microstrip line is constructed of a copper conductor and nylon phenolic board. The relative dielectric constant of the board material is 4.19, measured at 25 GHz, and its thickness is 0.4836 mm (19 mils). The line width is 0.635 mm (25 mils) and the line thickness is 0.071 mm (2.8 mils). Calculate the

a) characteristic impedance Z_0 of the microstrip line;
b) dielectric filling factor q;
c) dielectric attenuation constant α_d;
d) surface skin resistivity R_s of the copper conductor at 25 GHz; and
e) conductor attenuation constant α_c.

3-5 A microstrip line is made of a copper conductor 0.254 mm (10 mils) wide on a G-10 fiberglass-epoxy board 0.20 mm (8 mils) in height. The relative dielectric constant ε_r of the board material is 4.8, measured at 25 GHz. The microstrip line 0.035-mm (1.4 mils) thick is to be used for 10 GHz. Determine the

a) characteristic impedance Z_0 of the microstrip line;
b) surface resistivity R_s of the copper conductor;
c) conductor attenuation constant α_c;
d) dielectric attenuation constant α_d; and
e) quality factors Q_c and Q_d.

3-6 A gold parallel stripline has the following parameters:

Relative dielectric constant of teflon	$\varepsilon_{rd} = 2.1$
Strip width	$w = 26$ mm
Separation distance	$d = 5$ mm
Conductivity of gold	$\sigma_c = 4.1 \times 10^7$ ℧/m
Frequency	$f = 10$ GHz

Determine the

a) surface resistance of the gold strip;
b) characteristic impedance of the stripline; and
c) phase velocity.

3-7 A gold parallel stripline has the following parameters:

Relative dielectric constant of polyethylene	$\varepsilon_{rd} = 2.25$
Strip width	$w = 25$ mm
Separation distance	$d = 5$ mm

Calculate the

a) characteristic impedance of the stripline;
b) stripline capacitance;
c) stripline inductance; and
d) phase velocity.

3-8 A 50-Ω coplanar stripline has the following parameters:

Relative dielectric constant of alumina	$\varepsilon_{rd} = 10$
Strip width	$w = 4$ mm
Strip thickness	$t = 1$ mm
TEM-mode field intensities	$E_y = 3.16 \times 10^3 \sin\left(\frac{\pi x}{w}\right) e^{-j\beta z}$
	$H_x = 63.20 \sin\left(\frac{\pi x}{w}\right) e^{-j\beta z}$

Find the

a) average power flow; and
b) peak current in one strip.

3-9 A shielded stripline has the following parameters:

Relative dielectric constant of the insulator polyethylene	$\varepsilon_{rd} = 2.25$
Strip width	$w = 2$ mm
Strip thickness	$t = 0.5$ mm
Shield depth	$d = 4$ mm

Calculate the

a) K factor;
b) fringe capacitance; and
c) characteristic impedance.

3-10 A shielded stripline is made of a gold strip in a polystyrene dielectric insulator and has the following parameters:

Relative dielectric constant of polystyrene	$\varepsilon_{rd} = 2.56$
Strip thickness	$t = 0.7$ mm
Strip width	$w = 1.4$ mm
Shield depth	$d = 3.5$ mm

Determine the

a) K factor;
b) fringe capacitance; and
c) characteristic impedance.

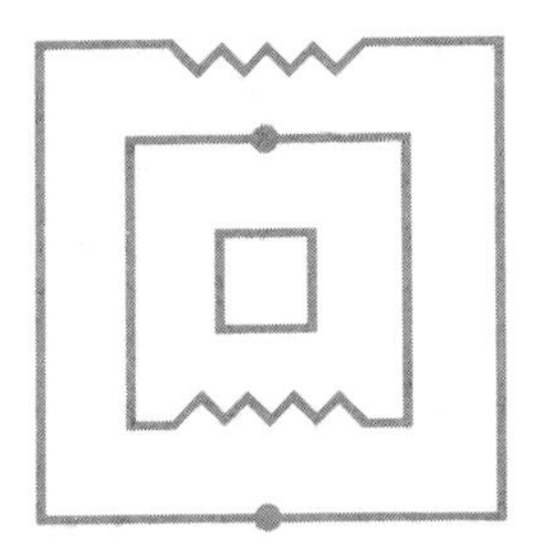

Chapter 4

Digital Transmission Lines

4-0 INTRODUCTION

A digital system uses the binary code to switch from the 0 state (ON, closed, or conducting state) to the 1 state (OFF, open, or nonconducting state). The switching time from one state to another must be very short, that is, in terms of nanoseconds or picoseconds. When information is transmitted by a binary code or a digital number, a pulse signal must be used. In general, digital or pulse transmission is very sensitive to the characteristic parameters of a transmission line.

A pulse can propagate along an ideal, long, and lossless line without distortion. However, its output waveform is delayed by line delay time, which we define as

$$T_d = \frac{\ell}{v_{ph}} \tag{4-0-1}$$

where ℓ is the length of the line (m), and v_{ph} is the phase velocity of the pulse (m/s).

4-1 PHASE VELOCITY AND GROUP VELOCITY

If a wave travels in a homogeneous lossless medium, all the frequency components have the same phase velocity without attenuation. Consequently, the wave always retains its original waveform. If the medium is lossy and frequency-dependent, however, the various frequency components have different phase velocities and rates of attenuation, and the wave changes its shape as it travels through the medium.

4-1-1 Phase Velocity

When a TEM wave of a single frequency travels through a lossless, frequency-independent, and vacuum-dielectric line, we define the phase velocity as

$$v_{ph} = \frac{\omega}{\beta} = \frac{1}{\sqrt{LC}}$$

$$= \frac{1}{\sqrt{\mu_0 \varepsilon_0}} = c \tag{4-1-1}$$

where β is the phase constant, and c is the velocity of light in free space.

If the line medium is lossy and frequency-dependent, the propagation constant is

$$\gamma = j\omega\sqrt{\mu\varepsilon}\sqrt{1 - j\frac{\sigma}{\omega\varepsilon}} \tag{4-1-2}$$

The factor $\sqrt{\varepsilon(1 - j(\sigma/\omega\varepsilon)}$ is a function of frequency. Thus if the loss tangent is very small, that is, $\sigma/(\omega\varepsilon) \ll 1$, we can write the attenuation constant and phase constant, respectively, as

$$\alpha = \frac{\sigma}{2}\sqrt{\frac{\mu}{\varepsilon}} \tag{4-1-3}$$

and

$$\beta = \omega\sqrt{\mu\varepsilon} \tag{4-1-4}$$

Finally, the phase velocity for a frequency-dependent dielectric line is

$$v_{ph} = \frac{1}{\sqrt{\mu\varepsilon}}$$

$$= \frac{c}{\sqrt{\varepsilon_r}} = \frac{c}{n} \tag{4-1-5}$$

where $n = \sqrt{\varepsilon_r}$ and is the refractive index of the medium. In a dispersive medium with no loss and $\varepsilon_r > 1$, the phase velocity of a traveling wave is always less than the velocity of light in vacuum.

4-1-2 Group Velocity

According to Fourier theory a dispersed wave contains not only the center carrier frequency but also the upper and lower sideband frequencies. Let $v_c(t)$ be the carrier wave and $v_s(t)$ be the signal that modulates the amplitude of the carrier wave:

$$v_c(t) = V_0 \sin(\omega_0 t)$$

and

$$v_s(t) = V_s \sin(\Delta\omega t)$$

where ω_0 is the carrier angular frequency, and $\Delta\omega$ is the signal angular frequency. If we let the modulation index or factor be

$$M = \frac{V_s}{V_0}$$

the modulated wave at $z = 0$ becomes

$$v(0, t) = V_0[1 + M \sin(\Delta\omega t)] \sin(\omega_0 t) \tag{4-1-6}$$

By applying the trigonometric identity, we obtain the following expression:

$$v(0, t) = V_0 \sin(\omega_0 t) + \frac{MV_0}{2}[\cos(\omega_0 t - \Delta\omega t) - \cos(\omega_0 t + \Delta\omega t)] \tag{4-1-7}$$

When the waves of the carrier frequency ω_0 and the side bands $(\omega_0 - \Delta\omega, \omega_0 + \Delta\omega)$ travel through a digital line, the carrier has a phase constant β_0; the upper and lower side bands have slightly different phase constants, $\beta_0 \pm \Delta\beta$, respectively. The traveling voltage wave of Eq (4-1-7) in the positive z direction is

$$v(z, t) = V_0 \sin(\omega_0 t - \beta_0 z) + \frac{MV_0}{2}\{\cos[(\omega_0 - \Delta\omega)t - (\beta_0 - \Delta\beta)z]$$
$$-\cos[(\omega_0 + \Delta\omega)t - (\beta_0 + \Delta\beta)z]\} \tag{4-1-8}$$

which we can also write as

$$v(z, t) = V_0[1 + M \sin(\Delta\omega t - \Delta\beta z)] \sin(\omega_0 t - \beta_0 z) \tag{4-1-9}$$

The factor $[1 + M \sin(\Delta\omega t - \Delta\beta z)]$ is the envelope of the modulated wave. For a constant phase point on the envelope, then, we have

$$\Delta\omega t - \Delta\beta z = \text{constant}$$

and we define the group velocity of the wave envelope as

$$\frac{dz}{dt} = \frac{\Delta\omega}{\Delta\beta} = v_{gr} \tag{4-1-10}$$

For a constant phase point on the signal, we have $\omega t - \beta z = \text{constant}$. Then the phase velocity of the signal wave is

$$\frac{dz}{dt} = \frac{\omega}{\beta} = v_{ph} \tag{4-1-11}$$

Finally, the product of the group velocity and phase velocity is

$$v_{gr}v_{ph} = c^2 \tag{4-1-12}$$

Figure 4-1-1 shows the relationship between the group velocity and phase velocity for a dispersed wave of two frequencies. In general, when the dielectric constant ε_r of a medium is greater than 1 and frequency-dependent, the phase velocity is less than the velocity of light in free space. As a result, the group velocity is larger than the velocity of light in vacuum.

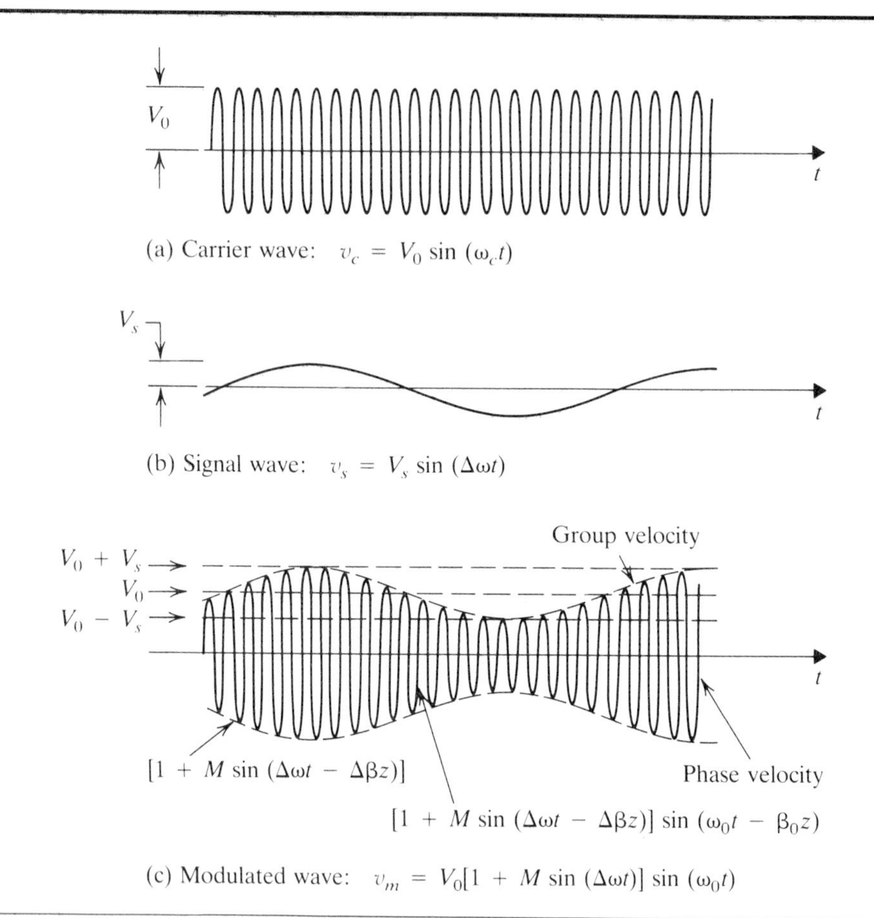

Figure 4-1-1 Amplitude modulation with a sinusoidal signal.

4-2 DISTORTION EFFECTS

A train of periodic pulses generally is composed of an infinite set of discrete frequency components from Fourier theory. The amplitude of the fundamental frequency term is much larger than the decreasing amplitudes of the harmonic frequency terms. The pulse phase distortion occurs when the various frequency components travel at different phase velocities. Amplitude dispersion arises when the frequency components suffer some irregular amplitude attenuations. Consequently, the output pulse waveform is distorted from the input pulse waveform.

Some dielectrics have a constant relative permittivity, but some have a frequency-dependent relative permittivity. Figure 4-2-1 shows plots of frequency-dependent relative permittivity for several commonly used dielectrics. You can see that the relative permittivity of the polyvinyl chloride varies from 4.80 at 1 kHz to 3 at 0.1 GHz. Table 4-2-1 lists the values of the constant relative permittivities of the dielectrics shown in Fig. 4-2-1.

Table 4-2-1
Relative permittivities of commonly used dielectrics

Material	Relative Permittivity, ε_r ($f \leq 10$ GHz)
Alumina	10
Beryllia	6
GaAs	13.10
Polyethylene	2.25
Polystyrene	2.56
Quartz	3.78
Teflon	2.10

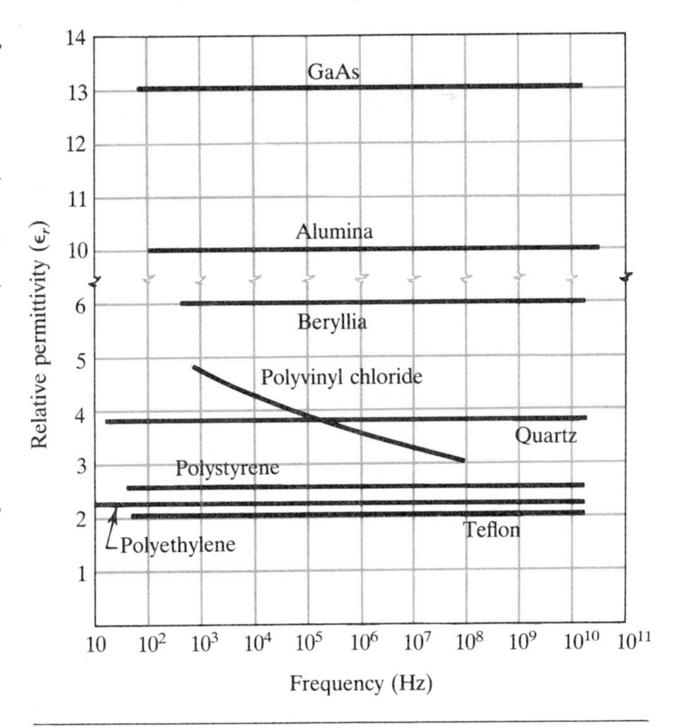

Figure 4-2-1 Relative permittivity versus frequency.

4-2-1 Distortionless Line and Inductive Loading

1. *Distortionless Line*

When a complex signal is impressed on a transmission line, the received wave will be of the same form as the transmitted wave only if all harmonic components of the original wave have no amplitude and phase distortions. The first condition requires that the attenuation constant α be independent of frequency, and therefore all components are attenuated uniformly. The second condition requires that the phase constant β be linearly proportional to frequency, and therefore the harmonics travel at equal phase velocity in all frequencies. In general, the attenuation constant α and phase constant β are related to the following expression:

$$\gamma = \alpha + j\beta = \sqrt{\mathbf{ZY}}$$
$$= \sqrt{(R + j\omega L)(G + j\omega C)} \qquad (4\text{-}2\text{-}1)$$

which ordinarily does not satisfy the two conditions. However, if the parameters R, L, G, and C meet the criterion

$$\frac{R}{L} = \frac{G}{C} \qquad (4\text{-}2\text{-}2)$$

a distortionless line would be obtained. If we substitute Eq. (4-2-2) into Eq. (4-2-1), we get

$$\gamma = \sqrt{RG}\sqrt{\left(1 + j\omega\frac{L}{R}\right)\left(1 + j\omega\frac{C}{G}\right)} = \sqrt{RG}\left(1 + j\omega\frac{C}{G}\right)$$

$$= \sqrt{RG} + j\omega\sqrt{RG}\frac{C}{G} = \sqrt{RG} + j\omega\sqrt{LC}$$

Then

$$\alpha + j\beta = \sqrt{RG} + j\omega\sqrt{LC} \tag{4-2-3}$$

Clearly, the attenuation constant α is independent of frequency, and the phase constant β varies linearly with frequency.

The phase velocity is

$$v_p = \frac{\omega}{\beta} = \frac{1}{\sqrt{LC}} \tag{4-2-4}$$

which is the velocity of light in free space for all frequencies and gives us a distortionless line. The lossless line is a special case of the distortionless line.

A transmission line normally has the following characteristics:

$$R \gg G \qquad L \gg C \quad \text{and} \quad \frac{R}{G} > \frac{L}{C}$$

which does not satisfy the criterion expressed as Eq. (4-2-2). Increasing G artificially will increase the attenuation and is unwise. Reducing R will require large sizes of wires. Reducing C by enlarging the spacing between the conductors could be done, but there are practical limitations on doing so. Thus the only practical way of satisfying Eq. (4-2-2) is to increase the inductance L by the process known as inductive loading.

2. *Inductive Loading*

On some types of transmission lines, the inductance L is increased above its normal value, which is called *inductive loading*. It is accomplished by inserting inductance coils at equal intervals or by wrapping the conductors with a high-permeability metal tape.

EXAMPLE 4-2-1 Inductive Loading

A digital line has the following constants:

$R = 10.0\ \Omega/\text{m}$
$L = 0.0036\ \text{H/m}$
$G = 0.2 \times 10^{-6}\ \mho/\text{m}$
$C = 0.0080 \times 10^{-6}\ \text{F/m}$

The line is to be equalized with inductors placed in series with the line at $\frac{1}{8}$-m intervals.

a) What should the inductance of each of the loading coils be?
b) With the line properly loaded, what should the terminal impedance be if the line is to be free from reflections?

Solutions:

a) The total inductance is

$$L = \frac{RC}{G} = \frac{(10)(0.0080 \times 10^{-6})}{0.2 \times 10^{-6}} = 0.4 \text{ H/m}$$

Thus each inductor should have

$$L = \frac{0.4 - 0.0036}{8} = 0.0496 \text{ H}$$

b)

$$Z_\ell = Z_0 = \sqrt{\frac{L}{C}} = \sqrt{\frac{0.4}{0.0080 \times 10^{-6}}} = 7070 \ \Omega$$

EXAMPLE 4-2-2 Computer Solution for Inductive Loading

```
PROGRAM    TELINE

010C       PROGRAM TO COMPUTE INDUCTANCE OF INDUCTIVE
020C+      LOADING FOR A TELEPHONE LINE
030C       TELINE
040        REAL L,N,LPM,LPC
050      1 READ 2,R,L,G,C,N
060        IF(R .LE. 0.0) STOP
070        G=G*0.000001
080        C=C*0.000001
090        LPM=R*C/G
100        LPC=(LPM-L)/N
110        ZL=SQRT(LPM/C)
120        PRINT 4, LPM,LPC,ZL
130      2 FORMAT(5F10.5)
140      4 FORMAT(//1H ,5X,14HLPM;HENRY/MILE,
150+       3X,14HLPC;HENRY/COIL,2X,7HZL;OHMS//
160+       5X,F10.5,7X,F10.5,6X,F10.5/)
190        GO TO 1
200        STOP
210        END
READY.
RUN

PROGRAM    TELINE

?    10.00000    0.00360    0.20000    0.00800    8.00000

        LPM;HENRY/MILE    LPC;HENRY/COIL   ZL;OHMS

           .40000            .04955        7071.06781

? 0.0

SRU        0.701 UNTS.

RUN COMPLETE.
```

4-2-2 Phase Distortion (Dispersion)

Dispersive media are those materials having characteristics that are dependent upon frequency. Historically, the phenomenon of frequency-dependent dispersion occurs in the refraction of light passing through a glass. A white light passes through a glass prism into its component colors as a result of the dispersive nature of glass. The violet light of higher frequency is bent more than the red light of lower frequency. In other words, the violet light travels more slowly than the red light, because the refractive index of violet light is larger than the refractive index of red light. We define the refractive index as

$$n = \sqrt{\varepsilon_r} \tag{4-2-5}$$

where ε_r is the dielectric constant.

Dispersion gives rise to the phase distortion of the wave. In a dispersive medium the group velocity can exceed the velocity of light, but the actual signal phase velocity must be less than the velocity of light. This condition results from the fact that

$$v_{gr} v_{ph} = c^2 \tag{4-2-6}$$

where v_{gr} is the group velocity, v_{ph} is the phase velocity, and c is the velocity of light in vacuum.

The group velocity can have almost any value from negative through positive to infinite. In a dispersionless medium the group velocity and the phase velocity are equal, or $v_{gr} = v_{ph} = c$. In general, the relative dielectric constants of dispersive media are functions of frequency, as previously shown in Fig. 4-2-1.

4-2-3 Amplitude Distortion (Attenuation)

A wave propagating in a dispersive line will undergo severe distortion because of a large attenuation constant α. Each frequency component will experience a different amount of attenuation along the line. If the line is very long, the output signal waveform will not duplicate the input signal waveform. Therefore the digital line must be limited to distances for which the total attenuation is not significant.

4-3 SKIN EFFECT

Skin effect is a phenomenon resulting from the frequency-dependence of current and/or magnetic field distribution in a conductor and also from the electric field arising from the current in the conductor. In general, skin effect shows the penetration of an electromagnetic wave into a conductor. From Eq. (0-5-61) the skin depth for a good conductor is

$$\delta = \frac{1}{\sqrt{\pi f \mu \sigma}} = \frac{1}{\alpha} = \frac{1}{\beta} \tag{4-3-1}$$

At very high frequencies the skin depth is extremely shallow. A piece of glass with an evaporated silver coating 5.40-μm thick is an excellent reflector at such frequencies.

Also, from Eq. (0-5-62) the classical surface resistance of a good conductor is related to the skin depth by

$$R_s = \frac{1}{\sigma\delta} = \sqrt{\frac{\omega\mu}{2\sigma}} \quad \Omega/\text{square} \tag{4-3-2}$$

The phase velocity of a wave in a good conductor is

$$v_{ph} = \omega\delta \tag{4-3-3}$$

However, at a very high frequency or a very low temperature, the mean-free-path of the conduction electrons between collisions becomes greater than the classical skin depth. This phenomenon causes a surface resistance that can be many times larger than the classical skin-effect calculation. In addition, the skin effect is also the source of a dispersion effect (phase distortion). Dispersion occurs when the phase velocity of a propagating wave is different for the various frequencies.

4-4 WAVE RESPONSES

If the relative dielectric constant of a medium is dependent upon frequency, the shunt capacitance of the transmission line is also a frequency-dependent parameter. As a result, the time constant of the equivalent circuit for the line is also frequency-dependent.

4-4-1 Pulse Responses

In digital transmission a pulse is commonly used to switch a circuit on or off. The rise-time constant of the equivalent circuit for the line is

$$\tau_r = \frac{0.35}{f_{3\text{dB}}} \text{ seconds} \tag{4-4-1}$$

where f is measured at the 3-dB level as indicated by the subscript, and 0.35 is a constant for the shape of the frequency-gain curve.

EXAMPLE 4-4-1 Rise Time of a Pulse Circuit

If the signal frequency is 10 GHz, determine the rise time of the pulse.

Solutions:

Using Eq. (4-4-1), we calculate the pulse rise time as

$$\tau_r = \frac{0.35}{10^{10}}$$

$$= 35 \text{ ps}$$

4-4-2 Step-Function Response

An open-circuited lossless line is depicted in Fig. 4-4-1. It shows a step function of $\mathbf{V}_g$ volts from a d-c generator with an internal impedance of 100 Ω, which is connected to an initially uncharged, open-circuited lossless line of $\mathbf{V}_0 = 50(\Omega)$ at $t = 0$. The voltage and current responses along the line can be described as follows:

1. Let T be the time required for the wave to travel once from the sending end to the receiving end. When the switch is closed at $t = 0$, the first incident voltage to the sending end of the line is

$$V_+ = \frac{V_g R_0}{Z_g + R_0} = \frac{V_g(50)}{100 + 50} = \frac{V_g}{3} \text{ volts} \tag{4-4-2}$$

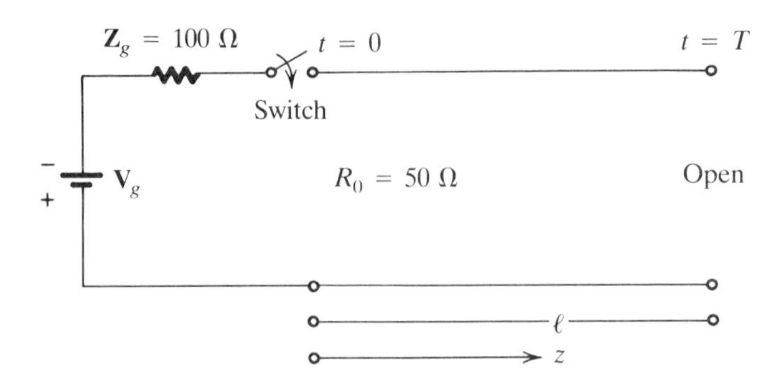

Figure 4-4-1 An open-circuited lossless line.

2. The first reflected voltage back from the receiving end is

$$V_- = V_+\Gamma_r = \frac{V_g}{3}\left(\frac{Z_r - R_0}{Z_r + R_0}\right) = \frac{V_g}{3}(1) = \frac{V_g}{3} \quad \text{volts} \tag{4-4-3}$$

3. The first reflected voltage from the sending end is

$$V_{+r} = V_-\Gamma_g = \frac{V_g}{3}\left(\frac{Z_g - R_0}{Z_g + R_0}\right) = \frac{V_g}{3}\left(\frac{1}{3}\right) = \frac{V_g}{9} \quad \text{volts} \tag{4-4-4}$$

4. The second reflected voltage from the receiving end is $V_g/9$, and so on.
5. The current along the line can be obtained by dividing the voltage by R_0 and changing the sign of Γ_r and Γ_g from positive to negative.
6. Figure 4-4-2 shows the zigzag diagrams for voltage and current on a lossless line.
7. The line voltage and current, respectively, are the sum of the two traveling waves, as shown in Fig. 4-4-3. Thus, as $t \to \infty$,

$$V_{\text{line}} = \frac{2V_g}{3}\left[1 + \frac{1}{3} + \left(\frac{1}{3}\right)^2 + \left(\frac{1}{3}\right)^3 + \cdots\right] = \frac{2V_g}{3}\left[\frac{1}{1 - \frac{1}{3}}\right] = V_g \tag{4-4-5}$$

and

$$I_{\text{line}} = \frac{V_g}{3R_0}\left[1 - 1 + \frac{1}{3} - \frac{1}{3} + \left(\frac{1}{3}\right)^2 - \left(\frac{1}{3}\right)^2 + \cdots\right] = 0 \tag{4-4-6}$$

which are shown in Fig. 4-4-3.
8. The sending-end voltage and current, respectively, are the sum of the two traveling waves at the sending end, as illustrated in Fig. 4-4-4.
9. The receiving-end voltage and current, respectively, are also the sum of the two traveling waves. Since the receiving end is an open circuit, the receiving-end voltage should be equal to the source voltage, and its current should be zero, as plotted in Fig. 4-4-5.

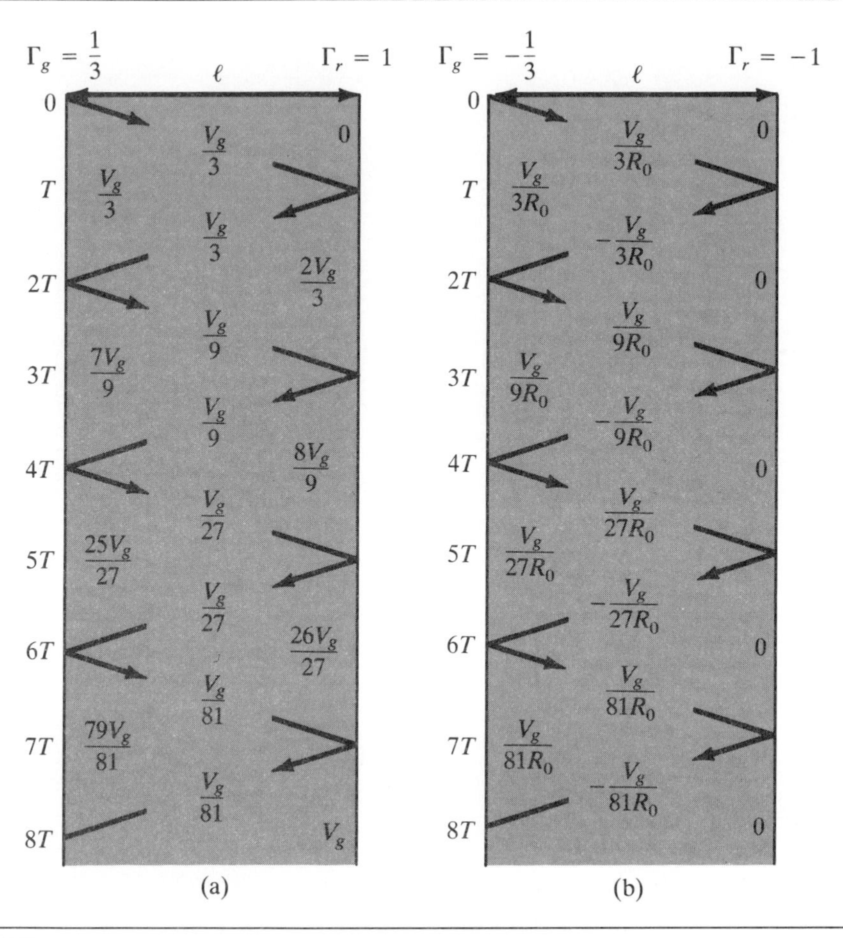

Figure 4-4-2 Zigzag diagrams of (a) voltage and (b) current.

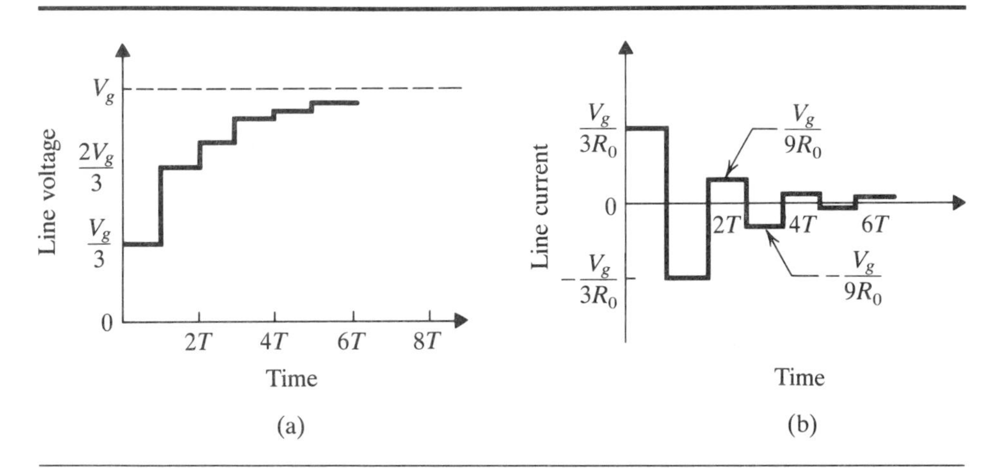

Figure 4-4-3 (a) Line voltage and (b) line current versus time.

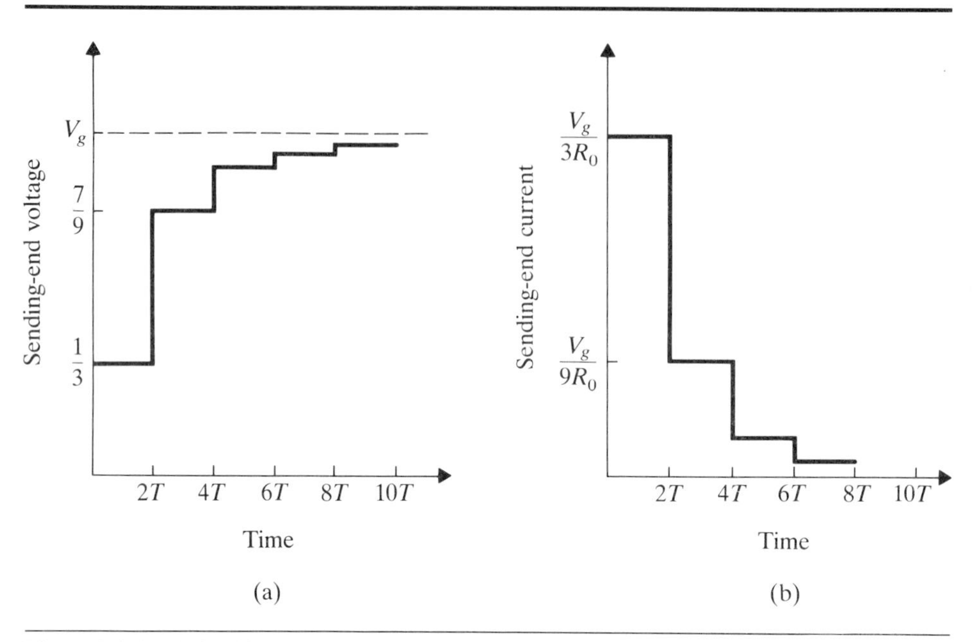

Figure 4-4-4 (a) Sending-end voltage and (b) sending-end current versus time.

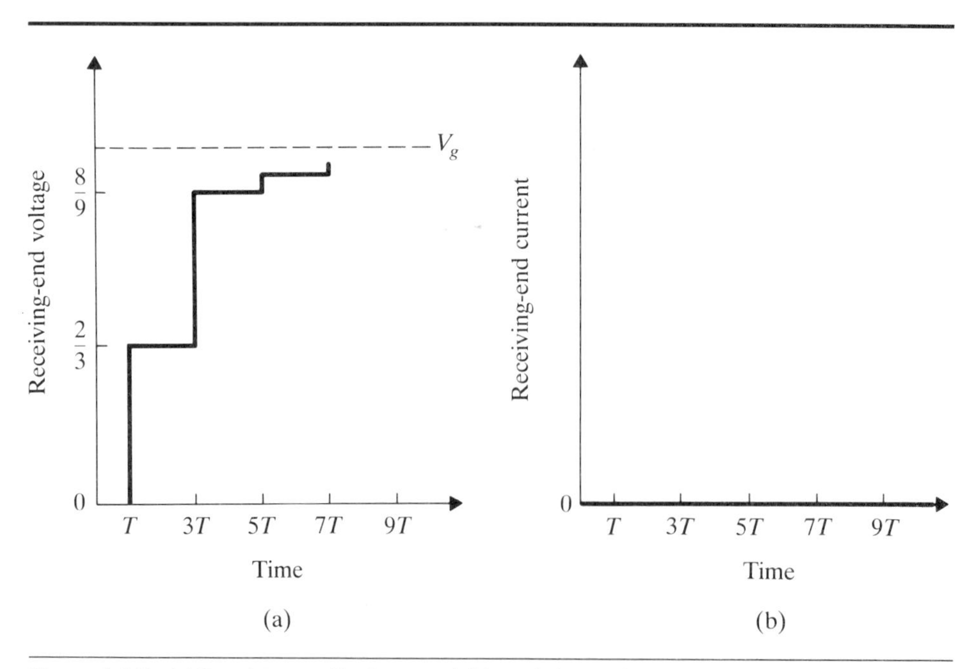

Figure 4-4-5 (a) Receiving-end voltage and (b) receiving-end current versus time.

EXAMPLE 4-4-2 Step-Function Response of a Digital Line

An open-circuited digital line has a characteristic impedance of 50 Ω and is connected to a step-function source of 9.9 V. The d-c generator has an output impedance of 100 Ω.

Calculate:

a) the first sending-end voltage;
b) the first receiving-end reflected voltage;
c) the final line voltage; and
d) the final line current.

Solutions:

a) Using Eq. (4-4-2), we obtain the first sending-end voltage:

$$V_+ = \frac{9.9}{3} = 3.3 \text{ V}$$

b) By using Eq. (4-4-3), we find that the first reflected voltage from the receiving end is

$$V_- = \frac{9.9}{3} = 3.3 \text{ V}$$

c) The final line voltage, from Eq. (4-4-5), is

$$V_{\text{line}} = 9.9 \text{ V}$$

d) The final line current, from Eq. (4-4-6), is

$$I_{\text{line}} = 0$$

EXAMPLE 4-4-3 Computer Solution for Line Voltage and Current of an Open-Circuited Line

```
PROGRAM    LINEVI

010C       PROGRAM TO COMPUTE VOLTAGE AND CURRENT ALONG
020C+      AN OPEN-CIRCUITED LINE
030C       LINEVI
040        REAL IL
050        RO=0.500E+02
060        ZG=0.100E+03
070C       LOAD IMPEDANCE SHOULD BE INFINITE AND IT IS
080C+      ASSUMED TO BE 10E+12
090        ZR=0.100E+14
100C       VG IS ASSUMED TO BE 99 VOLTS RMS
110        VG=0.990E+02
120        PRINT 4, RO,ZG,ZR,VG
130C       FIRST INCIDENT VOLTAGE TO SENDING END OF LINE
140        VPLUS=VG*RO/(ZG+RO)
150C       REFLECTION COEFFICIENT AT LOAD
160        GAMMAR=(ZR-RO)/(ZR+RO)
170C       FIRST REFLECTION VOLTAGE BACK FROM RECEIVING
180C+      END TOWARD THE SENDING END
190        VMINUS=VPLUS*GAMMAR
```

(continued)

EXAMPLE 4-4-3 *(continued)*

```
200C      REFLECTION COEFFICIENT AT GENERATOR
210       GAMMAG=(ZG-RO)/(ZG+RO)
220C      FIRST REFLECTED VOLTAGE FROM THE SENDING END
230C+     TOWARD THE RECEIVING END
240       VPLUSR=VMINUS*GAMMAG
250C      LINE VOLTAGE IS THE SUM OF TWO TRAVELING WAVES
260       VL1=0.0
262       DO 8 I=1,11
263       VLI=VPLUS*(GAMMAR*GAMMAG)**(I-1)
264     8 VL1=VL1+VLI
265       DO 9 J=1,10
266       VLJ=VPLUS*GAMMAR**J*GAMMAG**(J-1)
267     9 VL2=VL2+VLJ
268       VL=VL1+VL2
269C      LINE CURRENT IS THE SUM OF TWO TRAVELING WAVES
270       IL=VPLUS/RO*(1.0+(-GAMMAR)+(-GAMMAR)*(-GAMMAG)
272+        +(-GAMMAR)**2*(-GAMMAG))
280       PRINT 6,GAMMAG,GAMMAR,VPLUS,VMINUS,VPLUSR,VL,TL
290     4 FORMAT(/1H ,6X,"RO",10X,"ZG",10X,"ZR",10X,"VG"/
300+      1X,4E12.3/)
310     6 FORMAT(/1H ,3X,"GAMMAG",2X,"GAMMAR",2X,"VPLUS",
320+      2X,"VMINUS",2X,"VPLUSR",3X,"VL",7X,"IL"/
325+      1X,7F8.3/)
330       STOP
340       END
READY.

RUN

PROGRAM   LINEVI

        RO          ZG          ZR          VG
     .500E+02    .100E+03    .100E+14    .990E+02

   GAMMAG  GAMMAR   VPLUS  VMINUS  VPLUSR      VL       IL
    .333   1.000  33.000  33.000  11.000  98.999     .000

SRU       0.692 UNTS.

RUN COMPLETE.
BYE

F228F10   LOG OFF     18.31.50.
F228F10   SRU       1.406 UNTS.
```

4-5 SUPERCONDUCTING TRANSMISSION LINE

For an ideal line the phase velocity and attenuation constant of the signal wave are the same for all frequencies, and the waveforms at any point on an infinitely long line are identical in shape except for the decreasing amplitude from the source to the load. A superconducting line, however, is dispersionless for all frequency components up to 1 GHz.

A superconducting transmission line is made of superconductor. A superconductor is a special type of metal, such as lead, niobium, and tin, that can carry electric current without resistance at very low temperatures. The critical temperature T_c for a superconducting Josephson effect is close to 0 °K. According to the Bardeen–Cooper–Schrieffer (BCS) theory, below the critical temperature T_c some free electrons in the metal are formed into what are called Cooper pairs. These Cooper-pair electrons can carry a supercurrent without a voltage drop across the superconductor, because its resistance is 0. Superconducting lines offer a distinct advantage over ordinary lines. They usually give negligible distortion and attenuation and have a very fast rise time for a pulse over a substantial length of line. In digital circuit applications, superconducting line is an effective type of transmission line. The characteristic parameters of some commonly used superconductors are listed in Table 4-5-1.

Table 4-5-1 Characteristic parameters of some commonly used superconductors

Metal	Critical Temperature, T_c (°K)	Magnetic Intensity at 0 °K H_0 (A/m)	Penetration Depth at 0 °K d (Å)	Resistivity at °C ρ ($\Omega \cdot$ cm)
Aluminum	1.196	7.87×10^3	500	2.828 (20 °C)
Cadmium	0.560	2.39×10^3	1,300	7.540 (18 °C)
Indium	3.407	23.32×10^3	640	8.370 (0 °C)
Lead	7.175	63.82×10^3	390	19.800 (0 °C)
Mercury	4.153	32.79×10^3	380	95.783 (20 °C)
Molybdenum	0.920	7.80×10^3	—	5.330 (22 °C)
Niobium	9.250	154.70×10^3	—	14.500 (22 °C)
Tin	3.740	24.27×10^3	510	11.500 (20 °C)

4-5-1 Surface Impedance

Figure 4-5-1 shows a schematic diagram of a parallel superconducting line. The surface impedance per unit length of a parallel superconducting line is a function of tem-

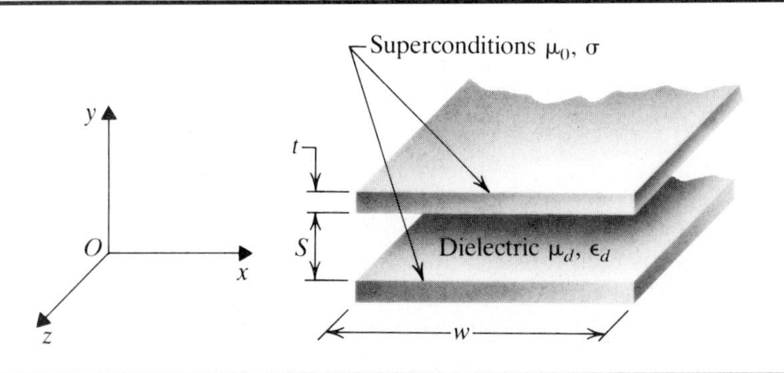

Figure 4-5-1 Schematic diagram of a parallel superconducting line.

perature, frequency, and magnetic field and is expressed [1] as

$$\mathbf{Z}_s = \frac{\mathbf{E}_z}{\mathbf{I}_x} = \frac{j\omega\mu_0 H_0}{H_0 w}\left(\frac{1}{d^2} + j\omega\mu_0\sigma\right)^{-1/2} \coth\left(\frac{t^2}{d^2} + j\omega\mu_0\sigma\right)^{1/2}$$

$$\mathbf{Z}_s = \frac{1}{w} j\omega\mu_0\left(\frac{1}{d^2} + j\omega\mu_0\sigma\right)^{-1/2} \coth\left(\frac{t}{d}\right) \quad \text{for } \frac{t^2}{d^2} \gg \omega\mu_0\sigma \tag{4-5-1}$$

and

$$\mathbf{Z}_s = \frac{1}{w} j\omega\mu_0\left(\frac{1}{d^2} + j\frac{2}{\delta^2}\right)^{-1/2} \coth\left(\frac{t}{d}\right) \quad \Omega/\text{unit length} \tag{4-5-2}$$

where

t = the superconductor thickness;
d = the superconducting penetration depth;
w = the superconductor width;
$\delta = \sqrt{2/(\omega\mu\sigma)}$ and is the skin depth;
μ_0 = the superconductor permeability;
σ = the superconductor conductivity; and
$\rho = 1/\sigma$ and is the temperature-dependent superconductor resistivity.

If the ordinary skin depth is much greater than the penetration depth ($\delta \gg d$), which is usually the case, the surface impedance per unit length in Eq. (4-5-2) becomes

$$\mathbf{Z}_s = \frac{j\omega\mu_0 d}{w} \coth\left(\frac{t}{d}\right) \quad \Omega/\text{unit length} \tag{4-5-3}$$

4-5-2 Characteristic Impedance

The characteristic impedance of a transmission line is the square root of the ratio of the series impedance and shunt admittance of the line, as described in Chapter 1. We can write the series impedance **Z** of a parallel superconducting line as

$$\mathbf{Z} = 2\mathbf{Z}_s + j\omega L = 2\mathbf{Z}_s + j\omega\frac{\mu_d S}{w} \tag{4-5-4}$$

where $L = \mu_d S/w$ and is the inductance per unit length of the parallel superconductor line; and S is the separation of two parallel superconductors. Substituting Eq. (4-5-3) into Eq. (4-5-4) yields the series impedance:

$$\mathbf{Z} = \frac{j\omega\mu_0}{w}\left[2d \coth\left(\frac{t}{d}\right) + \mu_{rd} S\right] \tag{4-5-5}$$

where $\mu_d = \mu_0\mu_{rd}$ has been replaced by $\mu_0\mu_{rd}$. When the superconductor thickness is much greater than the superconducting penetration depth ($t \gg d$) and the value of coth (t/d) approaches 1, the series impedance becomes a pure reactance, or

$$\mathbf{Z} = j\omega L = j\omega\mu_0\frac{S}{w}\left(\mu_{rd} + \frac{2d}{S}\right) \tag{4-5-6}$$

We can express the shunt admittance per unit length of a parallel superconducting line as

$$\mathbf{Y} = G + j\omega C = G + j\omega\varepsilon_d \frac{w}{S} \tag{4-5-7}$$

where $C = \varepsilon_d w/S$ and is the capacitance per unit length of the parallel superconducting line; and G is the conductance of the parallel superconducting line. Finally, the characteristic impedance of a parallel superconducting line is

$$\mathbf{Z}_0 = \sqrt{\frac{\mathbf{Z}}{\mathbf{Y}}} = \sqrt{\frac{j\omega\mu_0 \frac{S}{w}\left(\mu_{rd} + \frac{2d}{S}\right)}{G + j\omega\varepsilon_d \frac{w}{S}}} = \frac{120\pi}{\sqrt{\varepsilon_{rd}}} \frac{S}{w}\left(1 + \frac{2d}{S}\right)^{1/2} \quad \text{for } G \ll C \tag{4-5-8}$$

where $\mu_d = \mu_0\mu_{rd}$ and $\varepsilon_d = \varepsilon_0\varepsilon_{rd}$ are replaced. This characteristic impedance is a pure resistance that has a slightly increased value because of the wave's penetration into the superconductor.

4-5-3 Propagation Constant

We can express the propagation constant of a parallel superconducting line as

$$\begin{aligned}\gamma = \sqrt{\mathbf{Z}\mathbf{Y}} &= \sqrt{j\omega\mu_0 \frac{S}{w}\left(\mu_{rd} + \frac{2d}{S}\right) j\omega\varepsilon_d \frac{w}{S}} \\ &= j\omega\sqrt{\mu_0\varepsilon_0}\left(\mu_{rd}\varepsilon_{rd} + \frac{2\varepsilon_{rd}d}{S}\right)^{1/2}\end{aligned} \tag{4-5-9}$$

Equation (4-5-9) is based on the assumption that the conductance $G = 0$. Then the real part of the propagation constant vanishes, and the attenuation loss of a parallel superconducting line is negligible.

EXAMPLE 4-5-1 Characteristics of a Superconducting Line

An aluminum superconducting line has the following parameters:

Relative dielectric constant	$\varepsilon_{rd} = 10$
Superconductor thickness	$t = 5$ mils, or 0.127 mm
Superconductor width	$w = 50$ mils, or 1.270 mm
Superconductor penetration depth	$d = 500$ Å
Separation between two superconductors	$S = 21$ mils, or 0.5334 mm
Operating frequency	$f = 1$ GHz

Calculate:

a) the surface impedance;
b) the characteristic impedance; and
c) the propagation constant.

Solutions:

a) From Eq. (4-5-3) the surface impedance is

$$\mathbf{Z}_s = \frac{j\omega\mu_0 d}{w} \coth\left(\frac{t}{d}\right) = j\omega \frac{(4\pi \times 10^{-7})(500 \times 10^{-10})}{1.27 \times 10^{-3}} (1)$$

$$= j4.95 \times 10^{-11} \times 2\pi \times 10^9$$

$$= j0.311\ \Omega/\text{m}$$

b) Using Eq. (4-5-8), we obtain the characteristic impedance of the line:

$$\mathbf{Z}_0 = \frac{120\pi(0.5334)}{\sqrt{10}(1.270)} \left(1 + \frac{2(500 \times 10^{-10})}{533.4 \times 10^{-6}}\right)^{1/2}$$

$$= 50\ \Omega$$

c) We calculate the propagation constant from Eq. (4-5-9):

$$\gamma = \frac{j\omega}{3 \times 10^8} \left[10 + \frac{(2 \times 10)(500 \times 10^{-10})}{533.4 \times 10^{-6}}\right]^{1/2}$$

$$= j1.05 \times 10^{-8} \times 2\pi \times 10^9 = j65.97$$

and

$$\beta = 65.97 \text{ rad/m}$$

4-6 OPTICAL FIBER TRANSMISSION LINE

Optical fiber is one of the latest innovations for signal links in digital transmission technology. Transmission loss is extremely low, and major applications are in computer links, industrial automation, medical instruments, telecommunications, and military command systems. This type of transmission line uses the binary digital light signal to switch the circuit on or off. The information signal is then reproduced at the output port in binary digital form of 0 or 1. We will discuss optical fibers in Chapter 7.

REFERENCE

[1] Matick, R. E., *Transmission Lines for Digital and Communication Networks*. New York; McGraw-Hill, 1969, pp. 235, 242.

Problems

4-1 A lossless coaxial line has a characteristic impedance of 75 Ω and terminates in its characteristic impedance. The length of the line is ℓ. At $t = 0$, the sending end is energized by a generator of 20 V (rms) with an internal resistance of 75 Ω. The time required for a wave to travel the length of the line is T. Sketch

a) a space–time diagram of the reflections;

b) the sending-end voltage and current as functions of time; and
c) the receiving-end voltage and current as functions of time.

4-2 Repeat Problem 4-1 for a short-circuited load. Show that $I_r = V_g Y_0$.

4-3 A copper coaxial line has the following parameters:

Copper conductivity	$\sigma = 5.8 \times 10^7$ ℧/m
Relative dielectric constant	$\varepsilon_r = 2.25$
Frequency	$f = 1$ GHz

Calculate the

a) loss tangent;
b) attenuation and phase constants; and
c) phase velocity.

4-4 A copper line is operating at a frequency of 10 GHz. Determine the

a) skin depth;
b) surface resistance;
c) phase velocity; and
d) propagation constant.

4-5 A superconducting line has a characteristic impedance of 50 Ω and terminates in a short circuit. The line length is ℓ and a d-c source of 10 V (rms) is applied to the line by a generator at $t = 0$. The generator has an output impedance of 75 Ω, and the time required for the 10-V wave to travel the length of the line is T. Sketch

a) a space–time diagram of reflections;
b) the sending-end voltage and current as functions of time; and
c) the receiving-end voltage and current as functions of time.

4-6 A distortionless digital line is determined by the ratios of R, L, G, and C, as shown in Eq. (4-2-2). In order to eliminate the distortion in a superconducting line, it is necessary to adjust the L/C ratio to equal the R/G ratio. If the R/G ratio is 10^6, determine the value of L.

4-7 A tin superconducting line has the following parameters:

Relative dielectric constant of beryllia	$\varepsilon_{rd} = 6$
Superconducting thickness	$t = 0.25$ mm
Superconducting width	$w = 3.10$ mm
Superconductor penetration depth	$d = 510$ Å
Separation between two superconductors	$S = 1$ mm
Operating frequency	$f = 1$ GHz

Determine the

a) surface impedance;
b) characteristic impedance; and
c) propagation constant.

4-8 An indium superconducting line has the following parameters:

Relative dielectric constant of polystyrene	$\varepsilon_{rd} = 2.56$

Superconductor thickness	$t = 0.30$ mm
Superconductor width	$w = 7.07$ mm
Superconductor penetration depth	$d = 640$ Å
Separation between two superconductors	$S = 150$ mm
Operating frequency	$f = 1$ GHz

Calculate the

a) characteristic impedance of the line;
b) surface resistance of the superconductor; and
c) propagation constant.

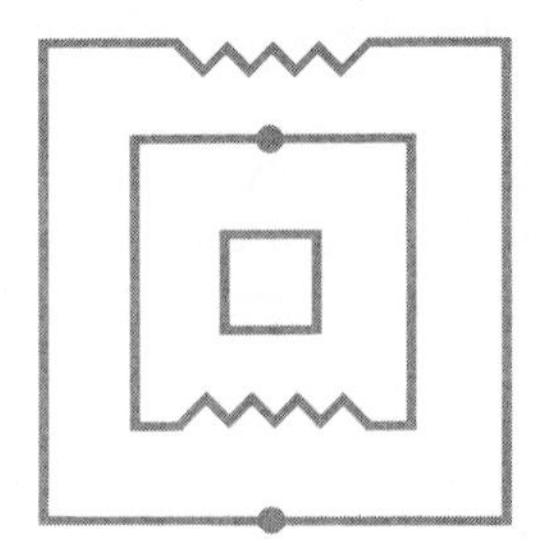

PART TWO

WAVEGUIDES

A waveguide commonly consists of a hollow metallic or dielectric tube of rectangular or circular shape used to guide an electromagnetic wave. Waveguides are used principally at frequencies in the microwave range; inconveniently large guides would be required to transmit RF power at the longer wavelengths.

In waveguides the electric and magnetic fields are confined to the space within the guides. Thus no power is lost through radiation, and even the dielectric loss is negligible, since the guides are normally air-filled. However, some power is lost as heat in the walls of the guides, but the loss is very small.

Several modes of electromagnetic waves can be propagated within a waveguide. These modes correspond to solutions of Maxwell's equations for the particular waveguides. A waveguide has a definite cutoff frequency for each allowed mode. If the frequency of the impressed signal is above the cutoff frequency for the mode, electromagnetic energy can be transmitted through the guide for that particular mode without attenuation. Otherwise, electromagnetic energy with a frequency below the cutoff frequency for that particular mode will be attenuated to a negligible value in a relatively short distance. The dominant mode in a particular guide is the mode having the lowest cutoff frequency. The dimensions of a guide should be chosen in such a way that, for a given input signal, only the energy of the dominant mode can be transmitted through the guide.

The process of solving waveguide problems consists of three steps.

Step 1: The desired wave equations are written in the form of either rectangular or cylindrical coordinate systems, whichever is suitable to the problem at hand.

Step 2: The boundary conditions are then applied to the wave equations obtained in step 1.

Step 3: The resultant equations usually are in the form of partial differential equations in either the time domain or the frequency domain. They can be solved by using the proper mathematical method.

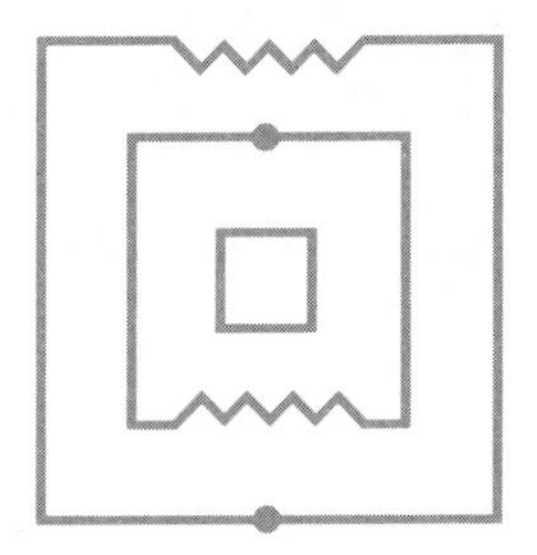

Chapter 5

Rectangular Waveguides

5-0 INTRODUCTION

A rectangular waveguide is a hollow metallic tube with a rectangular cross-section. The conducting walls of the guide confine the electromagnetic fields and thereby guide the electromagnetic wave. A number of distinct field configurations or modes can exist in waveguides. When the waves travel longitudinally down the guide, the plane waves are reflected from wall to wall. This condition produces a component of either an electric or a magnetic field in the direction of propagation of the resultant wave; therefore the wave is no longer a transverse electromagnetic (TEM) wave. Figure 5-0-1 shows that any uniform plane wave in a lossless guide may be resolved into TE and TM waves.

When the wavelength λ is in the direction of propagation of the incident wave, one component, λ_n, will be in the direction normal to the reflection plane, and another, λ_p, will be parallel to the plane. These components are

$$\lambda_n = \frac{\lambda}{\cos \theta} \tag{5-0-1}$$

and

$$\lambda_p = \frac{\lambda}{\sin \theta} \tag{5-0-2}$$

where θ is the angle of incidence, and λ is the wavelength of the impressed signal in an unbounded medium.

A plane wave in a waveguide resolves into two components: one standing wave in

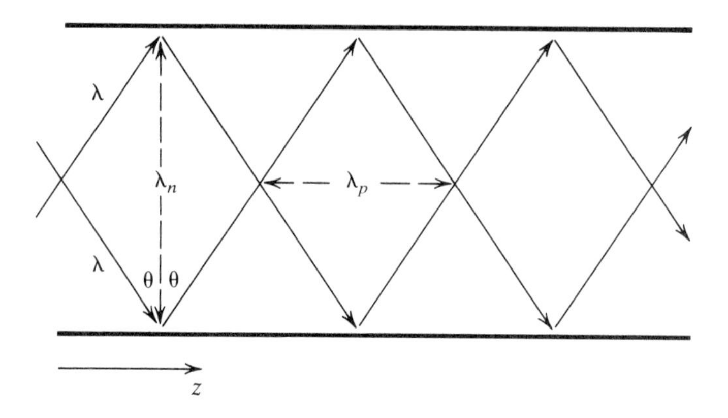

Figure 5-0-1 Plane wave reflected in a waveguide.

the direction normal to the reflecting walls of the guide; and one traveling wave in the direction parallel to the reflecting walls. In lossless waveguides the modes are classified as either transverse electric (TE) or transverse magnetic (TM). In rectangular guides the modes are designated TE_{mn} or TM_{mn}. The m denotes the number of half-waves of electric or magnetic intensity in the x direction, and n denotes the number of half-waves in the y direction, if we assume that the propagation of the wave is in the positive z direction.

In this chapter you will study wave modes, power transmission, and power losses in rectangular waveguides.

5-1 WAVE EQUATIONS IN RECTANGULAR COORDINATES

As we have stated previously, each wave equation has both a time-domain and a frequency-domain solution. However, to simplify the solution of wave equations in three dimensions—plus a time variable—we present only the sinusoidal steady-state, or frequency-domain, solution. We begin with a rectangular coordinate system as shown in Fig. 5-1-1.

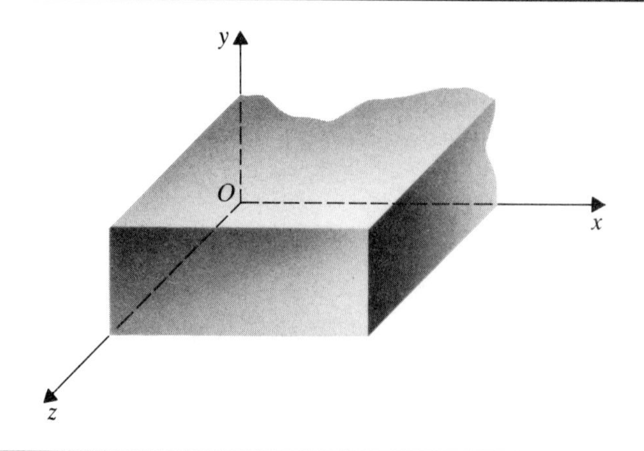

Figure 5-1-1 Rectangular coordinates.

Recall that the electric and magnetic wave equations in the frequency domain, Eqs. (0-3-27) and (0-3-28), are

$$\nabla^2 \mathbf{E} = \gamma^2 \mathbf{E} \tag{5-1-1}$$

and

$$\nabla^2 \mathbf{H} = \gamma^2 \mathbf{H} \tag{5-1-2}$$

where $\gamma = \sqrt{j\omega\mu(\sigma + j\omega\varepsilon)} = \alpha + j\beta$. These two equations are called the *vector-wave equations.*

The rectangular coordinates follow the usual right-hand system. The rectangular components of **E** or **H** satisfy the complex scalar-wave, or Helmholtz, equation:

$$\nabla^2 \psi = \gamma^2 \psi \tag{5-1-3}$$

The Helmholtz equation expressed in rectangular coordinates is

$$\frac{\partial^2 \psi}{\partial x^2} + \frac{\partial^2 \psi}{\partial y^2} + \frac{\partial^2 \psi}{\partial z^2} = \gamma^2 \psi \tag{5-1-4}$$

Equation (5-1-4) is a linear and nonhomogeneous partial differential equation in three dimensions. Using the method of separation of variables, we assume that the solution is in the form:

$$\psi = X(x)Y(y)Z(z) \tag{5-1-5}$$

where

$X(x)$ is a function of the x coordinate only;

$Y(y)$ is a function of the y coordinate only; and

$Z(z)$ is a function of the z coordinate only.

Substituting Eq. (5-1-5) into Eq. (5-1-4) and dividing the result by Eq. (5-1-5), we get

$$\frac{1}{X}\frac{d^2X}{dx^2} + \frac{1}{Y}\frac{d^2Y}{dy^2} + \frac{1}{Z}\frac{d^2Z}{dz^2} = \gamma^2 \tag{5-1-6}$$

Since the sum of the three terms on the left-hand side of Eq. (5-1-6) is a constant and each term is an independent variable, it follows that each term must be equal to a constant. If we let the three terms be k_x^2, k_y^2, and k_z^2, respectively, the separation equation becomes

$$-k_x^2 - k_y^2 - k_z^2 = \gamma^2 \tag{5-1-7}$$

Thus the general solution of each part of the differential equation, Eq. (5-1-6),

$$\frac{d^2X}{dx^2} = -k_x^2 X \tag{5-1-8}$$

$$\frac{d^2Y}{dy^2} = -k_y^2 Y \tag{5-1-9}$$

and

$$\frac{d^2Z}{dz^2} = -k_z^2 Z \tag{5-1-10}$$

will take the form:

$$X = A \sin (k_x x) + B \cos (k_x x) \tag{5-1-11}$$

$$Y = C \sin (k_y y) + D \cos (k_y y) \tag{5-1-12}$$

and

$$Z = E \sin (k_z z) + F \cos (k_z z) \tag{5-1-13}$$

And, finally, the total solution of the Helmholtz equation in rectangular coordinates is

$$\psi = [A \sin (k_x x) + B \cos (k_x x)][C \sin (k_y y) + D \cos (k_y y)][E \sin (k_z z) + F \cos (k_z z)] \tag{5-1-14}$$

As we noted previously, the convention for the propagation of the wave in the guide is in the positive z direction. However, we should note further that the propagation constant γ_g in the guide differs from the intrinsic propagation constant γ of the dielectric. Let

$$\gamma_g^2 = \gamma^2 + k_x^2 + k_y^2 = \gamma^2 + k_c^2 \tag{5-1-15}$$

where $k_c = \sqrt{k_x^2 + k_y^2}$ and is usually called the cutoff wave number. For a lossless dielectric, $\gamma^2 = -\omega^2 \mu\varepsilon$; then

$$\gamma_g = \pm j\sqrt{\omega^2 \mu\varepsilon - k_c^2} \tag{5-1-16}$$

The three cases for the propagation constant γ_g in the waveguide are:

Case 1—There will be no wave propagation (evanescence) in the guide if $\omega_c^2 \mu\varepsilon = k_c^2$ and $\gamma_g = 0$. This is the critical condition for cutoff of propagation. The cutoff frequency is

$$f_c = \frac{1}{2\pi\sqrt{\mu\varepsilon}}\sqrt{k_x^2 + k_y^2} \tag{5-1-17}$$

Case 2—The wave will be propagating in the guide if $\omega^2 \mu\varepsilon > k_c^2$ and

$$\gamma_g = \pm j\beta_g = \pm j\omega\sqrt{\mu\varepsilon}\sqrt{1 - \left(\frac{f_c}{f}\right)^2} \tag{5-1-18}$$

This means that the operating frequency must be above the cutoff frequency in order for a wave to propagate in the guide.

Case 3—The wave will be attenuated if $\omega^2 \mu\varepsilon < k_c^2$ and

$$\gamma_g = \pm \alpha_g = \pm \omega\sqrt{\mu\varepsilon}\sqrt{\left(\frac{f_c}{f}\right)^2 - 1} \tag{5-1-19}$$

This means that if the operating frequency is below the cutoff frequency, the wave will decay exponentially with respect to a factor of $-\alpha_g z$, and there will be no wave propagation because the propagation constant is a real quantity.

Therefore the solution to the Helmholtz equation in rectangular coordinates is

$$\psi = [A \sin (k_x x) + B \cos (k_x x)][C \sin (k_y y) + D \cos (k_y y)]e^{-j\beta_g z} \tag{5-1-20}$$

5-2 TM MODES IN RECTANGULAR WAVEGUIDES

The TM_{mn} modes in a rectangular guide are characterized by $\mathbf{H}_z = 0$. In other words, the z component of the electric field **E** must exist in order for energy to be transmitted in the guide. Consequently, the Helmholtz equation for **E** in the rectangular coordinates is

$$\nabla^2 E_z = \gamma^2 E_z \tag{5-2-1}$$

We can state one solution of the Helmholtz equation as

$$E_z = [A_m \sin(k_x x) + B_m \cos(k_x x)][C_n \sin(k_y y) + D_n \cos(k_y y)]e^{-j\beta_g z} \tag{5-2-2}$$

which we must determine according to the boundary conditions. Figure 5-2-1 shows the coordinates of a rectangular waveguide for TM modes.

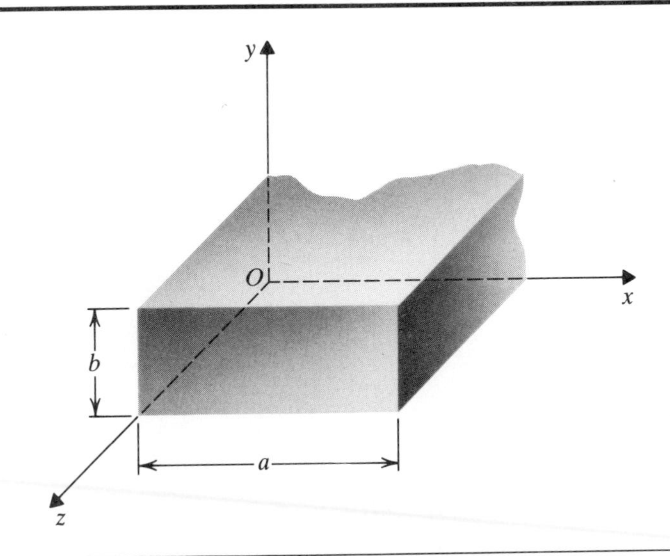

Figure 5-2-1 Coordinates of rectangular guide for TM modes.

The boundary conditions on E_z require that the field vanish at the waveguide walls, since the electric field E_z is zero on the conducting surface. That is, for $E_z = 0$ at $x = 0$ and a,

$$B_m = 0 \quad \text{and} \quad k_x a = m\pi \quad \text{or} \quad k_x = \frac{m\pi}{a}$$

and for $E_z = 0$ at $y = 0$ and b,

$$D_n = 0 \quad \text{and} \quad k_y b = n\pi \quad \text{or} \quad k_y = \frac{n\pi}{b}$$

Thus our solution in Eq. (5-2-2) reduces to

$$E_z = E_{0z} \sin\left(\frac{m\pi x}{a}\right) \sin\left(\frac{n\pi y}{b}\right) e^{-j\beta_g z} \tag{5-2-2a}$$

where

$$m = 1, 2, 3, \ldots,$$

and

$$n = 1, 2, 3, \ldots.$$

If $m = 0$ or $n = 0$ or $m = n = 0$, the field intensities all vanish: There is no TM_{01}, TM_{10}, or TM_{00} mode in a rectangular waveguide.

For a lossless dielectric, Maxwell's curl equations in frequency domain are

$$\nabla \times \mathbf{E} = -j\omega\mu\mathbf{H} \tag{5-2-3}$$

and

$$\nabla \times \mathbf{H} = j\omega\varepsilon\mathbf{E} \tag{5-2-4}$$

In rectangular coordinates, their components are

$$\frac{\partial E_z}{\partial y} - \frac{\partial E_y}{\partial z} = -j\omega\mu H_x \tag{5-2-5}$$

$$\frac{\partial E_x}{\partial z} - \frac{\partial E_z}{\partial x} = -j\omega\mu H_y \tag{5-2-6}$$

$$\frac{\partial E_y}{\partial x} - \frac{\partial E_x}{\partial y} = -j\omega\mu H_z \tag{5-2-7}$$

$$\frac{\partial H_z}{\partial y} - \frac{\partial H_y}{\partial z} = j\omega\varepsilon E_x \tag{5-2-8}$$

$$\frac{\partial H_x}{\partial z} - \frac{\partial H_z}{\partial x} = j\omega\varepsilon E_y \tag{5-2-9}$$

and

$$\frac{\partial H_y}{\partial x} - \frac{\partial H_x}{\partial y} = j\omega\varepsilon E_z \tag{5-2-10}$$

By substituting $\frac{\partial}{\partial z} = -j\beta_g$ and $H_z = 0$, into the preceding equations, we can simplify them as follows:

$$\frac{\partial E_z}{\partial y} + j\beta_g E_y = -j\omega\mu H_x \tag{5-2-11}$$

$$j\beta_g E_x + \frac{\partial E_z}{\partial x} = j\omega\mu H_y \tag{5-2-12}$$

$$\frac{\partial E_y}{\partial x} - \frac{\partial E_x}{\partial y} = 0 \tag{5-2-13}$$

$$\beta_g H_y = \omega\varepsilon E_x \tag{5-2-14}$$

$$-\beta_g H_x = \omega\varepsilon E_y \tag{5-2-15}$$

and

$$\frac{\partial H_y}{\partial x} - \frac{\partial H_x}{\partial y} = j\omega\varepsilon E_z \tag{5-2-16}$$

We can solve Eqs. (5-2-11)–(5-2-16) simultaneously for E_x, E_y, H_x, and H_y in terms of E_z. The resultant field equations for TM modes are

$$E_x = \frac{-j\beta_g}{k_c^2} \frac{\partial E_z}{\partial x} \tag{5-2-17}$$

$$E_y = \frac{-j\beta_g}{k_c^2} \frac{\partial E_z}{\partial y} \tag{5-2-18}$$

$$E_z = E_{0z} \sin\left(\frac{m\pi x}{a}\right) \sin\left(\frac{n\pi y}{b}\right) e^{-j\beta_g z} \qquad [(5\text{-}2\text{-}2a)]$$

$$H_x = \frac{j\omega\varepsilon}{k_c^2} \frac{\partial E_z}{\partial y} \tag{5-2-19}$$

$$H_y = \frac{-j\omega\varepsilon}{k_c^2} \frac{\partial E_z}{\partial x} \tag{5-2-20}$$

and

$$H_z = 0 \tag{5-2-21}$$

where $\beta_g^2 - \omega^2\mu\varepsilon = -k_c^2$. Differentiating Eq. (5-2-2a) with respect to x or y and substituting the results into Eqs. (5-2-17)–(5-2-21), we obtain a new set of TM_{mn}-mode field equations in a rectangular waveguide:

$$E_x = E_{0x} \cos\left(\frac{m\pi x}{a}\right) \sin\left(\frac{n\pi y}{b}\right) e^{-j\beta_g z} \tag{5-2-22}$$

$$E_y = E_{0y} \sin\left(\frac{m\pi x}{a}\right) \cos\left(\frac{n\pi y}{b}\right) e^{-j\beta_g z} \tag{5-2-23}$$

$$E_z = E_{0z} \sin\left(\frac{m\pi x}{a}\right) \sin\left(\frac{n\pi y}{b}\right) e^{-j\beta_g z} \qquad [(5\text{-}2\text{-}2a)]$$

$$H_x = H_{0x} \sin\left(\frac{m\pi x}{a}\right) \cos\left(\frac{n\pi y}{b}\right) e^{-j\beta_g z} \tag{5-2-24}$$

$$H_y = H_{0y} \cos\left(\frac{m\pi x}{a}\right) \sin\left(\frac{n\pi y}{b}\right) e^{-j\beta_g z} \tag{5-2-25}$$

and

$$H_z = 0 \tag{5-2-26}$$

We can derive the characteristic equations for TM modes in a rectangular waveguide in the following manner. The cutoff wave number k_c, as defined by Eq. (5-1-15) for the TM_{mn} modes, is

$$k_c = \sqrt{\left(\frac{m\pi}{a}\right)^2 + \left(\frac{n\pi}{b}\right)^2} = \omega_c\sqrt{\mu\varepsilon} \tag{5-2-27}$$

where a and b are in meters, and $\frac{m\pi}{a}$ and $\frac{n\pi}{b}$ were substituted for k_x and k_y, respectively.

The cutoff frequency, as defined in Eq. (5-1-17) for the TM_{mn} modes, is

$$f_c = \frac{1}{2\sqrt{\mu\varepsilon}} \sqrt{\frac{m^2}{a^2} + \frac{n^2}{b^2}} \tag{5-2-28}$$

The propagation constant, as defined in Eq. (5-1-16), becomes β_g for $\alpha_g = 0$ and is

$$\beta_g = \omega\sqrt{\mu\varepsilon}\sqrt{1 - \left(\frac{f_c}{f}\right)^2} \tag{5-2-29}$$

The phase velocity in the positive z direction for the TM_{mn} modes is

$$v_g = \frac{\omega}{\beta_g} = \frac{v_p}{\sqrt{1 - (f_c/f)^2}} \tag{5-2-30}$$

where $v_p = 1/\sqrt{\mu\varepsilon}$ and is the phase velocity in an unbounded dielectric.

The characteristic wave impedance of TM_{mn} modes in the guide, from Eqs. (5-2-14) or (5-2-15), is

$$Z_g = \frac{\beta_g}{\omega\varepsilon} = \eta\sqrt{1 - (f_c/f)^2} \tag{5-2-31}$$

where $\eta = \sqrt{\mu/\varepsilon}$ and is the intrinsic impedance in an unbounded dielectric. The wavelength λ_g in the guide for the TM_{mn} modes is

$$\lambda_g = \frac{\lambda}{\sqrt{1 - (f_c/f)^2}} \tag{5-2-32}$$

where $\lambda = v_p/f$ is the wavelength in an unbounded dielectric.

Equation (5-2-31) defines the wave impedance or field impedance of TM_{mn} modes in a rectangular waveguide. However, in practical engineering design, the characteristic impedance of the TM-mode waveguide is desirable for impedance matching. The characteristic impedance of a rectangular waveguide is defined as the ratio of the voltage over current in the guide. For example, the characteristic impedance of the TM_{mn}-mode rectangular waveguide can be expressed as

$$Z_{0g} = \frac{V}{I} = \frac{-bE_y}{aH_x} = \frac{b\eta}{a}\sqrt{1 - \left(\frac{f_c}{f}\right)^2} \tag{5-2-33}$$

EXAMPLE 5-2-1 TM-mode Wave Propagation in a Rectangular Waveguide

An air-filled rectangular waveguide is used to propagate a TM-mode wave at 20 GHz and the guide inside dimensions are $a = 2.286$ cm and $b = 1.016$ cm.

Calculate:

a) the lowest TM mode;
b) the cutoff frequency;
c) the phase constant;
d) the wave impedance in the guide; and
e) the phase velocity.

Solutions:

a) The lowest TM mode is TM_{11}.
b) The cutoff frequency is

$$f_c = \frac{3 \times 10^8}{2}\sqrt{\left(\frac{1}{2.286 \times 10^{-2}}\right)^2 + \left(\frac{1}{1.016 \times 10^{-2}}\right)^2}$$

$$= 16.2 \text{ GHz}$$

c) The phase constant is

$$\beta_g = \omega\sqrt{\mu_0\varepsilon_0}\sqrt{1-\left(\frac{f_c}{f}\right)^2} = \frac{2\pi \times 20 \times 10^9}{3 \times 10^8}\sqrt{1-\left(\frac{16.2}{20}\right)^2}$$

$$= 245.90 \text{ rad/m}$$

d) The wave impedance in the guide is

$$Z_g = \frac{\beta_g}{\omega\varepsilon_0} = \sqrt{\frac{\mu_0}{\varepsilon_0}}\sqrt{1-\left(\frac{f_c}{f}\right)^2} = 377\sqrt{1-\left(\frac{16.2}{20}\right)^2}$$

$$= 221.30\ \Omega$$

e) The phase velocity is

$$v_g = \frac{\omega}{\beta_g} = \frac{2\pi \times 20 \times 10^9}{245.90}$$

$$= 5.11 \times 10^8 \text{ m/s}$$

5-3 TE MODES IN RECTANGULAR WAVEGUIDES

We have previously assumed that waves are propagating in the positive z direction in a waveguide. The TE_{mn} modes in a rectangular guide are characterized by $E_z = 0$. In other words, the z component of the magnetic field H_z must exist in order for energy to be transmitted in the guide. Consequently, the Helmholtz equation for H in rectangular coordinates is

$$\nabla^2 H_z = \gamma^2 H_z \tag{5-3-1}$$

We can state one solution of the Helmholtz equation in the form:

$$H_z = \left[A_m \sin\left(\frac{m\pi x}{a}\right) + B_m \cos\left(\frac{m\pi x}{a}\right)\right]\left[C_n \sin\left(\frac{n\pi y}{b}\right) + D_n \cos\left(\frac{n\pi y}{b}\right)\right]e^{-j\beta_g z} \tag{5-3-2}$$

which we must determine according to the boundary conditions. Figure 5-3-1 shows the coordinates of a rectangular waveguide for TE modes.

Since $E_x = 0$, $\partial H_z/\partial y = 0$ at $y = (0, b)$; hence, $C_n = 0$. Since $E_y = 0$, $\partial H_z/\partial x = 0$ at $x = (0, a)$; hence, $A_m = 0$. The general conclusion is that the normal derivative of H_z must vanish at the conducting surfaces; that is,

$$\frac{\partial H_z}{\partial n} = 0 \qquad \text{at guide walls} \tag{5-3-3}$$

Therefore the magnetic field in the positive z direction is

$$H_z = H_{0z} \cos\left(\frac{m\pi x}{a}\right)\cos\left(\frac{n\pi y}{b}\right)e^{-j\beta_g z} \tag{5-3-4}$$

where H_{0z} is the amplitude constant.

Expanding the two curl equations, as shown in Eqs. (5-2-3) and (5-2-4), and

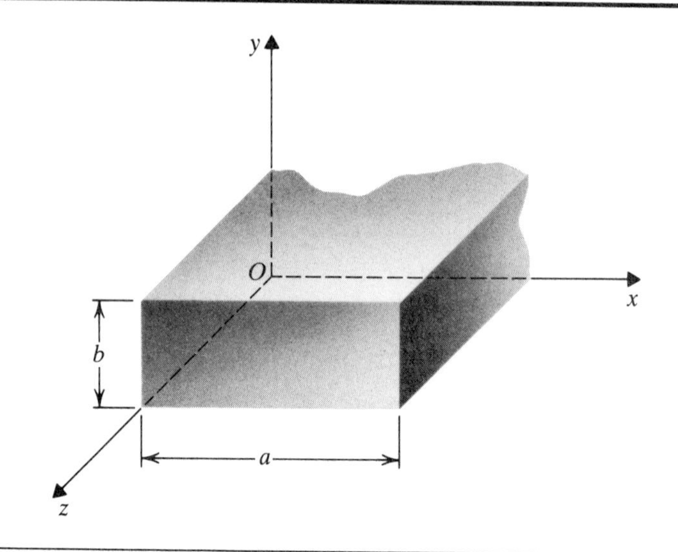

Figure 5-3-1 Coordinates of rectangular guide for TE modes.

substituting $\partial/\partial z = -j\beta_g$ and $E_z = 0$, yield the six field equations:

$$\beta_g E_y = -\omega\mu H_x \tag{5-3-5}$$

$$\beta_g E_x = \omega\mu H_y \tag{5-3-6}$$

$$\frac{\partial E_y}{\partial x} - \frac{\partial E_x}{\partial y} = -j\omega\mu H_z \tag{5-3-7}$$

$$\frac{\partial H_z}{\partial y} + j\beta_g H_y = j\omega\varepsilon E_x \tag{5-3-8}$$

$$-j\beta_g H_x - \frac{\partial H_z}{\partial x} = j\omega\varepsilon E_y \tag{5-3-9}$$

and

$$\frac{\partial H_y}{\partial x} - \frac{\partial H_x}{\partial y} = 0 \tag{5-3-10}$$

We can solve Eqs. (5-3-5)–(5-3-10) for E_x, E_y, H_x, and H_y in terms of H_z to obtain the TE-mode field equations:

$$E_x = \frac{-j\omega\mu}{k_c^2}\frac{\partial H_z}{\partial y} \tag{5-3-11}$$

$$E_y = \frac{j\omega\mu}{k_c^2}\frac{\partial H_z}{\partial x} \tag{5-3-12}$$

$$E_z = 0 \tag{5-3-13}$$

$$H_x = \frac{-j\beta_g}{k_c^2}\frac{\partial H_z}{\partial x} \tag{5-3-14}$$

$$H_y = \frac{-j\beta_g}{k_c^2}\frac{\partial H_z}{\partial y} \tag{5-3-15}$$

and

$$H_z = H_{0z} \cos\left(\frac{m\pi x}{a}\right) \cos\left(\frac{n\pi y}{b}\right) e^{-j\beta_g z} \qquad [(5\text{-}3\text{-}4)]$$

where $k_c^2 = \omega^2\mu\varepsilon - \beta_g^2$ has been replaced. Differentiating Eq. (5-3-4) with respect to x and y and then substituting the results into Eqs. (5-3-11)–(5-3-15), we obtain a new set of TE_{mn}-mode field equations in a rectangular waveguide:

$$E_x = E_{0x} \cos\left(\frac{m\pi x}{a}\right) \sin\left(\frac{n\pi y}{b}\right) e^{-j\beta_g z} \qquad (5\text{-}3\text{-}16)$$

$$E_y = E_{0y} \sin\left(\frac{m\pi x}{a}\right) \cos\left(\frac{n\pi y}{b}\right) e^{-j\beta_g z} \qquad (5\text{-}3\text{-}17)$$

$$E_z = 0 \qquad (5\text{-}3\text{-}18)$$

$$H_x = H_{0x} \sin\left(\frac{m\pi x}{a}\right) \cos\left(\frac{n\pi y}{b}\right) e^{-j\beta_g z} \qquad (5\text{-}3\text{-}19)$$

$$H_y = H_{0y} \cos\left(\frac{m\pi x}{a}\right) \sin\left(\frac{n\pi y}{b}\right) e^{-j\beta_g z} \qquad (5\text{-}3\text{-}20)$$

and

$$H_z = H_{0z} \cos\left(\frac{m\pi x}{a}\right) \cos\left(\frac{n\pi y}{b}\right) e^{-j\beta_g z} \qquad [(5\text{-}3\text{-}4)]$$

where

$m = 0, 1, 2, \ldots,$
$n = 0, 1, 2, \ldots,$ and
$m = n = 0$ excepted.

Some TE-mode–characteristic equations are identical to those of the TM modes, but some are different. For convenience, all are shown, as follows:

Cutoff frequency
$$f_c = \frac{1}{2\sqrt{\mu\varepsilon}} \sqrt{\frac{m^2}{a^2} + \frac{n^2}{b^2}} \qquad (5\text{-}3\text{-}21)$$

Phase constant
$$\beta_g = \omega\sqrt{\mu\varepsilon} \sqrt{1 - \left(\frac{f_c}{f}\right)^2} \qquad (5\text{-}3\text{-}22)$$

Wavelength
$$\lambda_g = \frac{\lambda}{\sqrt{1 - \left(\frac{f_c}{f}\right)^2}} \qquad (5\text{-}3\text{-}23)$$

Phase velocity
$$v_g = \frac{v_p}{\sqrt{1 - \left(\frac{f_c}{f}\right)^2}} \qquad (5\text{-}3\text{-}24)$$

Wave impedance
$$Z_g = \frac{E_x}{H_y} = -\frac{E_y}{H_x} = \frac{\omega\mu}{\beta_g} = \frac{\eta}{\sqrt{1 - \left(\frac{f_c}{f}\right)^2}} \qquad (5\text{-}3\text{-}25)$$

Table 5-3-1 Modes of $\frac{(f_c)_{mn}}{f_c}$ for $a \geq b$

a/b \ Modes f/f_{10}	TE_{10}	TE_{01}	TE_{11}, TM_{11}	TE_{20}	TE_{02}	TE_{21}, TM_{21}	TE_{12}, TM_{12}	TE_{22}, TM_{22}	TE_{30}
1	1	1	1.414	2	2	2.236	2.236	2.838	3
1.5	1	1.5	1.803	2	3	2.500	3.162	3.606	3
2	1	2	2.236	2	4	2.828	4.123	4.472	3
3	1	3	3.162	2	6	3.606	6.083	6.325	3
∞	1	∞	∞	2	∞	∞	∞	∞	3

Characteristic impedance $$Z_{0g} = \frac{b}{a} \frac{\eta}{\sqrt{1 - \left(\frac{f_c}{f}\right)^2}} \tag{5-3-26}$$

The cutoff frequency shown in Eq. (5-3-21) is a function of the modes and guide dimensions, so the physical size of the waveguide will determine the propagation of the modes. Table 5-3-1 shows the ratio of cutoff frequency of some modes with respect to that of the dominant mode in physical dimension terms.

In a rectangular guide the corresponding TE_{mn} and TM_{mn} modes are always degenerate. In a square guide the TE_{mn}, TE_{nm}, TM_{mn}, and TM_{nm} modes form a foursome of degeneracy. Rectangular guides are ordinarily of a size that only one mode will propagate, and their dimensions have a ratio of $a = 2b$. The mode with the lowest cutoff frequency in a particular guide is called the dominant mode. The dominant mode in a rectangular guide with $a > b$ is the TE_{10} mode. Each mode has a specific mode pattern (or field pattern).

Normally, all modes exist simultaneously in a given waveguide, and the situation is not very serious. Actually, only the dominant mode propagates, and the higher modes near the sources or discontinuities decay very fast.

EXAMPLE 5-3-1 Wave Propagation in Rectangular Waveguide

An air-filled rectangular waveguide has inside dimensions of 7 cm × 3.5 cm, as shown in Fig. 5-3-2.

Determine:

a) the dominant mode;
b) the cutoff frequency;
c) the phase velocity of the guided wave in the guide at a frequency of 3.5 GHz; and
d) the guide wavelength at the same frequency.

Solutions:

a) The dominant mode is the TE_{10} mode.

b) $$f_c = \frac{c}{2b} = \frac{3 \times 10^8}{2(7 \times 10^{-2})}$$

$$= 2.14 \text{ GHz}$$

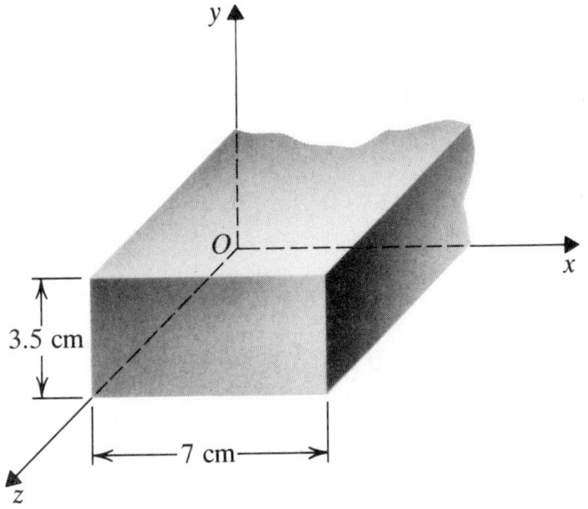

Figure 5-3-2 Wave propagation in a rectangular waveguide for Example 5-3-1.

c) $$v_g = \frac{c}{\sqrt{1-(f_c/f)^2}} = \frac{3 \times 10^8}{\sqrt{1-(2.14/3.5)^2}}$$

$$= 3.78 \times 10^8 \text{ m/s}$$

d) $$\lambda_g = \frac{\lambda_0}{\sqrt{1-(f_c/f)^2}} = \frac{3 \times 10^8/(3.5 \times 10^9)}{\sqrt{1-(2.14/3.5)^2}}$$

$$= 10.8 \text{ cm}$$

EXAMPLE 5-3-2 Computer Solution for TE_{01} Mode in Rectangular Waveguide

```
PROGRAM    GUIDE
010C       PROGRAM TO COMPUTE CUTOFF FREQUENCY, WAVE
020C+      VELOCITY AND WAVELENGTH FOR TE01 MODE IN
030C+      A RECTANGULAR WAVEGUIDE
040C       GUIDE
045        REAL LAMDAO,LAMDAG
050      1 READ 2, A,B,F
060        IF(A .LE. 0.0) STOP
070        C=0.30000E+09
080        LAMDAO=C/F
090        FC=C/(2.0*B)
100        VG=C/SQRT(1.0-(FC/F)**2)
105        LAMDAG=LAMDAO/SQRT(1.0-(FC/F)**2)
110        PRINT 4, FC,VG,LAMDAG
120      2 FORMAT(2F10.5, E12.5)
130      4 FORMAT(//1H ,22HCUTOFF FREQUENCY FC = ,
140+       1X,E12.5,1X,5HHERTZ/
```

(continued)

EXAMPLE 5-3-2 *(continued)*

```
150+     1X,22HWAVE VELOCITY     VG = ,E12.5,1X,5HM/SEC/
160+     1X,22HWAVE LENGTH   LAMDAG = ,E12.5,1X,5HMETER/)
170      GO TO 1
180      STOP
190      END
READY.
RUN

PROGRAM    GUIDE
?     0.03500    0.07000 0.35000E+10

CUTOFF FREQUENCY FC =     .21429E+10 HERTZ
WAVE VELOCITY     VG =   .37943E+09 M/SEC
WAVE LENGTH   LAMDAG =   .10841E+00 METER
?     0.0
SRU        0.709 UNTS.
RUN COMPLETE.
BYE
F228F10    LOG OFF      17.05.55.
F228F10    SRU        5.779 UNTS.
```

5-4 POWER TRANSMISSION AND POWER LOSSES IN RECTANGULAR WAVEGUIDES

5-4-1 Power Transmission in Rectangular Guides

We can calculate the power transmitted through a waveguide and the power loss in the guide walls by means of the complex Poynting theorem. We assume that the guide is terminated in such a way that there is no reflection from the receiving end or that the guide is very long, compared to the wavelength. From the Poynting theorem in Section 0-4, the power transmitted through a guide is

$$P_{tr} = \oint_s \mathbf{p} \cdot d\mathbf{s} = \oint_s \tfrac{1}{2}(\mathbf{E} \times \mathbf{H}^*) \cdot d\mathbf{s} \tag{5-4-1}$$

For a lossless dielectric, the time-averaged power flow through a rectangular guide is

$$P_{tr} = \frac{1}{2Z_g} \int_a |E|^2 \, da = \frac{Z_g}{2} \int_a |H|^2 \, da \tag{5-4-2}$$

where

$Z_g = E_x/H_y = -E_y/H_x$;
$|E|^2 = |E_x|^2 + |E_y|^2$; and
$|H|^2 = |H_x|^2 + |H_y|^2$.

For TE_{mn} modes the average power transmitted through a rectangular waveguide is

$$P_{tr} = \frac{\sqrt{1 - (f_c/f)^2}}{2\eta} \int_0^b \int_0^a (|E_x|^2 + |E_y|^2)\, dx\, dy \tag{5-4-3}$$

and for TM_{mn} modes the average power transmitted through a rectangular waveguide is

$$P_{tr} = \frac{1}{2\eta\sqrt{1 - (f_c/f)^2}} \int_0^b \int_0^a (|E_x|^2 + |E_y|^2)\, dx\, dy \tag{5-4-4}$$

where $\eta = \sqrt{\mu/\varepsilon}$ and is the intrinsic impedance in an unbounded dielectric.

5-4-2 Power Losses in Rectangular Guide

The three types of power losses in a rectangular waveguide are:

1. Losses by the signal with frequency below the cutoff frequency.
2. Losses in the dielectric.
3. Losses in the guide walls.

1. *Power Losses Due to Signal Frequency*

As we described in Section 5-1, when the impressed frequency is below the cutoff frequency, the signal is attenuated exponentially with respect to $-\alpha_g z$, and nonpropagation (or evanescence) occurs because the propagation constant is a real value. Thus waveguides with dimensions much smaller than the cutoff are often used as attenuators. Recall that the attenuation constant α_g for the TE_{mn} and TM_{mn} modes is expressed in Eq. (5-1-19) as

$$\alpha_g = \omega\sqrt{\mu\varepsilon}\sqrt{\left(\frac{f_c}{f}\right)^2 - 1} \tag{5-4-4a}$$

2. *Power Losses Due to Dielectric Attenuation*

Recall that in a low-loss dielectric (that is, $\sigma \ll \omega\varepsilon$) the propagation constant for a plane wave traveling in an unbounded lossy dielectric is given in Eq. (0-5-72) by

$$\alpha = \frac{\sigma}{2}\sqrt{\frac{\mu}{\varepsilon}} = \frac{\eta\sigma}{2} \tag{5-4-5}$$

The attenuations resulting from the low-loss dielectric in the rectangular waveguide are

$$\alpha_g = \frac{\sigma\eta}{2\sqrt{1 - (f_c/f)^2}} \qquad \text{for } TE_{mn} \text{ mode} \tag{5-4-6}$$

and

$$\alpha_g = \frac{\sigma\eta}{2}\sqrt{1 - (f_c/f)^2} \qquad \text{for } TM_{mn} \text{ mode} \tag{5-4-7}$$

If $f \gg f_c$, the attenuation constant in the guide approaches that for unbounded dielectric in Eq. (5-4-5). However, if the operating frequency is far below the cutoff frequency, or $f \ll f_c$, the attenuation constant becomes very large and nonpropagation occurs.

3. *Power Losses Due to the Guide Walls*

When the electric and magnetic intensities propagate through a lossy waveguide, we can write their magnitudes as

$$|E| = |E_{0z}|e^{-\alpha_g z} \tag{5-4-8}$$

and

$$|H| = |H_{0z}|e^{-\alpha_g z} \tag{5-4-9}$$

where E_{0z} and H_{0z} are the field intensities at $z = 0$. Note that for a low-loss guide the time-average power flow decreases proportionally to $e^{-2\alpha_g z}$. Hence,

$$P_{tr} = (P_{tr} + P_{loss})e^{-2\alpha_g z} \tag{5-4-10}$$

For $P_{loss} \ll P_{tr}$ and $2\alpha_g z \ll 1$,

$$\frac{P_{loss}}{P_{tr}} + 1 \doteq 1 + 2\alpha_g z \tag{5-4-11}$$

Finally,

$$\alpha_g = \frac{P_L}{2P_{tr}} \tag{5-4-12}$$

where P_L is the power loss per unit length. Consequently, the attenuation constant of the guide walls equals the ratio of the power loss per unit length to twice the power transmitted through the guide.

Since the electric and magnetic field intensities established at the surface of a low-loss guide wall decay exponentially with respect to the skin depth while the waves progress into the walls, we should define a surface resistance of the guide walls as

$$R_s \equiv \frac{\rho}{\delta} = \frac{1}{\sigma\delta} = \frac{\alpha_g}{\sigma} = \sqrt{\frac{\pi f \mu}{\sigma}} \quad \Omega/\text{square} \tag{5-4-13}$$

where

ρ is the resistivity of the conducting wall in $\Omega \cdot$ m;

σ is the conductivity in ℧/m; and

δ is the skin depth or depth of penetration in m.

We obtain the power loss per unit length of guide by integrating the power density over the surface of the conductor corresponding to unit length of the guide, or

$$P_L = \frac{R_s}{2}\int_s |H_t|^2\, ds \quad \text{W/unit length} \tag{5-4-14}$$

where H_t is the tangential component of magnetic intensity at the guide walls. Substituting Eqs. (5-4-2) and (5-4-14) in Eq. (5-4-12) yields

$$\alpha_g = \frac{R_s \displaystyle\int_s |H_t|^2\, ds}{2Z_g \displaystyle\int_a |H|^2\, da} \tag{5-4-15}$$

where

$$|H|^2 = |H_x|^2 + |H_y|^2, \text{ and } |H_t|^2 = |H_{tx}|^2 + |H_{ty}|^2.$$

EXAMPLE 5-4-1 TE_{10} Mode in Rectangular Waveguide

An air-filled waveguide with a cross-section of 2 cm × 1 cm transports energy in the TE_{10} mode at the rate of 0.5 hp. The impressed frequency is 30 GHz. What is the peak value of the electric field occurring in the guide? (See Fig. 5-4-1.)

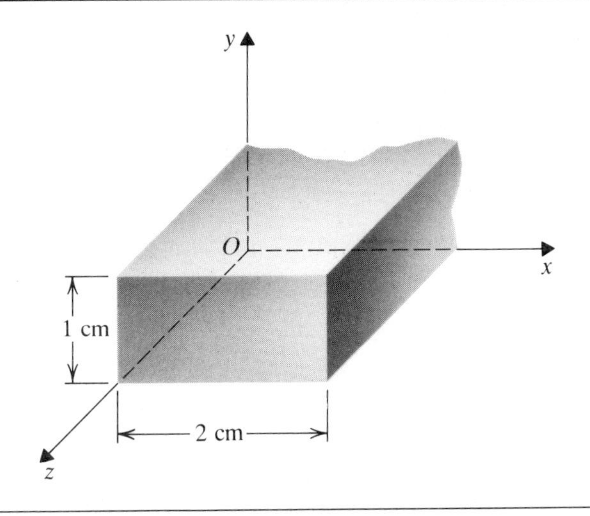

Figure 5-4-1 Rectangular waveguide for Example 5-4-1.

Solution:

The field components of the dominant mode TE_{10} may be obtained by substituting $m = 1$ and $n = 0$ in Eqs. (5-3-16)–(5-3-20) and (5-3-4). Then

$$E_x = 0 \qquad\qquad H_x = -\frac{E_{0y}}{Z_g}\sin\left(\frac{\pi x}{a}\right)e^{-j\beta_g z}$$

$$E_y = E_{0y}\sin\left(\frac{\pi x}{a}\right)e^{-j\beta_g z} \qquad\qquad H_y = 0$$

and and

$$E_z = 0 \qquad\qquad H_z = H_{0z}\cos\left(\frac{\pi x}{a}\right)e^{-j\beta_g z}$$

where $Z_g = \omega\mu_0/\beta_g$. We find the phase constant β_g from

$$\beta_g = \sqrt{\omega^2\mu_0\varepsilon_0 - \frac{\pi^2}{a^2}} = \pi\sqrt{\frac{(2f)^2}{c^2} - \frac{1}{a^2}}$$

$$= \pi\sqrt{\frac{4 \times 9 \times 10^{20}}{9 \times 10^{16}} - \frac{1}{4 \times 10^{-4}}}$$

$$= 193.5\pi, \text{ or } 608.81 \text{ rad/m}$$

The power delivered in the z direction by the guide is

$$P = \text{Re}\left[\tfrac{1}{2}\int_0^b\int_0^a (\mathbf{E}\times\mathbf{H}^*)\right]\cdot dx\,dy\,\mathbf{u}_z$$

$$= \tfrac{1}{2}\int_0^b\int_0^a \left\{\left[\mathbf{E}_{0y}\sin\left(\frac{\pi x}{a}\right)e^{-j\beta_g z}\mathbf{u}_y\right]\times\left[-\frac{\beta_g}{\omega\mu_0}\mathbf{E}_{0y}\sin\left(\frac{\pi x}{a}\right)e^{+j\beta_g z}\mathbf{u}_x\right]\right\}\cdot dx\,dy\,\mathbf{u}_z$$

$$= \tfrac{1}{2}E_{0y}^2\frac{\beta_g}{\omega\mu_0}\int_0^b\int_0^a\left\{\sin\frac{\pi x}{a}\right\}^2 dx\,dy$$

$$= \tfrac{1}{4}E_{0y}^2\frac{\beta_g}{\omega\mu_0}ab$$

Thus

$$373 = \frac{1}{4}E_{0y}^2\frac{193.5\pi\times 10^{-2}\times 2\times 10^{-2}}{2\pi\times 3\times 10^{10}\times 4\pi\times 10^{-7}}$$

and the peak value of electric intensity is

$$E_{0y} = 53.87\ \text{kV/m}$$

EXAMPLE 5-4-2 Computer Solution for Power Transmission in Rectangular Waveguide

```
PROGRAM   FIELD
010C      PROGRAM TO COMPUTE ELECTRIC FIELD INTENSITY
020C      FOR TE01 MODE IN A RECTANGULAR WAVEGUIDE
030C      FIELD
040     1 READ 2, A,B,PR,F
050       IF(A .LE. 0.0) STOP
060       PI=3.141593
070       W=2.0*PI*F
080       U0=4.0E-07*PI
090       E0=8.854E-12
100       BETAG=SQRT(W**2*U0*E0-(PI/B)**2)
110       E0X=SQRT(4.0*373.0*W*U0/(BETAG*A*B))
120       PRINT 4, E0X, BETAG
130     2 FORMAT(3F10.5, E12.5)
140     4 FORMAT(//1H ,20HPEAK VALUE OF E0X = ,F12.5,
150+      1X,15HVOLTS PER METER/
155+      1X,23HPHASE CONSTANT BETAG = ,F10.5,
157+      1X,13HRADIANS/METER/)
160       GO TO 1
170       STOP
180       END
READY.
RUN

PROGRAM   FIELD
?     0.01000   0.02000 373.00000 0.30000E+11
```

EXAMPLE 5-4-2 *(continued)*

```
 PEAK VALUE OF EOX =  53874.63424 VOLTS PER METER
 PHASE CONSTANT BETAG =  608.80925 RADIANS/METER
?    0.0
SRU      0.713 UNTS.
RUN COMPLETE.
BYE
F228F10    LOG OFF     18.47.07.
F228F10    SRU       1.000 UNTS.
```

5-5 EXCITATIONS OF MODES AND MODE PATTERNS IN RECTANGULAR WAVEGUIDES

5-5-1 Excitations of Modes in Rectangular Waveguides

In general, the field intensities of the desired mode in a waveguide can be established by means of a probe (monopole antenna) or loop-coupling device (loop antenna). A probe should be located so as to excite the electric field intensity of the mode and a coupling loop in such a way as to generate the magnetic field intensity for the desired mode. If two or more probes or loops are to be used, care must be taken to ensure the proper phase relationship between the currents in the various antennas. This relationship can be achieved by inserting additional lengths of transmission line in one or more of the antenna feeders. Impedance matching can be accomplished by varying the position and depth of the antenna in the guide or by using impedance matching stubs on the coaxial line feeding the waveguide. A device that excites a mode in the guide can also serve reciprocally as a receiver or collector of energy for that mode. The methods of excitation for various modes in rectangular waveguides are shown in Fig. 5-5-1.

In order to excite a TE_{10} mode in one direction of the guide, the two exciting antennas should be arranged in such a way that the field intensities cancel each other in one direction and reinforce in the other direction. Figure 5-5-2 shows an arrangement for launching a TE_{10} mode in one direction only. The two antennas are placed a quarter-wavelength apart and their phases are in time quadrature. Phasing is compensated by use of an additional quarter-wavelength section of line connected to the antenna feeders. The field intensities radiated by the two antennas are in phase opposition to the left of the antennas and cancel each other, whereas in the region to the right of the antennas the field intensities are in time phase and reinforce each other. The resulting wave propagates to the right in the guide.

Some higher modes are generated by discontinuities of the waveguide, such as obstacles, bends, and loads. However, the higher order modes are, in general, more highly attenuated than the corresponding dominant mode. However, the dominant mode tends to remain as a dominant wave, even when the guide is large enough to support the higher order modes.

5-5-2 Mode Patterns in Rectangular Waveguides

In general, the mode pattern is dependent on the electrical size of the guide and the impressed signal wavelength. As we described in Sections 5-2 and 5-3 for TE_{mn} and TM_{mn} modes in the rectangular guide, the integer m denotes the number of half-waves

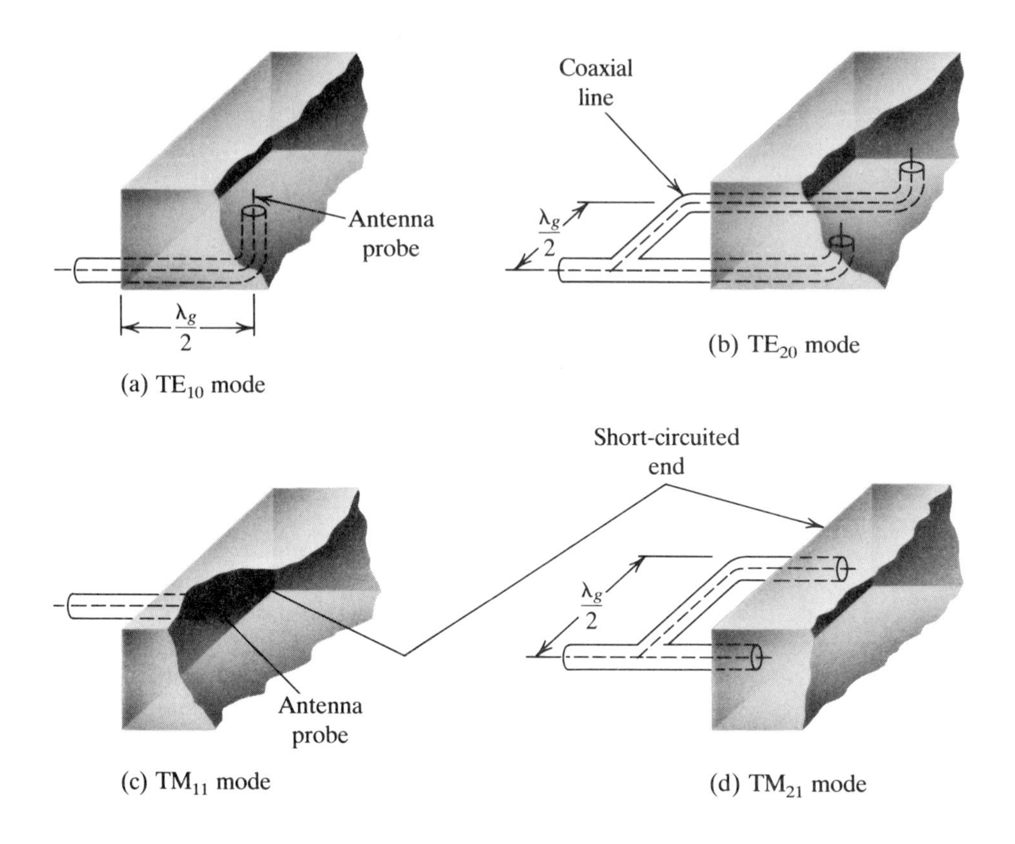

Figure 5-5-1 Methods of exciting various modes in rectangular waveguides.

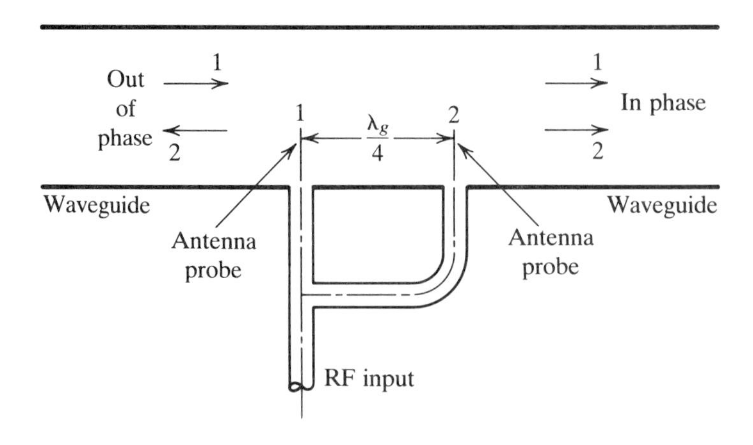

Figure 5-5-2 A method of launching a TE_{10} mode in one direction only.

of electric or magnetic intensity in the x direction, and n denotes the number of half-waves in the y direction if we assume that the direction of wave propagation is in the positive z direction. We can describe the orientations of the lines of electric and magnetic forces in the guide in the following ways:

1. The lines of electric force are oriented according to the field excited by the source and change periodically with the number of half-waves.
2. The lines of magnetic force encircle the lines of electric force or the displacement-current field in accordance with the right-hand rule.
3. Since the displacement-current field is given by

$$\mathbf{J}_d = \varepsilon \frac{\partial \mathbf{E}}{\partial t}$$

   the displacement-current field is maximum where the electric field is zero, and vice versa.
4. The lines of magnetic field are maximum where the electric field is maximum, and these lines are oriented to the direction of the wave propagation.

Figure 5-5-3 shows the mode patterns of various modes in a rectangular guide.

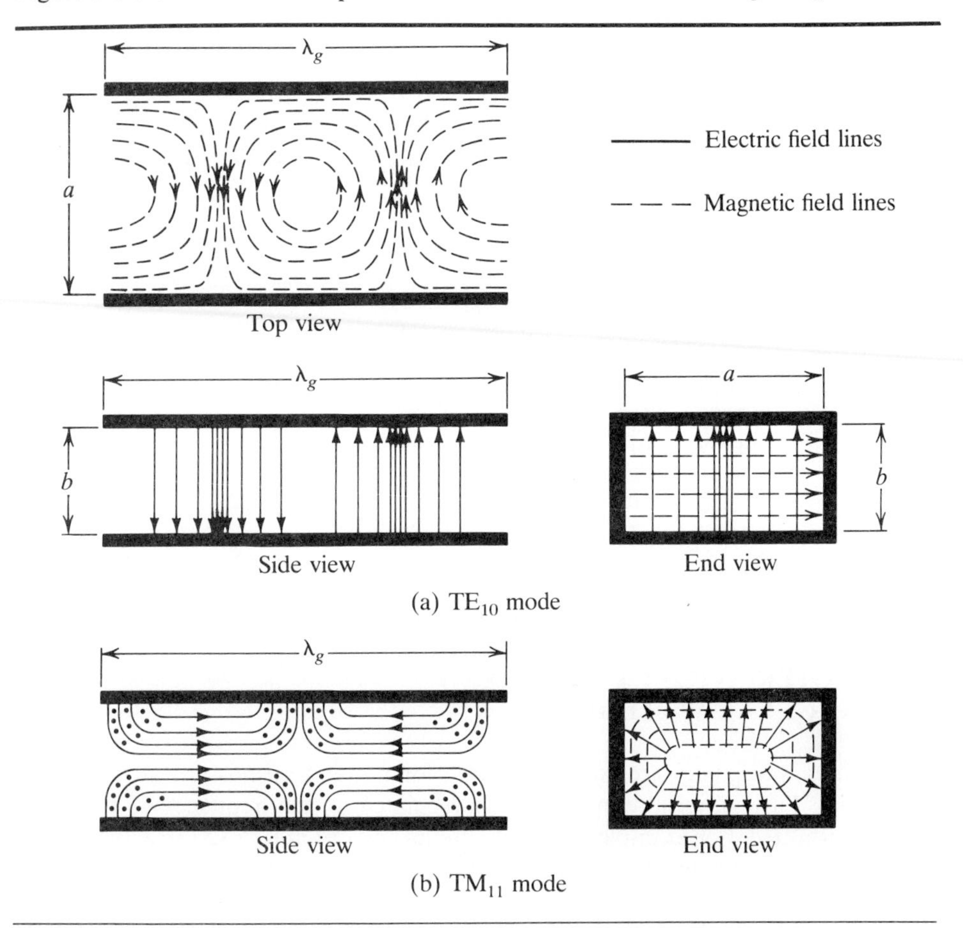

Figure 5-5-3 Field patterns of two different modes in infinitely long rectangular guides.

Table 5-6-1 Characteristics of standard rectangular waveguides

EIA Designation WR( )	Physical Dimensions: Inside in cm (in.) Width a	Height b	Outside in cm (in.) Width a	Height b	Cutoff Frequency for Air-filled Waveguide in GHz	Recommended Frequency Range for TE_{10} Mode in GHz
2300	58.420 (23.000)	29.210 (11.500)	59.055 (23.250)	29.845 (11.750)	0.257	0.32– 0.49
2100	53.340 (21.000)	26.670 (10.500)	53.973 (21.250)	27.305 (10.750)	0.281	0.35– 0.53
1800	45.720 (18.000)	22.860 (9.000)	46.350 (18.250)	23.495 (9.250)	0.328	0.41– 0.62
1500	38.100 (15.000)	19.050 (7.500)	38.735 (15.250)	19.685 (7.750)	0.394	0.49– 0.75
1150	29.210 (11.500)	14.605 (5.750)	29.845 (11.750)	15.240 (6.000)	0.514	0.64– 0.98
975	24.765 (9.750)	12.383 (4.875)	25.400 (10.000)	13.018 (5.125)	0.606	0.76– 1.15
770	19.550 (7.700)	9.779 (3.850)	20.244 (7.970)	10.414 (4.100)	0.767	0.96– 1.46
650	16.510 (6.500)	8.255 (3.250)	16.916 (6.660)	8.661 (3.410)	0.909	1.14– 1.73
510	12.954 (5.100)	6.477 (2.500)	13.360 (5.260)	6.883 (2.710)	1.158	1.45– 2.20
430	10.922 (4.300)	5.461 (2.150)	11.328 (4.460)	5.867 (2.310)	1.373	1.72– 2.61
340	8.636 (3.400)	4.318 (1.700)	9.042 (3.560)	4.724 (1.860)	1.737	2.17– 3.30
284	7.214 (2.840)	3.404 (1.340)	7.620 (3.000)	3.810 (1.500)	2.079	2.60– 3.95
229	5.817 (2.290)	2.908 (1.145)	6.142 (2.418)	3.233 (1.273)	2.579	3.22– 4.90
187	4.755 (1.872)	2.215 (0.872)	5.080 (2.000)	2.540 (1.000)	3.155	3.94– 5.99
159	4.039 (1.590)	2.019 (0.795)	4.364 (1.718)	2.344 (0.923)	3.714	4.64– 7.05
137	3.485 (1.372)	1.580 (0.622)	3.810 (1.500)	1.905 (0.750)	4.304	5.38– 8.17
112	2.850 (1.122)	1.262 (0.497)	3.175 (1.250)	1.588 (0.625)	5.263	6.57– 9.99

5-6 CHARACTERISTICS OF STANDARD RECTANGULAR WAVEGUIDES

Rectangular waveguides are commonly used for power transmission at microwave frequencies. Their physical dimensions are regulated by the frequency of the signal being transmitted. For example, at X-band frequencies from 8 to 12 GHz, the outside dimensions of a rectangular waveguide—designated as EIA WR(90) by the Electronic Industry Association—are 2.54 cm (1.0 in.) wide and 1.27 cm (0.5 in.) high, and its inside dimensions are 2.286 cm (0.90 in.) wide and 1.016 cm (0.40 in.) high. Table 5-6-1 contains a tabulation of the characteristics of the standard rectangular waveguides.

Table 5-6-1 *(continued)*

EIA Designation WR( )	Physical Dimensions Inside in cm (in.) Width a	Height b	Outside in cm (in.) Width a	Height b	Cutoff Frequency for Air-filled Waveguide in GHz	Recommended Frequency Range for TE_{10} Mode in GHz
90	2.286 (0.900)	1.016 (0.400)	2.540 (1.000)	1.270 (0.500)	6.562	8.20– 12.50
75	1.905 (0.750)	0.953 (0.375)	2.159 (0.850)	1.207 (0.475)	7.874	9.84– 15.00
62	1.580 (0.622)	0.790 (0.311)	1.783 (0.702)	0.993 (0.391)	9.494	11.90– 18.00
51	1.295 (0.510)	0.648 (0.255)	1.499 (0.590)	0.851 (0.335)	11.583	14.50– 22.00
42	1.067 (0.420)	0.432 (0.170)	1.270 (0.500)	0.635 (0.250)	14.058	17.60– 26.70
34	0.864 (0.340)	0.432 (0.170)	1.067 (0.420)	0.635 (0.250)	17.361	21.70– 33.00
28	0.711 (0.280)	0.356 (0.140)	0.914 (0.360)	0.559 (0.220)	21.097	26.40– 40.00
22	0.569 (0.224)	0.284 (0.112)	0.772 (0.304)	0.488 (0.192)	26.362	32.90– 50.10
19	0.478 (0.188)	0.239 (0.094)	0.681 (0.268)	0.442 (0.174)	31.381	39.20– 59.60
15	0.376 (0.148)	0.188 (0.074)	0.579 (0.228)	0.391 (0.154)	39.894	49.80– 75.80
12	0.310 (0.122)	0.155 (0.061)	0.513 (0.202)	0.358 (0.141)	48.387	60.50– 91.90
10	0.254 (0.100)	0.127 (0.050)	0.457 (0.180)	0.330 (0.130)	59.055	73.80–112.00
8	0.203 (0.080)	0.102 (0.040)	0.406 (0.160)	0.305 (0.120)	73.892	92.20–140.00
7	0.165 (0.065)	0.084 (0.033)	0.343 (0.135)	0.262 (0.103)	90.909	114.00–173.00
5	0.130 (0.051)	0.066 (0.026)	0.257 (0.101)	0.193 (0.076)	115.385	145.00–220.00
4	0.109 (0.043)	0.056 (0.022)	0.211 (0.083)	0.157 (0.062)	137.615	172.00–261.00
3	0.086 (0.034)	0.043 (0.017)	0.163 (0.064)	0.119 (0.047)	174.419	217.00–333.00

Note: EIA stands for Electronic Industry Association, and WR represents rectangular waveguide.

5-7 RECTANGULAR-CAVITY RESONATORS AND QUALITY FACTOR Q

In general, a cavity resonator is a metallic enclosure that confines electromagnetic energy. The stored electric and magnetic energies inside the cavity determine its equivalent inductance and capacitance. The energy dissipated by the finite conductivity of the cavity walls determines its equivalent resistance. In practice, rectangular-cavity (a), circular-cavity (b), and reentrant-cavity (c–g) resonators, as shown in Fig. 5-7-1, are commonly used.

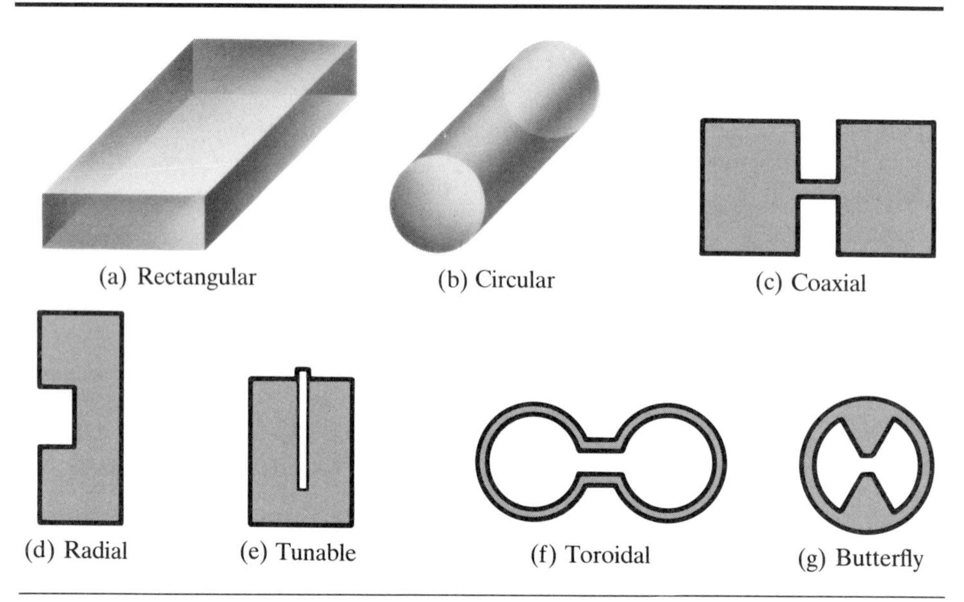

Figure 5-7-1 Various types of resonators.

Theoretically, a resonator has an infinite number of resonant modes, and each mode corresponds to a definite resonant frequency. When the frequency of an impressed signal is equal to a resonant frequency, a maximum amplitude of the standing wave occurs, and the peak energy stored in the electric field is equal to that of the magnetic field. The mode having the lowest resonant frequency is called the *dominant mode.*

5-7-1 Rectangular-Cavity Resonator

The electromagnetic fields inside a cavity should satisfy Maxwell's equations, subject to the boundary conditions that the electric field tangential to and the magnetic field normal to the metal walls must vanish. The geometry of a rectangular cavity is illustrated in Fig. 5-7-2.

The wave equations for a rectangular resonator should satisfy the boundary conditions of zero tangential **E** at four of the walls. We merely choose the harmonic functions in z to satisfy this condition at the remaining two (end) walls. Thus we can write the wave equations for the rectangular-cavity resonator from Eqs. (5-3-4) and (5-2-2a) as

$$H_z = H_{0z} \cos\left(\frac{m\pi x}{a}\right) \cos\left(\frac{n\pi y}{b}\right) \sin\left(\frac{p\pi z}{d}\right) \qquad \text{for } \mathrm{TE}_{mnp} \qquad (5\text{-}7\text{-}1)$$

where

$m = 0, 1, 2, 3, \ldots$ represents the number of the half-wave periodicity in the x direction;

$n = 0, 1, 2, 3, \ldots$ represents the number of the half-wave periodicity in the y direction; and

$p = 1, 2, 3, 4, \ldots$ represents the number of the half-wave periodicity in the z direction.

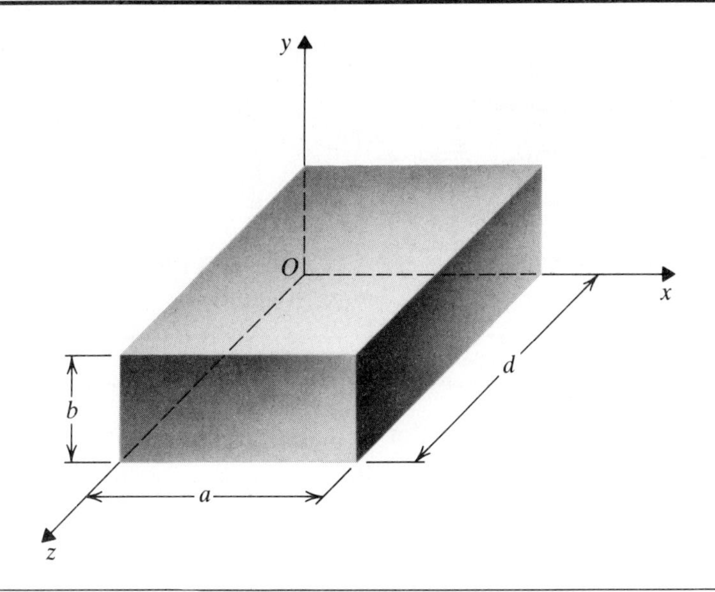

Figure 5-7-2 Coordinates of a rectangular cavity.

and

$$E_z = E_{0z} \sin\left(\frac{m\pi x}{a}\right) \sin\left(\frac{n\pi y}{b}\right) \cos\left(\frac{p\pi z}{d}\right) \quad \text{for } \mathrm{TM}_{mnp} \tag{5-7-2}$$

where

$m = 1, 2, 3, 4, \ldots;$
$n = 1, 2, 3, 4, \ldots;$ and
$p = 0, 1, 2, 3, \ldots.$

The separation equation for both the TE and TM modes is

$$k^2 = \left(\frac{m\pi}{a}\right)^2 + \left(\frac{n\pi}{b}\right)^2 + \left(\frac{p\pi}{d}\right)^2 \tag{5-7-3}$$

For a lossless dielectric, $k^2 = \omega^2\mu\varepsilon$; therefore the resonant frequency for the TE and TM modes is

$$f_r = \frac{1}{2\sqrt{\mu\varepsilon}\sqrt{(m/a)^2 + (n/b)^2 + (p/d)^2}} \tag{5-7-4}$$

and for $b < a < d$, the dominant mode is the TE_{101} mode.

In general, a straight-wire probe inserted at the position of maximum electric intensity is used to excite a mode, and a loop coupling placed at the position of maximum magnetic intensity is utilized to launch a specific mode. Figure 5-7-3 shows the methods of excitation for the rectangular resonator. The maximum amplitude of the standing wave occurs when the frequency of the impressed signal is equal to the resonant frequency.

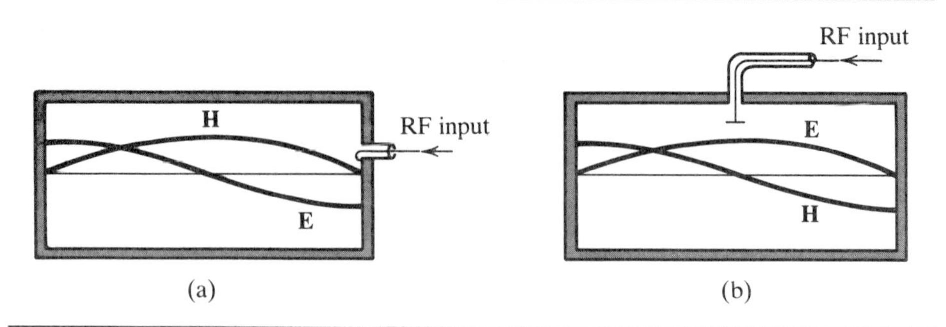

Figure 5-7-3 Methods of exciting a resonator.

5-7-2 The Q of a Cavity Resonator

The quality factor Q is a measure of the frequency selectivity of a resonant or antiresonant circuit and is defined as

$$Q \equiv 2\pi\left(\frac{\text{maximum energy stored}}{\text{energy dissipated per cycle}}\right)$$

$$= \frac{\omega w}{\langle p \rangle} \tag{5-7-5}$$

where w is the maximum energy stored, and $\langle p \rangle$ is the average power loss. At resonant frequency, the electric and magnetic energies are equal and in time quadrature. When the electric energy is at its maximum, the magnetic energy is zero, and vice versa. We calculate the total energy stored in the resonator by integrating the energy density over the volume of the resonator. That is,

$$w_e = \int_v \frac{\varepsilon}{2}|E|^2\,dv = w_m = \int_v \frac{\mu}{2}|H|^2\,dv = w \tag{5-7-6}$$

where $|E|$ and $|H|$ are the peak values of the field intensities. We can evaluate the average power loss in the resonator by integrating the power density as given in Eq. (0-5-64) over the inner surface of the resonator, or

$$P = \frac{R_s}{2}\int_s |H_t|^2\,da \tag{5-7-7}$$

where R_s represents the surface resistance of the resonator in Ω/square, and H_t represents the peak value of the tangential magnetic intensity. Substituting Eqs. (5-7-6) and (5-7-7) into Eq. (5-7-5) yields

$$Q = \frac{\omega\mu \int_v |H|^2\,dv}{R_s \int_s |H_t|^2\,da} \tag{5-7-8}$$

The peak value of the magnetic intensity is related to the tangential components by

$$|H|^2 = |H_t|^2 + |H_n|^2 \tag{5-7-9}$$

where $|H_n|$ is the peak value of the normal magnetic intensity. The value of $|H_t|^2$ at the resonator walls is approximately twice the value of $|H|^2$ averaged over the resonator volume. So the Q of a cavity resonator in Eq. (5-7-8) can be expressed approximately by

$$Q = \frac{\omega\mu(\text{volumes})}{2R_s(\text{surface areas})} \tag{5-7-10}$$

EXAMPLE 5-7-1 Rectangular-Cavity Resonator

An air-filled rectangular-cavity resonator has as inside dimensions $a = 1.295$ cm, $b = 0.648$ cm, and $d = 2.10$ cm.

Calculate:

a) the dominant mode;
b) the resonant frequency; and
c) the resonant frequency if the resonator is dielectrically loaded with $\varepsilon_r = 2.56$.

Solutions:

a) The dominant mode is TE_{101} because $b < a < d$.
b) The resonant frequency is

$$f_r = \frac{3 \times 10^8}{2}\sqrt{\left(\frac{1}{1.295 \times 10^{-2}}\right)^2 + \left(\frac{0}{b}\right)^2 + \left(\frac{1}{2.1 \times 10^{-2}}\right)^2}$$

$$= 1.5 \times 10^8 \times 90.72$$

$$= 13.61 \text{ GHz}$$

c) If the resonator is dielectrically loaded, its resonant frequency is

$$f_{r\varepsilon} = \frac{f_r}{\sqrt{\varepsilon_r}} = \frac{13.61 \times 10^9}{\sqrt{2.56}}$$

$$= 8.51 \text{ GHz}$$

Problems

5-1 An air-filled rectangular waveguide has dimensions of $a = 2.286$ cm and $b = 1.016$ cm. The signal frequency is 3 GHz. Calculate the following for the TE_{10}, TE_{01}, TE_{11}, and TM_{11} modes:

a) Cutoff frequency
b) Wavelength in the waveguide
c) Phase constant and phase velocity in the waveguide
d) Group velocity and wave impedance in the waveguide

5-2 Show that the TM_{01} and TM_{10} modes do not exist in a rectangular waveguide.

5-3 The dominant mode TE_{10} is propagated in a rectangular waveguide of dimensions $a = 6$ cm and $b = 3$ cm. The distance between maximum and minimum is 4.55 cm. Determine the signal frequency of the dominant mode.

5-4 A TE mode of 10 GHz is propagated in an air-filled rectangular waveguide. The magnetic field in the z direction is

$$H_z = H_0 \cos\left(\frac{\pi x}{a}\right) \cos\left(\frac{\pi y}{b}\right) \quad \text{A/m}$$

The phase constant is $\beta = 1.0367$ rad/cm, and the quantities x and y are expressed in cm. Determine the cutoff frequency f_c, phase velocity v_g, guided wavelength λ_g, and

$$H_y = jH_0 \cos\left(\frac{\pi x}{a}\right) \sin\left(\frac{\pi y}{a}\right) \quad \text{A/m}$$

5-5 A rectangular waveguide is designed to propagate the dominant mode TE_{10} at a frequency of 5 GHz. The cutoff frequency is 0.8 of the signal frequency. The ratio of the guide height to width is 0.5. The time-average power flowing through the guide is 1 kW. Determine the magnitudes of the electric and magnetic intensities in the guide and indicate where they occur in the guide.

5-6 An air-filled rectangular waveguide has dimensions of $a = 1.295$ cm and $b = 0.648$ cm. The guide transports energy in the dominant mode TE_{10} at a rate of 1 hp (746 Joules). If the frequency is 20 GHz, what is the peak value of electric field that occurs in the guide?

5-7 A rectangular waveguide is filled by dielectric material of $\epsilon_r = 9$ and has inside dimensions of 7 cm × 3.5 cm. It operates in the dominant TE_{10} mode. Determine the

a) cutoff frequency;
b) phase velocity in the guide at a frequency of 2 GHz; and
c) guided wavelength λ_g at the same frequency.

5-8 The electric field intensity of the dominant TE_{10} mode in a lossless rectangular waveguide is

$$\mathbf{E}_y = \mathbf{E}_{0y} \sin\left(\frac{\pi y}{a}\right) e^{-j\beta z} \qquad \text{for } f > f_c$$

Calculate the

a) magnetic field intensity $\mathbf{H}$;
b) cutoff frequency; and
c) time-average power transmitted.

5-9 An air-filled rectangular waveguide with dimensions of 3 cm × 1 cm operates in the TE_{10} mode at 10 GHz. The waveguide is perfectly matched and the maximum **E** field existing everywhere in the guide is 10^3 V/m. Determine the expressions for the waveguide voltage, current, and characteristic impedance.

5-10 The dominant mode TE_{10} is propagated in a rectangular waveguide of dimensions $a = 2.28$ cm and $b = 1$ cm. Assume an air dielectric with a breakdown gradient of 30 kV/cm and a frequency of 10 GHz. There are no standing waves in the guide. Determine the maximum average power that can be carried by the guide.

5-11 A rectangular waveguide is terminated in an unknown impedance at $z = 25$ cm. A dominant mode TE_{10} is propagated in the guide, and its VSWR is measured as 2.8 at a frequency of 8 GHz. The adjacent minima are located at $z = 9.46$ cm and $z = 12.73$ cm.

Determine the

a) value of the load impedance in terms of Z_0;
b) position closest to the load where an inductive window is placed in order to obtain a VSWR of unity; and
c) value of the window admittance.

5-12 A rectangular-cavity resonator has dimensions of $a = 5$ cm, $b = 2$ cm, and $d = 15$ cm. Calculate the

a) resonant frequency of the dominant mode for an air-filled cavity; and
b) resonant frequency of the dominant mode for a dielectric-filled cavity of $\varepsilon_r = 2.56$.

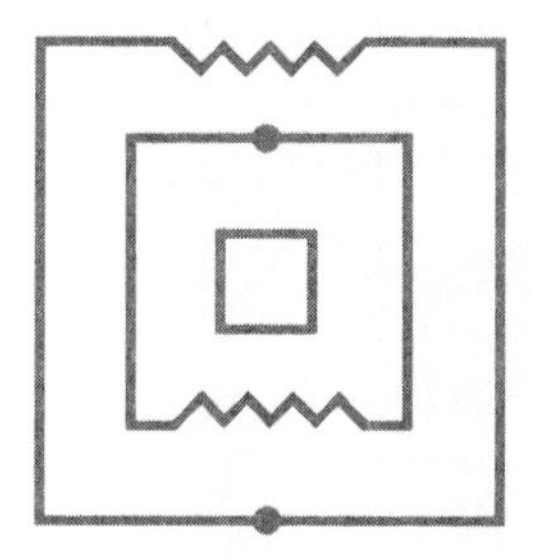

Chapter 6

Circular Waveguides

6-0 INTRODUCTION

A circular waveguide is a tubular, circular conductor. A plane wave propagating through a circular waveguide produces a transverse electric (TE) or transverse magnetic (TM) mode. Several other types of waveguides, such as elliptical and reentrant guides, also propagate electromagnetic waves. In this chapter you will study the wave modes, power transmission, and power losses in circular waveguides.

6-1 WAVE EQUATIONS IN CYLINDRICAL COORDINATES

As we explained in Chapter 5 for rectangular waveguides, only a sinusoidal steady-state or frequency-domain solution will be presented for circular waveguides. A cylindrical coordinate system is shown in Fig. 6-1-1.

The scalar Helmholtz equation in cylindrical coordinates is

$$\frac{1}{r}\frac{\partial}{\partial r}\left(r\frac{\partial \psi}{\partial r}\right) + \frac{1}{r^2}\frac{\partial^2 \psi}{\partial \phi^2} + \frac{\partial^2 \psi}{\partial z^2} = \gamma^2 \psi \tag{6-1-1}$$

Using the method of separation of variables, we obtain the solution in the form:

$$\psi = R(r)\Phi(\phi)Z(z) \tag{6-1-2}$$

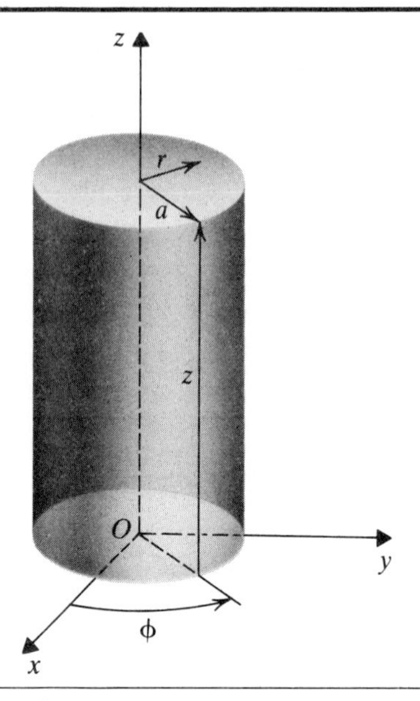

Figure 6-1-1 Cylindrical coordinates.

where

$R(r)$ represents the function of the r coordinate only;

$\Phi(\phi)$ represents the function of the ϕ coordinate only; and

$Z(z)$ represents the function of the z coordinate only.

When we substitute Eq. (6-1-2) into Eq. (6-1-1) and divide the result by Eq. (6-1-2), we get

$$\frac{1}{rR}\frac{d}{dr}\left(r\frac{dR}{dr}\right)+\frac{1}{r^2\Phi}\frac{d^2\Phi}{d\phi^2}+\frac{1}{Z}\frac{d^2Z}{dz^2}=\gamma^2 \tag{6-1-3}$$

Since the sum of the three independent terms is a constant, each of the three terms must be a constant. Let's set the third term equal to the constant γ_g^2:

$$\frac{d^2Z}{dz^2}=\gamma_g^2 Z \tag{6-1-4}$$

Then the solution of Eq. (6-1-4) is

$$Z=Ae^{-\gamma_g z}+Be^{\gamma_g z} \tag{6-1-5}$$

where γ_g is the propagation constant of the wave in the guide.

Inserting γ_g^2 for the third term of the left-hand side of Eq. (6-1-3) and multiplying the resultant by r^2, we obtain

$$\frac{r}{R}\frac{d}{dr}\left(r\frac{dR}{dr}\right)+\frac{1}{\Phi}\frac{d^2\Phi}{d\phi^2}-(\gamma^2-\gamma_g^2)r^2=0 \tag{6-1-6}$$

The second term is a function of ϕ only; hence, equating the second term to the constant $-n^2$ yields

$$\frac{d^2\Phi}{d\phi^2} = -n^2\Phi \tag{6-1-7}$$

The solution of Eq. (6-1-7) is a harmonic function:

$$\Phi = A_n \sin(n\phi) + B_n \cos(n\phi) \tag{6-1-8}$$

Replacing the ϕ term with $-n^2$ in Eq. (6-1-6) and multiplying through by R, we have

$$r\frac{d}{dr}\left(r\frac{dR}{dr}\right) + [(k_c r)^2 - n^2]R = 0 \tag{6-1-9}$$

This is the Bessel function of order n, in which

$$k_c^2 + \gamma^2 = \gamma_g^2 \tag{6-1-10}$$

Equation (6-1-10) is called the characteristic equation of the Bessel function. For a lossless guide, Eq. (6-1-10) reduces to

$$\beta_g = \pm\sqrt{\omega^2\mu\varepsilon - k_c^2} \tag{6-1-11}$$

The solution of Bessel's equation takes the form:

$$R = C_n J_n(k_c r) + D_n N_n(k_c r) \tag{6-1-12}$$

where $J_n(k_c r)$ is the nth-order Bessel function of the first kind, representing a standing wave of $\cos(k_c r)$ for $r < a$, as shown in Fig. 6-1-2; and $N_n(k_c r)$ is the nth-order Bessel function of the second kind, representing a standing wave of $\sin(k_c r)$ for $r > a$, as shown in Fig. 6-1-3. The value of $k_c r$ is the argument of the Bessel function.

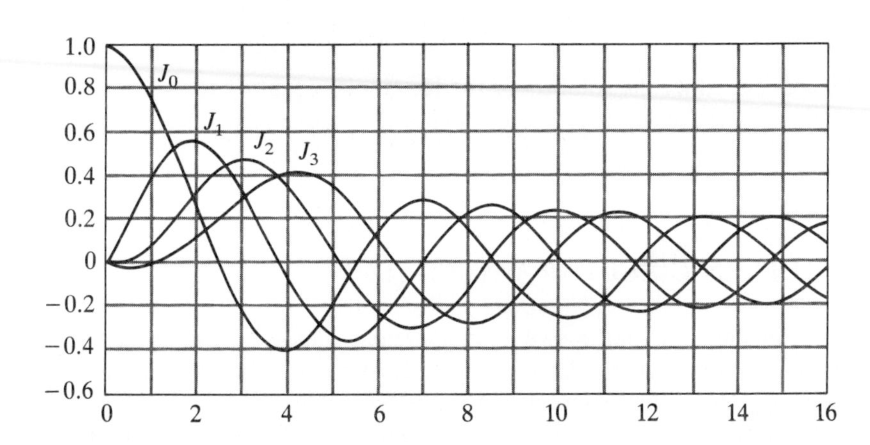

Figure 6-1-2 Bessel functions of the first kind.

Therefore the total solution of the Helmholtz equation in cylindrical coordinates becomes

$$\psi = [C_n J_n(k_c r) + D_n N_n(k_c r)][A_n \sin(n\phi) + B_n \cos(n\phi)]e^{\pm j\beta_g z} \tag{6-1-13}$$

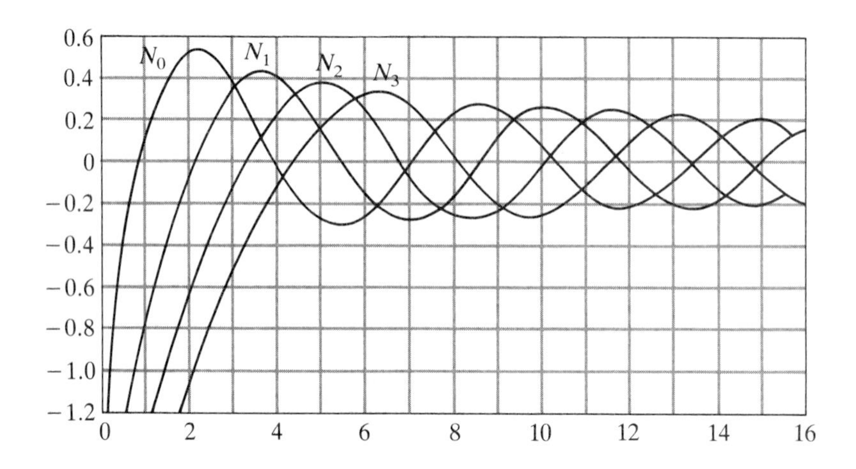

Figure 6-1-3 Bessel functions of the second kind.

At $r = 0$, however, $k_c r = 0$, the function N_n approaches infinity, and $D_n = 0$. Thus at $r = 0$ on the z axis, the field must be finite. Also, by manipulating trigonometric functions, we can convert the two sinusoidal terms to

$$(A_n \sin (n\phi) + B_n \cos (n\phi) = \sqrt{A_n^2 + B_n^2} \cos\left[n\phi + \tan^{-1}\left(\frac{A_n}{B_n}\right)\right]$$

$$= F_n \cos (n\phi) \tag{6-1-14}$$

Finally, we can reduce the solution of the Helmholtz equation to

$$\psi = \psi_0 \cos (n\phi) J_n(k_c r) e^{-j\beta_g z} \tag{6-1-15}$$

6-2 TE MODES IN CIRCULAR WAVEGUIDES

We commonly assume that the waves in a circular waveguide are propagating in the positive z direction. Figure 6-2-1 shows the coordinates of a circular guide.

The TE_{np} modes in the circular guide are characterized by $E_z = 0$, which means that the z component of the magnetic field H_z must exist in the guide in order for electromagnetic energy to be transmitted. A Helmholtz equation for H_z in a circular guide is

$$\nabla^2 H_z = \gamma^2 H_z \tag{6-2-1}$$

Its solution, as shown in Eq. (6-1-15), is

$$H_z = H_{0z} J_n(k_c r) \cos (n\phi) e^{-j\beta_g z} \tag{6-2-2}$$

which is subject to boundary conditions.

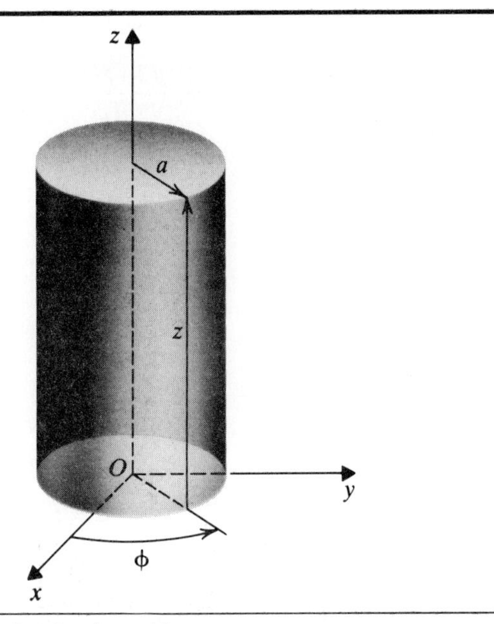

Figure 6-2-1 Coordinates of a circular guide.

For a lossless dielectric, Maxwell's curl equations in the frequency domain are

$$\nabla \times \mathbf{E} = -j\omega\mu\mathbf{H} \tag{6-2-3}$$

and

$$\nabla \times \mathbf{H} = -j\omega\varepsilon\mathbf{E} \tag{6-2-4}$$

In cylindrical coordinates, their components are

$$\frac{1}{r}\frac{\partial E_z}{\partial \phi} - \frac{\partial E_\phi}{\partial z} = -j\omega\mu H_r \tag{6-2-5}$$

$$\frac{\partial E_r}{\partial z} - \frac{\partial E_z}{\partial r} = -j\omega\mu H_\phi \tag{6-2-6}$$

$$\frac{1}{r}\frac{\partial}{\partial r}(rE_\phi) - \frac{1}{r}\frac{\partial E_r}{\partial \phi} = -j\omega\mu H_z \tag{6-2-7}$$

$$\frac{1}{r}\frac{\partial H_z}{\partial \phi} - \frac{\partial H_\phi}{\partial z} = j\omega\varepsilon E_r \tag{6-2-8}$$

$$-j\beta_g H_r - \frac{\partial H_z}{\partial r} = j\omega\varepsilon E_\phi \tag{6-2-9}$$

and

$$\frac{1}{r}\frac{\partial}{\partial r}(rH_\phi) - \frac{1}{r}\frac{\partial H_r}{\partial \phi} = j\omega\varepsilon E_z \tag{6-2-10}$$

When we replace the differentiation $\frac{\partial}{\partial z}$ *with* $(-j\beta_g)$ and the z component of electric field

E_z by 0, the TE-mode equations for a circular waveguide are

$$E_r = -\frac{j\omega\mu}{k_c^2}\frac{1}{r}\frac{\partial H_z}{\partial \phi} \qquad (6\text{-}2\text{-}11)$$

$$E_\phi = \frac{j\omega\mu}{k_c^2}\frac{\partial H_z}{\partial r} \qquad (6\text{-}2\text{-}12)$$

$$E_z = 0 \qquad (6\text{-}2\text{-}13)$$

$$H_r = \frac{-j\beta_g}{k_c^2}\frac{\partial H_z}{\partial r} \qquad (6\text{-}2\text{-}14)$$

$$H_\phi = \frac{-j\beta_g}{k_c^2}\frac{1}{r}\frac{\partial H_z}{\partial \phi} \qquad (6\text{-}2\text{-}15)$$

and

$$H_z = H_{0z}J_n(k_c r)\cos(n\phi)e^{-j\beta_g z} \qquad (6\text{-}2\text{-}16)$$

where $k_c^2 = \omega^2\mu\varepsilon - \beta_g^2$ has been replaced.

The boundary conditions require that the ϕ component of the electric field E_ϕ—which is tangential to the inner surface of the circular waveguide at $r = a$—must vanish or that the r component of the magnetic field H_r—which is normal to the inner surface of $r = a$—must vanish. Consequently,

$$E_\phi = 0 \text{ at } r = a, \therefore \left.\frac{\partial H_z}{\partial r}\right|_{r=a} = 0 \quad \text{or} \quad H_r = 0 \text{ at } r = a, \therefore \left.\frac{\partial H_z}{\partial r}\right|_{r=a} = 0$$

This requirement is equivalent to that expressed in Eq. (6-2-2) as

$$\left.\frac{\partial H_z}{\partial r}\right|_{r=a} = H_{0z}J_n'(k_c a)\cos(n\phi)e^{-j\beta_g z} = 0 \qquad (6\text{-}2\text{-}17)$$

Hence,

$$J_n'(k_c a) = 0 \qquad (6\text{-}2\text{-}18)$$

where J_n' indicates the derivative of J_n.

Since the J_n' functions are oscillatory, the $J_n'(k_c a)$ functions are also oscillatory. An infinite sequence of values of $(k_c a)$ satisfies Eq. (6-2-18). These points—the roots of Eq. (6-2-18)—correspond to the maxima and minima of the curves $J_n'(k_c a)$, as shown previously in Fig. 6-1-2. Table 6-2-1 contains a few roots of $J_n'(k_c a)$ for some lower order n values.

Table 6-2-1 The pth zeros of $J_n'(k_c a)$ for TE_{np} modes

p \ n	0	1	2	3	4	5
1	3.832	1.841	3.054	4.201	5.317	6.416
2	7.016	5.331	6.706	8.015	9.282	10.520
3	10.173	8.536	9.969	11.346	12.682	13.987
4	13.324	11.706	13.170			

We can write an expression for the permissible values of k_c as

$$k_c = \frac{X'_{np}}{a} \tag{6-2-19}$$

where X'_{np} is the argument of the Bessel function.

Substituting Eq. (6-2-2) into Eqs. (6-2-11)–(6-2-16) yields the complete field equations of the TE_{np} modes in circular waveguides:

$$E_r = E_{0r} J_n\left(\frac{X'_{np} r}{a}\right) \sin(n\phi) e^{-j\beta_g z} \tag{6-2-20}$$

$$E_\phi = E_{0\phi} J'_n\left(\frac{X'_{np} r}{a}\right) \cos(n\phi) e^{-j\beta_g z} \tag{6-2-21}$$

$$E_z = 0 \tag{6-2-22}$$

$$H_r = -\frac{E_{0\phi}}{Z_g} J'_n\left(\frac{X'_{np} r}{a}\right) \cos(n\phi) e^{-j\beta_g z} \tag{6-2-23}$$

$$H_\phi = \frac{E_{0r}}{Z_g} J_n\left(\frac{X'_{np} r}{a}\right) \sin(n\phi) e^{-j\beta_g z} \tag{6-2-24}$$

and

$$H_z = H_{0z} J_n\left(\frac{X'_{np} r}{a}\right) \cos(n\phi) e^{-j\beta_g z} \tag{6-2-25}$$

where

$Z_g = \frac{E_r}{H_\phi} = -\frac{E_\phi}{H_r}$ have been replaced for the wave impedance in the guide;

$n = 0, 1, 2, 3, \ldots,$ and

$p = 1, 2, 3, 4, \ldots$

The subscript n represents the number of full cycles of field variation in one revolution through 2π radians of ϕ. The subscript p indicates the number of zeros of E_ϕ; that is, $J'_n[(X'_{np} r)/a]$ indicates the zeros along the radial of a guide, but the zero on the axis is excluded if it exists.

We determine the mode propagation constant from Eqs. (6-1-11) and (6-2-19) for a lossless waveguide, and its phase constant is

$$\beta_g = \sqrt{\omega^2 \mu\varepsilon - \left(\frac{X'_{np}}{a}\right)^2} \tag{6-2-26}$$

The cutoff wave number of a mode is that number for which the mode propagation constant vanishes. Hence,

$$k_c = \frac{X'_{np}}{a} = \omega_c \sqrt{\mu\varepsilon} \tag{6-2-27}$$

The cutoff frequency for TE modes in a circular guide, then, is

$$f_c = \frac{X'_{np}}{2\pi a \sqrt{\mu\varepsilon}} \tag{6-2-28}$$

and the phase velocity for TE modes is

$$v_g = \frac{\omega}{\beta_g} = \frac{v_p}{\sqrt{1-(f_c/f)^2}} \tag{6-2-29}$$

where $v_p = 1/\sqrt{\mu\varepsilon} = c/\sqrt{\mu_r\varepsilon_r}$ and is the phase velocity in an unbounded dielectric.

The wavelength and wave impedance, respectively, for TE modes in a circular guide are

$$\lambda_g = \frac{\lambda}{\sqrt{1-(f_c/f)^2}} \tag{6-2-30}$$

and

$$Z_g = \frac{\omega\mu}{\beta_g} = \frac{\eta}{\sqrt{1-(f_c/f)^2}} \tag{6-2-31}$$

where

$\lambda = v_p/f$ and is the wavelength in an unbounded dielectric; and
$\eta = \sqrt{\mu/\varepsilon}$ and is the intrinsic impedance in an unbounded dielectric.

The characteristic impedance of the TE_{np}-mode circular waveguide is

$$Z_{0g} = \frac{aE_r}{2\pi a H_\phi} = \frac{\eta}{2\pi\sqrt{1-(f_c/f)^2}} \tag{6-2-32}$$

EXAMPLE 6-2-1 TE Mode in Circular Waveguide

A TE_{11} mode is propagating through a circular waveguide. The radius of the guide is 5 cm, and the guide contains an air dielectric. (See Fig. 6-2-2.)

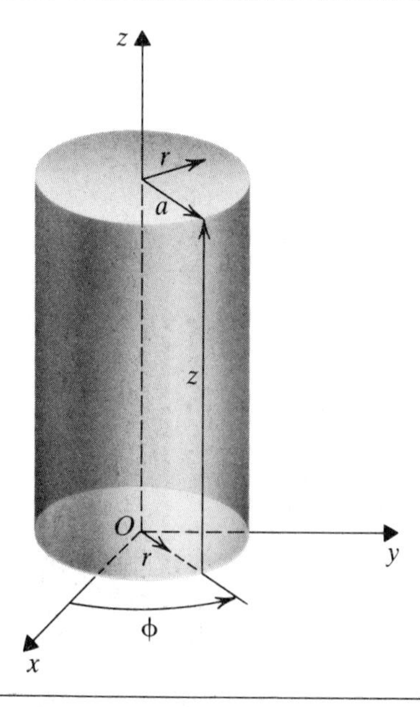

Figure 6-2-2 Circular waveguide diagram for Example 6-2-1.

Determine:

a) the cutoff frequency;
b) the wavelength λ_g in the guide for an operating frequency of 10 GHz; and
c) the wave impedance Z_g in the guide.

Solutions:

a) From Table 6-2-1 for the TE_{11} mode, $n = 1$ and $p = 1$, so $X'_{11} = 1.841 = k_c a$. The cutoff wave number is

$$k_c = \frac{1.841}{a} = \frac{1.841}{5 \times 10^{-2}}$$
$$= 36.82$$

The cutoff frequency is

$$f_c = \frac{k_c}{2\pi\sqrt{\mu_0 \varepsilon_0}} = \frac{36.82 \times 3 \times 10^8}{2\pi}$$
$$= 1.758 \times 10^9 \text{ Hz}$$

b) The phase constant in the guide is

$$\beta_g = \sqrt{\omega^2 \mu_0 \varepsilon_0 - k_c^2}$$
$$= \sqrt{(2\pi \times 10^{10})^2 \times 4\pi \times 10^{-7} \times 8.85 \times 10^{-12} - (36.82)^2}$$
$$= 50.7 \text{ rad/m}$$

The wavelength in the guide is

$$\lambda_g = \frac{2\pi}{\beta_g} = \frac{6.28}{50.7}$$
$$= 12.4 \text{ cm}$$

c) The wave impedance in the guide is

$$Z_g = \frac{\omega \mu_0}{\beta_g} = \frac{2\pi \times 10^{10} \times 4\pi \times 10^{-7}}{50.7}$$
$$= 465\ \Omega$$

EXAMPLE 6-2-2 Computer Solution for TE Mode in a Circular Waveguide

```
PROGRAM    LGUIDE

010C      PROGRAM TO COMPUTE CUTOFF FREQUENCY,WAVE IMPEDANCE,
020C+     AND WAVELENGTH FOR TE-ONE-ONE MODE IN CYLINDRICAL
030C+     WAVEGUIDE
040C      LGUIDE
050       DIMENSION TE(4,3)
055       INTEGER P
057       REAL KC,LAMBDA
060       DATA TE/3.832,1.841,3.054,4.201,7.016,5.331,6.706,
070+              8.015,10.173,8.536,9.969,11.346/
100     1 READ 2, N,P,A,F
```

(continued)

EXAMPLE 6-2-2 *(continued)*

```
110       IF((N .EQ. 0.0) .AND. (P .EQ. 0.0)) STOP
120       PRINT 4, N,P,A,F
130       N=N+1
140       PI=3.141593
150       WO=4E-07*PI
160       EO=8.854E-12
170       KC=TE(N,P)/A
180       FC=KC/(2.*PI*SQRT(UO*EO))
190       BETAG=SQRT((2.*PI*F)**2*UO*EO-KC**2)
200       LAMBDA=2.*PI/BETAG
210       ZG=2.*PI*F*UO/BETAG
220       PRINT 6, FC,BETAG,LAMBDA,ZG
230     2 FORMAT(2I1,F10.5,E10.3)
240     4 FORMAT(//1H ,24HWAVE MODE NUMBER   N,P = ,2I1/
250+      1X,24HRADIUS OF WAVEGUIDE A = ,F10.5,1X,6HMETERS/
260+      1X,24HOPERATING FREQUENCY F = ,E10.3,1X,5HHERTS/)
270     6 FORMAT(//1H ,23HCUTOFF FREQUENCY   FC = ,
280+      1X,E13.3,1X,5HHERTS/
290+      1X,23HPHASE CONSTANT BETAG = ,F10.5,1X,9HRADIANS/M/
300+      1X,23HWAVELENGTH    LAMBDA = ,F10.5,1X,6HMETERS/
310+      1X,23HWAVE IMPEDANCE    ZG = ,F10.5,1X,4HOHMS/)
320       GO TO 1
330       STOP
340       END
READY.
RUN

PROGRAM   LGUIDE

? 11    0.05000 0.300E+10

 WAVE MODE NUMBER   N,P = 11
 RADIUS OF WAVEGUIDE A =    .05000 METERS
 OPERATING FREQUENCY F =    .300E+10 HERTS

 CUTOFF FREQUENCY   FC =      .176E+10 HERTS
 PHASE CONSTANT BETAG =    50.96582 RADIANS/M
 WAVELENGTH    LAMBDA =      .12328 METERS
 WAVE IMPEDANCE    ZG =   464.76356 OHMS

? 00

SRU       0.781 UNTS.

RUN COMPLETE.
```

6-3 TM MODES IN CIRCULAR WAVEGUIDES

The TM_{np} modes in a circular guide are characterized by $H_z = 0$. However, the z component of the electric field E_z must exist in order for energy to be transmitted in the guide. Consequently, the Helmholtz equation for E_z in a circular waveguide is

$$\nabla^2 E_z = \gamma^2 E_z \tag{6-3-1}$$

Its solution, as shown in Eq. (6-1-15), is

$$E_z = E_{0z} J_n(k_c r) \cos(n\phi) e^{-j\beta_g z} \tag{6-3-2}$$

which is subject to boundary conditions. Boundary conditions require that the tangential component of the electric field E_z at $r = a$ vanish. Consequently,

$$J_n(k_c a) = 0 \tag{6-3-3}$$

Since $J_n(k_c r)$ functions are oscillatory, as shown previously in Fig. 6-1-2, there is an infinite number of roots of $J_n(k_c r)$. Table 6-3-1 contains a few roots of $J_n(k_c r)$ for some lower order n values.

Table 6-3-1 The pth zeros of $J_n(k_c r)$.

p \ n	0	1	2	3	4	5
1	2.405	3.832	5.136	6.380	7.588	8.771
2	5.520	7.106	8.417	9.761	11.065	12.339
3	8.645	10.173	11.620	13.015	14.372	
4	11.792	13.324	14.796			

For $H_z = 0$ and $\partial/\partial z = -j\beta_g$, the field equations in the circular guide, after expansion of $\nabla \times \mathbf{E} = -j\omega\mu\mathbf{H}$ and $\nabla \times \mathbf{H} = j\omega\varepsilon\mathbf{E}$, are

$$E_r = \frac{-j\beta_g}{k_c^2}\frac{\partial \mathbf{E}_z}{\partial r} \tag{6-3-4}$$

$$E_\phi = \frac{-j\beta_g}{k_c^2}\frac{1}{r}\frac{\partial \mathbf{E}_z}{\partial \phi} \tag{6-3-5}$$

$$E_z = E_{0z}J_n(k_c r)\cos(n\phi)e^{-j\beta_g z} \qquad [(6\text{-}3\text{-}2)]$$

$$H_r = \frac{j\omega\varepsilon}{k_c^2}\frac{1}{r}\frac{\partial E_z}{\partial \phi} \tag{6-3-6}$$

$$H_\phi = -\frac{j\omega\varepsilon}{k_c^2}\frac{\partial E_z}{\partial r} \tag{6-3-7}$$

and

$$H_z = 0 \tag{6-3-8}$$

where $k_c^2 = \omega^2\mu\varepsilon - \beta_g^2$ has been replaced.

Differentiating Eq. (6-3-2) with respect to z and substituting the result in Eqs. (6-3-4)–(6-3-8) yield the field equations of TM_{np} modes in circular waveguide:

$$E_r = E_{0r}J_n'\left(\frac{X_{np}r}{a}\right)\cos(n\phi)e^{-j\beta_g z} \tag{6-3-9}$$

$$E_\phi = E_{0\phi}J_n\left(\frac{X_{np}r}{a}\right)\sin(n\phi)e^{-j\beta_g z} \tag{6-3-10}$$

$$E_z = E_{0z}J_n\left(\frac{X_{np}r}{a}\right)\cos(n\phi)e^{-j\beta_g z} \tag{6-3-11}$$

$$H_r = \frac{E_{0\phi}}{Z_g}J_n\left(\frac{X_{np}r}{a}\right)\sin(n\phi)e^{-j\beta_g z} \tag{6-3-12}$$

$$H_\phi = \frac{E_{0r}}{Z_g}J_n'\left(\frac{X_{np}r}{a}\right)\cos(n\phi)e^{-j\beta_g z} \tag{6-3-13}$$

and

$$H_z = 0 \tag{6-3-14}$$

where

$$Z_g = \frac{E_r}{H_\phi} = -\frac{E_\phi}{H_r} = \frac{\beta_g}{\omega\varepsilon} \quad \text{and} \quad k_c = \frac{X_{np}}{a};$$

X_{np} = the argument of the Bessel function;
$n = 0, 1, 2, 3, \ldots$; and
$p = 1, 2, 3, 4, \ldots$.

Some of the TM-mode characteristic equations for the circular guide are identical to those for the TE mode, but some are different. For convenience, we show all of them here:

$$\beta_g = \sqrt{\omega^2\mu\varepsilon - \left(\frac{X_{np}}{a}\right)^2} \tag{6-3-15}$$

$$k_c = \frac{X_{np}}{a} = \omega_c\sqrt{\mu\varepsilon} \tag{6-3-16}$$

$$f_c = \frac{X_{np}}{2\pi a\sqrt{\mu\varepsilon}} \tag{6-3-17}$$

$$v_g = \frac{\omega}{\beta_g} = \frac{v_p}{\sqrt{1-(f_c/f)^2}} \tag{6-3-18}$$

$$\lambda_g = \frac{\lambda}{\sqrt{1-(f_c/f)^2}} \tag{6-3-19}$$

and

$$Z_g = \frac{\beta_g}{\omega\varepsilon} = \eta\sqrt{1-(f_c/f)^2} \tag{6-3-20}$$

Note that the dominant mode, or the mode of lowest cutoff frequency in a circular waveguide, is the TE_{11} mode, which has the smallest value of the product, $k_c a = 1.841$, as shown in Tables 6-2-1 and 6-3-1.

EXAMPLE 6-3-1 Wave Propagation in Circular Waveguide

An air-filled circular waveguide has a radius of 2 cm and is to carry energy at a frequency of 10 GHz. Find all the TE_{np} and TM_{np} modes for which energy transmission is possible.

Solution:

Since the physical dimension of the guide and the frequency of the wave remain constant, the value of $k_c a$ is also constant. Hence,

$$k_c a = (\omega_0\sqrt{\mu_0\varepsilon_0})a = \frac{2\pi \times 10^{10}}{3 \times 10^8} \times 2 \times 10^{-2}$$

$$= 4.18$$

Any mode that has a value of $(k_c a)$ less than or equal to 4.18 (that is, $k_c a \leqq 4.18$) will propagate the wave with the frequency of 10 GHz. The possible modes are:

$TE_{11}(1.841)$ $TM_{01}(2.405)$
$TE_{21}(3.054)$ $TM_{11}(3.832)$
$TE_{01}(3.832)$

EXAMPLE 6-3-2 Computer Solution for Wave Propagation in Circular Waveguide

```
PROGRAM    KCA

010C       PROGRAM TO COMPUTE POSSIBLE MODES IN
020C+      CYLINDRICAL WAVEGUIDE
030C       KCA
040        DIMENSION TE(4,3), TM(4,3)
050        INTEGER P
060        REAL KCA
070        DATA TE/3.832,1.841,3.054,4.207,7.016,5.331,
080+               6.706,8.015,10.173,8.536,9.969,11.346/,
090+            TM/2.405,3.832,5.136,6.380,5.520,7.106,
100+               8.417,9.761,8.654,10.173,11.620,13.015/
110        A=0.02
115        F=10.E+09
120        PI=3.141593
130        U0=4.0E-07*PI
140        E0=8.854E-12
150        KCA=(2.0*PI*F*SQRT(U0*E0))*A
155        PRINT 6, KCA
160        DO 4 P=1,3
170        DO 4 N=1,4
190        IF(TE(N,P).GT.KCA) GO TO 9
210     4  PRINT 8, TE(N,P)
212     9  DO 5 P=1,3
213        DO 5 N=1,4
214        IF(TM(N,P).GT.KCA) STOP
216     5  PRINT 10, TM(N,P)
218     6  FORMAT(1H ,"KCA          = ",F10.3)
220     8  FORMAT(1H ,"TE(N,P)      = ",F10.3)
224    10  FORMAT(1H ,"TM(N,P)      = ",F10.3)
250        STOP
260        END
READY.
RUN

PROGRAM    KCA

 KCA         =        4.192
 TE(N,P)     =        3.832
 TE(N,P)     =        1.841
 TE(N,P)     =        3.054
 TM(N,P)     =        2.405
 TM(N,P)     =        3.832

SRU        0.735 UNTS.

RUN COMPLETE.
```

6-4 TEM MODES IN CIRCULAR WAVEGUIDES

As we stated in Section 5-0, the transverse electromagnetic (TEM) mode, or transmission-line mode, is characterized by $E_z = H_z = 0$. This condition means that the electric and magnetic fields are transverse to the direction of wave propagation. The TEM mode cannot exist in hollow waveguides, since it requires two conductors, such as a coaxial transmission line or a two-open–wire line.

Analysis of the TEM mode illustrates a clearly analogous relationship between circuit theory and field theory. Figure 6-4-1 shows a coaxial line, which we use for purposes of the analysis.

Maxwell's curl equations in cylindrical coordinates

$$\nabla \times \mathbf{E} = -j\omega\mu\mathbf{H} \tag{6-4-1}$$

and

$$\nabla \times \mathbf{H} = j\omega\varepsilon\mathbf{E} \tag{6-4-2}$$

become

$$\beta_g E_r = \omega\mu H_\phi \tag{6-4-3}$$

$$\beta_g E_\phi = \omega\mu H_r \tag{6-4-4}$$

$$\frac{\partial}{\partial r}(rE_\phi) - \frac{\partial E_r}{\partial \phi} = 0 \tag{6-4-5}$$

$$\beta_g H_r = -\omega\varepsilon E_\phi \tag{6-4-6}$$

$$\beta_g H_\phi = \omega\varepsilon E_r \tag{6-4-7}$$

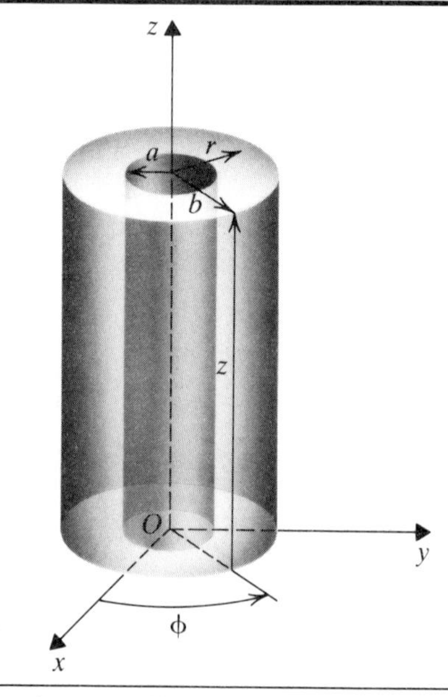

Figure 6-4-1 Coordinates of a coaxial line.

and

$$\frac{\partial}{\partial r}(rH_\phi) - \frac{\partial H_r}{\partial \phi} = 0 \tag{6-4-8}$$

where $\frac{\partial}{\partial r}$ is replaced by $-j\beta_g$ and $E_z = H_z = 0$. Substituting Eq. (6-4-4) into Eq. (6-4-6) yields the propagation constant of the TEM mode in a coaxial line:

$$\beta_g = \omega\sqrt{\mu\varepsilon} \tag{6-4-9}$$

which is the phase constant of the wave in a lossless transmission line with a dielectric.

When we compare Eq. (6-4-9) with the characteristic equation of the Helmholtz equation in cylindrical coordinates, Eq. (6-1-11),

$$\beta_g \pm \sqrt{\omega^2\mu\varepsilon - k_c^2} \tag{6-4-10}$$

it is evident that

$$k_c = 0 \tag{6-4-11}$$

which means that the cutoff frequency of the TEM mode in a coaxial line and an ordinary two-wire line is zero.

From Eq. (6-4-9) we can express the phase velocity of the TEM mode:

$$v_p = \frac{\omega}{\beta_g} = \frac{1}{\sqrt{\mu\varepsilon}} \tag{6-4-12}$$

which is the velocity of light in an unbounded dielectric.

We can find the wave impedance of the TEM mode from either Eqs. (6-4-3) and (6-4-6) or Eqs. (6-4-4) and (6-4-7):

$$\eta(\text{TEM}) = \sqrt{\frac{\mu}{\varepsilon}} \tag{6-4-13}$$

which is the wave impedance of the lossless transmission lines in a dielectric.

Ampere's law states that the line integral of **H** about any closed path is exactly equal to the current enclosed by that path; that is,

$$\oint \mathbf{H} \cdot d\ell = I = I_0 e^{-j\beta_g z} = 2\pi r H_\phi \tag{6-4-14}$$

where I is the complex current that must be supported by the center conductor of a coaxial line. This requirement clearly demonstrates that the TEM mode can only exist in a two-conductor system—not in the hollow waveguide, because the center conductor does not exist.

In summary, the properties of the TEM modes in a lossless medium are as follows:

1. Its cutoff frequency is zero.
2. Its transmission line is a two-conductor system.
3. Its wave impedance is the impedance in an unbounded dielectric.
4. Its propagation constant is the constant in an unbounded dielectric.
5. Its phase velocity is the velocity of light in an unbounded dielectric.

6-5 POWER TRANSMISSION AND POWER LOSSES IN CIRCULAR WAVEGUIDES

6-5-1 Power Transmission in Circular Guides and Coaxial Lines

In general, we can calculate the power transmitted through circular waveguides and coaxial lines by means of the complex Poynting theorem described in Section 0-4. For a lossless dielectric, we can express the time-average power transmitted through a circular guide as

$$\langle P_{tr} \rangle = \frac{1}{2Z_g} \int_0^{2\pi} \int_0^a [|E_r|^2 + |E_\phi|^2] r\, dr\, d\phi \tag{6-5-1}$$

and

$$\langle P_{tr} \rangle = \frac{Z_g}{2} \int_0^{2\pi} \int_0^a [|H_r|^2 + |H_\phi|^2] r\, dr\, d\phi \tag{6-5-2}$$

where $Z_g = E_r/H_\phi = -E_\phi/H_r$ and is the wave impedance in the guide; and a is the radius of the circular guide. Substituting Z_g for a particular mode in Eq. (6-5-1) yields the power transmitted by that mode through the guide.

For TE_{np} modes the average power transmitted through a circular guide is

$$\langle P_{tr} \rangle = \frac{\sqrt{1 - (f_c/f)^2}}{2\eta} \int_0^{2\pi} \int_0^a [|E_r|^2 + |E_\phi|^2] r\, dr\, d\phi \tag{6-5-3}$$

where $\eta = \sqrt{\mu/\varepsilon}$ and is the intrinsic impedance in an unbounded dielectric.

For TM_{np} modes the average power transmitted through a circular guide is

$$\langle P_{tr} \rangle = \frac{1}{2\eta\sqrt{1 - (f_c/f)^2}} \int_0^{2\pi} \int_0^a [|E_r|^2 + |E_\phi|^2] r\, dr\, d\phi \tag{6-5-4}$$

For the TEM mode in coaxial lines the average power transmitted through a coaxial line or a two-open–wire line is

$$\langle P_{tr} \rangle = \frac{1}{2\eta} \int_0^{2\pi} \int_0^a [|E_r|^2 + |E_\phi|^2] r\, dr\, d\phi \tag{6-5-5}$$

If we assume that the current carried by the center conductor of a coaxial line is

$$I_z = I_0 e^{-j\beta_g z} \tag{6-5-6}$$

we can express the magnetic intensity induced by the current around the center conductor by rearranging Ampere's law; that is,

$$H_\phi = \frac{I_0}{2\pi r} e^{-j\beta_g z} \tag{6-5-7}$$

The potential rise from the outer conductor to the center conductor is

$$V_r = -\int_b^a E_r\, dr = -\int_b^a \eta H_\phi\, dr$$

$$= \frac{I_0 \eta}{2\pi} \ln\left(\frac{b}{a}\right) e^{-j\beta_g z} \tag{6-5-8}$$

The characteristic impedance of a coaxial line is

$$Z_0 = \frac{V}{I} = \frac{\eta}{2\pi} \ln\left(\frac{b}{a}\right) \tag{6-5-9}$$

where $\eta = \sqrt{\mu/\varepsilon}$ is the intrinsic impedance in an unbounded dielectric.

From Eq. (6-5-5) we can express the power transmitted by TEM mode in a coaxial line:

$$\langle P_{tr} \rangle = \frac{1}{2\eta} \int_0^{2\pi} \int_a^b |\eta H_\phi|^2 r \, dr \, d\phi = \frac{\eta I_0^2}{4\pi} \ln\left(\frac{b}{a}\right) \tag{6-5-10}$$

Substituting $|V_r|$ from Eq. (6-5-8) into Eq. (6-5-10) yields

$$\langle P_{tr} \rangle = \tfrac{1}{2} V_0 I_0 \tag{6-5-11}$$

Thus the power transmission derived from the Poynting theory is the same as that derived from the circuit theory for an ordinary transmission line.

6-5-2 Power Losses in Circular Guides and Coaxial Lines

The theory and equations presented in Sections 5-2 and 5-3 for the TE and TM modes in rectangular waveguides are applicable to the TE and TM modes in circular guides. The power losses for the TEM mode in coaxial lines can be calculated from the transmission-line theory by

$$P_L = 2\alpha P_{tr} \tag{6-5-12}$$

where

P_L represents the power loss per unit length;

$\langle P_{tr} \rangle$ represents the transmitted power; and

α is the attenuation constant;

For low-loss conductor, the magnitude of the attenuation constant is

$$\alpha = \frac{1}{2}\left(R\sqrt{\frac{C}{L}} + G\sqrt{\frac{L}{C}}\right) \tag{6-5-13}$$

6-6 EXCITATION OF MODES AND MODE PATTERNS IN CIRCULAR WAVEGUIDES

6-6-1 Excitation of Modes in Circular Waveguides

As we indicated in Sections 6-2 and 6-3, TE modes have no electric-field z component and TM modes have no magnetic-intensity z component. If a device is inserted in a circular guide in such a way that it excites only a z component of electric intensity, the wave propagating through the guide will be of the TM modes; however, if a device is placed in a circular guide in such a manner that only the z component of magnetic intensity exists, the traveling wave will be of the TE modes. The methods of excitation for two modes in circular waveguides are shown in Fig. 6-6-1.

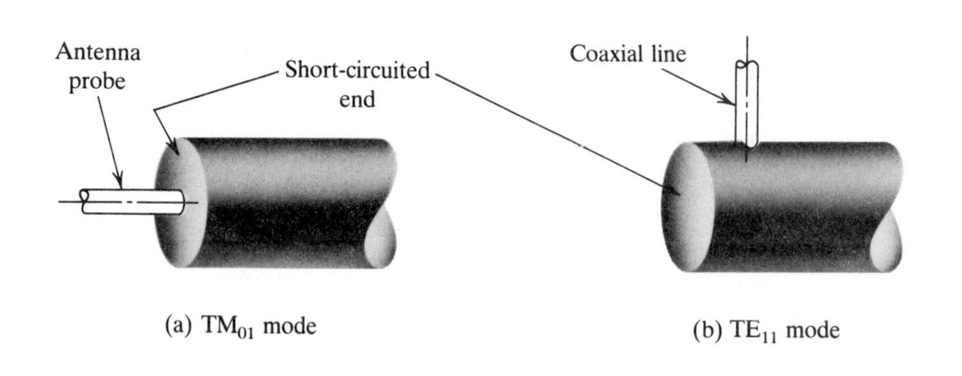

Figure 6-6-1 Methods of exciting two different modes in circular waveguides.

A common way used to excite the TM modes in a circular guide is with a coaxial line, as shown in Fig. 6-6-2. At the end of the coaxial line a large magnetic intensity is present in the ϕ direction of wave propagation. The magnetic field from the coaxial line will excite the TM modes in the guide. However, when the guide is connected to the source by a coaxial line, discontinuity at the junction increases the standing-wave ratio on the line and eventually decreases the amount of power transmitted. Placing a tuning device around the junction in order to suppress the reflection is often necessary.

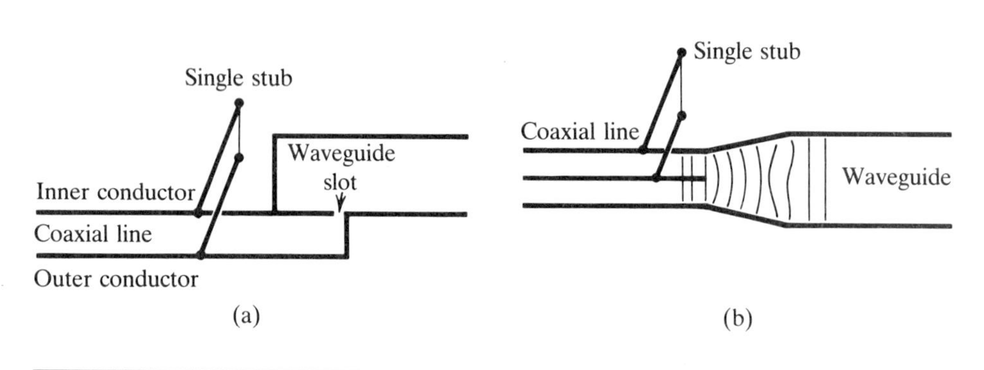

Figure 6-6-2 Methods of exciting waveguides from coaxial lines.

6-6-2 Mode Patterns in Circular Waveguides

The orientations of electric and magnetic-force lines described in this section are applicable to the TE and TM modes in circular waveguides. Figure 6-6-3 shows the mode patterns of two modes in circular guides.

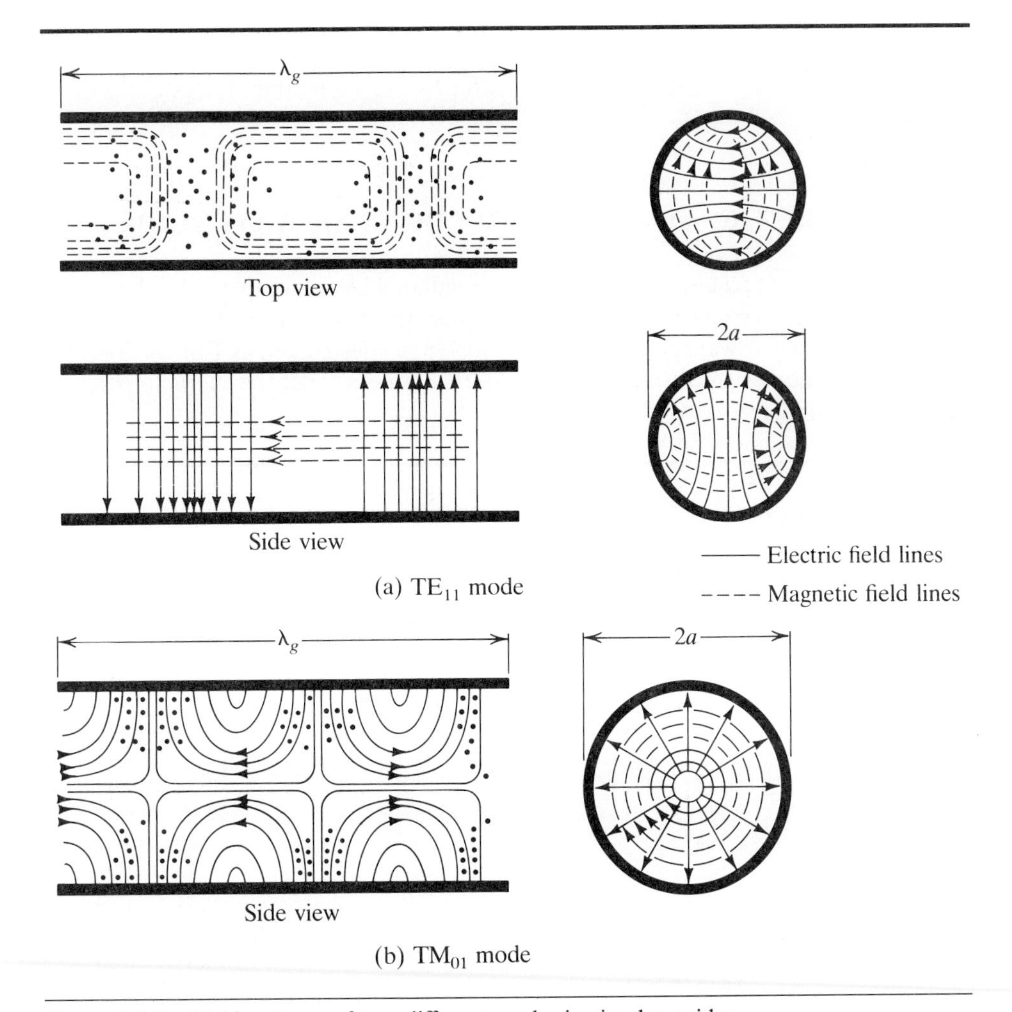

Figure 6-6-3 Field patterns of two different modes in circular guides.

EXAMPLE 6-5-1 Coaxial Line

A coaxial line is used to transmit a TEM-mode wave and its dimensions are $a = 1$ cm and $b = 2$ cm. The dielectric between the two conductors is $\varepsilon_r = 2.56$ and the current I_0 is 500 mA.

Calculate:

a) the characteristic impedance of the line; and
b) the transmitted power.

Solutions:

a) The characteristic impedance is

$$Z_0 = \frac{\eta}{2\pi}\ln\left(\frac{b}{a}\right) = \frac{377}{\sqrt{2.56} \times 2\pi}\ln\left(\frac{2}{1}\right) = 37.5 \times 0.693 = 26\ \Omega$$

b) The transmitted power is

$$\langle P_{tr} \rangle = \frac{\eta I_0^2}{4\pi} \ln\left(\frac{b}{a}\right) = \frac{\eta}{2\pi} \ln\left(\frac{b}{a}\right) \frac{I_0^2}{2} = 26 \times \frac{(0.5)^2}{2} = 3.25 \text{ W}$$

6-7 CHARACTERISTICS OF STANDARD CIRCULAR WAVEGUIDES

The inner diameter of a circular waveguide is regulated by the frequency of the signal being transmitted. For example, at X-band frequencies of from 8 to 12 GHz, the inner diameter of a circular waveguide is 2.383 cm (0.938 in.) designated as EIA WC(94) by the Electronic Industry Association. Table 6-7-1 contains a list of the characteristics of the standard circular waveguides.

Table 6-7-1 Characteristics of standard circular waveguides

EIA Designation WC( )	Inside diameter $2a$ cm (in.)	Cutoff Frequency for Air-Filled Waveguide (GHz)	Recommended Frequency Range for TE_{11} Mode (GHz)
992	25.184 (9.915)	0.698	0.80– 1.10
847	21.514 (8.470)	0.817	0.94– 1.29
724	18.377 (7.235)	0.957	1.10– 1.51
618	15.700 (6.181)	1.120	1.29– 1.76
528	13.411 (5.280)	1.311	1.51– 2.07
451	11.458 (4.511)	1.534	1.76– 2.42
385	9.787 (3.853)	1.796	2.07– 2.83
329	8.362 (3.292)	2.102	2.42– 3.31
281	7.142 (2.812)	2.461	2.83– 3.88
240	6.104 (2.403)	2.880	3.31– 4.54
205	5.199 (2.047)	3.381	3.89– 5.33
175	4.445 (1.750)	3.955	4.54– 6.23
150	3.810 (1.500)	4.614	5.30– 7.27
128	3.254 (1.281)	5.402	6.21– 8.51
109	2.779 (1.094)	6.326	7.27– 9.97
94	2.383 (0.938)	7.377	8.49– 11.60
80	2.024 (0.797)	8.685	9.97– 13.70
69	1.748 (0.688)	10.057	11.60– 15.90
59	1.509 (0.594)	11.649	13.40– 18.40
50	1.270 (0.500)	13.842	15.90– 21.80
44	1.113 (0.438)	15.794	18.20– 24.90
38	0.953 (0.375)	18.446	21.20– 29.10
33	0.833 (0.328)	21.103	24.30– 33.20
28	0.714 (0.281)	24.620	28.30– 38.80
25	0.635 (0.250)	27.683	31.80– 43.60
22	0.556 (0.219)	31.617	36.40– 49.80
19	0.478 (0.188)	36.776	42.40– 58.10
17	0.437 (0.172)	40.227	46.30– 63.50
14	0.358 (0.141)	49.103	56.60– 77.50
13	0.318 (0.125)	55.280	63.50– 87.20
11	0.277 (0.109)	63.462	72.70– 99.70
9	0.239 (0.094)	73.552	84.80–116.00

Note: EIA stands for Electronic Industry Association, and WC represents Circular Waveguide

6-8 CIRCULAR-CAVITY RESONATOR

A circular-cavity resonator is a circular waveguide with two ends closed by metal walls, as depicted in Fig. 6-8-1. The wave function in the circular resonator should satisfy Maxwell's equations, subject to the same boundary conditions that we described for the rectangular-cavity resonator. We merely choose the harmonic functions in z to satisfy the boundary conditions at the two end walls. These conditions can be achieved if

$$H_z = H_{0z} J_n\left(\frac{X'_{np}r}{a}\right) \cos(n\phi) \sin\left(\frac{q\pi z}{d}\right) \quad \text{for } \mathrm{TE}_{npq} \tag{6-8-1}$$

where

$n = 0, 1, 2, 3, \ldots$ is the number of the periodicity in the ϕ direction;
$p = 1, 2, 3, 4, \ldots$ is the number of zeros of the field in the radial direction;
$q = 1, 2, 3, 4, \ldots$ is the number of half-waves in the axial direction;
J_n is the Bessel function of the first kind; and
H_{0z} is the amplitude of the magnetic field.

and

$$E_z = E_{0z} J_n\left(\frac{X_{np}r}{a}\right) \cos(n\phi) \cos\left(\frac{q\pi z}{d}\right) \quad \text{for } \mathrm{TM}_{npq} \tag{6-8-2}$$

where

$n = 0, 1, 2, 3, \ldots;$
$p = 1, 2, 3, 4, \ldots;$
$q = 0, 1, 2, 3, \ldots;$ and
E_{0z} is the amplitude of the electric field.

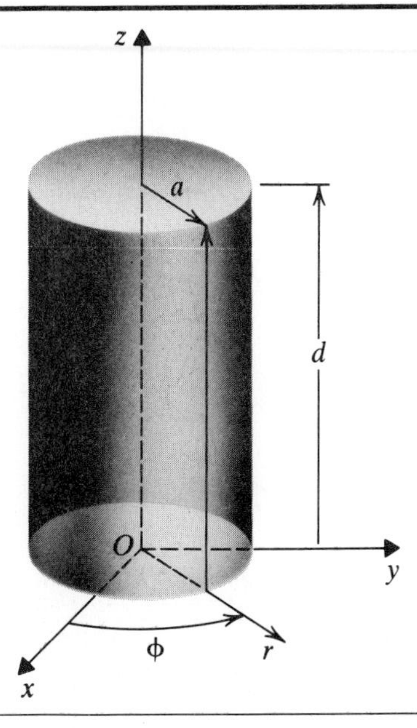

Figure 6-8-1 A circular resonator.

The separation equations for the TE and TM modes are

$$k^2 = \left(\frac{X'_{np}}{a}\right)^2 + \left(\frac{q\pi}{d}\right)^2 \qquad \text{TE modes} \tag{6-8-3}$$

and

$$k^2 = \left(\frac{X_{np}}{a}\right)^2 + \left(\frac{q\pi}{d}\right)^2 \qquad \text{TM modes} \tag{6-8-4}$$

Substituting $k^2 = \omega^2\mu\varepsilon$ in Eqs. (6-8-3) and (6-8-4) yields the resonant frequencies for the TE and TM modes:

$$f_r = \frac{1}{2\pi\sqrt{\mu\varepsilon}}\sqrt{\left(\frac{X'_{np}}{a}\right)^2 + \left(\frac{q\pi}{d}\right)^2} \qquad \text{TE modes} \tag{6-8-5}$$

and

$$f_r = \frac{1}{2\pi\sqrt{\mu\varepsilon}}\sqrt{\left(\frac{X_{np}}{a}\right)^2 + \left(\frac{q\pi}{d}\right)^2} \qquad \text{TM modes} \tag{6-8-6}$$

Note that for $2a > d$, the TM_{110} mode is dominant and that for $2a \le d$, the TE_{111} mode is dominant.

EXAMPLE 6-8-1 Circular-Cavity Resonator

An air-filled circular-cavity resonator has an inner radius $a = 2.383$ cm and height $d = 4.854$ cm.

Determine:

a) the resonant (or dominant) mode;
b) the resonant frequency; and
c) the resonant frequency if the resonator is dielectrically loaded with $\varepsilon_r = 2.25$.

Solutions:

a) The resonant mode is TE_{111} mode because $2a < d$.
b) The resonant frequency is

$$f_r = \frac{3 \times 10^8}{2\pi}\sqrt{\left(\frac{1.841}{2.383 \times 10^{-2}}\right)^2 + \left(\frac{1 \times \pi}{4.854 \times 10^{-2}}\right)^2}$$

$$= 4.81 \text{ GHz}$$

c) If the resonator is dielectrically loaded with $\varepsilon_r = 2.25$, the resonant frequency is

$$f_{re} = \frac{f_r}{\sqrt{\varepsilon_r}} = \frac{4.81 \times 10^9}{\sqrt{2.25}} = 3.21 \text{ GHz}$$

Problems

6-1 An air-filled circular waveguide is to be operated at a frequency of 6 GHz and is to have dimensions such that $f_c = 0.8f$ for the dominant mode. Determine the

a) diameter of the guide;
b) wavelength λ_g; and
c) the phase velocity v_g.

6-2 An air-filled circular waveguide of 1.191 cm inside radius is operated in the TE_{01} mode.

a) Calculate the cutoff frequency.
b) If the guide is to be filled with a dielectric material of $\varepsilon_r = 2.25$, to what size must its radius be changed in order to maintain the same cutoff frequency?

6-3 An air-filled circular waveguide has a radius of 1.012 cm and is to carry energy at a frequency of 10 GHz. Find all TE and TM modes for which transmission is possible.

6-4 A TE_{11} wave is propagating through a circular waveguide. The diameter of the guide is 11.458 cm, and the guide is air-filled. Find the

a) cutoff frequency;
b) wavelength λ_g in the guide for a frequency of 3 GHz; and
c) wave impedance in the guide.

6-5 An air-filled circular waveguide has a diameter of 1 cm and is to carry energy at a frequency of 10 GHz. Determine all TE_{nm} modes for which transmission is possible.

6-6 A circular waveguide has a cutoff frequency of 9 GHz in the dominant mode. Determine the

a) inside diameter of the guide if it is air-filled; and
b) inside diameter of the guide if the guide is dielectric-filled. The dielectric constant $\varepsilon_r = 4$.

6-7 A circular waveguide has a radius of 3 cm. The waveguide is to be made resonant for a TE_{01} mode at 10 GHz by placing two perfectly conducting plates at its two ends. Determine the minimum distance in cm between the two plates.

6-8 A coaxial resonator is constructed of a section of coaxial line 5 cm long. The resonator is filled with dielectric material ($\varepsilon_r = 9$) and open circuited at both ends. The radius of the inner conductor is 1 cm and the radius of the outer conductor is 2.5 cm. Determine the

a) resonant frequency of the resonator; and
b) resonant frequency of the same cavity with one end open and the other shorted.

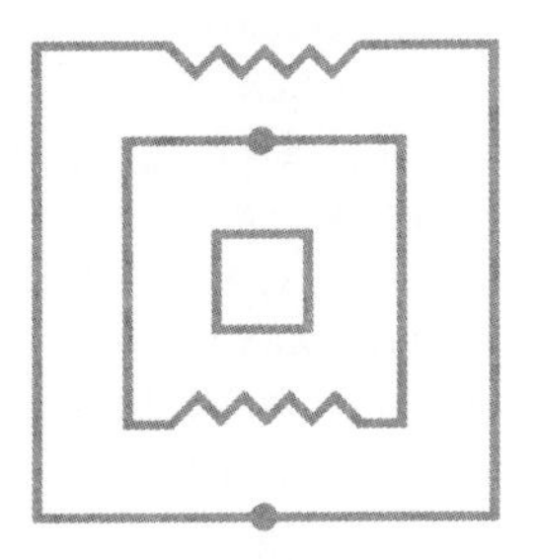

Chapter 7

Optical-Fiber Waveguides

7-0 INTRODUCTION

The optical fiber is one of the latest additions to signal links in submillimeter-wave technology. Transmission loss is extremely low, which makes the optical fiber extremely valuable for applications involving computer links, industrial automation, medical instruments, telecommunications, and military command and control systems. In this chapter, we discuss the operational mechanisms, modes, structures, and characteristics of optical-fiber cables and systems.

7-1 OPTICAL FIBERS

An optical-fiber cable is an assembly of several optical fibers. The optical fiber is simply a circular dielectric waveguide system. The center part consists of a core with a larger refractive index; the outer part consists of a cladding with a smaller refractive index. If a beam of electromagnetic energy impinges on this system through one end face of the fiber, the light beam will travel through the fiber and emerge from the other end face.

The basic difference in signal transmission between a metal transmission line and an optical fiber lies in the carriers. In a metal line the carriers are electrons; in an optical fiber the carriers are photons. The advantages of optical fibers over metal lines in long-distance communication systems are the low-transmission loss and attractive bandwidths. Installed optical-fiber cables have shown losses in the neighborhood of only 4 dB/km at wavelengths of 0.82 to 0.85 μm. Several experiments have demonstrated that optical fiber 30 km or longer had a loss of less than 0.7 dB/km at wavelengths near 1.3 μm. In data communication systems a bit rate of 50 Mbit/s for a repeater span of at least 10 km was achieved for a multimode fiber.

Light-wave transmission by optical fiber has reached a fully commercial stage, with carrier wavelengths in the range of 0.82 to 0.85 μm. In the future operational carrier wavelengths of 1.2 to 1.6 μm are expected. The key components needed for a light-wave system are optical fibers, light-wave sources [lasers or light-emitting diodes (LEDs)], and light-wave detectors (avalanche photodiodes or p-i-n photodiodes). Optical-fiber cables have been used in underwater intercontinental communication systems, intercity and metropolitan telephone systems, data communication systems, computer and switching-link systems, automation process systems, and military command and control systems.

7-1-1 Materials and Fabrication

1. *Materials*

Three major materials are used in the production of optical fibers:

Silica fibers. Basically, silica fibers are made of silicon dioxide (SiO_2) along with other metal oxides to establish a difference in refractive index between the core and cladding. Various dopants—such as TiO_2, Al_2O_3, GeO_2, and P_2O_5—have been used to increase or decrease the refractive index of silica. The cladding has a lower refractive index than the core, but its coefficient of thermal expansion is higher.

Glass fibers. Glass fibers are made of compound glasses with low melting temperatures and long-term chemical stability.

Plastic fibers. Plastic fibers consist of plastics having higher attenuation losses than silica and glass. They are commonly used for short-distance links in computers.

2. *Fabrications*

Four fabrication processes are used to make optical fibers:

Outside vapor-phase oxidation (OVPO) process. Developed by the Corning Glass Works, the OVPO process was used to deposit high-silica glass from vapor-phase sources. When the mandrel is removed, the "soot" tube is sintered and subsequently drawn to a fiber.

Modified chemical vapor deposition (MCVD) process. The MCVD process was invented by Bell Laboratories. Glass of the desired composition is deposited inside a silica tube, layer by layer, using a flame-hydrolysis process, over the length of a mandrel. When deposition is complete, the material is sintered and the mandrel removed. The consolidated tube is then fused and collapsed simultaneously into a perform rod.

Vapor-phase axial deposition (VAD) process. Introduced by the Nippon Telegraph and Telephone Public Corporation, the VAD process is similar to the OVPO process, except that deposition occurs at the end of a growing "soot" cylinder.

Plasma chemical vapor deposition (PCVD) process. The Philip Company pioneered the PCVD process, which uses microwave plasma to excite reactants and deposit vitreous material directly inside a silica tube.

All four processes fabricate fibers by using compound SiO_2 (silica) with relatively small amounts of germanium (Ge), phosphorus (P), and sometimes boron (B) as dopants to alter the refractive index and lower the working temperature somewhat below that for pure SiO_2.

7-1-2 Physical Structures

As noted, an optical fiber is simply a circular dielectric waveguide system. It consists of a central core and an outside cladding. The core is a cylinder of transparent dielectric rod with a higher refractive index n_1, and the cladding is a second dielectric sheathing, or covering (usually glass fused to the core) with a lower refractive index n_2. Optical fibers generally are classified as follows:

1. Monomode step-index fiber, which has a core radius of 1 to 16 μm and a cladding radius of 50 to 100 μm.
2. Multimode step-index fiber, which has a core radius of 25 to 60 μm and a cladding radius of 50 to 150 μm.
3. Multimode graded-index fiber, which has a core radius of 10 to 35 μm and a cladding radius of 50 to 80 μm.

Figure 7-1-1 shows diagrams of these three types of optical fiber with their index profiles.

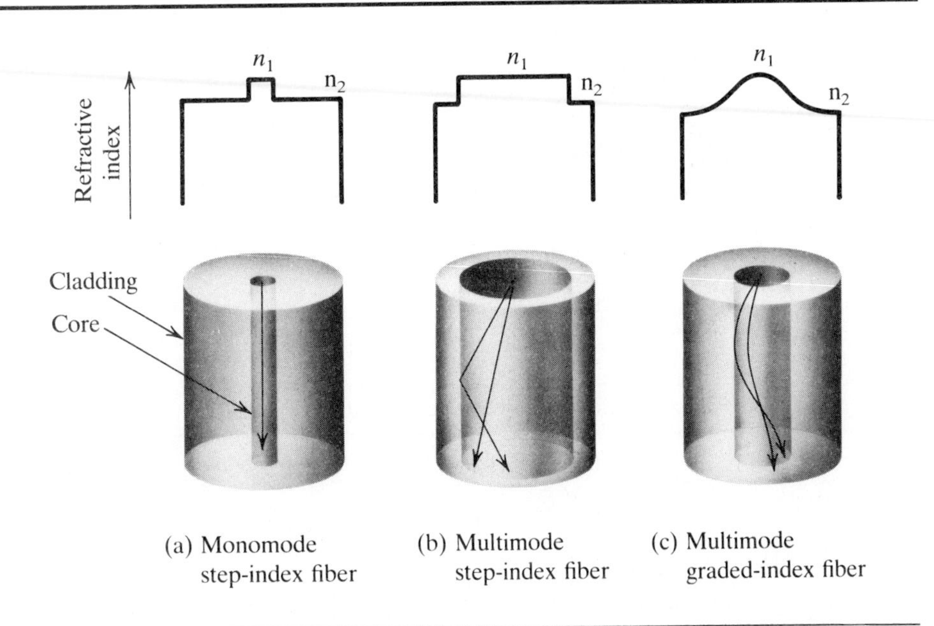

Figure 7-1-1 Physical structures of the three types of optical fibers.

An optical-fiber cable consists of several optical fibers encased in a protective material. Many different types of optical-fiber cables are available. Figure 7-1-2 shows several typical cable configurations.

One fiber cable may contain hundreds of fibers for high-capacity channels. Figure 7-1-3 shows a cable that consists of 12 stacked ribbons, each containing 12 optical fibers.

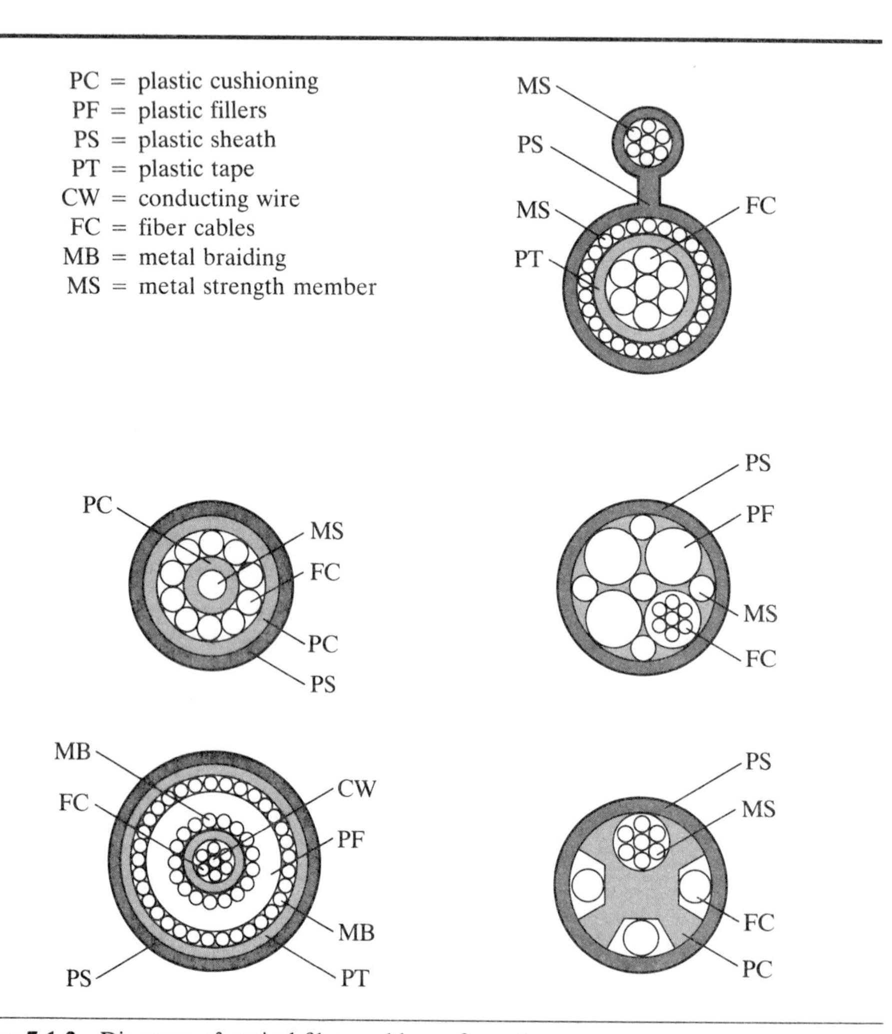

Figure 7-1-2 Diagrams of optical-fiber–cable configurations.

SOURCE: AFTER G. R. AND H. A. ELION [1]; REPRINTED WITH PERMISSION OF THE MARCEL DEKKER COMPANY, INC.

7-1-3 Losses

In an ideal case, we consider the fiber core to be a perfectly transparent material and the cladding a shield that confines the electromagnetic energy to the core region by total reflection. Otherwise, we would have to deal with absorption, transmission, scattering, and microbending losses.

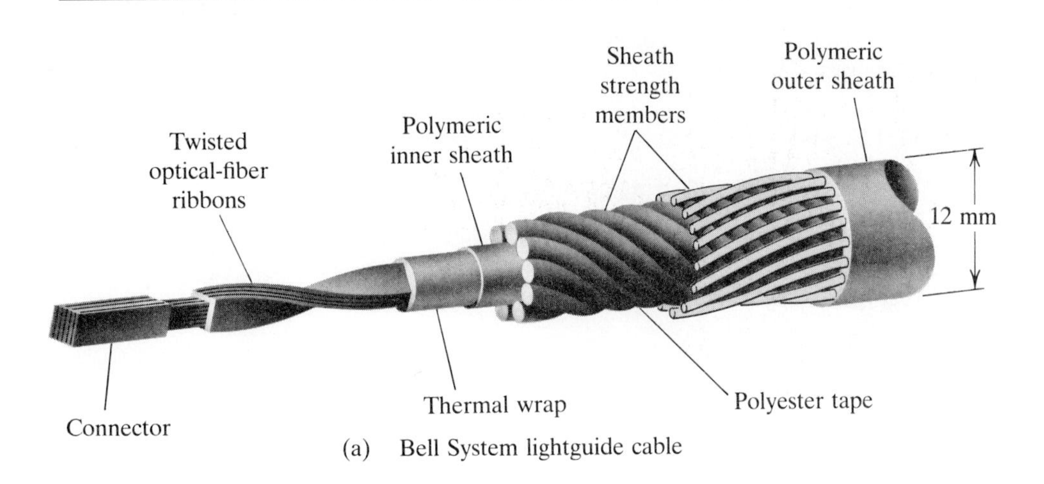

(a) Bell System lightguide cable

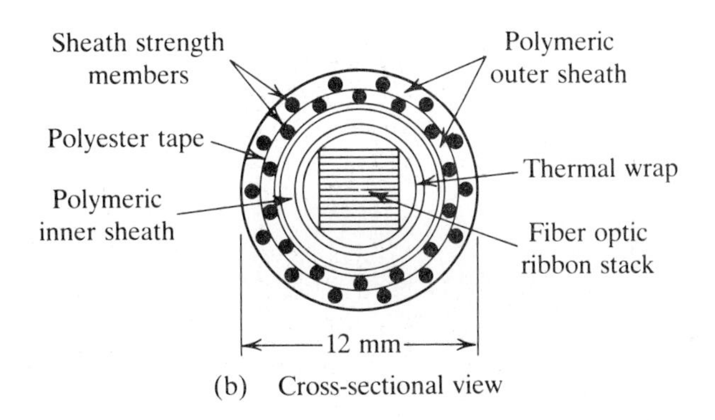

(b) Cross-sectional view

Figure 7-1-3 Optical-fiber cable with 144 fibers.

SOURCE: AFTER I. JACOBS AND S. E. MILLER [2]; © 1977 IEEE.

1. *Absorption loss.* If the core material is not perfectly transparent because of an impurity, the absorption loss for a length ℓ of the light path through the core glass is

$$\text{Absorption loss} = e^{-\alpha\ell} \qquad (7\text{-}1\text{-}1)$$

   where

   α is the absorption coefficient per unit length;

   $\ell = n_1(n_1^2 - \sin^2\theta_m)^{-1/2}$ and is the light-path length; and

   θ_m is the maximum acceptance angle, as defined in Section 7-2-3.

2. *Scattering loss.* Fiber-waveguide scattering losses are caused mainly by geometric irregularities at the core–cladding interface.
3. *Microbending loss.* The microbending of an optical fiber causes radiation losses, if the fiber is distorted at joints separated as much as 1 mm longitudinally.

In particular, plastic optical fibers have higher absorption losses than silica and glass fibers. Therefore they are used primarily for short-distance applications, such as in computer links.

EXAMPLE 7-1-1 Absorption Loss of an Optical Fiber

An optical fiber has the following parameters:

Maximum acceptance angle	$\theta_m = 30^\circ$
Core refractive index	$n_1 = 1.40$
Absorption coefficient	$\alpha = 0.47$ Np/km

Calculate:

a) the light-path length; and
b) the absorption loss.

Solutions:

a) The light-path length ℓ is

$$\ell = \frac{1.40}{\sqrt{(1.40)^2 - \sin^2(30^\circ)}} = \frac{1.40}{1.31} = 1.07 \text{ km}$$

b) The absorption loss is

$$\begin{aligned} \text{Loss} &= e^{-\alpha\ell} = e^{-0.47(1.07)} = e^{-0.50} = 0.61 \\ &= 10 \log(0.61) = -2.20 \text{ dB/km} \end{aligned}$$

7-1-4 Characteristics

In the twenty-first century microwave electronic devices and circuits will be heavily based on electrooptics. The characteristics of optical fibers are:

1. *Small size and light weight.* An optical-fiber cable is much smaller and lighter in weight than a copper-wire line. These features are especially important for underwater cables and overcrowded transmission lines in general.
2. *Low cost and low loss.* The costs of optical-fiber cables and their installation are much lower than for metal cables. The transmission loss of fibers is very low at 4 dB/km, contrasted to that of coaxial lines at 150 dB for 0.5 km at 1 GHz or that of microstrip lines at 150 dB for 0.4 km at 10 GHz.
3. *Immunity of interference.* An optical-fiber cable does not generate or receive any electrical or electromagnetic noise or interference.
4. *High reliability and durability.* Optical-fiber cables are safe to use in any explosive environment and eliminate the hazards of short circuits in wire lines.
5. *High bandwidth.* An optical-fiber cable can carry more channels than either coaxial or microstrip lines.
6. *High security capability.* Optical-fiber cables are immune to electrical or electromagnetic noise or grounding, such as cross-talk or jamming. Such cables increase the security of data transmission, which is very important in military communications and in civilian applications in which privacy is a major concern.

7-2 OPERATIONAL MECHANISMS OF OPTICAL FIBERS

We can analyze light-wave propagation in an optical fiber with a nondissipative core of radius a and refractive index n_1 imbedded in a nondissipative cladding medium of radius b and refractive index n_2 by solving Maxwell's field equations for specific boundary conditions. The resultant waves in an optical fiber are neither transverse electric (TE) nor transverse magnetic (TM), but hybrid. In other words, there is no nonzero component of both electric **E** and magnetic **H** waves in the direction of propagation. A cylindrical coordinate system (r, ϕ, z) for an optical fiber is shown in Fig. 7-2-1.

The optical properties of a material are usually characterized by a complex refractive index N, which contains two constants: the refractive index n and the extinction index k. That is,

$$N = n - jk \tag{7-2-1}$$

The extinction index k is related to the exponential decay of the wave as it passes through the medium. For nondissipative optical fibers, $k = 0$. The refractive index n is defined as the ratio of the velocity of light in vacuum to the velocity of light in a particular medium. For nonmagnetic media the refractive index is

$$n = \frac{v_0}{v_\varepsilon} = \frac{\sqrt{\mu_0 \varepsilon}}{\sqrt{\mu_0 \varepsilon_0}} = \sqrt{\varepsilon_r} \tag{7-2-2}$$

where ε_r is the relative dielectric constant of the medium. Typical values of n are 1.00 for air, 1.50 for polyethylene, 1.60 for polystyrene, and 8.94 for fresh water.

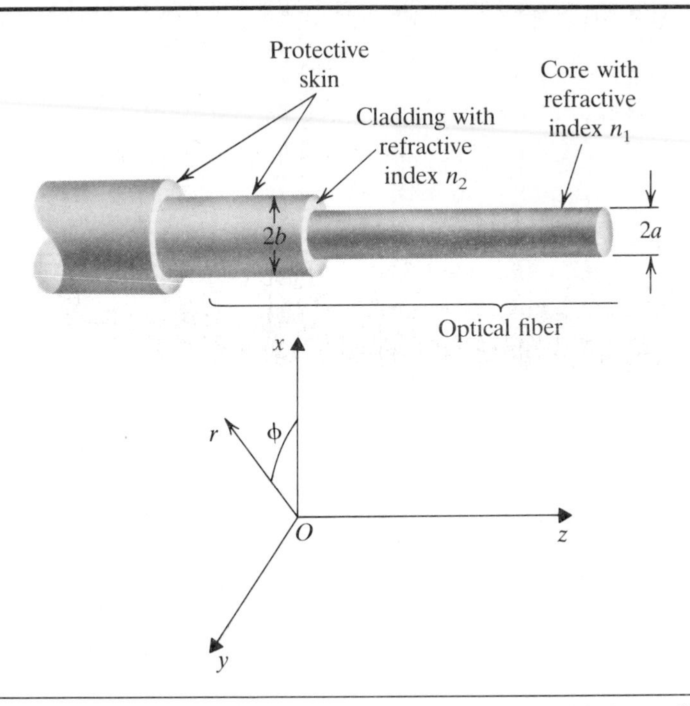

Figure 7-2-1 Schematic diagram of an optical fiber and its cylindrical coordinate system (r, ϕ, z).

7-2-1 Wave Equations

As shown in Fig. 7-2-1, the optical-fiber waveguide consists of a core of higher refractive index n_1 and radius a surrounded by a cladding of lower refractive index n_2 and radius b. Both regions are assumed to be perfect insulators with free-space magnetic permeability μ_0. Such a structure can support an infinite number of modes, but for specific values of n_1, n_2, a, and b, only a finite number are waveguide modes that have localized fields in the vicinity of the core. The other unbound modes correspond, for example, to light striking the core on one side, passing through the core, and emerging from the other side.

A waveguide mode is a coherent distribution of light, which is localized in the core by total internal reflection and which propagates through the guide with a well-defined phase velocity. We choose the z axis of the cylindrical coordinate system (r, ϕ, z) to be along the fiber core axis. We can then express the z components of the electric field **E** and magnetic field **H** parallel to the fiber guide as follows, assuming that the exponential factor $e^{j(\omega t - \beta_g z)}$ is implied. In the core region

$$E_z = AJ_n(kr)\cos(n\phi) \tag{7-2-3}$$

and

$$H_z = BJ_n(kr)\sin(n\phi) \tag{7-2-4}$$

and in the cladding region

$$E_z = CH_n(\chi r)\cos(n\phi) \tag{7-2-5}$$

and

$$H_z = DH_n(\chi r)\sin(n\phi) \tag{7-2-6}$$

where

$J_n(kr)$ is the nth-order Bessel function of the first kind;
$H_n(\chi r)$ is the nth-order Hankel function of the first kind;
k is the transverse propagation constant in the core region;
χ is the transverse propagation constant in the cladding region;
$k^2 = \beta_1^2 - \beta_g^2$ and is the separation equation in the core region;
$\chi^2 = \beta_g^2 - \beta_2^2$ and is the separation equation in the cladding region;
$\beta_g = \omega\sqrt{\mu_0\varepsilon_0}$ and is the phase constant in free space;
$\beta_1 = \omega\sqrt{\mu_1\varepsilon_1}$ and is the phase constant in the core region; and
$\beta_2 = \omega\sqrt{\mu_2\varepsilon_2}$ and is the phase constant in the cladding region.

The Bessel function J_n exhibits oscillatory behavior for real k, as do the sinusoidal functions. Hence, this function represents a cylindrical standing wave in the core region for $r < a$. The Hankel function H_n represents a traveling wave for real χ in the cladding region for $r > a$, as do the exponential functions. For a nondissipative medium, the Hankel function becomes the modified Bessel function. In order for electric and magnetic fields to be evanescent in the cladding region, χ must be imaginary.

From Eqs. (7-2-3) and (7-2-4) the ϕ components of the field equations in the core region are

$$E_\phi = \left[A\frac{j\beta_g n}{k^2 r}J_n(kr) + B\frac{j\omega\mu_1}{k^2}J_n'(kr)\right]\sin(n\phi) \tag{7-2-7}$$

and

$$H_\phi = -\left[A\frac{j\omega\varepsilon_1}{k^2}J_n'(kr) + B\frac{j\beta_g n}{k^2 r}J_n(kr)\right]\cos(n\phi) \tag{7-2-8}$$

Similarly, from Eqs. (7-2-5) and (7-2-6) the ϕ components of the field equations for the cladding region are

$$E_\phi = -\left[C\frac{j\beta_g n}{\chi^2 r}H_n(\chi r) + D\frac{j\omega\mu_2}{\chi^2}H_n'(\chi r)\right]\sin(n\phi) \tag{7-2-9}$$

and

$$H_\phi = \left[C\frac{j\omega\varepsilon_2}{\chi^2}H_n'(\chi r) + D\frac{j\beta_g n}{\chi^2 r}H_n(\chi r)\right]\cos(n\phi) \tag{7-2-10}$$

The primes on J_n and H_n refer to differentiation with respect to their arguments (kr) and (χr), respectively.

The tangential field components should be continuous at the interface $r = a$; therefore

$$AJ_n(ka) = CH_n(\chi a) \tag{7-2-11}$$

$$BJ_n(ka) = DH_n(\chi a) \tag{7-2-12}$$

$$A\frac{j\beta_g n}{k^2 a}J_n(ka) + B\frac{j\omega\mu_1}{k^2}J_n'(ka) = -C\frac{j\beta_g n}{\chi^2 a}H_n(\chi a) - D\frac{j\omega\mu_2}{\chi^2}H_n'(\chi a) \tag{7-2-13}$$

and

$$A\frac{j\omega\varepsilon_1}{k^2}J_n'(ka) + B\frac{j\beta_g n}{k^2 a}J_n(ka) = -C\frac{j\omega\varepsilon_2}{\chi^2}H_n'(\chi a) - D\frac{j\beta_g n}{\chi^2 a}H_n(\chi a) \tag{7-2-14}$$

When we have determined constants A, B, C, and D, we can finally derive the electric and magnetic field equations from Maxwell's equations.

EXAMPLE 7-2-1 Wave Propagation in an Optical Fiber

A monomode step-index fiber has the following parameters:

Carrier wavelength	$\lambda = 0.82\ \mu\text{m}$
Carrier frequency	$f = 3.66 \times 10^{14}$ Hz
Core radius	$a = 2\ \mu\text{m}$
Core refractive index	$n_1 = 1.50$
Cladding radius	$b = 80\ \mu\text{m}$
Cladding refractive index	$n_2 = 1.35$

Calculate:

a) the phase constant in the core region;
b) the phase constant in the cladding region;
c) the phase constant in free space;
d) the transverse propagation constant in the core region;
e) the transverse propagation constant in the cladding region;
f) the value of the Bessel function $J_1(ka)$; and
g) the value of the Hankel function $H_1(\chi b)$.

Solutions:

a) The phase constant in the core region is

$$\beta_1 = \omega\sqrt{\mu_1\varepsilon_1} = \omega\frac{n_1}{c} = 2\pi \times 3.66 \times 10^{14}\left(\frac{1.5}{3 \times 10^8}\right)$$

$$= 1.15 \times 10^7 \text{ rad/m}$$

b) The phase constant in the cladding region is

$$\beta_2 = 2\pi \times 3.66 \times 10^{14}\left(\frac{1.35}{3 \times 10^8}\right)$$

$$= 1.03 \times 10^7 \text{ rad/m}$$

c) The phase constant in free space is

$$\beta_g = \frac{2\pi \times 3.66 \times 10^{14}}{3 \times 10^8}$$

$$= 7.67 \times 10^6 \text{ rad/m}$$

d) The transverse propagation constant in the core region is

$$k = \sqrt{(1.15 \times 10^7)^2 - (7.67 \times 10^6)^2}$$

$$= 8.6 \times 10^6 \text{ rad/m}$$

e) The transverse propagation constant in the cladding region is

$$\chi = \sqrt{(7.67 \times 10^6)^2 - (1.03 \times 10^7)^2}$$

$$= j6.9 \times 10^6 \text{ rad/m}$$

f) The Bessel function $J_1(ka)$ is

$$J_1(ka) = J_1(8.6 \times 10^6 \times 2 \times 10^{-6}) = J_1(17.20)$$

$$= -0.12814$$

g) The Hankel function $H_1(\chi b)$ is

$$H_1(\chi b) = H_1(j6.9 \times 10^6 \times 80 \times 10^{-6})$$

$$= H_1(j552)$$

The Hankel function vanishes in the cladding region because its argument is imaginary.

7-2-2 Wave Modes and Cutoff Wavelengths

1. *Wave Modes*

Recall that for metallic circular waveguides the wave modes are usually designated TE_{np} and TM_{np}, meaning transverse electric and transverse magnetic waves, respectively. However, in dielectric circular waveguides only the cylindrically symmetric ($n = 0$) modes are either transverse electric (TE_{0p}) or transverse magnetic (TM_{0p}). The other modes are all hybrid, in which the electric field E_z and the magnetic field H_z coexist.

The designation of hybrid modes is based on the relative contributions of waves E_z and H_z to a transverse component of the field at some reference point. If electric wave E_z makes the larger contribution, the mode is considered to be E-like and is designated EH_{np}. Conversely, if magnetic wave H_z makes the dominant contribution, the mode is designated HE_{np}. The method of designation is arbitrary, because it does not depend on any particular component of the chosen field, the reference point, or how far the wavelength is from the cutoff. The use of two letters, EH and HE, in the designation merely identifies the hybrid nature of these modes. The subscripts on the EH_{np} and HE_{np} modes refer to the nth order of the Bessel function and the pth rank, where the rank gives the successive solutions of the boundary-condition equation involving the Bessel function of the first kind J_n.

2. *Cutoff Wavelengths*

For the cutoff condition, $J_n(ka) = 0$, which means that the limit of the argument of the Hankel function (χa) approaches zero [4]. Then the separation equation in the cladding region, as shown in Eq. (7-2-6), becomes

$$\beta_g^2 = \omega^2 \mu_2 \varepsilon_2 \tag{7-2-15}$$

and the separation equation in the core region is

$$k^2 = \omega^2 \mu_1 \varepsilon_1 - \omega^2 \mu_2 \varepsilon_2 \tag{7-2-16}$$

If we let the argument $(k_{np}a)$ of the Bessel function of the first kind be X_{np},

$$k_{np} = \frac{X_{np}}{a} \tag{7-2-17}$$

The cutoff condition is

$$X_{np} = \frac{2\pi a}{\lambda_0}(n_1^2 - n_2^2)^{1/2} \tag{7-2-18}$$

and we can express the free-space cutoff wavelength as

$$\lambda_0 = \frac{2\pi a}{X_{np}}(n_1^2 - n_2^2)^{1/2} \tag{7-2-19}$$

For a core refractive index n_1, a cladding refractive index n_2, and a core radius a, the free-space cutoff wavelength for a monomode EH_{01} operation is

$$\lambda_{0c} = \frac{2\pi a}{2.405}(n_1^2 - n_2^2)^{1/2} \tag{7-2-20}$$

The cutoff parameter X_{np} (or $K_{np}a$) is usually called the V number of the fiber. The number of propagating modes in the step-index fiber is proportional to its V number.

EXAMPLE 7-2-2 Cutoff Wavelength of Monomode Optical Fibers

A monomode optical fiber has the following parameters:

Core radius	$a = 5\ \mu\text{m}$
Core refractive index	$n_1 = 1.40$
Cladding refractive index	$n_2 = 1.05$

Calculate the cutoff wavelength for the monomode EH_{01} operation.

Solution:

The argument of the Bessel function of the first kind for the EH_{01} mode is 2.405, so the cutoff wavelength is

$$\lambda_{0c} = \frac{2\pi \times 5 \times 10^{-6}}{2.405}[(1.40)^2 - (1.05)^2]^{1/2}$$

$$= 12.11\ \mu\text{m}$$

Thus any signal with a wavelength larger than 12.11 μm will be cut off.

The wave modes capable of propagating in an optical fiber are those for which X_{np} are less than the values determined by Eq. (7-2-18). Because X_{np} forms an increasing sequence for a fixed n and increasing p or for a fixed p and increasing n, the number of allowed modes increases with the square of the radius a. That is, the total number of modes for an optical fiber [5] is

$$\text{Modes} = \frac{16}{\lambda_0^2}(n_1^2 - n_2^2)a^2 \tag{7-2-21}$$

The electromagnetic field is guided only partially within the core region, whereas outside the core the electromagnetic field is evanescent in a direction normal to propagation. Among the electromagnetic wave modes there is one, namely, the HE_{11} mode, that has no cutoff wavelength. Only this HE_{11} mode can propagate in an optical fiber for a wavelength greater than the highest cutoff wavelength of the other modes.

The modes of operation for optical fibers are commonly classified as single mode, such as monomode step-index fiber, and multiple modes, such as multimode step-index and multimode graded-index fiber. Figure 7-2-2 shows a plot of Bessel functions used for determining cutoff conditions for optical-fiber modes.

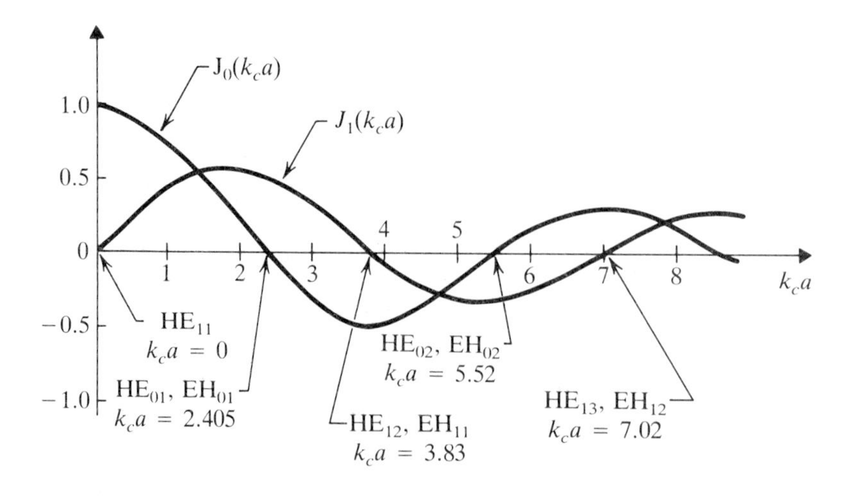

Figure 7-2-2 Cutoff conditions for optical-fiber modes.

7-2-3 Total Internal Reflection and Numerical Aperture

1. *Total Internal Reflection*

If light wave in medium 1 propagates into medium 2, Snell's law states that

$$\frac{\sin \phi_2}{\sin \phi_1} = \frac{\beta_1}{\beta_2} = \frac{v_2}{v_1} = \sqrt{\frac{\varepsilon_1}{\varepsilon_2}} = \frac{n_1}{n_2} \quad \text{for } \mu_1 = \mu_2 = \mu_0 \tag{7-2-22}$$

where

ϕ_1 is the incident angle in medium 1;
ϕ_2 is the transmission angle in medium 2;
β_1 is the incident phase constant in medium 1;
β_2 is the transmission phase constant in medium 2;
v_1 is the wave velocity in medium 1;
v_2 is the wave velocity in medium 2;
ε_1 is the dielectric permittivity of medium 1;
ε_2 is the dielectric permittivity of medium 2;
n_1 is the refractive index of medium 1; and
n_2 is the refractive index of medium 2.

The total reflection occurs at $\phi_2 = 90°$, resulting in the incident angle in medium 1 of

$$\phi_1 = \phi_c = \sin^{-1}\left(\frac{n_2}{n_1}\right) \tag{7-2-23}$$

The angle specified by Eq. (7-2-23) is called the *critical incident angle* for total reflection. A wave incident on the interface of the core and cladding in an optical fiber at an angle equal to or greater than the critical angle will be totally reflected. There is a real critical angle only if $n_1 > n_2$. Thus the total internal reflection occurs only if the wave propagates from the core toward the cladding, because the value of $\sin \phi_c$ must be less than or equal to 1.

2. *Numerical Aperture (NA)*

In an optical-fiber waveguide, the light wave is incident on one end face of the fiber. Figure 7-2-3 shows a diagram for the light wave impinging on a fiber.

Before going on, we need to define several terms used in optical-fiber cables and systems:

1. The *acceptance angle* (or incident acceptance angle) is defined as any angle measured from the longitudinal center line to the maximum acceptance angle of an incident ray that will be accepted for transmission along a fiber.
2. The *acceptance cone* is a cone having an angle equal to twice the acceptance angle.
3. The *critical angle* (or incident critical angle) is the smallest angle made by a meridional ray in an optical fiber that can be totally reflected from the innermost interface. Thus it determines the maximum acceptance angle at which a meridional ray can be accepted for transmission along a fiber.

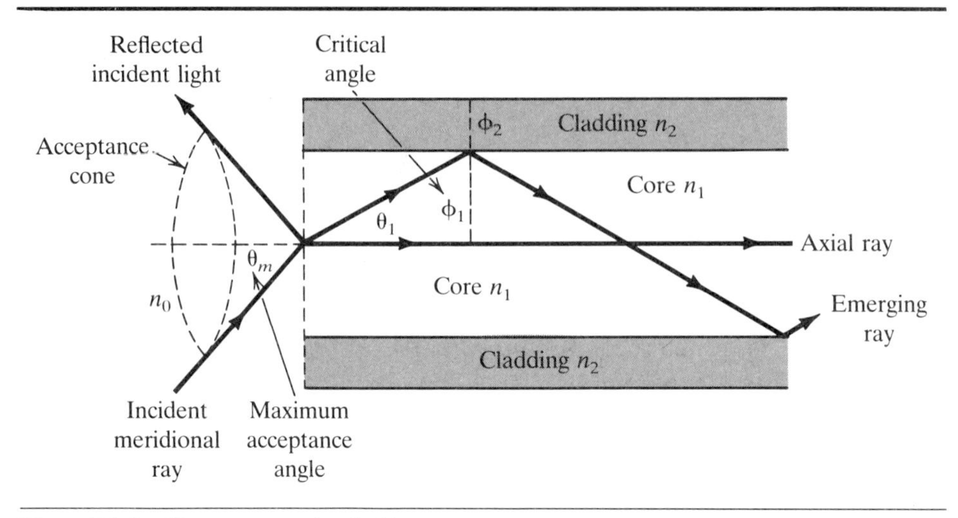

Figure 7-2-3 Wave propagation in fiber.

4. The *maximum acceptance angle* depends on the refractive indexes of the two media that determine the critical angle.
5. A *numerical aperture* (NA) is defined as a number that expresses the light-gathering power of a fiber.

For an optical fiber the most important parameter is the numerical aperture, which is expressed as

$$NA = \sin \theta \tag{7-2-24}$$

where θ represents the acceptance angle. According to Snell's law,

$$n_0 \sin \theta = n_1 \sin \theta_1 = n_1 \cos \phi_1 \tag{7-2-25}$$

and

$$n_1 \sin \phi_1 = n_2 \sin \phi_2 \tag{7-2-26}$$

For $n_1 > n_2$, total reflection occurs at

$$\sin \phi_1 > \frac{n_2}{n_1} \tag{7-2-27}$$

Then

$$\cos \phi_1 < \left(1 - \frac{n_2^2}{n_1^2}\right)^{1/2} \tag{7-2-28}$$

When we substitute Eq. (7-2-27) into Eq. (7-2-26), using the trigonometric identity, we get

$$n_0 \sin \theta < (n_1^2 - n_2^2)^{1/2} \tag{7-2-29}$$

or

$$\sin \theta < \frac{1}{n_0}(n_1^2 - n_2^2)^{1/2} \tag{7-2-30}$$

Equation (7-2-30) determines the incident acceptance angle of a meridional ray for total internal reflection in an optical fiber. It follows, then, that the maximum acceptance angle is

$$\sin \theta_m = \frac{1}{n_0}(n_1^2 - n_2^2)^{1/2} \qquad \text{for } n_1^2 < (n_2^2 + n_0^2) \tag{7-2-31}$$

and

$$\sin \theta_m = 1.0 \qquad \text{for } n_1^2 > (n_2^2 + n_0^2) \tag{7-2-32}$$

Normally, the optical fiber is immersed in free space or in a vacuum, so $n_0 = 1.0$. Then Eqs. (7-2-31) and (7-3-32), respectively, become

$$\sin \theta_m = (n_1^2 - n_2^2)^{1/2} \qquad \text{for } n_1^2 < (n_2^2 + 1) \tag{7-2-33}$$

and

$$\sin \theta_m = 1.0 \qquad \text{for } n_1^2 > (n_2^2 + 1) \tag{7-2-34}$$

Equation (7-2-33) is a measure of the numerical aperture of an optical fiber; that is,

$$NA = (n_1^2 - n_2^2)^{1/2} \tag{7-2-35}$$

The magnitude of the numerical aperture also indicates the acceptance of impinging light, degree of openness, light-gathering capability, and acceptance cone. The light-gathering power for a meridional ray in an optical fiber is

$$P = (NA)^2 = n_1^2 - n_2^2 \tag{7-2-36}$$

EXAMPLE 7-2-3 Characteristics of a Multimode Step-Index Fiber

A multimode step-index fiber has the following parameters:

Core refractive index	$n_1 = 1.54$
Cladding refractive index	$n_2 = 1.49$
Air refractive index	$n_0 = 1.00$

Calculate:

a) the critical angle for total reflection in the core region;
b) the maximum incident angle at the interface between air and the fiber end; and
c) the light-gathering power.

Solutions:

a) Using Eq. (7-2-23), we find that the critical angle is

$$\phi_c = \sin^{-1}\left(\frac{1.49}{1.54}\right)$$

$$= 75.36^\circ$$

b) Then from Eq. (7-2-31), we obtain the maximum incident angle:

$$\theta_m = \sin^{-1}\left[\frac{(1.54)^2 - (1.49)^2}{1}\right]^{1/2}$$

$$= \sin^{-1}(0.39)$$

$$= 22.79^\circ$$

c) And from Eq. (7-2-36), we determine the light-gathering power:

$$P = (1.54)^2 - (1.49)^2$$
$$= 0.15$$

7-2-4 Light-Gathering Power

In an optical-fiber system, we consider the light source to be a Lambertian source, that is, one having an intensity of

$$I = I_0 \cos \theta \tag{7-2-37}$$

where I_0 is the light intensity in the normal direction; and θ is the angle between the normal to the source and the direction of measurement.

The unit of light intensity, or luminous flux, is called the *lumen*. The eyes of most people are most sensitive to the green color of wavelength 5550 Å. At the peak of the relative luminosity curve 5550° Å, 1 W of radiant flux of monochromatic radiation corresponds to 685 lumens of luminous flux, or light flux. Thus for the normal eye, 1 lumen is equivalent to 1/685 W (or 0.00146 W) of monochromatic light of green color of wavelength 5550 Å.

If the source emitting light flux is a point source, its light intensity in a particular direction is defined as the number of lumens per unit solid angle radiated in that direction. The unit solid angle (or steradian) is defined as the angle subtended at the center of a sphere of radius r by that portion of its surface area equal to r^2. The unit of light intensity I is 1 lumen (lm) per steradian (sr), also called 1 candle. Since the total solid angle subtended by a point source is 4π sr, a point source of light intensity of 1 candle emitting uniformly in all directions radiates 4π lm.

Another quantity frequently used in practice for light intensity is a measure of the illumination of a surface. The units are lumens per unit area, that is, lm/m^2 or lm/ft^2, depending upon the unit of area chosen. In engineering practice, illumination is measured in a unit called the *foot-candle*, which is the illumination produced when the light from a point source of 1 candle falls normally on a surface at a distance of 1 foot. The foot-candle, then, is numerically equal to 1 lm/ft^2. In the MKS system, the unit of illumination is lumens per square meter (lm/m^2), and this unit is also called the *lux*. Since 1 $\text{ft}^2 = 0.093\ \text{m}^2$, it follows that an illumination of 1 lm/ft^2 or 1 foot-candle is approximately equal to 10 lm/m^2 or 10 lux.

Figure 7-2-4 shows a diagram for the light output from a Lambertian source through a spherical surface of radius r.

The amount of light collected by a fiber that is normal to a Lambertian source is defined by the dielectric boundary of the fiber or the maximum acceptance angle θ_m:

$$\text{Light} = \int_0^{\theta_m} I_0 \cos \theta \cdot dA = \int_0^{\theta_m} I_0 \cos \theta \cdot 2\pi \sin \theta \, d\theta \tag{7-2-38}$$
$$= \pi I_0 \sin^2 \theta_m$$

The total light emitted by the Lambertian source is πI_0, and the fraction of light collected by the fiber is the light-collecting efficiency, or

$$\eta_m = \sin^2 \theta_m = (NA)^2 \qquad \text{for } NA < 1.0 \tag{7-2-39}$$

and

$$\eta_m = 1.0 \qquad \text{for } NA = 1.0$$

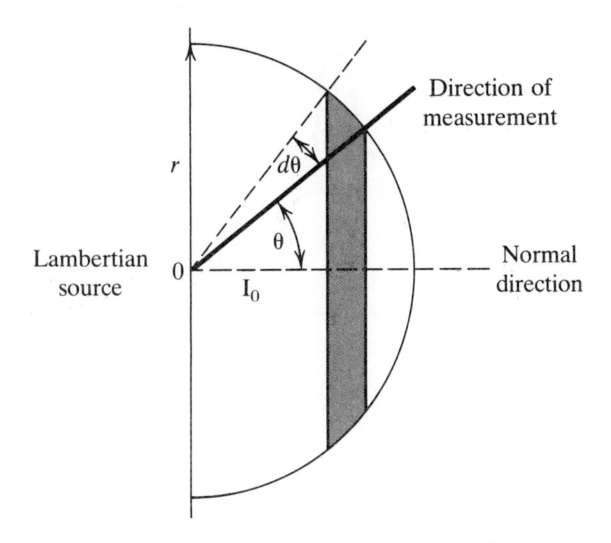

Figure 7-2-4 Light from a Lambertian source.

In practice, the light-collecting efficiency is much less than 1 because of many factors, such as the angled-end–face effect, the fiber-curvature effect, and the varying-diameter effect.

The light-gathering power of an optical fiber [3] is

$$P = n_0^2 - \frac{2}{\pi}\Bigg\{(n_1^2 - n_2^2)^{1/2}[n_0^2 - (n_1^2 - n_2^2)]^{1/2} + [n_0^2 - 2(n_1^2 - n_0^2)] \cos^{-1}\left(\frac{n_1^2 - n_2^2}{n_0^2}\right)^{1/2}\Bigg\} \tag{7-2-40}$$

or

$$P = 1 - \frac{2}{\pi}\{(NA)[1 - (NA)^2]^{1/2} + [1 - 2(NA)^2] \cos^{-1}(NA)\} \qquad \text{for } n_0 = 1.0 \tag{7-2-41}$$

where $(NA)^2 = n_1^2 - n_2^2$ is the meridional light-gathering power.

7-3 STEP-INDEX FIBERS

A step-index fiber is one in which there is an abrupt change in refractive indexes between the core and cladding. Two modes can exist in a step-index fiber: monomode fiber and multimode fiber. If the core radius is very small compared to the wavelength of the light source (say, of the order of 1 to 16 μm) only a single mode is propagated. If the core radius is large enough (say, 30 μm) multimodes coexist.

7-3-1 Monomode Step-Index Fibers

A monomode step-index fiber is a low-loss optical fiber with a very small core. The fiber requires a laser source for the input signals because of the very small acceptance

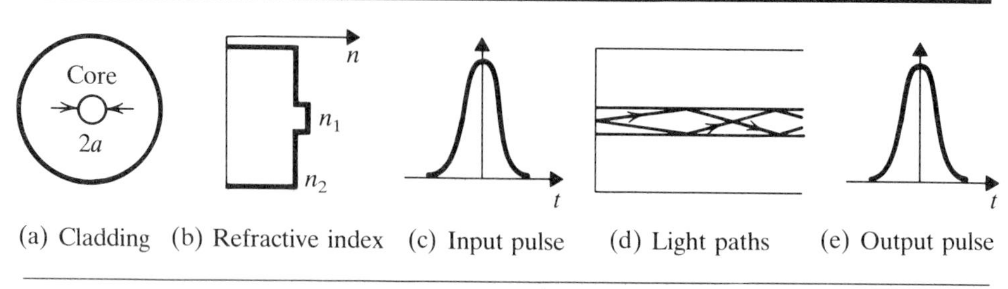

Figure 7-3-1 Diagrams for monomode step-index fiber.

SOURCE: AFTER T. G. GIALLORENZI [10]; © 1978 IEEE.

aperture (or acceptance cone). When the small core radius approaches the wavelength of the light source, only a single mode is propagated. In practice, we find that the core radius is from 1 to 16 μm, and the differential in refractive indexes between the core and cladding is about 0.6 percent. The bit rate is from 20 Mbit/s-km to 19 Gbit/s-km. This fiber is ideally suited for long-distance and high-bandwidth applications, such as telecommunication systems.

According to Eq. (7-2-38), the magnitude of X_{np} required for a monomode operation in a step-index fiber must be less than 2.405. That is,

$$X_{np} = \frac{2\pi a}{\lambda_0}(n_1^2 - n_2^2)^{1/2} < 2.405 \tag{7-3-1}$$

Figure 7-3-1 shows the diagrams for a monomode step-index fiber: light path and signal waveform profiles.

EXAMPLE 7-3-1 Determination of Core Radius of a Monomode Optical Fiber for a Single Mode

A monomode optical fiber has the following parameters:

Core refractive index	$n_1 = 1.51$
Cladding refractive index	$n_2 = 1.49$
Light source—Nd^{3+}: YAG laser	$\lambda_0 = 1.064\ \mu\text{m}$

Determine the core radius that will support only a single-mode operation.

Solution:

From Eq. (7-3-1), we can calculate the maximum core radius:

$$a \le \frac{2.405(1.064 \times 10^{-6})}{2\pi\sqrt{(1.51)^2 - (1.49)^2}}$$

$$a \le 1.66\ \mu\text{m}$$

7-3-2 Multimode Step-Index Fibers

A multimode step-index fiber has a large core and large numerical aperture, so it can couple efficiently to a light source LED. The core radius is from 25 to 60 μm and the differential in refractive indexes is from 1 to 10 percent. The bit rate is under 100 Mbit/

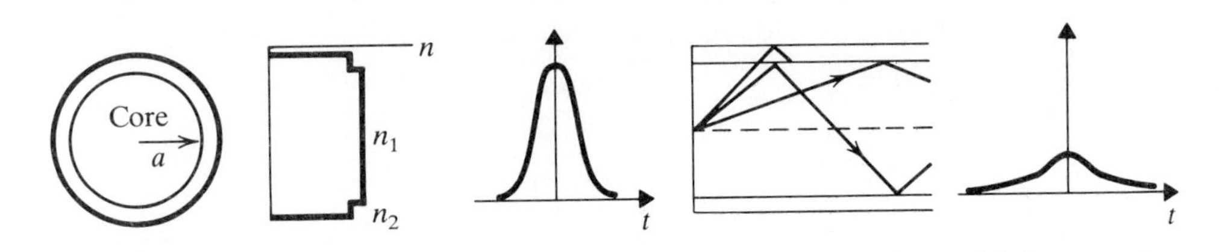

Figure 7-3-2 Diagrams for multimode step-index fiber.

SOURCE: AFTER T. G. GIALLORENZI [10]; © 1978 IEEE.

s-km. This fiber is suitable for low-bandwidth, short-distance, and low-cost applications, such as links in computers. Figure 7-3-2 shows the diagrams for a multimode step-index fiber: light path and signal waveform profiles.

EXAMPLE 7-3-2 Calculations for a Multimode Step-Index Fiber

A multimode step-index fiber has the following parameters:

Core refractive index	$n_1 = 1.45$
Numerical aperture	$NA = 0.35$
V number	$X_{np} = 100$
Wavelength of LED source	$\lambda_0 = 0.87\ \mu m$

Calculate:

a) the cladding refractive index n_2;
b) the core radius a; and
c) the cladding radius b.

Solutions:

a) The cladding refractive index is

$$n_2 = (n_1^2 - NA^2)^{1/2} = \sqrt{(1.45)^2 - (0.35)^2}$$
$$= 1.41$$

b) The core radius is obtained from Eq. (7-3-1) as

$$a = \frac{\lambda_0 X_{np}}{2\pi(NA)} = \frac{0.87 \times 100}{6.283 \times 0.35}$$
$$= 39.56\ \mu m$$

c) In order for electric and magnetic fields to be evanescent in the cladding region, the propagation constant χ of the Hankel function must be imaginary. That is,

$$\chi = j\sqrt{\beta_2^2 - \beta_g^2} = j\sqrt{\omega^2 \mu_2 \varepsilon_2 - \omega^2 \mu_0 \varepsilon_0}$$
$$= j\frac{2\pi}{\lambda_0}\sqrt{n_2^2 - 1} = j\frac{6.2832}{0.87}\sqrt{(1.4)^2 - 1}$$
$$= j7.18$$

From Appendix K we see that the argument of the Hankel function would be $4X_{np} = 400$. The cladding radius, then, is

$$b = \frac{400}{7.18}$$
$$= 55.71\ \mu\text{m}$$

7-4 GRADED-INDEX FIBER

A graded-index fiber is one in which the refractive index of the core region is decreased monotonically from the center and converged into a flat configuration at the cladding region. Much of the longer path lengths are through the lower index of refractive material and so the increased velocity over this part of the path compensates somewhat for the longer path lengths; the spread in group velocity between modes is not as great as for the step-index fiber. In order to meet bandwidth requirements for telecommunication applications, multimode fibers are usually fabricated with graded-index profiles to reduce intermodal dispersion. The graded-index fiber has a relatively large core, so it can support multimode operation. This type of fiber is suitable for high-bandwidth and medium-distance applications. Table 7-4-1 contains a list of some typical parameters for graded-index fibers. Figure 7-4-1 shows the diagrams for a multimode graded-index fiber: light path and signal waveform profiles.

Table 7-4-1 Parameters of graded-index fiber

Parameters	Values
Bit rate	140–1000 Mbit/s-km
Core radius	10–35 μm
Cladding radius	50–80 μm
Losses	2–5 dB/km
Numerical aperture (NA)	0.15–0.25
Number of modes at 0.9 μm	140–900
Pulse dispersion/mode	0.1–4.0 ns/km
Refractive index n_1	1.47–1.50
Deviation index Δ	0.7–30%

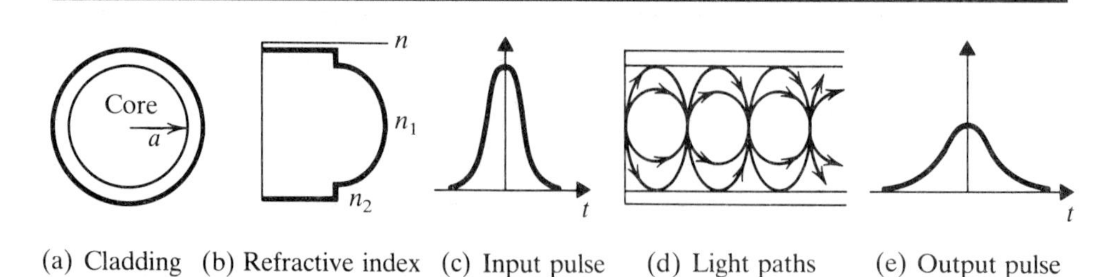

(a) Cladding (b) Refractive index (c) Input pulse (d) Light paths (e) Output pulse

Figure 7-4-1 Diagrams for graded-index fiber.

SOURCE: AFTER T. G. GIALLORENZI [10]; © 1978 IEEE.

7-4-1 Refractive-Index Profiles

In a multimode graded-index fiber the refractive index [6] is

$$n(r) = n_1\left[1 - 2\Delta\left(\frac{r}{a}\right)^{\alpha}\right]^{1/2} \quad \text{for } r \le a \tag{7-4-1}$$

and

$$n(r) = n_1(1 - 2\Delta)^{1/2} \quad \text{for } r > a \tag{7-4-2}$$

where

n_1 is the refractive index at the center of the core region;

$\Delta = \dfrac{n_1 - n_2}{n_2} \doteq \dfrac{n_1 - n_2}{n_1} \doteq \dfrac{n_1^2 - n_2^2}{2n_1^2}$ and is the deviation index; and

α is a parameter that is also called the power-law coefficient between 1 and ∞.

Figure 7-4-2 shows the cross-section of a circular symmetric index profile in a multimode fiber. All profiles reach a constant cladding index n_2 at $r = a$, whereas the index is n_1 for $r = 0$ at the core center. The core profile has a cone shape or a nearly parabolic shape for $\alpha = 1$ and becomes convergent to the form of the step profile for $\alpha = \infty$. Figure 7-4-3 illustrates some index profiles, as defined by Eq. (7-4-1), for the small deviation index Δ.

The total number of modes in a graded-index fiber is

$$N = \left(\frac{2\pi a}{\lambda_0} n_1\right)^2 \Delta\left(\frac{\alpha}{\alpha + 2}\right) \tag{7-4-3}$$

For a parabolic profile at $\alpha = 2$, the number of modes in a graded-index fiber is equal to one-half the number that exists in a step-index fiber ($\alpha = \infty$). That is,

$$N\text{ (parabolically graded)} = \frac{N\text{ (step)}}{2} \tag{7-4-4}$$

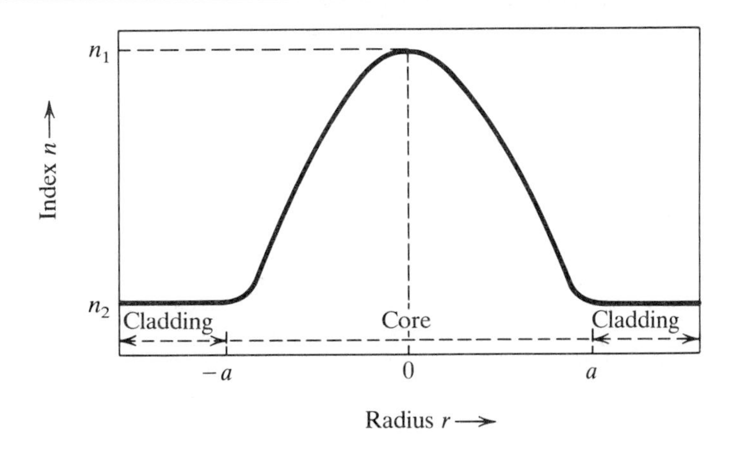

Figure 7-4-2 Cross-sectional diagram for a graded-refractive index.

SOURCE: FROM D. GLOGE AND A. J. MARCATILI [11]; REPRINTED BY PERMISSION OF THE BELL SYSTEM, AT&T, INC.

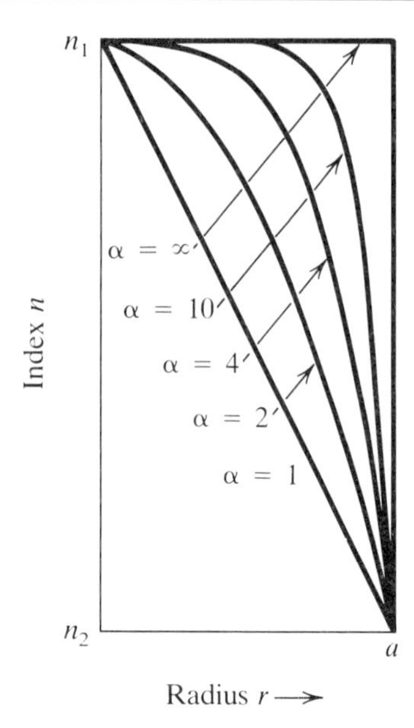

Figure 7-4-3 Graded-index profiles.

SOURCE: FROM D. GLOGE AND A. J. MARCATILI [11]; REPRINTED BY PERMISSION OF THE BELL SYSTEM, AT&T, INC.

We can then express the total number of modes in a graded-index fiber as

$$N = \frac{V^2}{2}\left(\frac{\alpha}{\alpha + 2}\right) \tag{7-4-5}$$

where $V = X_{np}$ is defined by Eq. (7-3-1).

EXAMPLE 7-4-1 Number of Modes in a Graded-Index Fiber

A graded-index fiber has the following parameters:

Core radius	$a = 20\ \mu m$
Power-law coefficient	$\alpha = 2$
Deviation index	$\Delta = 0.03$
Core refractive index	$n_1 = 1.52$
Wavelength of LED light source	$\lambda_0 = 0.87\ \mu m$

Calculate:

a) the total number of modes for the graded-index fiber;
b) the number of modes for a step-index profile; and
c) the graded refractive index.

Solutions:

a) From Eq. (7-4-3), we find that the number of modes for the graded-index fiber is

$$N = \left[\frac{6.283 \times 20 \times 10^{-6}}{0.87 \times 10^{-6}} \times 1.52\right]^2 (0.03)\left(\frac{2}{2+2}\right) = 723 \text{ modes}$$

b) From Eq. (7-4-4), we obtain the number of modes for a step-index profile:

$$N = 2 \times 723$$
$$= 1446 \text{ modes}$$

c) From Eq. (7-4-2), we determine that the graded refractive index is

$$n_2 = 1.52 \times (1 - 2 \times 0.03)^{1/2}$$
$$= 1.47$$

7-4-2 Wave Patterns

Graded-index fibers offer multimode propagation in a relatively large-core fiber coupled with low-mode dispersion. A typical bit rate is 1400 Mbit/s over a link 8 to 10 km long in conjunction with a GaAs light source, whereas a step-index fiber provides an upper limit of perhaps 10 to 20 Mbit/s-km. There are three wave patterns in a graded-index fiber, as shown in Fig. 7-4-4:

1. *Center ray.* Since the index distribution is graded, with a high index in the center surrounded by a steadily decreasing index, the medium behaves similarly to a series of lenses. A ray traversing the center of the medium follows the axis, traveling in high-index material all the way but traversing the shortest possible physical path from end to end.
2. *Meridional ray.* A ray leaving the center of the fiber at an angle to the axis is bent by the index profile (lenslike structure), curves after some characteristic distance, and crosses the axis again. It continues in a sinusoidal path to the end and thus travels a greater distance than the center ray, but much of its path is in the low-index material. This ray is called the meridional ray; that is, the ray passes through the axis of the fiber while being reflected internally.

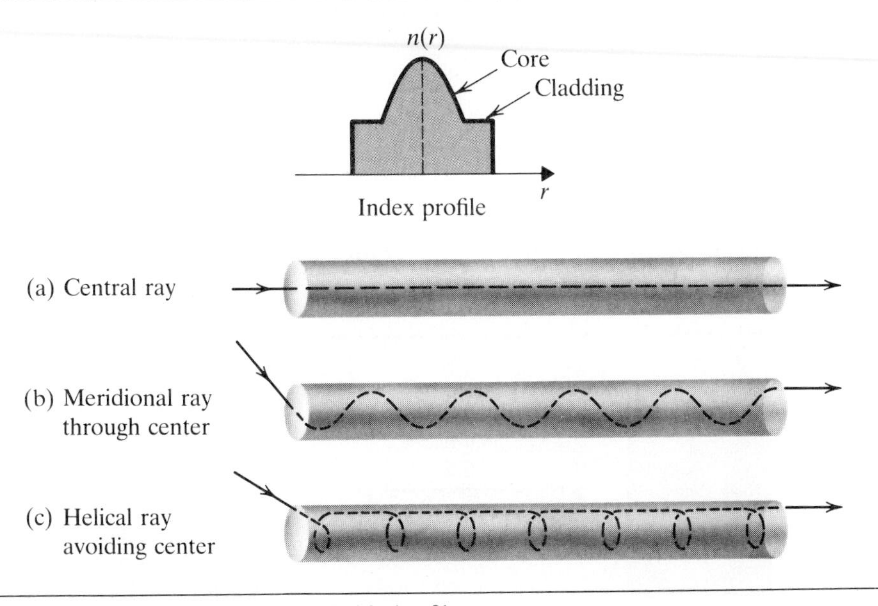

Figure 7-4-4 Wave patterns in graded-index fibers.

SOURCE: FROM J. E. MIDWINTER [7]; REPRINTED BY PERMISSION OF JOHN WILEY & SONS, INC.

3. *Skew ray* (*helical ray*). The third ray is a helical wave, formed when light is launched at a skew angle along the surface of a constant-radius cylinder at some intermediate radius and index. Once again the length of the physical path is longer than that taken by the center ray, but it traverses lower index material. This ray is called a skew, or helical, ray because it never intersects the axis of the fiber while being reflected internally.

A ray of light is defined as the rectilinear path in a homogeneous medium along which light is propagated. If we have a luminous object, choose one point on it, and draw a straight line from the point in the direction of the propagation of the light, we will have drawn a representation of a *ray of light*. Obviously, we can draw an infinite number of rays of light from a point source of light. We call this collection of rays, or a cone, of light, from a point source a *pencil of rays*. If the luminous object is an extended object, we can draw a pencil of rays from each point of the luminous object. We call the collection of pencils of rays from all the points on the object a *beam of light*.

7-5 OPTICAL-FIBER COMMUNICATION SYSTEMS

An optical-fiber communication system is essentially a dielectric-filled circular waveguide with some electronic components and circuits, as shown in Fig. 7-5-1.

An incoming binary-digital signal switches the light source laser or LED on and off. When the light is on, the light impinges on one end of the optical fiber, passes through it by total internal reflection at the core-cladding interface, and emerges at the far end. Then the emerged light impinges on the photodetector, which converts the photon light to electron-hole pairs in the detector, and a current pulse is generated. Consequently, the receiver circuits will reproduce the information signal in a binary-digital form of 0 and 1. The design of an optical-fiber communication system is based on four requirements: (1) the data rate (bandwidth); (2) the signal-to-noise ratio (SNR); (3) the distance between terminals; and (4) the type of source information (digital or analog).

7-5-1 Light Sources

A light source for optical-fiber communication systems must have certain characteristics, such as a long life, high efficiency, low cost, high capacity, and sufficient output power. There are two such light sources:

LED sources. Because LED (light emitting diodes) emits light randomly in all directions from its junction, the light is incoherent. Therefore the transmission

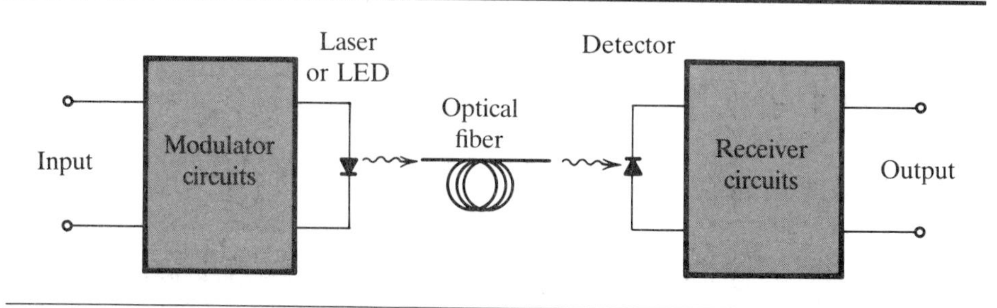

Figure 7-5-1 Diagram of an optical-fiber communication system.

of LED-generated signals inherently involves multimodes. The GaAs LED is an adequate choice for data links in the wavelength range of 0.82–0.85 μm.

Laser sources. The laser (light amplification by stimulated emission of radiation) beam is coherent and an ideal light source for monomode fibers. Because the light is very narrow, it can be coupled efficiently into the fibers and it reduces the effect of the intrinsic chromatic dispersion.

The power needed for optical fiber communication is usually no more than a few milliwatts. At this power level the fibers are linear. When the power reaches the level of 100 mW, however, single-mode fibers begin to exhibit nonlinearity. A laser power of about 2.5 W marks the onset of nonlinearity for multimode fibers.

Generally gas lasers (such as CO_2 and He–Ne) are too large, expensive, and inefficient for optical-fiber communication. At a wavelength of 1.06 μm, the Nd^{3+}: YAG (neodymium: yttrium–aluminum–garnet) laser is one of the most useful light sources for single-mode fibers. Ternary AlGaAs and InGaAs lasers are commonly used in the 0.82–0.85 μm range. Figure 7-5-2 shows a light-source AlGaAs laser coupled to a glass fiber.

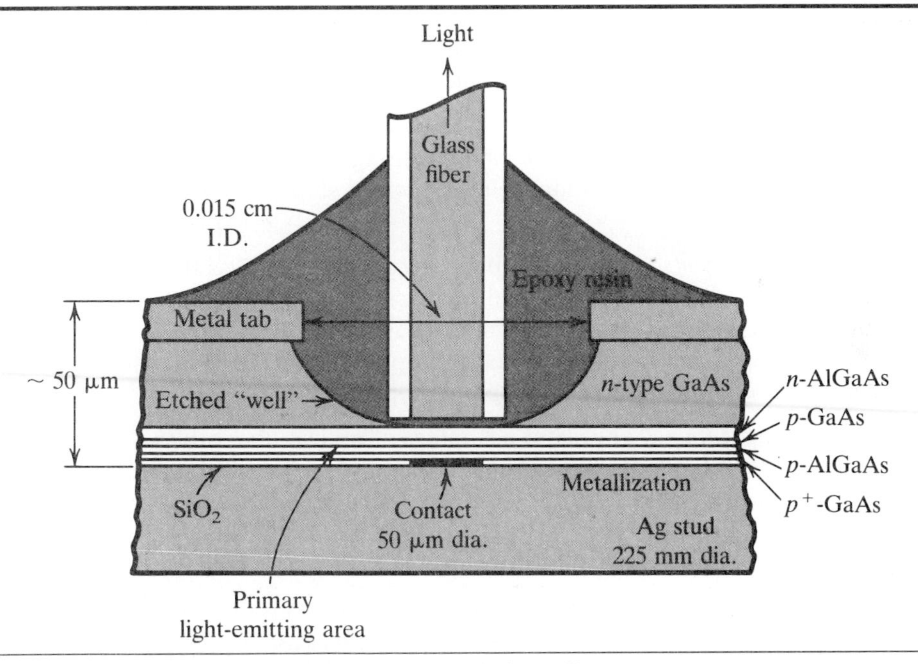

Figure 7-5-2 Diagram of an AlGaAs laser coupled to a fiber.

SOURCE: FROM C. A. BURRUS AND B. I. MILLER [8]; REPRINTED BY PERMISSION OF OPTICS COMMUNICATIONS CO.

7-5-2 Light Detectors

The functions of light detectors in an optical-fiber communication system is to detect the optical signal from the fiber and convert it into an electrical signal. The GaAs photodetector is a suitable choice for wavelengths of 0.82 to 0.85 μm. The PIN photodetectors, such as InGaAs and InGaAsP, are commonly used in the 1.3–1.5 μm region. At the receiver side, a low-noise GaAs MESFET preamplifier is usually used immediately after the photodetector for a better signal-to-noise ratio at the output terminal.

7-5-3 Applications

We can make several assumptions about the communication applications of multimode optical fibers:

1. The index profile of the fibers is circularly symmetric about the core axis.
2. The core diameter is large enough (say, 100 wavelengths) that we can expect many modes to propagate.
3. The index difference is small enough that we can consider the modes to be transverse electromagnetic (TEM).
4. The index variation is very small over distances of a wavelength, and thus we can make the local plane-wave approximation.

Table 7-5-1 lists the applications and characteristics of the three major optical-fiber systems [9]. As we have previously noted, optical-fiber cables are used primarily in computer, industrial automation, medical instrument, military, and telecommunications systems.

Computer applications. Computer terminals and their internal links require a very high data rate (up to several Gbit/s · km). Optical fibers can easily meet this requirement. Because the fibers offer freedom from electromagnetic interference and from grounding problems, the signal quality is excellent.

Industrial automation. Optical fibers are used in industrial automation in the areas of process control, discrete manufacturing automation, and transportation. Transportation includes airways, shipping, highways, railways, and so on.

Medical instruments. Historically, the medical profession was the first to use the optical fiber—for the diagnosis and treatment of disease. In the late 1930s a guided light was used to provide illumination for simple inspection instruments. Medical instruments that can be made of optical fibers are the cardioscope, colonoscope, endoscope, gastroscope, and opthalmoscope.

Military applications. The major advantage of optical-fiber cables is the absence of cross-talk, RF interference, grounding problems, and outside jamming. Such fibers also greatly increase security. Consequently, optical fibers are commonly used as communication links in aircraft, missiles, submarines, ocean-surface vessels, and military command and control systems.

Telecommunications. The primary use of optical fibers is in telephone and telegraph. Optical-fiber telecommunication systems can solve many inherent problems that occur in wire systems, such as ringing, echoes, and cross-talk. Such fibers are commonly used in voice telephones, video phones, telegraph services, and various news broadcast systems.

EXAMPLE 7-5-1 Design Example: Choosing a Light Source

The choice of a light source for a specific optical-fiber system is an essential part of the design process. The output power of a light source is usually expressed in terms of its input current. In practice, the light source is terminated in a fiber connector. For a LED light source and a 50-μm core fiber, the coupling loss may be as high as 20 dB, whereas for a laser and the same fiber, the coupling loss can be as low as 3 dB.

Table 7-5-1 Applications and characteristics of three major optical fibers.

	Monomode Step-Index Fiber	Multimode Graded-Index Fiber	Multimode Step-Index Fiber
	Cladding Core Protective plastic coating		
Source	Requires laser	Laser or LED	Laser or LED
Bandwidth	Very very large >3 GHz-km	Very large 200 MHz to 3 GHz-km	Large <200 MHz-km
Splicing	Very difficult due to small core	Difficult but doable	Difficult but doable
Example of application	Submarine cable system	Telephone trunk between central offices	Data links
Cost	Less expensive	Most expensive	Least expensive

SOURCE: AFTER A. H. CHERIN [9]; REPRINTED BY PERMISSION OF MCGRAW-HILL BOOK COMPANY.

An optical-fiber system is to be designed to meet the following specifications:

LED source to fiber coupling loss	20 dB
Laser source to fiber coupling loss	3 dB
Fiber length	4 km
Fiber loss	4 dB/km
Number of splices	1 splice/0.5 km
Splice loss	0.5 dB/splice
Fiber-to-detector coupling loss	0.1 dB
Minimum power required for receiver	−40 dBm

Solutions:

a) Selection of LED source.

1. Choose the LED 1A83 (ASEA-HAFO shown in Appendix L as the light source; it has an output power of 10 mW from a drive current of 0.1 A. So

$$P_{in} = 10 \text{ mW} = 10 \text{ dBm}$$

2. Then the total link loss is

$$\text{Loss} = 20 \text{ dB} + 4(4 \text{ dB}) + 8(0.5 \text{ dB}) + 0.1 \text{ dB} = 40.1 \text{ dB}$$

3. The power available at the receiver is

$$P_{avg} = 10 \text{ dBm} - 40.1 \text{ dB} = -30.1 \text{ dBm}$$

4. The power margin at the receiver is

$$P = -30.1 \text{ dBm} - (-40 \text{ dBm}) = 9.9 \text{ dBm}$$
$$= 9.772 \text{ mW}$$

b) Selection of laser source.

1. A CW-laser diode has an output power of 10 mW from a drive current of 0.2 A. So

$$P_{in} = 10 \text{ mW} = 10 \text{ dBm}$$

2. The total link loss is

$$\text{Loss} = 3 \text{ dB} + 4(4 \text{ dB}) + 8(0.5 \text{ dB}) + 0.1 \text{ dB} = 23.1 \text{ dB}$$

3. The power available at the receiver is

$$P_{avg} = 10 \text{ dBm} - 23.1 \text{ dB} = -13.1 \text{ dBm}$$

4. The power margin at the receiver is

$$P = -13.1 \text{ dBm} - (-40 \text{ dBm}) = 26.9 \text{ dBm}$$
$$= 489.8 \text{ mW}$$

REFERENCES

[1] Elion, G. R., and Elion, H. A., *Fiber Optics in Communications Systems*. New York: Marcel Dekker, 1978, p. 35.

[2] Jacobs, Ira, and Miller, S. E., "Optical Transmission of Voice and Data," *IEEE Spectrum*, vol. 14, no. 2, p. 39, February 1977.

[3] Potter, R. J., et al., *J. Opt. Soc. Am.*, 53:256 (1963).

[4] Schelkunoff, S. A., *Electromagnetic Waves*. New York: D. Van Nostrand, 1964, p. 247.

[5–7] Midwinter, John E., *Optical Fibers for Transmission*. New York: John Wiley & Sons, 1979, pp. 108, 83, 82.

[8] Burrus, C. A., and Miller, B. I., "Small Area, Double Heterostructure AlGaAs Electroluminescent Diode Source for Optical-Fiber Transmission Lines," *Opt. Common.*, 4:307 (1971).

[9] Cherin, Allen H., *An Introduction to Optical Fibers*. New York: McGraw-Hill, 1983, p. 3.

[10] Giallorenzi, Thomas G., "Optical Communications Research and Technology: Fiber Optics, *Proc. IEEE*, vol. 66, no. 7 (July 1978).

[11] Gloge, D., and Marcatili, A. J., "Multimode Theory of Graded-Core Fibers," *Bell Technical Journal*, vol. 52, no. 9, November 1973, pp. 1564 and 1568, Figs. 1 and 5.

SUGGESTED READINGS

1. Allan, W. B., *Fiber Optics: Theory and Practice*. London: Plenum Press, 1973.
2. Cherin, Allen H., *An Introduction to Optical Fibers*. New York: McGraw-Hill, 1983.
3. Elion, G. R., and Elion, H. A., *Fiber Optics in Communications Systems*. New York: Marcel Dekker, 1978.
4. Howes, M. J., and Morgan, D. V. (Eds.), *Optical Fibre Communications*. New York: John Wiley & Sons, 1980.
5. *IEEE Journal of Quantum Electronics*, vol. QE-18, no. 4. Special issue on optical guided wave technology, April 1982.
6. *IEEE Proc.*, vol. 68, no. 6. Special issue on optical communications, June 1980.
7. *IEEE Proc.*, vol. 61, no. 12. Special issue on optical fibers, December 1973.
8. *IEEE Proc.*, vol. 68, no. 10. Special issue on optical-fiber communications, October 1980.
9. *IEEE Proc.*, vol. 66, no. 7. Special issue on optical-fiber light paths, July 1978.
10. *IEEE Trans. on Electron Devices*, vol. ED-29, no. 9. Special issue on optical fibers, September 1982.
11. *IEEE Trans. on Microwave Theory and Techniques*, vol. MTT-30, no. 4. Special issue on optical guided wave technology, April 1982.
12. Kao, Charles K., *Optical Fiber Systems: Technology, Design, and Applications*. New York: McGraw-Hill, 1982.
13. Midwinter, John E., *Optical Fibers for Transmission*. New York: John Wiley & Sons, 1979.

Problems

7-1 The absorption loss of an optical fiber is described by Eq. (7-1-1), where ℓ is the light-path length. Verify that $\ell = n_1/(n_1^2 - \sin^2 \theta_{\max})^{1/2}$.

7-2 The numerical aperture of an optical fiber is defined as $NA = \sin \theta$, where θ is the acceptance angle. Verify Eq. (7-2-28).

7-3 An optical fiber has $n_1 = 1.30$ and $n_2 = 1.02$. Calculate the meridional light-gathering power.

7-4 The wave equations in the core region ($r < a$) of an optical fiber are represented by the Bessel functions, as shown in Eqs. (7-2-3) and (7-2-4). From these two equations and Maxwell's equations, derive Eqs. (7-2-7) and (7-2-8).

7-5 The wave equations in the cladding region ($r > a$) of an optical fiber are represented by the Hankel functions, as shown in Eqs. (7-2-5) and (7-2-6). From these two equations and Maxwell's equations, derive Eqs. (7-2-9) and (7-2-10).

7-6 For cutoff wavelengths, the Hankel function H_n must vanish at the interface between the core and cladding ($r = a$). Derive the free-space cutoff wavelength Eq. (7-2-19).

7-7 The cladding refractive index n_2 of an optical fiber is 1.10, and the core radius a is 30λ. Determine the wave component $K_v(u)$ in the cladding region for a distance r of 31λ.

7-8 Derive the equation of the maximum acceptance angle, as shown in Eq. (7-2-31), from Fig. 7-2-2.

7-9 A multimode step-index fiber has the following parameters:

Carrier wavelength	$\lambda = 0.85\ \mu m$
Carrier frequency	$f = 3.53 \times 10^{14}$ Hz
Core radius	$a = 30\ \mu m$
Core refractive index	$n_1 = 1.48$
Cladding radius	$b = 60\ \mu m$
Cladding refractive index	$n_2 = 1.20$

Calculate the

a) phase constant in the core region;
b) phase constant in the cladding region;
c) phase constant in free space;
d) transverse propagation constant in the core region;
e) transverse propagation constant in the cladding region;
f) value of the Bessel function $J_1(ka)$; and
g) value of the Hankel function $H_1(\chi b)$.

7-10 A step-index fiber operates in a single mode EH_{01}. Its radius a is 5 μm, and its refractive indexes are $n_1 = 1.15$ and $n_2 = 1.08$. Determine the cutoff wavelength and frequency.

7-11 A multimode step-index fiber has a core radius a of 40 μm and a cladding radius b of 50 μm. The refractive index n_1 of the core is 1.40, and the refractive differential between n_1 and n_2 is 10%. Calculate the

a) light-gathering power;
b) cutoff wavelength; and
c) number of modes.

7-12 Describe the characteristics and applications of the monomode and multimode step-index fibers.

7-13 A multimode step-index fiber has the following parameters:

Core radius	$a = 40\ \mu m$
Core refractive index	$n_1 = 1.52$
Cladding refractive index	$n_2 = 1.48$
Wavelength of LED light source	$\lambda = 0.87\ \mu m$

Determine the

a) numerical aperture of the fiber;

b) maximum acceptance angle of the fiber;
c) maximum number of modes; and
d) light-gathering power.

7-14 A multimode step-index fiber has the following parameters:

Core refractive index	$n_1 = 1.50$
Numerical aperture	$NA = 0.37$
V number	$X_{np} = 120$
Wavelength of Nd^{3+}:YAG light source	$\lambda = 1.06\ \mu m$

Calculate the

a) cladding refractive index n_2;
b) core radius a; and
c) cladding radius b.

7-15 A graded-index fiber has $n_1 = 150$ and $n_2 = 1.05$. Determine the refractive index at $r = a/2$ for $\alpha = 4$.

7-16 A graded-index fiber can support multiple modes. Its core radius a is 30 μm, and its refractive indexes are $n_1 = 1.50$ and $n_2 = 1.05$. Determine the number of modes.

7-17 Describe the characteristics and applications of multimode graded-index fibers.

7-18 Derive the equation of the total number of modes for a graded-index fiber, as shown in Eq. (7-4-5), from Eq. (7-4-3).

7-19 A graded-index fiber has the following parameters:

Core radius	$a = 30\ \mu m$
Core refractive index	$n_1 = 1.50$
Deviation index	$\Delta = 0.02$
Power law coefficient	$\alpha = 2$
Wavelength of light source	$\lambda_0 = 1.06\ \mu m$

Determine the

a) number of modes for the graded-index fiber;
b) graded refractive index; and
c) number of modes for a step-index fiber.

7-20 An optical-fiber digital line has a LED source of 20 mW and the following parameters:

LED source-to-fiber coupling loss	15 dB
Fiber length	2 km
Fiber loss	4 dB/km
Fiber-to-detector coupling loss	0.1 dB
Minimum power required for receiving	−30 dBm

Calculate the

a) total line loss;
b) available power; and
c) power margin at the receiver.

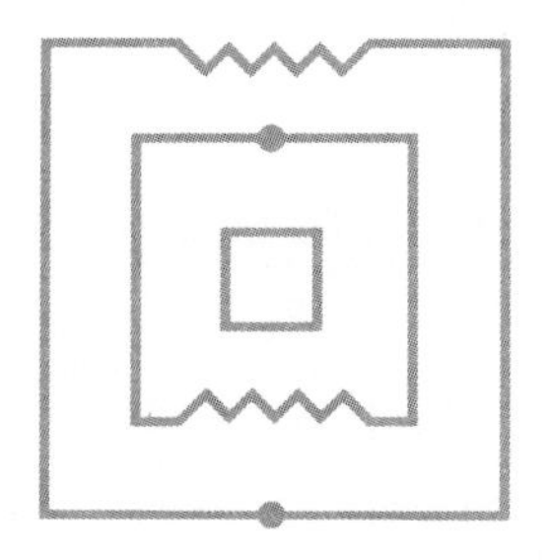

Chapter 8

Dielectric Planar Waveguides

8-0 INTRODUCTION

A dielectric planar waveguide consists of two parallel dielectric plates but is different from a rectangular waveguide because its top and bottom plates are not connected. This type of waveguide is similar to a two-conductor line or hollow tube, so it can support the TE and TM modes, as well as the TEM modes. In addition, recent advances in monolithic microwave integrated circuit (MMIC) technology make a coplanar waveguide preferable for monolithic circuits.

The planar waveguide has a cutoff frequency, as does the hollow waveguide. In general, the planar waveguide can generate, propagate, combine, detect, deflect, divide, modulate, demodulate, and switch optical guided waves. This type of waveguide is commonly used in microwave integrated circuits for signal transmission. In this chapter we discuss several widely used dielectric planar waveguides, such as infinite parallel-plate waveguide, dielectric-slab waveguide, coplanar waveguide, thin film-on-conductor waveguide, and thin film-on-dielectric waveguide, which are shown schematically in Fig. 8-0-1.

8-1 PARALLEL-PLATE WAVEGUIDES

A parallel-plate waveguide is composed of two perfect conducting plates separated by a dielectric medium having constitutive parameters of μ_d and ε_d, as shown in Fig. 8-1-1.

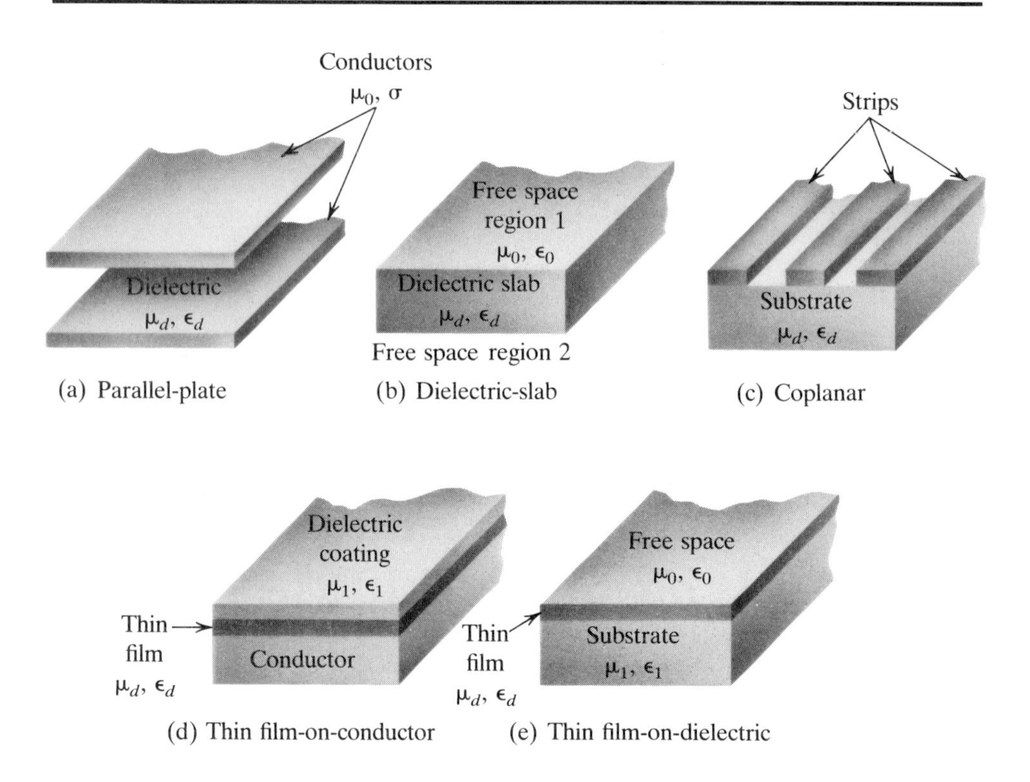

Figure 8-0-1 Schematic diagrams of commonly used dielectric planar waveguides.

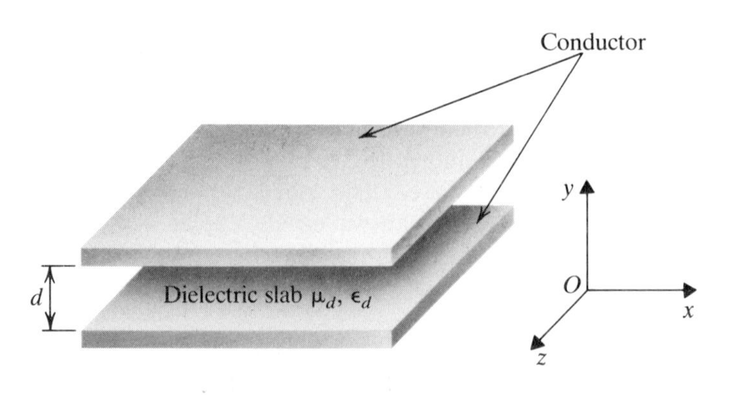

Figure 8-1-1 Schematic diagram of a parallel-plate waveguide.

We assume that the two plates are infinite in the x and z directions. In Section 3-1 we described the characteristics of a TEM wave propagating along a finite parallel stripline. The field behaviors of the TEM mode have characteristics similar to those of an unbounded dielectric medium.

8-1-1 TM Waves along the Plates

An infinite parallel-plate waveguide can support TM and TE waves. Let's suppose that TM waves propagate in the positive z direction. Then there are no field variations in the x direction, because the width in the x direction is infinite.

Also, the fringing field effects outside the edges in the x direction are negligible for $x \gg d$, since we are assuming that the plates extend infinitely in the x and z directions. For the TM mode ($H_z = 0$), we expressed the electric-field component as

$$E_z(y, z) = E_{z0}(y)e^{-\gamma z} \tag{8-1-1}$$

The TM wave equation is

$$\nabla^2 E_z - \gamma^2 E_z = 0 \tag{8-1-2}$$

or

$$\frac{\partial^2 E_z}{\partial y^2} + \frac{\partial^2 E_z}{\partial z^2} - \gamma^2 E_z = 0 \tag{8-1-3}$$

where $\gamma^2 = -\omega^2 \mu_d \varepsilon_d$. The general solution of Eq. (8-1-2) is

$$E_z(y, z) = E_{z0} \sin\left(\frac{n\pi y}{d}\right) e^{-\gamma_z z} \tag{8-1-4}$$

where

$\gamma_z = [(n\pi/d)^2 - \omega^2 \mu_d \varepsilon_d]^{1/2}$ is the wave propagation constant; and
$n = 0, 1, 2, 3, \ldots$

When $\gamma_z = 0$, the cutoff frequency is

$$f_c = \frac{n}{2d\sqrt{\mu_d \varepsilon_d}} = \frac{nc}{2d\sqrt{\varepsilon_{rd}}} \qquad \text{for } \mu_d = \mu_0 \tag{8-1-5}$$

and the cutoff wavelength is

$$\lambda_c = \frac{2d}{n} \tag{8-1-6}$$

For propagation, $(n\pi/d)^2 < \omega^2 \mu_d \varepsilon_d$, the propagation constant is imaginary. That is,

$$\gamma_z = j\beta_z = j\sqrt{\omega^2 \mu_d \varepsilon_d - (n\pi/d)^2} = j\omega\sqrt{\mu_d \varepsilon_d}\left[1 - \left(\frac{f_c}{f}\right)^2\right]^{1/2} \tag{8-1-7}$$

Finally, the field equations for the TM mode in an infinite parallel-plate waveguide are

$$E_z(y, z) = E_{z0} \sin\left(\frac{n\pi y}{d}\right) e^{-j\beta_z z} \tag{8-1-8}$$

$$E_y = \frac{-j\beta_z}{k} E_{z0} \cos\left(\frac{n\pi y}{d}\right) e^{-j\beta_z z} \tag{8-1-9}$$

and

$$H_x = \frac{j\omega\varepsilon}{k} E_{z0} \cos\left(\frac{n\pi y}{d}\right) e^{-j\beta_z z} \tag{8-1-10}$$

8-1-2 TE Waves Along the Plates

We assume that there are no field variations in the x direction. For the TE mode ($E_z = 0$), the magnetic-field component is

$$H_z(y, z) = H_{z0}(y) e^{-\gamma z} \tag{8-1-11}$$

The wave equation is

$$\nabla^2 H_z - \gamma^2 H_z = 0 \tag{8-1-12}$$

or

$$\frac{\partial^2 H_z}{\partial y^2} + \frac{\partial^2 H_z}{\partial z^2} - \gamma^2 H_z = 0 \tag{8-1-13}$$

Similarly, the general solution of Eq. (8.12) is

$$H_z(y, z) = H_{z0} \cos\left(\frac{n\pi y}{d}\right) e^{-j\beta_z z} \tag{8-1-14}$$

where

$$\beta_z = \omega\sqrt{\mu_d \varepsilon_d}\left[1 - \left(\frac{f_c}{f}\right)^2\right]^{1/2}; \text{ and} \tag{8-1-15}$$

$n = 1, 2, 3, \ldots$

The cutoff frequency is

$$f_c = \frac{n}{2d\sqrt{\mu_d \varepsilon_d}} \tag{8-1-16}$$

and the cutoff wavelength is

$$\lambda_c = \frac{2d}{n} \tag{8-1-17}$$

The phase velocity of the wave in the waveguide is

$$v_{ph} = \frac{\omega}{\beta_z} = \frac{1}{\sqrt{\mu_d \varepsilon_d}}\left[1 - \left(\frac{f_c}{f}\right)^2\right]^{-1/2} \tag{8-1-18}$$

The final field equations for TE modes in a parallel-plate waveguide are

$$E_x(y, z) = \frac{j\omega\mu}{k} H_{z0} \sin\left(\frac{n\pi y}{d}\right) e^{-j\beta_z z} \tag{8-1-19}$$

$$H_y(y, z) = \frac{j\beta_z}{k} H_{z0} \sin\left(\frac{n\pi y}{d}\right) e^{-j\beta_z z} \tag{8-1-20}$$

and

$$H_z(y, z) = H_{z0} \cos\left(\frac{n\pi y}{d}\right) e^{-j\beta_z z} \tag{8-1-21}$$

8-1-3 Attenuation and Intrinsic Impedance

The intrinsic wave propagation constant in a medium is

$$\gamma = \sqrt{j\omega\mu(\sigma + j\omega\varepsilon)} = j\omega\sqrt{\mu\varepsilon}\left(1 + \frac{\sigma}{j\omega\varepsilon}\right)^{1/2} \tag{8-1-22}$$

The propagation constant caused by losses in the dielectric at frequencies above the cutoff frequency f_c, then, is

$$\begin{aligned}\gamma_z &= \sqrt{\left(\frac{n\pi}{d}\right)^2 - \omega^2\mu_d\left(\varepsilon_d + \frac{\sigma_d}{j\omega\varepsilon_d}\right)} \\ &= \left[\left(\frac{n\pi}{d}\right)^2 - \omega^2\mu_d\varepsilon_d + \frac{j\omega\sigma_d\mu_d}{\varepsilon_d}\right]^{1/2} \\ &= \frac{\sigma}{2}\sqrt{\frac{\mu_d}{\varepsilon_d}}\left[1 - \left(\frac{f_c}{f}\right)^2\right]^{-1/2} + j\omega\sqrt{\mu_d\varepsilon_d}\left[1 - \left(\frac{f_c}{f}\right)^2\right]^{1/2}\end{aligned} \tag{8-1-23}$$

where $f_c = \dfrac{n}{2d\sqrt{\mu_d\varepsilon_d}}$, as defined in Eq. (8-1-5).

Finally, the attenuation constant in the dielectric for the TM mode is

$$\alpha_d = \frac{\sigma}{2}\sqrt{\frac{\mu_d}{\varepsilon_d}}\left[1 - \left(\frac{f_c}{f}\right)^2\right]^{-1/2} \quad \text{Np/m} \tag{8-1-24}$$

For the TE mode, the attenuation constant is

$$\alpha_d = \frac{\sigma n}{2}\left[1 - \left(\frac{f_c}{f}\right)^2\right]^{-1/2} \quad \text{Np/m} \tag{8-1-25}$$

When $n = 0$, $E_z = 0$, and only the transverse components H_x and E_y exist. As we described in Section 6-4, the TEM mode is the dominant mode in a parallel-plate waveguide, and its cutoff frequency is 0.

The intrinsic impedance of one conducting plate is

$$\begin{aligned}\eta &= \sqrt{\frac{j\omega\mu}{\sigma + j\omega\varepsilon}} = \sqrt{\frac{j\omega\mu}{\sigma}} \quad \text{for } \sigma \gg \omega\varepsilon \\ &= \sqrt{\frac{\omega\mu}{\sigma}}\,\underline{/45^\circ} = (1 + j)\sqrt{\frac{\omega\mu}{2\sigma}} \\ &= (1 + j)\frac{1}{\sigma\delta} = (1 + j)R_s\end{aligned} \tag{8-1-26}$$

The expression $R_s = \sqrt{\omega\mu/(2\sigma)}$ is called the *skin effect*, or surface resistance, in ohms per square.

EXAMPLE 8-1-1 Characteristics of an Infinite Parallel-Plate Waveguide

An infinite parallel-plate waveguide has the following parameters:

Operating frequency	$f = 10$ GHz
Separation distance	$d = 8$ mm
Conductivity of gold plate	$\sigma_{Au} = 4.1 \times 10^7$ ℧/m
Relative dielectric constant of GaAs insulator	$\varepsilon_{rd} = 13.10$
Conductivity of GaAs substrate	$\sigma_d = 1.17 \times 10^{-2}$ ℧/m

Calculate:

a) the cutoff frequency for the TM mode;
b) the phase velocity for the TM wave;
c) the intrinsic impedance of the plate; and
d) the attenuation constant for the TM mode.

Solutions:

a) Using Eq. (8-1-5), we obtain the cutoff frequency:

$$f_c = \frac{1 \times 3 \times 10^8}{2 \times 8 \times 10^{-3}\sqrt{13.1}}$$

$$= 5.18 \text{ GHz}$$

b) We calculate the phase velocity from Eq. (8-1-18) as

$$v_{ph} = \frac{3 \times 10^8}{\sqrt{13.1}}\left[1 - \left(\frac{5.18}{10}\right)^2\right]^{-1/2}$$

$$= 0.97 \times 10^8 \text{ m/s}$$

c) We get the intrinsic impedance from Eq. (8-1-26):

$$\eta = (1 + j)\left(\frac{2\pi \times 10 \times 10^9 \times 4\pi \times 10^{-7}}{2 \times 4.1 \times 10^7}\right)^{1/2}$$

$$= 0.031 + j0.031$$

d) The attenuation is computed from Eq. (8-1-24) as

$$\alpha_d = \frac{1.17 \times 10^{-2}}{2}\left(\frac{377}{\sqrt{13.1}}\right)\left[1 - \left(\frac{5.18}{10}\right)^2\right]^{-1/2}$$

$$= 0.713 \text{ Np/m} = 6.20 \text{ dB/m}$$

8-2 DIELECTRIC-SLAB WAVEGUIDES

Thin dielectric-slab waveguides are commonly used in optical integrated circuits for signal transmission. Figure 8-2-1 shows schematically a dielectric-slab waveguide. The thin dielectric slab of thickness d, permittivity ε_d, and permeability μ_d is situated in two free-space regions of μ_0 and ε_0.

We assume that there is no field dependence on the x coordinate, that the dielectric is lossless, and that the waves are propagating in the positive z direction. As for the wave propagation in a rectangular waveguide, we can also analyze the wave propagation in a dielectric-slab waveguide as TE and TM modes.

8-2-1 TE Waves Along a Dielectric-Slab Waveguide

For the TE mode ($E_z = 0$) and no x dependency, the wave equation for a dielectric-slab waveguide is

$$\frac{\partial^2 H_z(y)}{\partial y^2} + k_y^2 H_z(y) = 0 \tag{8-2-1}$$

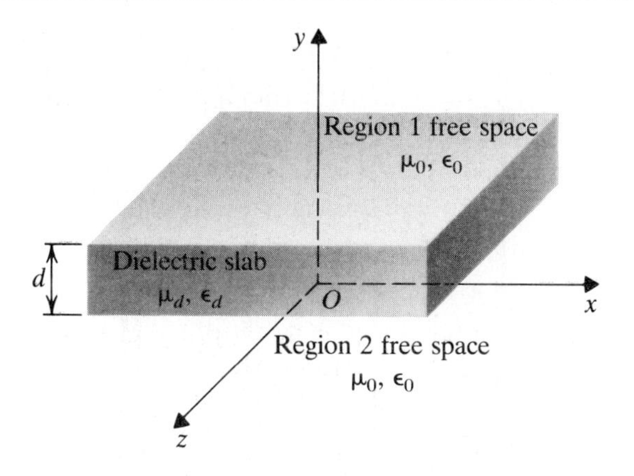

Figure 8-2-1 Schematic diagram of a dielectric-slab waveguide.

where $k_y^2 = \omega^2 \mu_d \varepsilon_d - \beta^2$ and is the characteristic equation for the dielectric slab; and k_y is the wave number in the y coordinate. We can write the solution of Eq. (8-2-1) as

$$H_z(y) = H_o \sin(k_y y) + H_e \cos(k_y y) \qquad \text{for } |y| \le \frac{d}{2} \tag{8-2-2}$$

where

sin $(k_y y)$ is the odd function (unsymmetric waves) of the coordinate y;
cos $(k_y y)$ is the even function (symmetric waves) of the coordinate y; and
k_y is the wave number in the y coordinate.

In free-space region 1 ($y > d/2$) and free-space region 2 ($y < -d/2$), the waves decay exponentially. That is,

$$H_z(y) = H_1 e^{-\alpha(y - d/2)} \qquad \text{for } y \ge \frac{d}{2} \tag{8-2-3}$$

and

$$H_z(y) = H_2 e^{-\alpha(y + d/2)} \qquad \text{for } y \le -\frac{d}{2} \tag{8-2-4}$$

where $\alpha = \sqrt{\beta^2 - \omega^2 \mu_0 \varepsilon_0}$ is the characteristic equation in free-space regions 1 and 2.

1. *Odd TE Modes*

The odd TE modes are expressed by the function:

$$H_z(y) = H_o \sin(k_y y) \tag{8-2-5}$$

For a dielectric slab, $|y| \le d/2$, we can write the field equations, after applying the boundary conditions, as

$$E_x(y) = \frac{-j\omega\mu_d}{k_y} H_o \cos(k_y y) \tag{8-2-6}$$

$$H_y(y) = \frac{-j\beta}{k_y} H_o \cos(k_y y) \tag{8-2-7}$$

and

$$H_z(y) = H_o \sin(k_y y) \tag{8-2-8}$$

For free-space region 1 ($y \geq d/2$), we can express the field equations, after applying the boundary conditions, as

$$E_x(y) = \frac{-j\omega\mu_0}{\alpha} H_o \sin\left(k_y \frac{d}{2}\right) e^{-\alpha(y-d/2)} \tag{8-2-9}$$

$$H_y(y) = \frac{-j\beta}{\alpha} H_o \sin\left(k_y \frac{d}{2}\right) e^{-\alpha(y-d/2)} \tag{8-2-10}$$

and

$$H_z(y) = H_o \sin\left(k_y \frac{d}{2}\right) e^{-\alpha(y-d/2)} \tag{8-2-11}$$

For free-space region 2 ($y \leq -d/2$), we can express the field equations, after applying the boundary conditions, as

$$E_x(y) = \frac{-j\omega\mu_0}{\alpha} H_o \sin\left(k_y \frac{d}{2}\right) e^{\alpha(y+d/2)} \tag{8-2-12}$$

$$H_y(y) = \frac{-j\beta}{\alpha} H_o \sin\left(k_y \frac{d}{2}\right) e^{\alpha(y+d/2)} \tag{8-2-13}$$

and

$$H_z(y) = -H_o \sin\left(k_y \frac{d}{2}\right) e^{\alpha(y+d/2)} \tag{8-2-14}$$

We can obtain the characteristic equation between α and k_y by equating Eq. (8-2-6) and Eq. (8-2-9) at $y = d/2$:

$$\frac{\alpha}{k_y} = \frac{\mu_0}{\mu_d} \tan\left(k_y \frac{d}{2}\right) \qquad \text{for odd TE modes} \tag{8-2-15}$$

a) Surface Impedance

The surface impedance of the dielectric-slab waveguide (looking down from the top surface of the dielectric slab) is

$$Z_s = \left.\frac{E_x}{H_z}\right|_{-y} = -j\frac{\omega\mu_0}{\alpha} \qquad \text{for odd TE modes} \tag{8-2-16}$$

We can also define the surface impedance from the bottom surface of the dielectric slab, looking up, as

$$Z_s = \left.\frac{-E_x}{H_z}\right|_{+y} = -j\frac{\omega\mu_0}{\alpha} \qquad \text{for odd TE modes} \tag{8-2-17}$$

These expressions indicate that an odd TE surface wave can be supported by a capacitive surface.

b) Cutoff Frequency

By adding α and k_y, as in Eqs. (8-2-1) and (8-2-4), we have

$$\alpha^2 + k_y^2 = \omega^2(\mu_d\varepsilon_d - \mu_0\varepsilon_0) \tag{8-2-18}$$

We can then express the attenuation constant as

$$\alpha = [\omega^2(\mu_d\varepsilon_d - \mu_0\varepsilon_0) - k_y^2]^{1/2} \tag{8-2-19}$$

When the phase constant β approaches $\omega\sqrt{\mu_0\varepsilon_0}$, as shown in Eq. (8-2-4), the attenuation α becomes 0, which means that the waves are no longer bound to the dielectric-slab region. In other words, the waves are cut off, and the cutoff condition is

$$\begin{aligned}\alpha = 0 &= \frac{\mu_0}{\mu_d} k_c \tan\left(k_y \frac{d}{2}\right) \\ &= \tan\left(\frac{\omega_c d}{2}\sqrt{\mu_d\varepsilon_d - \mu_0\varepsilon_0}\right)\end{aligned} \tag{8-2-20}$$

where $k_c = k_y = \omega_c\sqrt{\mu_d\varepsilon_d - \mu_0\varepsilon_0}$ has been replaced; and

$$\pi f_c d\sqrt{\mu_d\varepsilon_d - \mu_0\varepsilon_0} = (n-1)\pi \qquad \text{where } n = 1, 2, 3, \ldots$$

Therefore the cutoff frequency for odd TE modes is

$$f_{co} = \frac{n-1}{d\sqrt{\mu_d\varepsilon_d - \mu_0\varepsilon_0}} \tag{8-2-21}$$

If $n = 1$, $f_c = 0$, which means that the lowest order, odd TE mode can propagate along a dielectric-slab waveguide for any thickness d. As the frequency increases, the attenuation constant α also increases and the TE waves will be confined to a region closer to the dielectric slab.

EXAMPLE 8-2-1 Characteristics of Odd TE Waves in Dielectric-Slab Waveguide

An odd TE wave propagates through a dielectric-slab waveguide that has the following parameters:

Relative dielectric constant of alumina slab	$\varepsilon_{rd} = 10$
Slab thickness	$d = 15$ mm
Frequency	$f = 10$ GHz
Integer	$n = 2$

Calculate:

a) the phase constant β;
b) the attenuation constant α;
c) the surface impedance Z_s; and
d) the cutoff frequency.

Solutions:

a) The phase constant is

$$\begin{aligned}\beta &= \sqrt{\omega^2\mu_d\varepsilon_d - \left(\frac{n\pi}{d}\right)^2} \\ &= \left[(2\pi \times 10^{10})^2 \times 4\pi \times 10^{-7} \times 8.854 \times 10^{-12} \times 10 - \left(\frac{2\pi}{15 \times 10^{-3}}\right)^2\right]^{1/2} \\ &= [43.93 \times 10^4 - 17.55 \times 10^4]^{1/2} \\ &= 5.14 \times 10^2 \text{ rad/m}\end{aligned}$$

b) The attenuation constant is

$$\alpha = \left[(2\pi \times 10^{10})^2 \times 4\pi \times 10^{-7} \times 8.854 \times 10^{-12} \times (10-1) - \left(\frac{2\pi}{15 \times 10^{-3}}\right)^2\right]^{1/2}$$

$$= [39.54 \times 10^4 - 17.55 \times 10^4]^{1/2}$$

$$= 469 \text{ Np/m}$$

c) The surface impedance is

$$Z_s = -j\frac{2\pi \times 10^{10} \times 4\pi \times 10^{-7}}{469}$$

$$= -j168.35\ \Omega$$

d) The cutoff frequency is

$$f_c = \frac{n-1}{d\sqrt{\mu_0 \varepsilon_0 (\varepsilon_{rd} - 1)}}$$

$$= \frac{(2-1)(3 \times 10^8)}{15 \times 10^{-3}\sqrt{10-1}}$$

$$= 6.67 \text{ GHz}$$

---

2. *Even TE Modes*

The even TE modes are expressed by the function:

$$H_z(y) = H_e \cos(k_y y) \tag{8-2-22}$$

For a dielectric slab, $|y| \leq d/2$, we can write the field equations, after applying the boundary conditions, as

$$E_x(y) = \frac{j\omega\mu_d}{k_y} H_e \sin(k_y y) \tag{8-2-23}$$

$$H_y(y) = \frac{j\beta}{k_y} H_e \sin(k_y y) \tag{8-2-24}$$

and

$$H_z(y) = H_e \cos(k_y y) \tag{8-2-25}$$

For free-space region 1 ($y \geq d/2$), we can express the field equations, after applying the boundary conditions, as

$$E_x(y) = \frac{-j\omega\mu_0}{\alpha} H_e \cos\left(k_y \frac{d}{2}\right) e^{-\alpha(y-d/2)} \tag{8-2-26}$$

$$H_y(y) = \frac{-j\beta}{\alpha} H_e \cos\left(k_y \frac{d}{2}\right) e^{-\alpha(y-d/2)} \tag{8-2-27}$$

and

$$H_z(y) = H_e \cos\left(k_y \frac{d}{2}\right) e^{-\alpha(y-d/2)} \tag{8-2-28}$$

For free-space region 2 ($y \leq d/2$), we can express the field equations, after applying the boundary conditions, as

$$E_x(y) = \frac{j\omega\mu_0}{\alpha} H_e \cos\left(k_y \frac{d}{2}\right) e^{\alpha(y+d/2)} \tag{8-2-29}$$

$$H_y(y) = \frac{j\beta}{\alpha} H_e \cos\left(k_y \frac{d}{2}\right) e^{\alpha(y+d/2)} \tag{8-2-30}$$

and

$$H_z(y) = H_e \cos\left(k_y \frac{d}{2}\right) e^{\alpha(y+d/2)} \tag{8-2-31}$$

The characteristic equation between α and k_y is

$$\frac{\alpha}{k_y} = -\frac{\mu_0}{\mu_d} \cot\left(k_y \frac{d}{2}\right) \qquad \text{for even TE modes} \tag{8-2-32}$$

a) Surface Impedance

The surface impedance of a dielectric-slab waveguide (either looking down from the top surface or looking up from the bottom surface) is

$$Z_s = \frac{E_x}{H_z}\bigg|_{-y} = \frac{(-j\omega\mu_o)/\alpha}{1} = -j\frac{\omega\mu_0}{\alpha} \qquad \text{for even TE modes} \tag{8-2-33}$$

or

$$Z_s = \frac{-E_x}{H_z}\bigg|_{+y} = \frac{-(+j\omega\mu_0)/\alpha}{1} = -j\frac{\omega\mu_0}{\alpha} \qquad \text{for even TE modes} \tag{8-2-34}$$

These expressions indicate that an even TE surface wave can be supported by a capacitive surface.

b) Cutoff Frequency

We can find the cutoff frequency by using the same method as before:

$$\alpha = 0 = -\frac{\mu_0}{\mu_d} k_c \cot\left(k_y \frac{d}{2}\right)$$

$$= -\frac{\mu_0}{\mu_d} k_c \cot\left(\frac{\omega_c d}{2}\sqrt{\mu_d\varepsilon_d - \mu_0\varepsilon_0}\right) \tag{8-2-35}$$

where $\pi f_c d\sqrt{\mu_d\varepsilon_d - \mu_0\varepsilon_0} = [n - (1/2)]\pi$; $n = 1, 2, 3, \ldots$. Therefore the cutoff frequency for even TE modes is

$$f_{ce} = \frac{n - 1/2}{d\sqrt{\mu_d\varepsilon_d - \mu_0\varepsilon_0}} \qquad \text{even TE modes} \tag{8-2-36}$$

8-2-2 TM Waves Along a Dielectric-Slab Waveguide

The analysis of TM waves along a dielectric-slab waveguide is similar to that of TE waves. For TM modes ($H_z = 0$) and without x dependence, the wave equation for a dielectric-slab waveguide is

$$\frac{d^2E_z(y)}{dy^2} + k_y^2 E_z(y) = 0 \tag{8-2-37}$$

where $k_y^2 = \gamma^2 + \omega^2 \mu_d \varepsilon_d$ is the characteristic equation for the dielectric slab. We can write the solution of Eq. (8-2-37) as

$$E_z(y) = E_o \sin(k_y y) + E_e \cos(k_y y) \tag{8-2-38}$$

For free-space regions 1 ($y \geq d/2$) and 2 ($y \leq -d/2$) the waves decay exponentially. That is,

$$E_z(y) = E_1 e^{-\alpha(y-d/2)} \qquad \text{for } y \geq d/2 \tag{8-2-39}$$

$$E_z(y) = E_2 e^{-\alpha(y+d/2)} \qquad \text{for } y \leq d/2 \tag{8-2-40}$$

where $\alpha = \sqrt{\beta^2 - \omega^2 \mu_0 \varepsilon_0}$ is the characteristic equation in free-space regions 1 and 2.

1. *Odd TM Modes*

The odd TM modes are expressed by the function:

$$E_z(y) = E_o \sin(k_y y) \tag{8-2-41}$$

For a dielectric slab, $|y| \leq d/2$, we can express the field equations, after applying the boundary conditions, as

$$E_y(y) = \frac{-j\beta}{k_y} E_o \cos(k_y y) \tag{8-2-42}$$

$$E_z(y) = E_o \sin(k_y y) \tag{8-2-43}$$

and

$$H_x(y) = \frac{j\omega\varepsilon_d}{k_y} E_o \cos(k_y y) \tag{8-2-44}$$

For free-space region 1 ($y \geq d/2$), we can write the field equations, after applying the boundary conditions, as

$$E_y(y) = \frac{-j\beta}{\alpha} E_o \sin\left(k_y \frac{d}{2}\right) e^{-\alpha[y-d/2]} \tag{8-2-45}$$

$$E_z(y) = E_o \sin\left(k_y \frac{d}{2}\right) e^{-\alpha[y-d/2]} \tag{8-2-46}$$

and

$$H_x(y) = \frac{j\omega\varepsilon_0}{\alpha} E_o \sin\left(k_y \frac{d}{2}\right) e^{-\alpha[y-d/2]} \tag{8-2-47}$$

For free-space region 2 ($y \leq -d/2$), we can write the field equations, after applying the boundary conditions, as

$$E_y(y) = \frac{-j\beta}{\alpha} E_o \sin\left(k_y \frac{d}{2}\right) e^{\alpha[y+d/2]} \tag{8-2-48}$$

$$E_z(y) = -E_o \sin\left(k_y \frac{d}{2}\right) e^{\alpha[y+d/2]} \tag{8-2-49}$$

and

$$H_x(y) = \frac{j\omega\varepsilon_0}{\alpha} E_o \sin\left(k_y \frac{d}{2}\right) e^{\alpha[y+d/2]} \tag{8-2-50}$$

We can obtain the characteristic equation between α and k_y from Eqs. (8-2-44) and (8-2-47):

$$\frac{\alpha}{k_y} = \frac{\varepsilon_0}{\varepsilon_d} \tan\left(k_y \frac{d}{2}\right) \tag{8-2-51}$$

The cutoff frequency is

$$f_{co} = \frac{n - 1}{d\sqrt{\mu_d \varepsilon_d - \mu_0 \varepsilon_0}} \tag{8-2-52}$$

and the surface impedance is

$$\begin{aligned} Z_s &= \left.\frac{E_z(y)}{H_x(y)}\right|_y \\ &= \frac{-1}{(j\omega\varepsilon_0)/\alpha} \\ &= j\frac{\alpha}{\omega\varepsilon_0} \end{aligned} \tag{8-2-53}$$

Equation (8-2-53) indicates that an odd TM surface wave can be supported by an inductive surface.

2. *Even TM Modes*

The even TM modes are described by the field equation:

$$E_z(y) = E_e \cos(k_y y) \tag{8-2-54}$$

For a dielectric slab ($|y| \leq d/2$), we can express the field equations, after applying the boundary conditions, as

$$E_y(y) = \frac{j\beta}{k_y} E_e \sin(k_y y) \tag{8-2-55}$$

$$E_y(y) = E_e \cos(k_y y) \tag{8-2-56}$$

and

$$H_x(y) = \frac{-j\omega\varepsilon_d}{k_y} E_e \sin(k_y y) \tag{8-2-57}$$

For free-space region 1 ($y \geq d/2$), we can write the field equations, after applying the boundary conditions, as

$$E_y(y) = \frac{-j\beta}{\alpha} E_e \cos\left(k_y \frac{d}{2}\right) e^{-\alpha(y - d/2)} \tag{8-2-58}$$

$$E_z(y) = E_e \cos\left(k_y \frac{d}{2}\right) e^{-\alpha(y - d/2)} \tag{8-2-59}$$

and

$$H_x(y) = \frac{j\omega\varepsilon_0}{\alpha} E_e \cos\left(k_y \frac{d}{2}\right) e^{-\alpha(y - d/2)} \tag{8-2-60}$$

For free-space region 2 ($y \leq -d/2$), we can write the field equations, after applying the boundary conditions, as

$$E_y(y) = \frac{j\beta}{\alpha} E_e \cos\left(k_y \frac{d}{2}\right) e^{\alpha(y+d/2)} \tag{8-2-61}$$

$$E_z(y) = E_e \cos\left(k_y \frac{d}{2}\right) e^{\alpha(y+d/2)} \tag{8-2-62}$$

and

$$H_x(y) = \frac{-j\omega\varepsilon_0}{\alpha} E_e \cos\left(k_y \frac{d}{2}\right) e^{\alpha(y+d/2)} \tag{8-2-63}$$

We can obtain the characteristic equation between α and k_y from Eqs. (8-2-60) and (8-2-63):

$$\frac{\alpha}{k_y} = -\frac{\varepsilon_0}{\varepsilon_d} \cot\left(k_y \frac{d}{2}\right) \tag{8-2-64}$$

The cutoff frequency is

$$f_{ce} = \frac{n - 1/2}{d\sqrt{\mu_d \varepsilon_d - \mu_0 \varepsilon_0}} \tag{8-2-65}$$

and the surface impedance is

$$\begin{aligned} Z_s &= \left. \frac{-E_z(y)}{H_x(y)} \right|_{-y} \\ &= \frac{-1}{(j\omega\varepsilon_0)/\alpha} \\ &= j\frac{\omega\varepsilon_0}{\alpha} \end{aligned} \tag{8-2-66}$$

Equation (8-2-66) indicates that an even TM surface wave can be supported by an inductive surface.

EXAMPLE 8-2-2 Characteristics of Even TM Waves in Dielectric-Slab Waveguides

An even TM wave propagates through a dielectric-slab waveguide that has the following parameters:

Relative dielectric constant of the beryllia slab	$\varepsilon_{rd} = 6$
Slab thickness	$d = 15$ mm
Frequency	$f = 9$ GHz
Integer	$n = 1$

Calculate:

a) the phase constant β;
b) the attenuation constant α;
c) the surface impedance Z_s; and
d) the cutoff frequency.

Solutions:

a) The phase constant is

$$\beta = [\omega^2 \mu_d \varepsilon_d - k_y^2]^{1/2}$$

$$= \left[(2\pi \times 9 \times 10^9)^2 \times 4\pi \times 10^{-7} \times 8.854 \times 10^{-12} \times 6 - \left(\frac{\pi}{15 \times 10^{-3}}\right)^2\right]^{1/2}$$

$$= [21.36 \times 10^4 - 4.387 \times 10^4]^{1/2}$$

$$= 412 \text{ rad/m}$$

b) The attenuation constant is

$$\alpha = [\beta^2 - \omega^2 \mu_0 \varepsilon_0]^{1/2}$$

$$= [(412)^2 - (2\pi \times 9 \times 10^9)^2 \times 4\pi \times 10^{-7} \times 8.854 \times 10^{-12}]^{1/2}$$

$$= 367 \text{ Np/m}$$

c) The surface impedance is

$$Z_s = j\frac{\omega\varepsilon_0}{\alpha} = j\frac{2\pi \times 9 \times 10^9 \times 8.854 \times 10^{-12}}{367}$$

$$= j1.36 \times 10^{-3}\ \Omega$$

d) From Eq. (8-2-65), we obtain the cutoff frequency:

$$f_{ce} = \frac{n - 1/2}{d\sqrt{\mu_d\varepsilon_d - \mu_0\varepsilon_0}} = \frac{n - 1/2}{d\sqrt{\mu_0\varepsilon_0(\varepsilon_{rd} - 1)}} = \frac{(1 - 1/2) \times 3 \times 10^8}{15 \times 10^{-3}\sqrt{6 - 1}}$$

$$= 4.46 \text{ GHz}$$

8-3 COPLANAR WAVEGUIDES

A coplanar waveguide (CPW) consists of a center strip of thin metallic film deposited on the surface of a dielectric substrate, with two ground electrodes parallel to and separate from the center strip on the same surface, as shown in Fig. 8-3-1. A coplanar waveguide permits easy connection of external elements, such as active devices and capacitors. This type of waveguide is also ideal for connecting various elements in

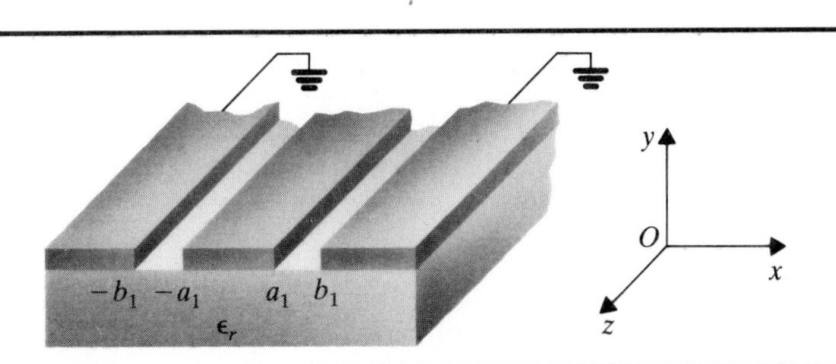

Figure 8-3-1 Schematic diagram of a coplanar waveguide.

monolithic microwave integrated circuits (MMICs) built on one semi-insulating substrate. The future solid-state airborne phased-array radar system requires a large number of identical circuits, and the coplanar waveguide is a suitable component for that system.

8-3-1 Characteristic Impedance

Because of the high dielectric constant of the substrate, most of the RF energy is stored in the dielectric; the loading effect of the grounded cover is negligible if it is more than two slot-widths from the center strip. We can calculate the characteristic impedance $\mathbf{Z}_0$ of a coplanar waveguide fabricated on a dielectric half-plane—with relative dielectric constant ε_r—as a function of the ratio a_1/b_1, where $2a_1$ is the width of the center strip, and $2b_1$ is the distance between two ground electrodes. We use a zero-order quasi-static approximation and transform the dielectric half-plane $\mathbf{Z}_1$ to the interior of a rectangle in the $\mathbf{Z}$ plane as shown in Fig. 8-3-2.

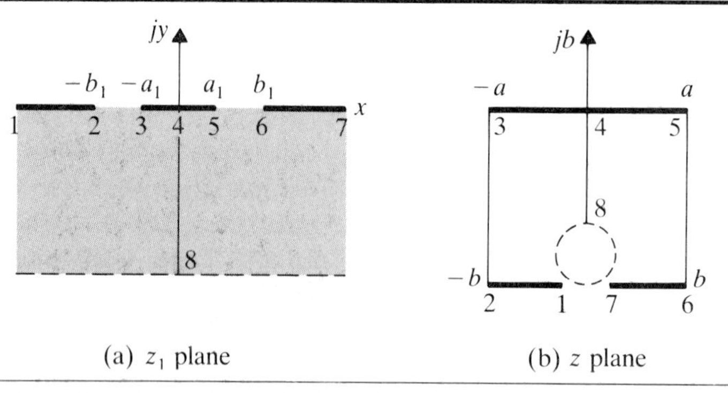

Figure 8-3-2 Transformation of mapping for a CPW.

The characteristic impedance of a coplanar waveguide (CPW) was developed by several research workers [1]. The transformation equation is

$$\frac{dZ}{dZ_1} = \frac{A}{\sqrt{(Z_1^2 - a_1^2)(Z_1^2 - b_1^2)}} \tag{8-3-1}$$

where

A is a constant;

$2a_1$ is the center-strip width; and

$2b_1$ is the distance between two ground electrodes.

We evaluate the ratio a/b of the rectangle in the Z plane, as shown in Fig. 8-3-2(b), by multiplying both sides of Eq. (8-3-1) by dZ_1 and carrying out the integration:

$$a + jb = \int_0^{b_1} \frac{A\,dZ_1}{\sqrt{(Z_1^2 - a_1^2)(Z_1^2 - b_1^2)}} \tag{8-3-2}$$

Equation (8-3-2) is one form of an elliptic integral. We can conveniently express the ratio a/b in terms of tabulated complete elliptic integrals:

$$\frac{a}{b} = \frac{K(k)}{K'(k)} \tag{8-3-3}$$

where

$k = a_1/b_1$ and is the modulus of the first kind of integral;

$K(k)$ is the complete elliptic integral of the first kind of modulus k[2];

$k' = (1 - k^2)^{1/2}$ and is the complementary modulus; and

$K'(k) = K(k')$ and is the complete integral of the first kind of complementary modulus k'.

By assuming a semi-infinite dielectric, in parallel with a half-space of air, we obtain the equivalent static capacitance per unit length for a pure TEM-mode wave propagating in the line:

$$C = \varepsilon_0(\varepsilon_r + 1)\frac{2a}{b} = 2\varepsilon_0(\varepsilon_r + 1)\frac{K(k)}{K'(k)} \tag{8-3-4}$$

where ε_r is the relative dielectric constant of the substrate. The phase velocity of the propagating wave is approximately related to the effective dielectric constant by

$$v_{ph} = \left(\frac{2}{\varepsilon_r + 1}\right)^{1/2} c \tag{8-3-5}$$

where $c = 3 \times 10^8$ m/s and is the velocity of light in vacuum. The characteristic impedance of a coplanar waveguide then becomes

$$\mathbf{Z}_0 = \frac{1}{cv_{ph}} \tag{8-3-6}$$

Figure 8-3-3 shows the curves for the characteristic impedance $\mathbf{Z}_0$ of a coplanar waveguide as a function of the ratio a_1/b_1, with the relative dielectric constant ε_r of the substrate as a parameter [1].

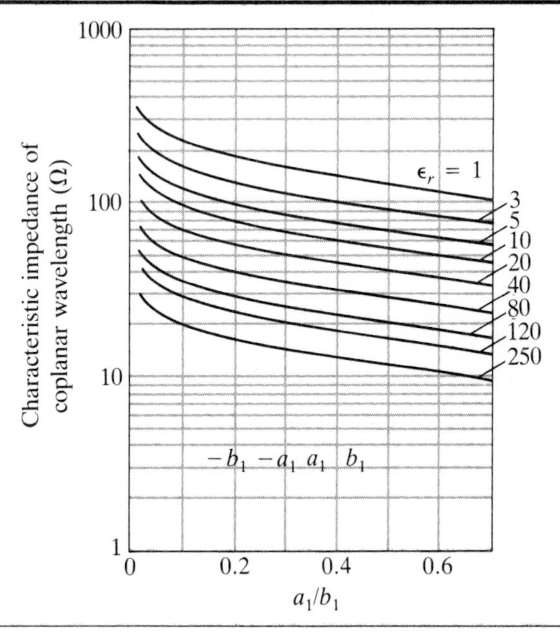

Figure 8-3-3 Characteristic impedance $\mathbf{Z}_0$ of CPW as a function of the ratio a_1/b_1, with the relative constant ε_r as a parameter.

EXAMPLE 8-3-1 Characteristic Impedance of a Coplanar Waveguide

A coplanar waveguide has the following parameters:

Relative dielectric constant of substrate	$\varepsilon_r = 13.10$
Ratio of center-strip width over the width from the center line of the center strip to the first edge of the ground strip	$a_1/b_1 = 0.70$

Calculate:

a) the capacitance of the CPW;
b) the phase velocity of the propagating wave; and
c) the characteristic impedance of the CPW.

Solutions:

a) By using Eq. (8-3-4), we obtain the capacitance of the CPW:

$$C = \varepsilon_0(\varepsilon_r + 1)\frac{2a}{b} = 2\varepsilon_0(\varepsilon_r + 1)\frac{K(k^2)}{K'(k^2)}$$

$$= 2\varepsilon_0(\varepsilon_r + 1)\frac{K(k^2)}{K(k'^2)}$$

$$= 2 \times 8.854 \times 10^{-12} \times (13.1 + 1) \times \frac{1.7457}{1.8626}$$

$$= 237 \text{ pF/m}$$

b) We find the phase velocity from Eq. (8-3-5):

$$v_{ph} = \left(\frac{2}{\varepsilon_r + 1}\right)^{1/2} c = \left(\frac{2}{13.1 + 1}\right)^{1/2} (3 \times 10^8)$$

$$= 1.13 \times 10^8 \text{ m/s}$$

c) Finally, we calculate the characteristic impedance from Eq. (8-3-6):

$$\mathbf{Z}_0 = \frac{1}{Cv_{ph}} = \frac{1}{237 \times 10^{-12} \times 1.13 \times 10^8}$$

$$= 37.34\ \Omega$$

8-3-2 Radiation Losses

When the distance between the ground electrodes approaches one wavelength of the signal wave, radiation becomes a problem. Because a coplanar waveguide is, in essence, an "edge-coupled" structure with a high concentration of charge and current near the strip edges, the losses tend to be higher than for a microstrip line. Another loss is ohmic loss, which is caused by the surface resistance of the conducting strip.

8-3-3 Applications

In microwave integrated circuits (MICs), wire bonds have always caused serious reliability and reproducibility problems. However, in monolithic microwave integrated circuits (MMICs), all active and passive circuit elements, that is, components and interconnections, are formed into the same surface of a semi-insulating substrate. Thus the coplanar waveguide structure meets MMIC design requirements and is widely used in the batch processing of hundreds of circuits per wafer of substrate.

8-4 THIN FILM-ON-CONDUCTOR WAVEGUIDES

A thin film-on-conductor waveguide is made of a thin film of lossless dielectric (μ_d, ε_d) of thickness d on a conducting ground plane, as shown in Fig. 8-4-1. The thin film is protected by a lossless dielectric coating (μ_1, ε_1). We assume that both the widths and lengths of the coating and ground plane extend to infinity. This type of dielectric thin-film waveguide is commonly used in microwave integrated circuits (MICs).

We also assume that the wave is propagating in the positive z direction and that there is no field dependence on the x coordinate. We can describe the wave propagation in a dielectric waveguide as TE and TM modes.

8-4-1 TM Modes in a Dielectric Thin Film

For TM modes ($H_z = 0$) and x independence, the wave equation and its solution, respectively, are

$$\frac{d^2E_z(y)}{dy^2} + k_y^2 E_z(y) = 0 \tag{8-4-1}$$

and

$$E_z(y) = E_o \sin(k_y y) + E_e \cos(k_y y) \tag{8-4-2}$$

The boundary condition at $y = 0$ requires that the tangential E field vanish, so

$$E_z(y)\Big|_{y=0} = 0\Big|_{y=0} + E_e\Big|_{y=0} = 0 \Rightarrow E_e = 0$$

The wave equation then is an odd function only:

$$E_z(y) = E_o \sin(k_y y) \tag{8-4-3}$$

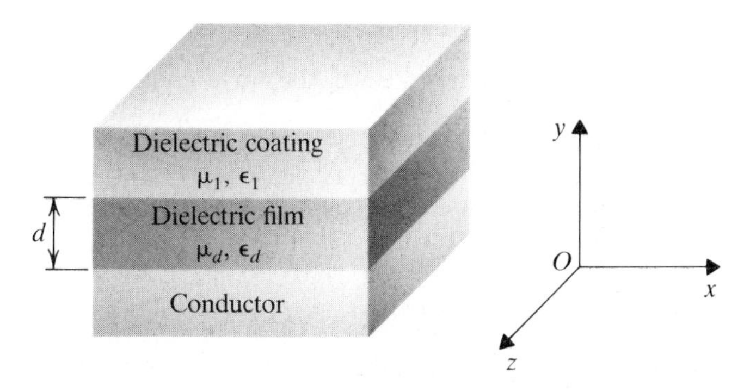

Figure 8-4-1 Schematic diagram of a thin film-on-conductor waveguide.

For the thin-film region ($0 \le y \le d$), the field equations are

$$E_y(y) = -\frac{j\beta}{k_y} E_o \cos(k_y y) \tag{8-4-4}$$

$$E_z(y) = E_o \sin(k_y y) \tag{8-4-5}$$

and

$$H_x(y) = \frac{j\omega\varepsilon_d}{k_y} E_o \cos(k_y y) \tag{8-4-6}$$

For the dielectric coating region ($y \ge d$), the field equations are

$$E_y(y) = -\frac{j\beta}{\alpha} E_o \sin(k_y y) e^{-\alpha(y-d)} \tag{8-4-7}$$

$$E_z(y) = E_o \sin(k_y y) e^{-\alpha(y-d)} \tag{8-4-8}$$

and

$$H_x(y) = \frac{j\omega\varepsilon_1}{\alpha} E_o \sin(k_y y) e^{-\alpha(y-d)} \tag{8-4-9}$$

where $\alpha = \sqrt{\beta^2 - \omega^2\mu_1\varepsilon_1}$. The characteristic equation between α and k_y at $y = d$ is

$$\frac{j\omega\varepsilon_d}{k_y} E_o \cos(k_y d) = \frac{j\omega\varepsilon_1}{\alpha} E_o \sin(k_y d) \tag{8-4-10}$$

or

$$\frac{\alpha}{k_y} = \frac{\varepsilon_1}{\varepsilon_d} \tan(k_y d) \tag{8-4-11}$$

The cutoff frequency is

$$f_{co} = \frac{n-1}{d\sqrt{\mu_d\varepsilon_d - \mu_1\varepsilon_1}} \tag{8-4-12}$$

where $n = 1, 2, 3, \ldots$

8-4-2 TE Modes in a Dielectric Thin Film

For TE modes ($E_z = 0$) and x independence, the wave equation and its solution, respectively, are

$$\frac{d^2 H_z(y)}{dy^2} + k_y^2 H_z(y) = 0 \tag{8-4-13}$$

and

$$H_z(y) = H_o \sin(k_y y) + H_e \cos(k_y y) \tag{8-4-14}$$

The boundary condition at $y = 0$ requires the continuity of the tangential H field to be

$$H_z(y)\Big|_{y=0} = 0 + H_e$$

The wave equation, then, is an even function only:

$$H_z(y) = H_e \cos(k_y y) \tag{8-4-15}$$

For the thin-film region ($0 \le y \le d$), the field equations are

$$E_x(y) = \frac{j\omega\mu_d}{k_y} H_e \sin(k_y y) \tag{8-4-16}$$

$$H_y(y) = \frac{j\beta}{k_y} H_e \sin(k_y y) \tag{8-4-17}$$

and

$$H_z(y) = H_e \cos(k_y y) \tag{8-4-18}$$

For the dielectric coating region ($y \ge d$), the field equations are

$$E_x(y) = \frac{-j\omega\mu_1}{\alpha} H_e \cos(k_y y) e^{-\alpha(y-d)} \tag{8-4-19}$$

$$H_y(y) = \frac{-j\beta}{\alpha} H_e \cos(k_y y) e^{-\alpha(y-d)} \tag{8-4-20}$$

and

$$H_z(y) = H_e \cos(k_y y) \tag{8-4-21}$$

where $\alpha = \sqrt{\omega^2(\mu_d\varepsilon_d - \mu_1\varepsilon_1) - k_y^2}$. The characteristic equation between α and k_y at $y = d$ is

$$\frac{j\omega\mu_d}{k_y} H_e \sin(k_y d) = \frac{-j\omega\mu_1}{\alpha} H_e \cos(k_y d) \tag{8-4-22}$$

or

$$\frac{\alpha}{k_y} = -\frac{\mu_1}{\mu_d} \cot(k_y d) \tag{8-4-23}$$

The cutoff frequency requires a zero propagation constant; that is,

$$\alpha = 0 = \cot\left(\omega_c d\sqrt{\mu_d\varepsilon_d - \mu_1\varepsilon_1}\right) \tag{8-4-24}$$

and the cutoff frequency is

$$f_{ce} = \frac{n + 1/2}{2d\sqrt{\mu_d\varepsilon_d - \mu_1\varepsilon_1}} \tag{8-4-25}$$

EXAMPLE 8-4-1 Cutoff Frequency of a Thin Film-on-Conductor Waveguide

A thin film-on-conductor waveguide has the following parameters:

Thin-film relative dielectric constant	$\varepsilon_{dr} = 1.50$
Film-coating relative dielectric constant	$\varepsilon_{1r} = 1.498$
Permeability	$\mu_d = \mu_1 = \mu_0$
Thickness of thin film	$d = 1\ \mu m$
Integer	$n = 1$

Calculate the cutoff frequency.

Solution:

By using Eq. (8-4-25), we obtain the cutoff frequency:

$$f_{ce} = \frac{1 + 1/2}{2 \times 10^{-6}\sqrt{4\pi \times 10^{-7} \times 8.854 \times 10^{-12}(1.5 - 1.498)}}$$
$$= 5.12 \times 10^{15} \text{ Hz}$$

8-5 THIN FILM-ON-DIELECTRIC WAVEGUIDES

A thin-film waveguide can be easily fabricated on a dielectric substrate in microwave integrated circuits (MICs). Figure 8-5-1 shows a dielectric thin film of thickness d, with permeability μ_d and permittivity ε_d, situated on a dielectric substrate of μ_1 and ε_1. We assume that the waveguide dimensions extend to infinity in both the x and z directions. The wave propagations of either the TE or TM modes are similar to those in a dielectric-slab waveguide, as discussed in Section 8-2.

8-5-1 TE Waves Along Thin Film

The wave equation for TE modes along thin film is

$$\frac{d^2H_z(y)}{dy^2} + k_y^2 H_z(y) = 0 \qquad (8\text{-}5\text{-}1)$$

where $k_y^2 = \omega^2\mu_d\varepsilon_d - \beta^2$. In the free-space region ($y \geq d$) and the substrate region ($y \leq 0$), the waves decay exponentially:

$$H_z(y) = H_0 e^{-\alpha(y-d)} \qquad \text{for } y \geq d \qquad (8\text{-}5\text{-}2)$$

and

$$H_z(y) = H_1 e^{+\alpha(y-0)} \qquad \text{for } y \leq 0 \qquad (8\text{-}5\text{-}3)$$

where

$\alpha^2 = \beta^2 - \omega^2\mu_0\varepsilon_0$ for the free-space region; and
$\alpha^2 = \beta^2 - \omega^2\mu_1\varepsilon_1$ for the substrate region.

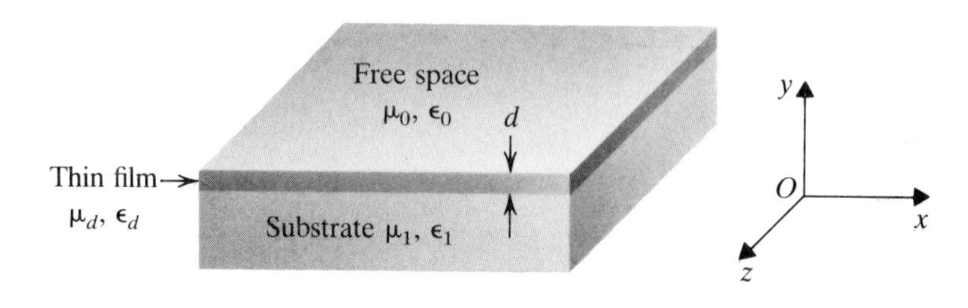

Figure 8-5-1 Schematic diagram of a thin film-on-dielectric waveguide.

8-5-2 TM Waves Along Thin Film

The wave equation for TM modes along thin film is

$$\frac{d^2E_z(y)}{dy^2} + k_y^2 E_z(y) = 0 \tag{8-5-4}$$

where $k_y^2 = \omega^2\mu_d\varepsilon_d - \beta^2$. In the free-space region $(y \geq d)$ and the substrate region $(y \leq 0)$, the waves must decay exponentially:

$$E_z(y) = E_0 e^{-\alpha(y-d)} \qquad \text{for } y \geq d \tag{8-5-5}$$

and

$$E_z(y) = E_1 e^{+\alpha(y-0)} \qquad \text{for } y \leq 0 \tag{8-5-6}$$

where

$\alpha^2 = \beta^2 - \omega^2\mu_0\varepsilon_0$ for the free-space region; and
$\alpha^2 = \beta^2 - \omega^2\mu_1\varepsilon_1$ for the substrate region.

EXAMPLE 8-5-1 Attenuation Constants of a Dielectric Thin Film-on-Dielectric Waveguide

A dielectric thin film-on-dielectric waveguide has the following parameters:

Operating frequency	$f = 18$ GHz
Relative dielectric constant of thin film	$\varepsilon_{rd} = 10$
Relative dielectric constant of substrate	$\varepsilon_{r1} = 2.1$
Film thickness	$d = 3$mm

Calculate:

a) the attenuation constants for the TE and TM modes in the free-space region; and
b) the attenuation constants for the TE and TM modes in the substrate region.

Solutions:

a) The attenuation constant in free space for the TE and TM modes is

$$\alpha = \sqrt{\beta^2 - \omega^2\mu_0\varepsilon_0} = \sqrt{\omega^2\mu_0\varepsilon_0(\varepsilon_{rd} - 1) - k_y^2}$$

$$= \left[(2\pi \times 18 \times 10^9)^2 \times 4\pi \times 10^{-7} \times 8.854 \times 10^{-12}(10-1) - \left(\frac{\pi}{3 \times 10^{-3}}\right)^2\right]^{1/2}$$

$$= [1.28 \times 10^6 - 1.10 \times 10^6]^{1/2}$$

$$= 4.24 \times 10^2 \text{ Np/m}$$

b) The attenuation constant in the substrate region for the TE and TM modes is

$$\alpha = \left[(2\pi \times 18 \times 10^9)^2 \times 4\pi \times 10^{-7} \times 8.854 \times 10^{-12} \times (10-2.1) - \left(\frac{\pi}{3 \times 10^{-3}}\right)^2\right]^{1/2}$$

$$= [1.124 \times 10^6 - 1.10 \times 10^6]^{1/2}$$

$$= 4.90 \times 10^2 \text{ Np/m}$$

REFERENCES

[1] Wen, Cheng P., "Coplanar Waveguide: A Surface Strip Transmission Line Suitable for Nonreciprocal Gyromagnetic Devices Applications." *IEEE Trans.*, vol. MTT-17, no. 12, December 1969, pp. 1087–1090.

[2] Jahnke, E., and Emde, F., *Tables of Functions with Formulae and Curves*, 4th ed. New York: Dover, 1945.

Problems

8-1 An infinite gold parallel-plate waveguide has the following parameters:

Separation distance	$d = 10$ mm
Gold conductivity	$\sigma_{Au} = 4.1 \times 10^7$ ℧/m
Relative dielectric constant of alumina	$\varepsilon_{rd} = 10$
Operating frequency	$f = 10$ GHz
Integer	$n = 1$
Conductivity of GaAs substrate	$\sigma_d = 1.17 \times 10^{-2}$ ℧/m

Calculate the

a) cutoff frequency for the TE wave;
b) phase velocity for the TE wave;
c) attenuation constant for the TE wave; and
d) intrinsic impedance for the plate.

8-2 An infinite gold parallel-plate waveguide has the following parameters:

Separation distance	$d = 12$ mm
Gold conductivity	$\sigma_{Au} = 4.1 \times 10^7$ ℧/m
Relative dielectric constant of beryllia	$\varepsilon_{rd} = 6$
Operating frequency	$f = 9$ GHz
Integer	$n = 1$
Conductivity of GaAs substrate	$\sigma_d = 1.17 \times 10^{-2}$ ℧/m

Calculate the

a) cutoff frequency for TM mode;
b) phase velocity for TM mode;
c) intrinsic impedance of the plate; and
d) attenuation constant of TM mode.

8-3 An even TE wave propagates through a dielectric-slab waveguide having the following parameters:

Relative dielectric constant of GaAs insulator	$\varepsilon_{rd} = 13.1$
Slab thickness	$d = 12$ mm
Operating frequency	$f = 10$ GHz
Integer	$n = 1$

Calculate the

a) phase constant β;
b) attenuation constant α;
c) surface impedance Z_s; and
d) cutoff frequency f_c.

8-4 An odd TM wave propagates through a dielectric-slab waveguide that has the following parameters:

Relative dielectric constant of alumina	$\varepsilon_{rd} = 10$
Slab thickness	$d = 10$ mm
Operating frequency	$f = 11$ GHz
Integer	$n = 2$

Determine the

a) phase constant β;
b) attenuation constant α;
c) surface impedance Z_s; and
d) cutoff frequency.

8-5 A coplanar waveguide has the following parameters:

Relative dielectric constant of alumina	$\varepsilon_{rd} = 10$
Dimension ratio	$a_1/b_1 = 0.33$

Determine the

a) capacitance of the CPW;
b) phase velocity of the propagating wave; and
c) characteristic impedance of the CPW.

8-6 A coplanar waveguide has the following parameters:

Relative dielectric constant of beryllia	$\varepsilon_{rd} = 6$
Dimension ratio	$a_1/b_1 = 0.4$

Determine the

a) capacitance of the CPW;
b) phase velocity of the propagating wave; and
c) characteristic impedance of the CPW.

8-7 A dielectric thin film-on-conductor waveguide has the following parameters:

Operating frequency	$f = 18$ GHz
Relative dielectric constant of thin film	$\varepsilon_{rd} = 10$
Relative dielectric constant of film coating	$\varepsilon_{r1} = 1.54$
Film thickness	$d = 6$ mm
Integer	$n = 2$

Find the

a) cutoff frequency for the TM wave; and
b) attenuation constant for the TM wave.

8-8 A dielectric thin film-on-conductor waveguide has the following parameters:

Operating frequency	$f = 20$ GHz
Relative dielectric constant of thin film	$\varepsilon_{rd} = 6$
Relative dielectric constant of film coating	$\varepsilon_{r1} = 1.50$
Film thickness	$d = 6$ mm
Integer	$n = 1$

Find the

a) cutoff frequency for the TE mode; and
b) attenuation constant for the TE mode.

8-9 A dielectric thin film-on-dielectric waveguide has the following parameters:

Operating frequency	$f = 18$ GHz
Relative dielectric constant of thin film	$\varepsilon_{rd} = 13.1$
Relative dielectric constant of the substrate	$\varepsilon_{r1} = 1.50$
Film thickness	$d = 3.75$ mm
Integer	$n = 1$

Find the

a) attenuation constant for the TE and TM waves in the free-space region; and
b) attenuation constant for the TE and TM waves in the substrate region.

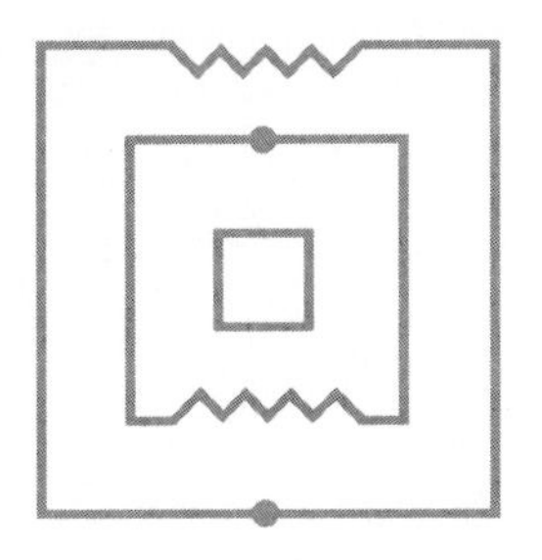

PART THREE

ANTENNAS

The antenna is a device used for radiating or receiving electromagnetic waves. The first antennas were constructed by Heinrich Rudolf Hertz in 1885. They were used in experiments that demonstrated the existence of electromagnetic waves. In 1896, Guglielmo Marconi built a large antenna to send telegraph signals over long distances. Since the days of Hertz and Marconi, antennas have been used widely for electromagnetic energy transmission in communication, radar, navigation, radio astronomy, and instrumentation systems. The purpose of this part is to present (1) antenna parameters and characteristics; (2) antenna structures; and (3) electromagnetic energy transmission systems.

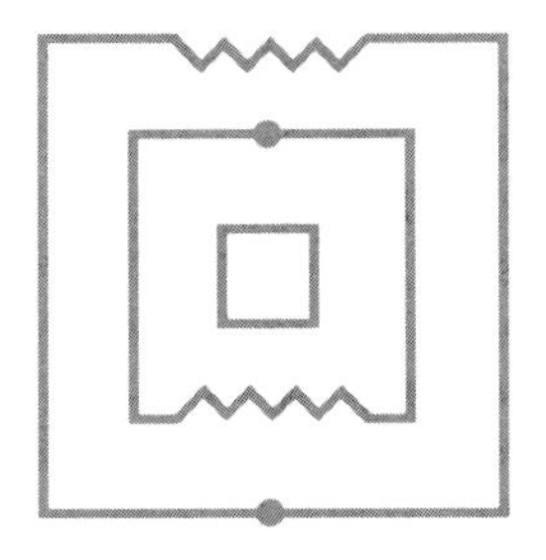

Chapter 9

Antenna Parameters and Characteristics

9-0 INTRODUCTION

Several antenna parameters often are used to describe the characteristics of antenna performance. These parameters are usually classified as radiation pattern, antenna aperture, antenna gain, antenna efficiency, antenna beamwidth, antenna directivity, antenna impedance, near field, and far field. The objective of this chapter is to describe these antenna parameters and characteristics.

9-1 FIELD EQUATIONS FOR THE SHORT-WIRE ANTENNA

One of the simplest antennas is the one-dimensional wire antenna. This type of antenna is well-understood, both theoretically and experimentally, and commonly used. It includes the short-wire antenna, short-dipole antenna, half-wave dipole antenna, short-monopole antenna, quarter-wave monopole antenna, and whip antenna. In this section, we derive the field equations for a short-wire antenna in order to use them to illustrate near and far fields in Section 9-2.

A linear antenna may consist of several elementary short conductors connected either in series or in parallel. Thus it is important for us to analyze first the radiation properties of a short-wire antenna. Figure 9-1-1 shows a shirt-wire antenna in both spherical and rectangular coordinates.

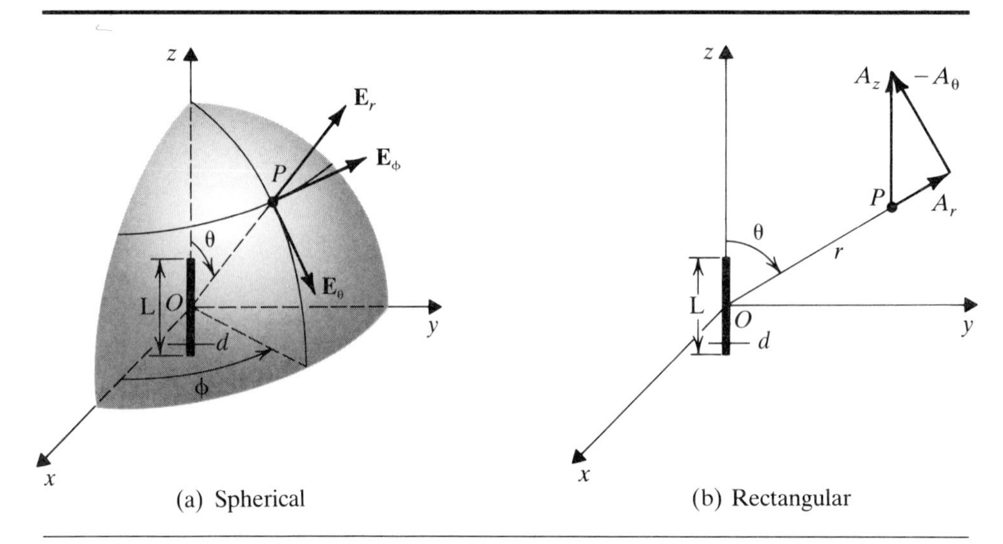

Figure 9-1-1 Coordinates for the short-wire antenna.

We assume (1) that the length of a wire antenna is much shorter than the wavelength, that is, $L \ll \lambda$; and (2) the diameter of the wire is very small compared to its length, that is, $d \ll L$. This type of antenna is usually called the short-wire, or the elemental, antenna. Because the wire is sufficiently short compared to the wavelength of the signal, we can consider the current flowing through the wire to be uniform over its entire length. We also assume that this uniform current flows in the positive z direction and express it as

$$\mathbf{I} = I_0 e^{j\omega t} \mathbf{a}_z \tag{9-1-1}$$

where I_0 is the magnitude of the current.

We can determine the radiation-field equation at the observation point P from a short-wire antenna by the retarded vector potential. The retarded magnetic vector potential at point P is

$$\mathbf{A} = \int \frac{\mu[\mathbf{I}]\,dz}{4\pi r} \mathbf{a}_z \tag{9-1-2}$$

where $[\mathbf{I}] = I_0 e^{j\omega(t - r/c)}$, and is the retarded current; and c is the velocity of light in free space.

Since $d \ll L \ll \lambda$, the magnetic vector potential is simply

$$\mathbf{A} = \frac{\mu[\mathbf{I}]L}{4\pi r} \mathbf{a}_z \tag{9-1-3}$$

With no loss of generality, we can omit the harmonic time variation and simplify Eq. (9-1-3):

$$A_z = \frac{\mu I_0 L}{4\pi r} e^{-j\beta r} \tag{9-1-4}$$

where $\beta = \dfrac{2\pi}{\lambda} = \dfrac{\omega}{c} = \omega\sqrt{\mu_0 \varepsilon_0}$.

You can see from Fig. 9-1-1(a) that the spherical components of the retarded magnetic vector potential are

$$A_r = A_z \cos\theta = \frac{\mu I_0 L}{4\pi r} \cos\theta \, e^{-j\beta r} \tag{9-1-5a}$$

and

$$A_\theta = -A_z \sin\theta = \frac{-\mu I_0 L}{4\pi r} \sin\theta \, e^{-j\beta r} \tag{9-1-5b}$$

We determine the magnetic field intensity **H** at point P from Maxwell's equation:

$$\mathbf{H} = \frac{1}{\mu}\nabla \times \mathbf{A} = \frac{1}{\mu r^2 \sin\theta}\begin{vmatrix} \mathbf{a}_r & r\mathbf{a}_\theta & r\sin\theta\,\mathbf{a}_\phi \\ \dfrac{\partial}{\partial r} & \dfrac{\partial}{\partial \theta} & 0 \\ A_r & rA_\theta & 0 \end{vmatrix} \tag{9-1-6}$$

Equation (9-1-6) shows that there is no ϕ variation of the retarded potential. Consequently, expanding Eq. (9-1-6) yields the magnetic field intensities:

$$H_r = H_\theta = 0 \tag{9-1-7a}$$

and

$$H_\phi = \frac{I_0 L}{4\pi}\left(\frac{j\beta}{r} + \frac{1}{r^2}\right)\sin\theta \, e^{-j\beta r} \tag{9-1-7b}$$

We can determine the electric field intensity **E** by using

$$\mathbf{E} = \frac{1}{j\omega\varepsilon}\nabla \times \mathbf{H} \tag{9-1-8}$$

Substituting Eq. (9-1-7) into Eq. (9-1-8) yields the electric field intensities:

$$E_\phi = 0 \tag{9-1-9a}$$

$$E_r = \frac{I_0 L}{2\pi j\omega\varepsilon}\left(\frac{j\beta}{r^2} + \frac{1}{r^3}\right)\cos\theta \, e^{-j\beta r} \tag{9-1-9b}$$

and

$$E_\theta = \frac{I_0 L}{4\pi j\omega\varepsilon}\left(\frac{(j\beta)^2}{r} + \frac{j\beta}{r^2} + \frac{1}{r^3}\right)\sin\theta \, e^{-j\beta r} \tag{9-1-9c}$$

EXAMPLE 9-1-1 Short-Wire Antenna

A short-wire antenna has the following parameters:

Antenna length	$L = 3$ cm
Current magnitude	$I_0 = 500$ mA
Operating frequency	$f = 1$ GHz

Determine:

a) the electric and magnetic fields $\mathbf{E}_r$, $\mathbf{E}_\theta$, and $\mathbf{H}_\phi$ at $r = 5\lambda$ in air; and
b) the intrinsic impedance of the medium in the r direction.

Solutions:

a) The electric and magnetic fields are

$$E_r = \frac{I_0 L}{2\pi j\omega\varepsilon_0}\left(\frac{j\beta}{r^2} + \frac{1}{r^3}\right)\cos\theta\, e^{-j\beta r}$$

$$= \frac{0.5 \times 3 \times 10^{-2}}{j2\pi \times 2\pi \times 10^9 \times 8.854 \times 10^{-12}}\left(\frac{j2\pi/0.30}{1.5^2} + \frac{1}{1.5^3}\right)\cos\theta\, e^{-j10\pi}$$

$$= \frac{0.043}{j}(j9.31 + 0.296)\cos\theta\, e^{-j10\pi}$$

$$= 0.40\cos\theta\, e^{-j10\pi}$$

$$E_\theta = \frac{I_0 L}{4\pi j\omega\varepsilon_0}\left(\frac{(j\beta)^2}{r} + \frac{j\beta}{r^2} + \frac{1}{r^3}\right)\sin\theta\, e^{-j\beta r}$$

$$= \frac{0.0215}{j}(-292.43 + j9.31 + 0.296)\sin\theta\, e^{-j10\pi}$$

$$= j6.29\sin\theta\, e^{-j10\pi}$$

and

$$H_\phi = \frac{I_0 L}{4\pi}\left(\frac{j\beta}{r} + \frac{1}{r^2}\right)\sin\theta\, e^{-j\beta r}$$

$$= \frac{0.5 \times 3 \times 10^{-2}}{4\pi}\left(\frac{j2\pi/0.3}{1.5} + \frac{1}{1.5^2}\right)\sin\theta\, e^{-j10\pi}$$

$$= 0.12 \times 10^{-2}(j13.96 + 0.44)\sin\theta\, e^{-j10\pi}$$

$$= j1.60 \times 10^{-2}\sin\theta\, e^{-j10\pi}$$

b) The intrinsic impedance is

$$\eta_0 = \frac{E_\theta}{H_\phi} = \frac{6.29}{1.6 \times 10^{-2}} = 377\ \Omega$$

9-2 NEAR FIELD AND FAR FIELD

The space around an antenna is divided into two regions: the one next to the antenna is called the *antenna region*; the one farther away is called the *outer region*. The outer region, in turn, is subdivided into two zones: the one nearer the antenna is called the *near field* or *Fresnel zone*; the one farther away is called the *far field* or *Fraunhofer zone*. The boundary between the two is commonly considered to be at

$$R = \frac{2D^2}{\lambda} \qquad (9\text{-}2\text{-}1)$$

where D is the largest dimension of the antenna in wavelength. Figure 9-2-1 shows the regions and zones around an antenna.

The magnetic-field and electric-field equations, Eqs. (9-1-7) and (9-1-9) contain terms varying as $1/r$, $1/r^2$, and $1/r^3$. These three terms represent three distinct fields.

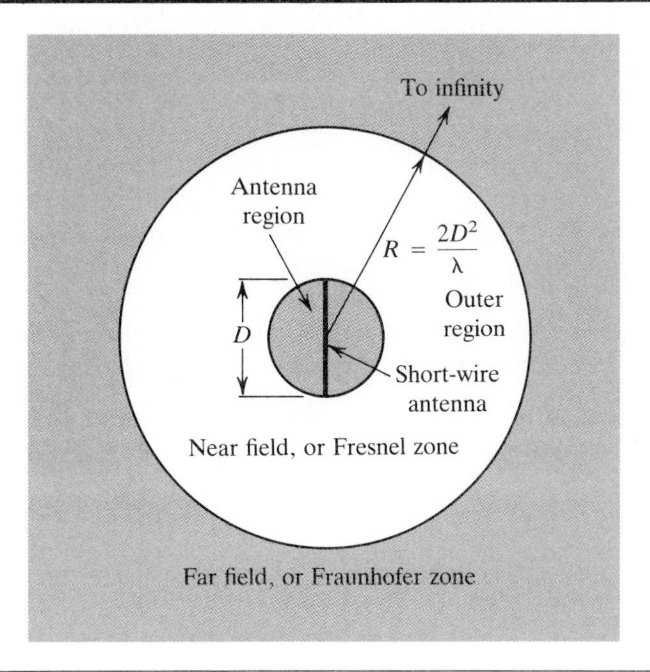

Figure 9-2-1 Near and far fields around an antenna.

9-2-1 The Electrostatic Field

At points very close to the current element, the $(1/r^3)$ terms are dominant. These terms correspond to the electric-dipole field produced by two point charges of equal magnitude and opposite sign, separated by a distance that is small compared to the distance to the observation point. These terms also represent energy stored in a capacitive field, which does not contribute to the radiated power. The field equations are

$$E_r = \frac{I_0 L}{2\pi j\omega\varepsilon r^3} \cos\theta \, e^{-j\beta r} \tag{9-2-2a}$$

and

$$E_\theta = \frac{I_0 L}{4\pi j\omega\varepsilon r^3} \sin\theta \, e^{-j\beta r} \tag{9-2-2b}$$

9-2-2 The Induction Field

At some distance from the source, the $(1/r^2)$ terms are predominant. These terms correspond to the induction field of the d-c element expressed by the Biot–Savard law from low-frequency theory. Equation (9-1-7b) shows that the boundary between the induction- and radiation-field components of magnetic intensity is at a distance of

$$r = \frac{1}{\beta} = \frac{\lambda}{2\pi} \simeq \frac{\lambda}{6} \tag{9-2-3}$$

That is, the distance is approximately $\frac{1}{6}$ wavelength from the antenna. In this region, the electric and magnetic fields are in phase quadrature. Energy is stored in the field during one quarter of a cycle and returned to the circuit during the next. The field equations for the induction region are

$$E_r = \frac{I_0 L \beta}{2\pi\omega\varepsilon r^2} \cos\theta \, e^{-j\beta r} \tag{9-2-4a}$$

and

$$H_\phi = \frac{I_0 L}{4\pi r^2} \sin\theta \, e^{-j\beta r} \tag{9-2-4b}$$

9-2-3 The Radiation Field

As the distance r increases, the $(1/r)$ terms of the electric field and the magnetic field rapidly decrease and soon approach 0 at very large r. At distance

$$R \geq \frac{2D^2}{\lambda} \tag{9-2-5}$$

the $(1/r)$ terms predominate. This condition is the so-called radiation field associated with high frequency. The far-field equations are

$$E_\theta = \frac{jI_0 L \beta}{4\pi r} \eta_0 \sin\theta \, e^{-j\beta r} \tag{9-2-6a}$$

and

$$H_\phi = \frac{jI_0 L \beta}{4\pi r} \sin\theta \, e^{-j\beta r} \tag{9-2-6b}$$

Equations (9-2-6a) and (9-2-6b) represent uniform plane waves in free space. These waves are always in time phase and mutually perpendicular to each other. The intrinsic impedance of free space is defined as

$$\eta_0 = \frac{E_\theta}{H_\phi} = \sqrt{\frac{\mu_0}{\varepsilon_0}} = 377\ \Omega \tag{9-2-7}$$

This relationship is valid only for the far field. The relationship of E/H in the near field is unknown and therefore is unmeasurable.

The near field of any antenna is composed of two regions: the inductive region and the Fresnel zone. The inductive field is considered to be significant up to about 1 wavelength from the antenna. The Fresnel zone (or interference region) is considered to begin at 1 wavelength from the antenna and to extend to the beginning of the far field. For shipboard radar antenna, the near field is significant and cannot be ignored. In this region, the far-field equations are not applicable. Within the near field, the power density is essentially constant and independent of the distance from the antenna. In any event, the maximum power density in the near field is approximately equal to four times the power density at the antenna aperture.

EXAMPLE 9-2-1 Induction-Field and Radiation-Field Equations

A short-wire antenna has the following parameters:

Antenna length	$L = 5$ cm
Current magnitude	$I_0 = 0.5$ A
Operating frequency	$f = 1$ GHz

Determine:

a) the induction field equations at $r = 0.15\lambda$;
b) the radiation field equations at $r = 2\lambda$; and
c) the intrinsic impedance of air for the radiation field.

Solutions:

a) The induction-field equations are

$$E_r = \frac{I_0 L\beta}{2\pi\omega\varepsilon_0 r^2} \cos\theta\, e^{-j\beta r}$$

$$= \frac{0.5 \times 5 \times 10^{-2} \times 2\pi/0.3}{(2\pi)^2 \times 10^9 \times 8.854 \times 10^{-12}(0.15 \times 0.3)^2} \cos\theta\, e^{-j45^\circ}$$

$$= 739.73 \cos\theta\, e^{-j45^\circ}$$

and

$$H_\phi = \frac{I_0 L}{4\pi r^2} \sin\theta\, e^{-j\beta r} = \frac{0.5 \times 5 \times 10^{-2}}{4\pi(0.15 \times 0.3)^2} \sin\theta\, e^{-j45^\circ}$$

$$= 0.98 \sin\theta\, e^{-j45^\circ}$$

b) The radiation-field equations are

$$E_\theta = \frac{jI_0 L\beta}{4\pi r} \eta_0 \sin\theta\, e^{-j\beta r} = j\frac{0.5 \times 5 \times 10^{-2} \times 2\pi/0.3}{4\pi \times 2 \times 0.3} 377 \sin\theta\, e^{-j720^\circ}$$

$$= j26.18 \sin\theta\, e^{-j720^\circ}$$

and

$$H_\phi = j\frac{I_0 L\beta}{4\pi r} \sin\theta\, e^{-j\beta r} = j\frac{0.5 \times 5 \times 10^{-2} \times 2\pi/0.3}{4\pi \times 2 \times 0.3} \sin\theta\, e^{-j720^\circ}$$

$$= j6.94 \times 10^{-2} \sin\theta\, e^{-j720^\circ}$$

c) The intrinsic impedance of air is

$$\eta_0 = \frac{E_\theta}{H_\phi} = \frac{26.18}{6.94 \times 10^{-2}}$$

$$= 377\ \Omega$$

9-3 POWER PATTERN, FIELD PATTERN, AND GROUND EFFECT

9-3-1 Power Pattern

Antenna radiation pattern is a measure of how the power density flowing through a sphere of large radius varies with the angles θ and ϕ. The power density of an electromagnetic plane wave in free space is defined as

$$P_d = \tfrac{1}{2}\mathrm{Re}(\mathbf{E} \times \mathbf{H}^*) \quad \mathrm{W/m^2} \tag{9-3-1}$$

where

> **E** is the peak value of the electric field intensity in volts per meter;
>
> **H** is the peak value of the magnetic field intensity in amperes per meter;
>
> **H*** is the complex conjugate of **H**; and
>
> Re indicates the real part of a complex quantity.

There are two ways to measure the radiation power pattern: (1) the absolute power pattern, that is, the power density at a specific distance from the antenna expressed explicitly in W/m^2; and (2) the relative power pattern, in which the power density often is expressed in relative terms, that is, with respect to the maximum value occurring at the peak point of the main lobe. The relative power pattern is called the *normalized power pattern.*

Antenna radiation patterns often are classified according to their shapes. Figure 9-3-1 shows three power patterns. The isotropic pattern has equal power density in all directions. This pattern often is used as a reference, although it cannot be achieved in practice. The omnidirectional pattern has equal power density in planes passing through the antenna in each direction. The pencil-beam pattern has a relatively narrow main lobe of circular cross-section.

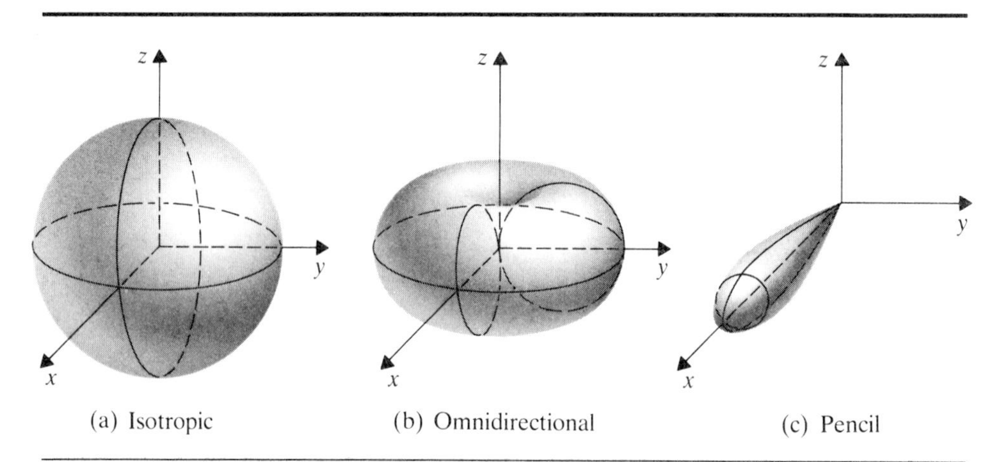

Figure 9-3-1 Antenna power patterns.

The energy radiated by an antenna spreads as the electromagnetic plane waves propagate through free space. The time rate of energy flow per unit area is the Poynting vector, a power density. The graph of power density at a constant radius as a function of either θ or ϕ is called a *power pattern.* Substituting Eqs. (9-2-6a) and (9-2-6b) into Eq. (9-3-1) yields the power-pattern equation for a short-wire antenna:

$$\mathbf{P}_d = \frac{1}{2}\text{Re}(\mathbf{E} \times \mathbf{H}^*) = \frac{\eta_0}{2}\left(\frac{I_0 L\beta}{4\pi r}\right)^2 \sin^2\theta \quad \text{W/m}^2 \tag{9-3-2a}$$

where

$$\eta_0 = \sqrt{\frac{\mu_0}{\varepsilon_0}} \qquad \text{and} \qquad \beta = \omega\sqrt{\mu_0\varepsilon_0}$$

If we denote the maximum power density at the peak of the main beam by P_{dm}, the relative power pattern in the y–z plane equals the square of the relative field pattern.

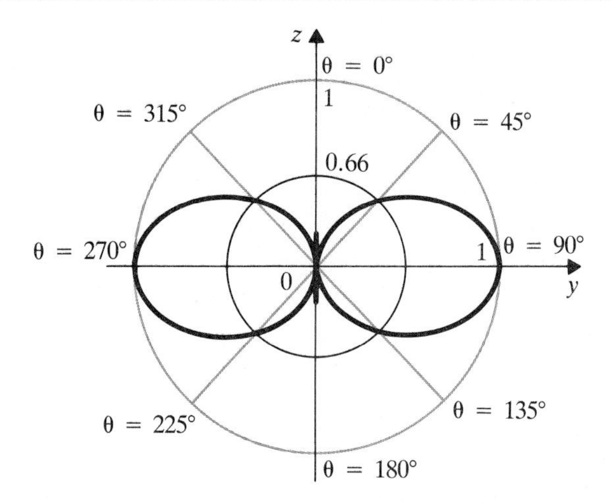

Figure 9-3-2 Polar power pattern of short-wire antenna.

Thus

$$\frac{P_d}{P_{dm}} = \left(\frac{E}{E_m}\right)^2 = \sin^2\theta \tag{9-3-2b}$$

In practice, power patterns often are expressed in terms of relative power. The relative power pattern of Eq. (9-3-2b) is shown in Figs. 9-3-2 and 9-3-3.

We can obtain the total power P_T flowing through a spherical surface of radius r by integrating that spherical surface:

$$\begin{aligned} P_T &= \int_s P_d\,da = \int_{\theta=0}^{2\pi}\int_{\theta=0}^{\pi} P_d r^2 \sin\theta\,d\theta\,d\phi \\ &= \frac{\eta_0(I_0 L\beta)^2}{16\pi}\int_0^{\pi}\sin^3\theta\,d\theta = \frac{\eta_0}{12\pi}(I_0 L\beta)^2 \end{aligned} \tag{9-3-3}$$

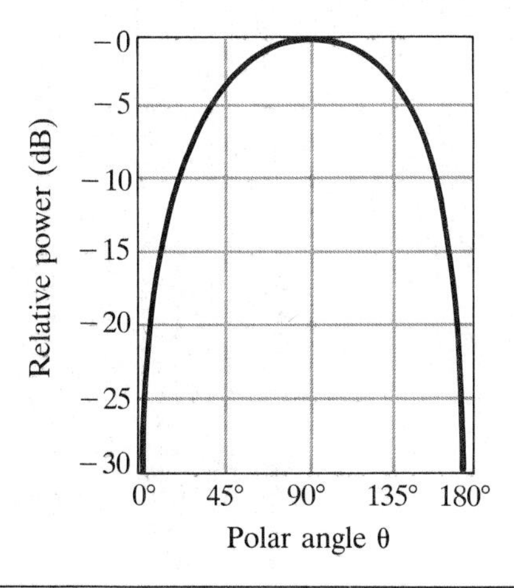

Figure 9-3-3 Rectangular power pattern of short-wire antenna.

Equation (9-3-3) expresses the average power radiated by a short-wire antenna. Assuming no losses in the antenna, this power also equals the power delivered to the antenna. Therefore the total power radiated equals the square of the rms current I flowing through the antenna times the radiation resistance R_r, or

$$
\begin{aligned}
P_T &= I^2 R_r = \left(\frac{I_0}{\sqrt{2}}\right)^2 R_r \\
&= \frac{\eta_0}{12\pi}(I_0 L\beta)^2
\end{aligned}
\tag{9-3-4}
$$

The radiation resistance of a short-wire antenna, then, is

$$R_r = 80\pi^2\left(\frac{L}{\lambda}\right)^2 \quad \Omega \tag{9-3-5}$$

where $\eta_0 = 377$ and $\beta = 2\pi/\lambda$ have been replaced.

EXAMPLE 9-3-1 Computer Plot for Relative Radiation Power Pattern of Short-Wire Antenna, as Expressed by Eq. (9-3-2a):

$$
\begin{aligned}
P_d &= \frac{1}{2}\,\text{Re}(\mathbf{E} \times \mathbf{H}^*) \\
&= \frac{\eta_0}{2}\left(\frac{I_0 L\beta}{4\pi r}\right)^2 \sin^2\theta \quad \text{W/m}^2
\end{aligned}
$$

```
PROGRAM    RADIA

010C       PROGRAM TO PLOT RELATIVE RADIATION POWER
020C+      PATTERN OF SHORT-WIRE ANTENNA
030C       RADIA
040        DIMENSION K(60)
050        DATA K/52*1H /
060        PRINT 3
070        DO 1 LL=1,361,10
080        A=LL-1
090        KP=40*SIN(A*3.141593/180.0)**2 + 12
100        K(12)=1H.
110        K(KP)=1H*
120        N=A
125        KPP=KP-12
130        PRINT 2,N,KPP,(K(JJ),JJ=4,52)
140      1 K(KP)=1H
150      2 FORMAT(I3,I7,60A1)
160      3 FORMAT(" DEG MAGNITUDE",40(1H.))
170        STOP
180        END
READY.
RUN
```

EXAMPLE 9-3-1 *(continued)*

```
PROGRAM   RADIA

 DEG MAGNITUDE........................................
  0      0        *
 10      1        .*
 20      4        .   *
 30     10        .         *
 40     16        .               *
 50     23        .                      *
 60     30        .                             *
 70     35        .                                  *
 80     38        .                                     *
 90     39        .                                      *
100     38        .                                     *
110     35        .                                  *
120     29        .                            *
130     23        .                      *
140     16        .               *
150      9        .        *
160      4        .   *
170      1        .*
180      0        *
190      1        .*
200      4        .   *
210     10        .         *
220     16        .               *
230     23        .                      *
240     30        .                             *
250     35        .                                  *
260     38        .                                     *
270     39        .                                      *
280     38        .                                     *
290     35        .                                  *
300     29        .                            *
310     23        .                      *
320     16        .               *
330      9        .        *
340      4        .   *
350      1        .*
360      0        *

SRU      0.910 UNTS.

RUN COMPLETE.
BYE

F228F10   LOG OFF    19.48.06.
F228F10   SRU       3.786 UNTS.
```

EXAMPLE 9-3-2 Radiation Resistance of a Short-Wire Antenna

A short-wire antenna has a length L of 0.1λ. Calculate its radiation resistance.

Solution:

The radiation resistance is

$$R_r = 80\pi^2\left(\frac{L}{\lambda}\right)^2 = 80\pi^2\left(\frac{0.1\lambda}{\lambda}\right)^2$$
$$= 7.89\ \Omega$$

EXAMPLE 9-3-3 Computer Plot for Radiation Resistance of Short-Wire Antenna, as Expressed by Eq. (9-3-5):

$$R_r = 80\pi^2\left(\frac{L}{\lambda}\right)^2\ \Omega$$

```
PROGRAM    RADRES

010C        PROGRAM TO PLOT RADIATION RESISTANCE OF
020C+       OF SHORT-WIRE ANTENNA
030C        THE VALUE OF L IS IN LAMBDA
040C        RADRES
050         DIMENSION LINE(41)
060         REAL L,LAMBDA
070         DO 4 J=1,41
080       4 LINE(J)=1H.
090         PRINT 6, LINE
100       6 FORMAT(/1H ,"    L    RESISTANCE",41A1/
105+        1X,"(LAMBDA) (OHMS)"/)
110         DO 8 J=1,41
120       8 LINE(J)=1H
130         LINE(2)=1H.
140         PI=3.141593
150         L=0.0
155         LAMBDA=1.0
160         DELTAL=0.05
170         DO 12 I=1,41
180         RR=80.0*(PI*L/LAMBDA)**2
182         N=IFIX(RR*0.01+3.0)
184         LINE(N)=1HX
190         PRINT 10, L,RR,LINE
200      10 FORMAT(2X,F5.2,1X,F10.3,41A1)
210         LINE(N)=1H
220         LINE(2)=1H.
230         L=L+DELTAL
240      12 CONTINUE
250         STOP
260         END
READY.
RUN
```

EXAMPLE 9-3-3 *(continued)*

```
PROGRAM    RADRES

    L     RESISTANCE......................................
 (LAMBDA)  (OHMS)

   0.00      0.000 .X
    .05      1.974 .X
    .10      7.896 .X
    .15     17.765 .X
    .20     31.583 .X
    .25     49.348 .X
    .30     71.061 .X
    .35     96.722 .X
    .40    126.331 . X
    .45    159.888 . X
    .50    197.392 . X
    .55    238.844 .  X
    .60    284.245 .  X
    .65    333.593 .   X
    .70    386.889 .   X
    .75    444.132 .    X
    .80    505.324 .     X
    .85    570.463 .     X
    .90    639.551 .      X
    .95    712.586 .       X
   1.00    789.569 .       X
   1.05    870.499 .        X
   1.10    955.378 .         X
   1.15   1044.204 .          X
   1.20   1136.979 .           X
   1.25   1233.701 .            X
   1.30   1334.371 .             X
   1.35   1438.989 .              X
   1.40   1547.554 .               X
   1.45   1660.068 .                X
   1.50   1776.529 .                 X
   1.55   1896.938 .                  X
   1.60   2021.295 .                    X
   1.65   2149.600 .                     X
   1.70   2281.853 .                      X
   1.75   2418.054 .                        X
   1.80   2558.202 .                         X
   1.85   2702.298 .                           X
   1.90   2850.342 .                            X
   1.95   3002.334 .                              X
   2.00   3158.274 .                               X

SRU        0.709 UNTS.

RUN COMPLETE.
```

9-3-2 Field Pattern

The radiation pattern of an antenna can also be plotted in terms of the electric or magnetic field intensity. The time-average power density flowing at the far field in free space can be expressed in terms of either electric intensity or magnetic intensity as

$$\langle P_d \rangle = \frac{1}{2\eta_0}|E|^2 = \frac{\eta_0}{2}|H|^2 \tag{9-3-6}$$

where

$$|E|^2 = |E_\theta|^2 + |E_\phi|^2$$
$$|H|^2 = |H_\theta|^2 + |H_\phi|^2$$

and

$$\eta_0 = \sqrt{\frac{\mu_0}{\varepsilon_0}} = 377\ \Omega$$

When the field intensity is expressed in V/m, it is an absolute field pattern; when the field intensity is expressed in units relative to its maximum value in some direction, it is a relative field pattern. Since the power is proportional to the square of the field intensity, the relative power pattern is equal to the square of the relative field pattern. That is,

$$P_d = \frac{|E|^2}{|E_m|^2} = \frac{|H|^2}{|H_m|^2} \tag{9-3-7}$$

where $|E_m|^2 = |E_{\theta m}|^2 + |E_{\phi m}|^2$ are the maximum values of E, $|H_m|^2 = |H_{\theta m}|^2 + |H_{\phi m}|^2$ are the maximum values of H, and

$$\frac{|E_\theta|}{|E_{\theta m}|} = \sin\theta \tag{9-3-7a}$$

Thus the normalized relative power pattern, as indicated in Eq. (9-3-2b) for a short-wire antenna is

$$P_d = \sin^2\theta \tag{9-3-7b}$$

The field and power patterns for a short-wire antenna are shown in Fig. 9-3-4.

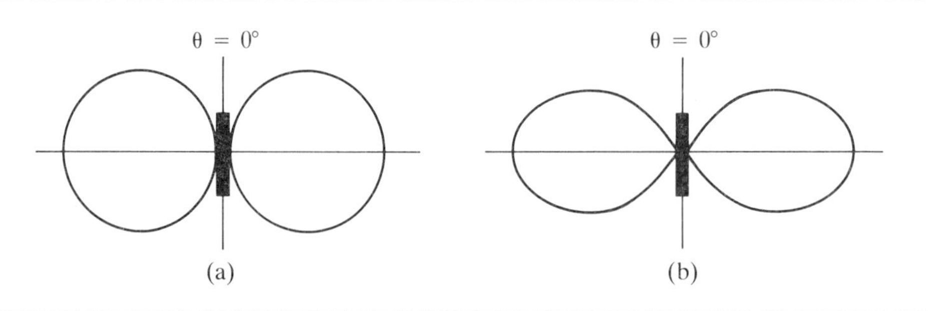

Figure 9-3-4 (a) Field pattern and (b) power pattern.

EXAMPLE 9-3-4 Computer Plot for an Electric or Magnetic Field Pattern of Short-Wire Antenna, as Expressed by

$$E_\theta = \frac{jI_0 L\beta}{4\pi r} \eta_0 \sin\theta\, e^{-j\beta r} \tag{9-2-6a}$$

and

$$H_\phi = \frac{jI_0 L\beta}{4\pi r} \sin\theta\, e^{-j\beta r} \tag{9-2-6b}$$

```
PROGRAM   EFIELD
010C     PROGRAM TO PLOT ELECTRIC OR MAGNETIC FIELD
020C+    PATTERN OF SHORT-WIRE ANTENNA
030C     EFIELD
040      DIMENSION LINE(41)
050      DO 6 J=1,41
060    6 LINE(J)=1H.
070      PRINT 7, LINE
080    7 FORMAT(/1H ,"ANGLE MAGNITUDE",41A1)
090      DO 8 J=1,41
100    8 LINE(J)=1H
110      LINE(20)=1H.
120      DTHETA=10.0*3.141593/180.0
130      THETA=0.0
140      ANGLE=0.0
150      DO 10 J1=1,37
160            ETHETA=SIN(THETA)
170            JS=19.0*(ETHETA+1.0)+1.5
180            LINE(JS)=X
190            PRINT 9, ANGLE, SIN(THETA), LINE
200          9 FORMAT(1X,F5.1,F7.3,41A1)
210            LINE(JS)=1H
220            LINE(20)=1H.
230            THETA=THETA+DTHETA
240            ANGLE=ANGLE+10.0
250   10 CONTINUE
260      STOP
270      END
READY.
RUN
```

(continued)

EXAMPLE 9-3-4 *(continued)*

```
PROGRAM    EFIELD

 ANGLE MAGNITUDE.....................................................
   0.0   0.000                         .
  10.0    .174                         .  .
  20.0    .342                         .     .
  30.0    .500                         .         .
  40.0    .643                         .           .
  50.0    .766                         .              .
  60.0    .866                         .               .
  70.0    .940                         .                 .
  80.0    .985                         .                  .
  90.0   1.000                         .                  .
 100.0    .985                         .                  .
 110.0    .940                         .                 .
 120.0    .866                         .               .
 130.0    .766                         .              .
 140.0    .643                         .           .
 150.0    .500                         .        .
 160.0    .342                         .     .
 170.0    .174                         .  .
 180.0   -.000                         .
 190.0   -.174                      .  .
 200.0   -.342                   .     .
 210.0   -.500               .         .
 220.0   -.643             .           .
 230.0   -.766          .              .
 240.0   -.866         .               .
 250.0   -.940 .                       .
 260.0   -.985.                        .
 270.0  -1.000.                        .
 280.0   -.985.                        .
 290.0   -.940 .                       .
 300.0   -.866         .               .
 310.0   -.766          .              .
 320.0   -.643             .           .
 330.0   -.500                .        .
 340.0   -.342                   .     .
 350.0   -.174                      .  .
 360.0    .000                         .

SRU          0.792 UNTS.

RUN COMPLETE.
```

9-3-3 Ground Effect

The presence of the ground affects the performance of high-frequency antennas in different ways. Antenna efficiency can be reduced because of losses resulting from the finite conductivity of the ground. The vertical plane radiation pattern of either

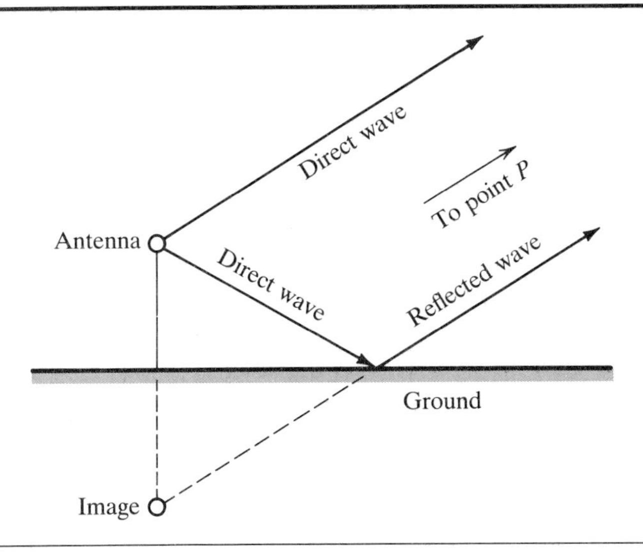

Figure 9-3-5 Ground effect on radiation pattern.

horizontally or vertically polarized antennas can be severely modified by the additive and subtractive interference between the direct waves and the reflected waves from the ground, as shown in Fig. 9-3-5.

At certain radiation angles, the direct and reflected waves arrive at a remote point in space in time phase, and the resultant field intensity is the sum of the wave components. Under perfect ground conditions, the ground-reflected wave is equal in magnitude to the direct wave. For a proper antenna height, a gain of 6 dB can be achieved at these radiation angles. At other radiation angles, the two wave components may be out of phase by 180°, and nulls are created in the radiation pattern. Figure 9-3-6

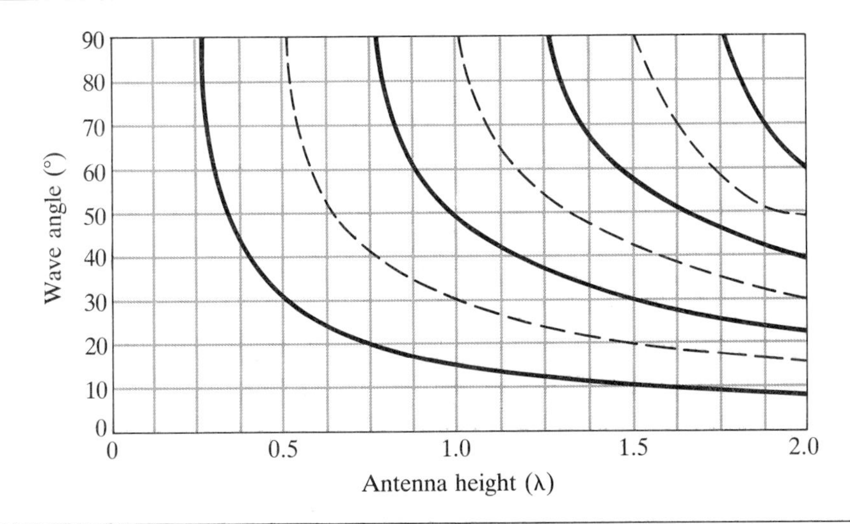

Figure 9-3-6 Angles of ground reflection maxima and minima for antenna heights up to 2 wavelengths.

shows the approximate radiation angles at which maxima and nulls will occur for various heights of horizontal and vertical antennas. The solid lines indicate minima and the dashed lines show maxima.

There are many ways to simulate the ground surface to be a perfect ground. The most effective method is to use a perfectly conducting screen over the ground; the reflection loss from the ground is reduced to 0, or a minimum, depending upon the conductivity of the metal screen.

9-4 ANTENNA BEAMWIDTH AND BANDWIDTH

9-4-1 Antenna beamwidth

The major lobes of a radiation pattern are those in which the radiation is maximum. Lobes of lesser radiation intensity are called minor lobes. We normally define the beamwidth of a directive antenna as the angular width, in degrees, of the major lobe between the two directions at which the relative radiated power equals one-half its value at the peak of the lobe. At these half-power points, the field intensity is 0.707 times its maximum, or 3 dB below the peak of the lobe. The front-to-back ratio is the ratio of magnitudes between the major and minor lobes. Figure 9-4-1 shows a major lobe with a beamwidth of 45 degrees.

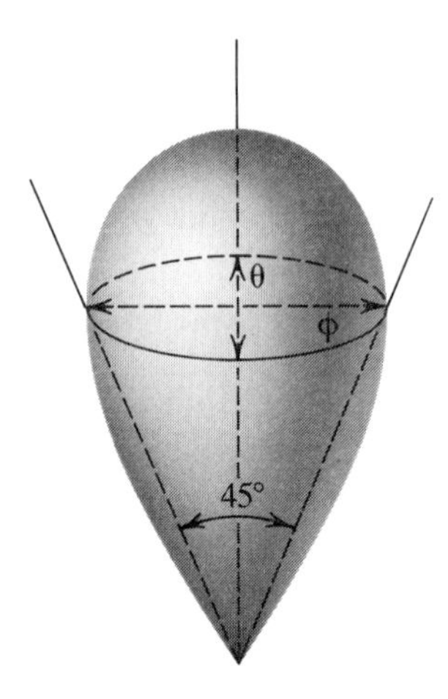

Figure 9-4-1 Antenna beamwidth.

We may also express the beamwidth of an antenna in dB down. For instance, if the beamwidths are −10 dB in width or −16 dB in width, the radiated power at the half-power point has decreased to $\frac{1}{10}$ or $\frac{1}{40}$ of its maximum value.

9-4-2 Antenna Bandwidth

The bandwidth of an antenna is defined as a frequency band over which the antenna electrical characteristics remain within acceptable limits. The electrical characteristics that may limit the bandwidth in a given application include antenna gain, directivity, efficiency, beamwidth, impedance, and standing-wave ratio.

Transmission lines, baluns, antenna couplers, or the antenna itself can be damaged if the antenna system is operated for transmitting electromagnetic energy outside its specified frequency band. Of course, no damage can result from reception, and many antennas can satisfactorily operate as receivers at frequencies outside their normally specified bandwidths.

The only type of antenna that offers the frequency-independent characteristics is the log-periodic antenna. This type of antenna not only makes patterns that are nearly frequency independent, but also offers the possibility of obtaining a particular variation of vertical pattern with frequency. This condition can be achieved by simply inclining the horizontally polarized log-periodic curtain at a proper angle with respect to the ground.

In general, the antenna input impedance is not a constant pure resistance. The impedance also has a reactance component that varies over the frequency range, in addition to the variation of the resistance component. This type of input impedance cannot be matched perfectly over a wide frequency band. Transmission-line theory has shown that, if a maximum bandwidth is to be obtained, the reflection coefficient must remain nearly constant over a bandwidth. Some studies have been made regarding the maximum attainable bandwidth for an impedance curve on the Smith chart that can be represented by a simple RLC circuit. Thus it is possible to predict the maximum bandwidth that can be obtained for an allowable value of the standing-wave ratio.

9-5 ANTENNA GAIN, DIRECTIVITY, AND EFFICIENCY

9-5-1 Antenna Gain

The power gain of an antenna is an important measure of its performance in an electronic system. The antenna power gain G is defined as

$$\text{Gain} \equiv \frac{\text{Maximum radiation intensity}}{\text{Maximum radiation intensity from a reference antenna with the same input power}}$$

or

$$g = \frac{\eta P_{d,\,\max}}{\langle P_d \rangle} \tag{9-5-1}$$

where η represents the efficiency of the antenna.

A commonly used reference antenna is the isotropic radiator, which we assume radiates energy uniformly in all directions. An isotropic antenna cannot be physically made, but we use it as a convenient reference, since we can easily calculate the radiation intensity for any input power. Other antennas often used as references are the half-wave dipole antenna, the quarter-wave monopole antenna, and the short-dipole antenna.

The antenna power gain is a function of direction and polarization; it represents a maximum power density in a specific direction.

If a lossless isotropic antenna has an input power P_{in}, its average radiation intensity $\langle P_d \rangle$ is

$$\langle P_d \rangle = \frac{P_{\text{in}}}{4\pi} \quad \text{W/unit solid angle} \tag{9-5-2}$$

where 4π represents the total solid angle of a sphere, and the unit of a solid angle is usually expressed in steradians (sr). Note that we can express the unit of power density in either W/m^2 or W/sr. The interchanging factor is r^2, where r is the spherical radius. Hence, we can write the power gain of an antenna as

$$g = \frac{4\pi P_{d,\max}\eta}{P_{\text{in}}} \qquad \text{(9-5-3)}$$

where g represents the numerical quantity of antenna power gain, and $P_{d,\max}$ represents the maximum power density radiated in a specified direction.

We can express the power gain of an antenna in decibels by taking 10 times the logarithm of g to the base 10. That is,

$$G = 10 \log (g) \quad \text{dB} \qquad \text{(9-5-4)}$$

where G stands for the antenna power gain in dB. If the radiation pattern is a field pattern, we should express the gain of an antenna as

$$G = 20 \log (g_f) \quad \text{dB} \qquad \text{(9-5-5)}$$

9-5-2 Antenna Directivity

All antennas exhibit directive effects. The property of radiating more strongly in some directions than in others is called the *directivity* of the antenna. For an isotropic antenna, the field intensity is the same everywhere on the equi-spherical–surface distance from the source. For an actual antenna radiating constant power, the directive pattern will result in greater power density in certain directions and less in the others. The ratio of the maximum power density to the average power density taken over the entire sphere is the numerical measure of the directivity of the antenna. That is,

$$\text{Directivity} = \frac{\text{Maximum power density}}{\text{Average power density}}$$

or

$$d = \frac{P_{d,\max}}{\langle P_d \rangle} \qquad \text{(9-5-6)}$$

The requirements of antenna gain can primarily be satisfied by sufficient antenna directivity, that is, radiation in the desired direction at the expense of radiation in all other directions. Directivity can usually be obtained by reducing the elevation beamwidth or the azimuthal beamwidth in order to achieve the required directive gain. If an omnidirectional azimuthal pattern is desired, the directive gain can be obtained only by reducing the vertical plane beamwidth.

Since total radiation power equals the average power density multiplied by the solid angle of 4π, we can express the directivity as

$$d = \frac{P_{d,\max}}{P_T/(4\pi)} = \frac{4\pi}{P_T/P_{d,\max}} \qquad \text{(9-5-7)}$$

where d represents the numerical quantity of antenna directivity.

Note that the ratio of the total radiated power to the power density per unit solid angle is the beam area in radians squared at half-points for a specific major lobe. Since the radiation pattern is a relative power pattern, we can write an expression for the beam area:

$$B = \frac{P_T}{P_{d,\max}} = \frac{\iint P_d \, d\Omega}{P_{d,\max}} = \iint \frac{P_d}{P_{d,\max}} \sin\theta \, d\theta \, d\phi \qquad \text{(9-5-8)}$$

where

$d\Omega = \sin\theta\, d\theta\, d\phi$ and is an element of solid angle;

P_d is the power density, in W/sr;

$P_{d,\max}$ is the maximum power density, in W/unit sr; and

P_T is the total power, in W.

For a unidirectional pencil-like power pattern, we can approximate the beam area:

$$B \simeq \theta\phi \quad \text{rad}^2 \tag{9-5-9}$$

where θ and ϕ are the half-power beamwidths in radians. Substituting Eq. (9-5-9) into Eq. (9-5-7) yields the directivity:

$$d = \frac{4\pi}{\theta\phi} \tag{9-5-10}$$

We can also express the directivity as

$$d = \frac{41{,}253}{\theta^\circ\phi^\circ} \tag{9-5-11}$$

where θ° and ϕ° are the half-power beamwidths in degrees.

Using Eq. (9-3-2a), we can obtain the maximum power density:

$$P_{d,\max} = \frac{\eta_0}{2}\left(\frac{I_0 L\beta}{4\pi r}\right)^2 \quad \text{W/m}^2 \tag{9-5-12}$$

Then from Eq. (9-3-3), we can find the average power density:

$$\langle P_d \rangle = \frac{\text{Total power}}{\text{Spherical surface}} = \frac{\eta_0 (I_0 L\beta)^2/(12\pi)}{4\pi r^2} = \frac{\eta_0}{3}\left(\frac{I_0 L\beta}{4\pi r}\right)^2 \tag{9-5-13}$$

Therefore the directivity of a short-wire antenna is

$$d = \frac{P_{d,\max}}{\langle P_d \rangle} = \frac{3}{2} \tag{9-5-14}$$

We can also express the directivity of an antenna in dB; that is,

$$D = 10 \log(d) \quad \text{dB} \tag{9-5-15}$$

where D represents the antenna directivity in dB.

9-5-3 Antenna Radiation Efficiency

The radiation efficiency of an antenna is the ratio of the radiation resistance to the total resistance of the antenna system. The total resistance includes radiation resistance, resistance in conductors and dielectrics, resistance of loading coils (if used), and resistance of the ground system. That is,

$$\eta = \frac{R_r}{R_r + R_c} \tag{9-5-16}$$

where R_r is the radiation resistance of the antenna, and R_c is the total lossy resistance. We discuss this relationship in detail in Section 10-6. Except for a perfectly efficient antenna, however, the antenna directivity is related to antenna gain by

$$g = \eta d \tag{9-5-17}$$

or $G = [\eta] + D$(dB), where $[\eta]$ stands for the antenna efficiency in dB. Many microwave antennas, such as the horn antenna, are very efficient ($\eta = 1$), whereas dipole and spiral antennas are much less efficient ($\eta \simeq 0.75$ and 0.60, respectively).

A transmitting antenna usually should be as efficient as possible. Most transmitting antennas of resonant dimensions and standing at least a quarter-wave above ground have efficiencies approaching 100 percent. The efficiency of a communication receiving antenna does not need to be high, because atmospheric, galactic, or man-made noise is usually much greater than receiver noise. An inefficient antenna will attenuate the received signal and noise equally. The essential requirement for determining the efficiency of a receiving antenna is that sufficient signal and noise be delivered to the receiver, so that the predetection signal-to-noise ratio is not determined by the amount of receiver noise.

9-6 MAXIMUM POWER TRANSFER AND EFFECTIVE APERTURE

9-6-1 Maximum Power Transfer

If a receiving antenna is placed in the electromagnetic field in free space, the antenna will intercept some energy and deliver it to the load. However, since the power density from the source is spreading through free space inversely with the square of the distance, the receiving antenna collects only a very small portion of the power transmitted from the source. Therefore maximizing the receiving power in the receiving system is of primary importance. Figure 9-6-1 shows a receiving antenna in free space and its equivalent circuit, where

$\mathbf{Z}_\ell = R_\ell + jX_\ell$ and is the load impedance;

$\mathbf{Z}_a = R_a + jX_a$ and is the antenna impedance; and

V and I are the rms voltage and rms current in the antenna induced by the passing electromagnetic wave.

In general, the antenna resistance consists of a radiation resistance R_r and an ohmic-loss resistance R_c, which in turn is the combination of the copper, dielectric, and mismatching losses. That is,

$$R_a = R_r + R_c \tag{9-6-1}$$

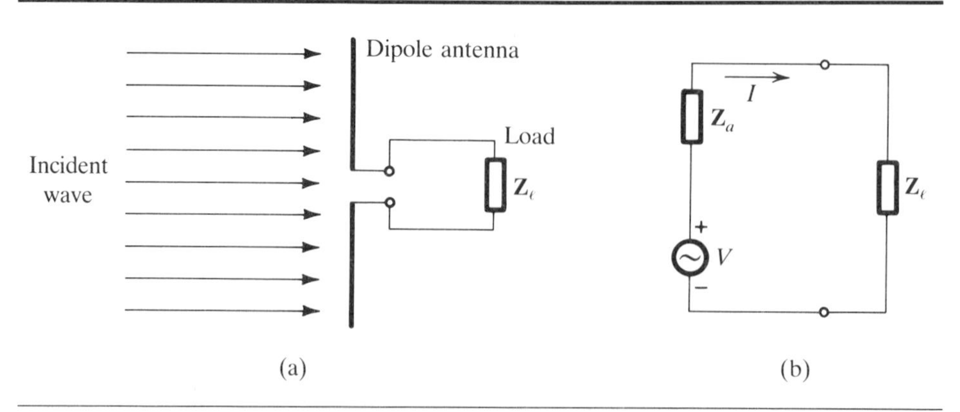

Figure 9-6-1 (a) Receiving antenna and (b) its equivalent circuit.

From the circuit theory for maximum power transfer, the reactances in the antenna and load must be conjugate, and their resistances must be equal. Usually, $R_c \ll R_r$. Thus

$$X_a = -X_\ell \tag{9-6-2}$$

and

$$R_r = R_\ell \tag{9-6-3}$$

Therefore the maximum power delivered by a lossless antenna to the load is

$$\begin{aligned} P_m &= I^2 R_\ell \\ &= \left(\frac{V}{R_r + R_\ell}\right)^2 R_\ell = \frac{V^2}{4R_\ell} \quad \text{W} \end{aligned} \tag{9-6-4}$$

We can obtain the antenna efficiency from the ordinary definition, or

$$\begin{aligned} \eta &= \frac{P_{\text{out}}}{P_{\text{in}}} \\ &= \frac{V^2 R_\ell/(R_r + R_\ell + R_c)^2}{V^2/(4R_r)} = \frac{4R_r R_\ell}{(R_r + R_\ell + R_c)^2} \end{aligned} \tag{9-6-5}$$

If the antenna is matched, $R_r + R_c = R_\ell$, the antenna efficiency becomes

$$\eta = \frac{R_r}{R_r + R_c} \tag{9-6-6}$$

As you can see, when $R_c \ll R_r$, the antenna efficiency approaches 100 percent, which is the normal case.

9-6-2 Effective Aperture

The different types of antenna apertures include the effective aperture, scattering aperture, loss aperture, and physical aperture. Here we describe only the effective aperture. The effective aperture of an antenna is defined as

$$\text{Effective aperture} = \frac{\text{Power absorbed by load}}{\text{Power density of incident wave}}$$

or

$$A_e = \frac{P}{P_d} \tag{9-6-7}$$

If the impedances are matched, the received power is a maximum and the effective aperture is also a maximum. From the field theory, we can express the induced voltage in terms of the electric intensity by

$$V = \int_\ell \mathbf{E} \cdot d\ell = EL \quad \text{V(rms)} \tag{9-6-8}$$

where L is the length of a short-dipole antenna.

The power density P_d of the incident wave at the antenna is

$$P_d = \frac{E^2}{\eta_0} = \frac{E^2}{120\pi} \tag{9-6-9}$$

Therefore the maximum effective aperture of a short-dipole antenna is

$$A_{em} = \frac{P}{P_d} = \frac{(EL)^2/(4R_r)}{E^2/\eta_0} = \frac{\eta_0 L^2}{4R_r}$$

$$= \frac{120\ L^2}{4[20\pi^2(L/\lambda)^2]} = \frac{3\lambda^2}{2\pi} \qquad (9\text{-}6\text{-}10)$$

where $R_r = 20\pi^2 L^2/\lambda^2$ was substituted for a short-dipole antenna resistance.

Furthermore, the gains G_1 and G_2 of two antennas are related to their directivities, maximum effective apertures, and effective apertures by the relationship:

$$\frac{G_1}{G_2} = \frac{\eta_1}{\eta_2}\frac{D_1}{D_2} = \frac{\eta_1 A_{em1}}{\eta_2 A_{em2}} = \frac{A_{e1}}{A_{e2}} \qquad (9\text{-}6\text{-}11)$$

An isotropic source has a directivity or gain of 1, so its maximum effective aperture is

$$A_{em} = \frac{D_1}{D_2} A_{em2} = \frac{1}{3/2}\left(\frac{3\lambda^2}{8\pi}\right) = \frac{\lambda^2}{4\pi} \qquad (9\text{-}6\text{-}12)$$

We can conclude therefore that the directivity of any antenna is equal to $4\pi/\lambda^2$ times its maximum effective aperture. That is,

$$D = \frac{4\pi}{\lambda^2} A_{em} \qquad (9\text{-}6\text{-}13)$$

Alternatively, since $A = \eta A_m$, the effective aperture of any antenna equals the effective aperture of an isotropic source multiplied by the gain of the antenna. That is,

$$A_e = \frac{\lambda^2}{4\pi} g \qquad (9\text{-}6\text{-}14)$$

Equation (9-6-14), however, is valid only for lossy antenna.

The effective aperture of large-opening antennas is of the same order of magnitude as the physical aperture. For pyramidal horns, the effective aperture is approximately 50 percent of the physical area of an optimum horn and may be as high as 80 percent of the physical area of a horn with a very long flare length. For parabolic-reflector antennas, the effective aperture is normally between 50 and 65 percent of the physical area. Table 9-6-1 shows the values of the effective aperture, directivity, gain, and radiation resistance for a number of simple antennas.

1. *Friis Transmission Formula*

The concept of effective aperture is useful in determining the free-space attenuation between two antennas. Figure 9-6-2 shows a transmitting antenna and a receiving antenna in free space, separated by distance of R. For simplicity we assume that the antennas and the transmission lines are lossless. The power density available to the receiving antenna is

$$P_d = \frac{P_t g_t}{4\pi R^2} \quad \text{W/m}^2 \qquad (9\text{-}6\text{-}15)$$

The power received by the receiver is expressed by

$$P_r = P_d A_r g_r = \frac{P_t g_t}{4\pi R^2} A_r g_r = \frac{P_t g_t}{4\pi R^2} A_{er} \quad \text{W} \qquad (9\text{-}6\text{-}16)$$

Table 9-6-1 Antenna Aperture, Directivity, Gain, and Radiation Resistance

Antenna Type	Effective Aperture	Directivity (Numeric)	Gain (dB)	Radiation Resistance (Ω)
Isotropic radiator	$\frac{\lambda^2}{4\pi}$	1	0	
Very short wire	$\frac{3\lambda^2}{8\pi}$	1.5	1.76	$80\pi^2\left(\frac{L}{\lambda}\right)^2$
Very short dipole	$\frac{3\lambda^2}{2\pi}$	1.5	1.76	$20\pi^2\left(\frac{L}{\lambda}\right)^2$
Half-wave dipole	$\frac{1.64\lambda^2}{4\pi}$	1.64	2.15	73
Long wire antenna	$\frac{\lambda^2}{4\pi}$	1	0	$30\pi L\omega\sqrt{\mu\varepsilon}$
Quarter-wave dipole above perfect ground plane		3.28	5.15	
Short monopole antenna	$\frac{3\lambda^2}{8\pi}$	3.00	4.77	$10\pi^2\left(\frac{L}{\lambda}\right)^2$
Quarter-wave monopole antenna	$\frac{1.64\lambda^2}{2\pi}$	3.30	5.19	36.5

Substituting Eq. (9-6-14) for the effective aperture of the transmitting antenna we get

$$\frac{P_r}{P_t} = \frac{A_{et}A_{er}}{\lambda^2 R^2} \tag{9-6-17}$$

which is the Friis transmission formula. Note that Eq. (9-6-16) is valid only in the far field or the Fraunhofer zone. That is,

$$R \geq \frac{2D^2}{\lambda} \tag{9-6-18}$$

where D is the largest linear dimension of either of the two antennas.

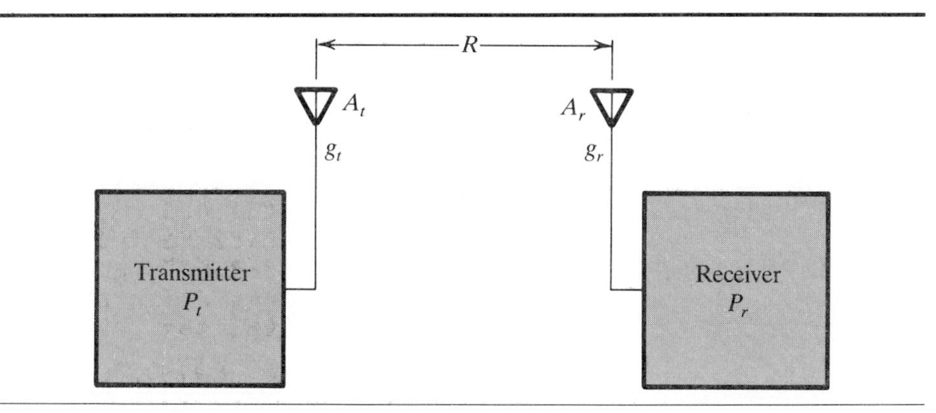

Figure 9-6-2 Free-space transmission by two antennas.

The effective aperture of the receiving antenna is

$$A_{er} = \frac{\lambda^2}{4\pi} g_r \tag{9-6-19}$$

so we can express the received power as

$$P_r = P_t\,(\text{dBW}) + G_t + G_r - 20 \log\left(\frac{4\pi R}{\lambda}\right) \quad \text{dBW} \tag{9-6-20}$$

The last term of Eq. (9-6-20) is the transmission loss in free space, which we can find from the nomograph of space attenuation in Fig. 9-6-3. For example, if the wavelength of a signal is 0.01 m and the range is 30 m, the free-space attenuation is about 92 dB.

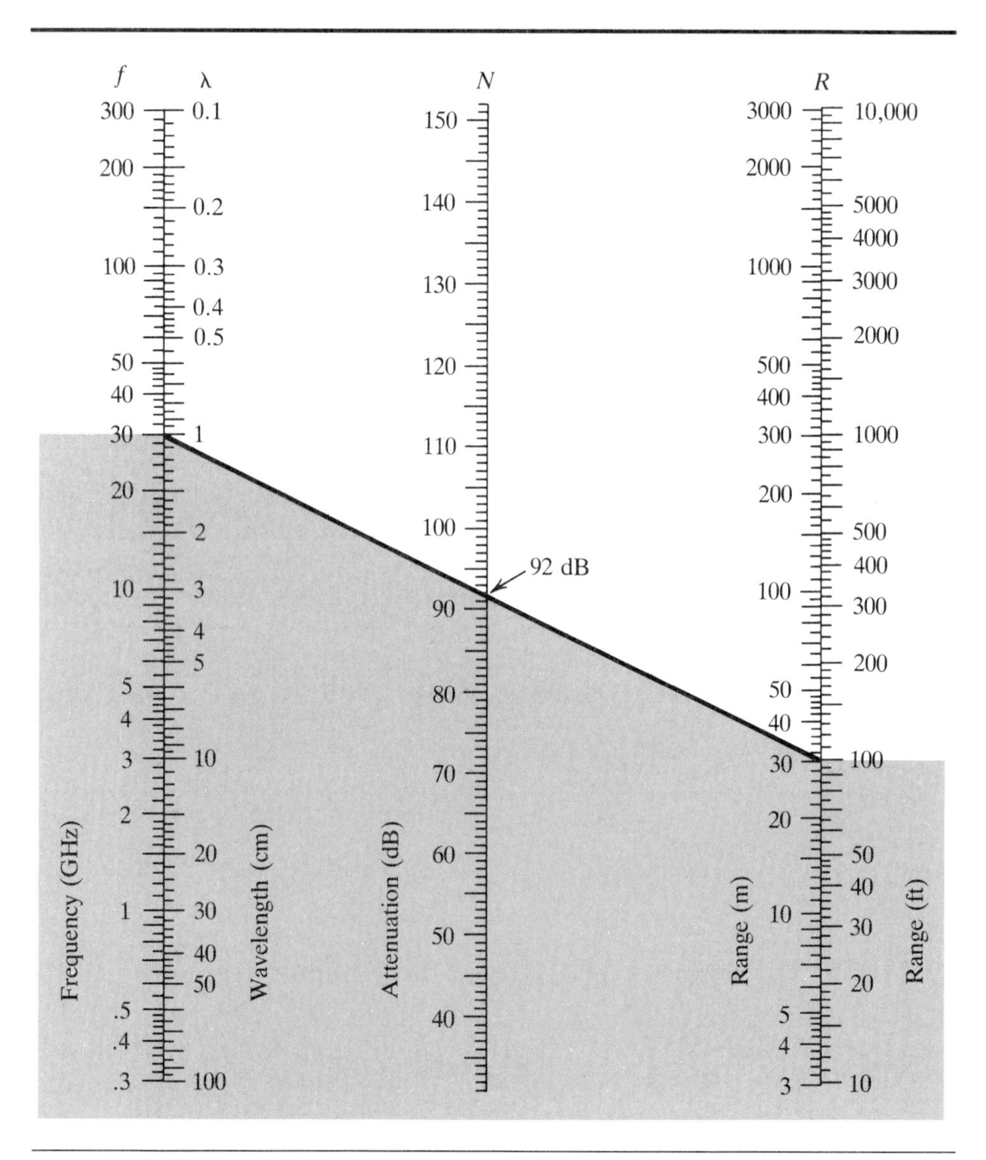

Figure 9-6-3 Nomograph of free-space attenuation.

EXAMPLE 9-6-1 Antenna Power Gain and Antenna Effective Area

An antenna has a directivity of 6 dB and an efficiency of 90% at an operating frequency of 3 GHa.

Calculate:

a) the antenna gain in dB; and
b) the antenna effective area.

Solutions:

a) The antenna power gain is

$$G = \eta d = [\eta] + \mathrm{D} = 0.98 \times 3.981 = 3.901 = 5.9 \text{ dB}$$
$$= -0.1 + 6 = 5.9 \text{ dB}$$

b) The antenna effective area is

$$A_e = \frac{\lambda^2}{4\pi} g = \frac{(0.1)^2}{4\pi}(3.981)$$
$$= 3.17 \times 10^{-3} \text{ m}^2$$

9-7 POLARIZATION

We define the polarization of an antenna as the locus of the tip of the time-varying electric field vector of the radiation from the antenna in the direction of the main beam. The polarization depends on the relative magnitude and phase of the orthogonal components of the electric field. The radiated wave can be linearly, circularly, or elliptically polarized. Figure 9-7-1 shows the vertical polarization of an electric wave against the horizontal polarization of a magnetic wave.

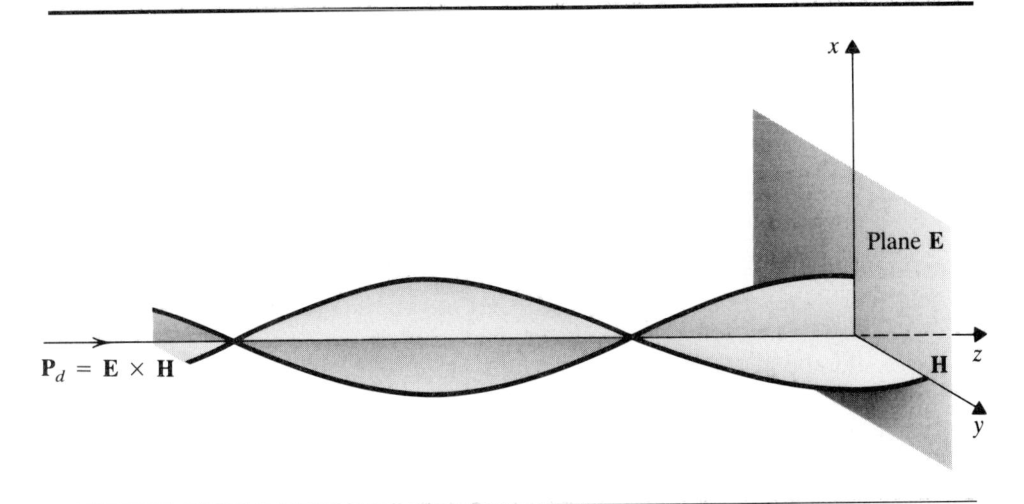

Figure 9-7-1 Vertical polarization of **E** versus horizontal polarization of **H**.

The vertical polarization of an electric field **E** is preferable in the following situations:

1. Surface wave propagation is needed.
2. A broadband omnidirectional radiation pattern is required.
3. Shipboard antennas are used for communication.
4. Shore-based stations are used for communication.

The horizontal polarization of an electric field **E** is desirable in the following situations:

1. A notable unidirectional antenna is needed.
2. Ground conductivity is sufficiently low.
3. Man-made noise is a problem.
4. High gain is required.

The characteristics of the ionosphere influence the polarization in many ways. Because of the presence of the earth's magnetic field, the incident wave from the transmitting antenna to the ionosphere splits into an ordinary and extraordinary waves, which propagate independently with different polarizations. However, the waves arriving at the receiving station can be considered to be circularly polarized. Therefore either vertical or horizontal polarization can be received, regardless of the polarization in the transmitting wave.

9-7-1 Duty Cycle

If the power delivered to the transmitting antenna is pulse modulated, the peak power density at the receiving antenna will exceed the average power density by a factor related to the duty cycle. The duty cycle (DC) is defined as the ratio of pulse duration to repetition period; that is,

$$\text{Duty cycle} = \frac{\text{Pulse duration}}{\text{Repetition period}}$$

or

$$DC = \frac{T_p}{T_r} = T_p f_r \tag{9-7-1}$$

and also as

$$\text{Duty cycle} = \frac{\text{Average power}}{\text{Peak power}}$$

or

$$DC = \frac{P_{\text{avg}}}{P_{\text{pk}}}$$

Figure 9-7-2 shows a pulse and its repetition period.

If the duration of a pulse is 1 μs and its repetition period is 2000 μs, which is equivalent to a pulse repetition frequency of 500 Hz, the duty cycle is 0.0005. If the peak power is 100 kW, the average power is 50 W.

We can also express the duty cycle in dB, or

$$DC\ (\text{dB}) = 10 \log\left(\frac{T_p}{T_r}\right) \tag{9-7-2}$$

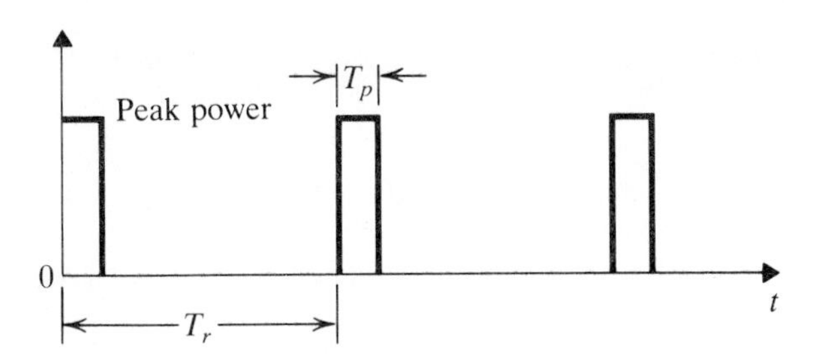

Figure 9-7-2 Pulse and its repetition period.

A symmetrical square wave would have a duty cycle of -3 dB. If the transmitted power P_{tr} is measured in terms of average power, the peak power density at distance R from the source is

$$P_d = \frac{P_t g_t}{4\pi R^2} \cdot DC \tag{9-7-3}$$

We can express Eq. (9-7-3) in dB as

$$P_d = -11 \text{ dB} - 20 \log (R) + P_t \text{ (dBW)} + G_t \text{ (dB)} + DC \text{ (dB)} \quad \text{dBW/m}^2 \tag{9-7-4}$$

For convenience, we can use the nomograph of the duty cycle shown in Fig. 9-7-3 (on the following page) to determine the DC in dB. The procedure is as follows:

1. Obtain the pulse width (PW) in μs and the pulse recurrence frequency (PRF) in pulses/s for the signal.
2. Lay a straight edge through point a for PW and point b for PRF.
3. Read the scale at point c in dB.
4. Example: PW = 0.5 μs and PRF = 2000 pulses/s; $DC = -30$ dB.

EXAMPLE 9-7-1 Antenna Receiving Power

A transmitting antenna has a power gain of 8 dB and transmits 2-kW peak power at a frequency of 1 GHz. The duty cycle is 0.003.

Determine:

a) the average power transmitted;
b) the peak power density at a distance of 100 m; and
c) the received power by an antenna with a gain of 10 dB at a distance of 100 m.

Solutions:

a) The average power is

$$P_{avg} = P_{pk} DC = 2 \times 10^3 \times 0.003 = 6 \text{ W}$$

b) The peak power density at 100 m is

$$P_d = \frac{P_t y_t}{4\pi R^2} DC = \frac{2 \times 10^3 \times 6.31}{4\pi (100)^2} \times 0.003 = 30 \text{ kW/m}^2$$

c) The power received is

$$P_r = P_t\,(\text{dBW}) + G_t + G_r - 20\log\left(\frac{4\pi R}{\lambda}\right)$$

$$= 33\ \text{dBW} + 8 + 10 - 20\log\left(\frac{4\pi \times 100}{0.3}\right)$$

$$= 51 - 72.44 = -21.44\ \text{dBW} \quad \text{or} \quad 8.56\ \text{dBm}$$

$$= 7.10\ \text{mW}$$

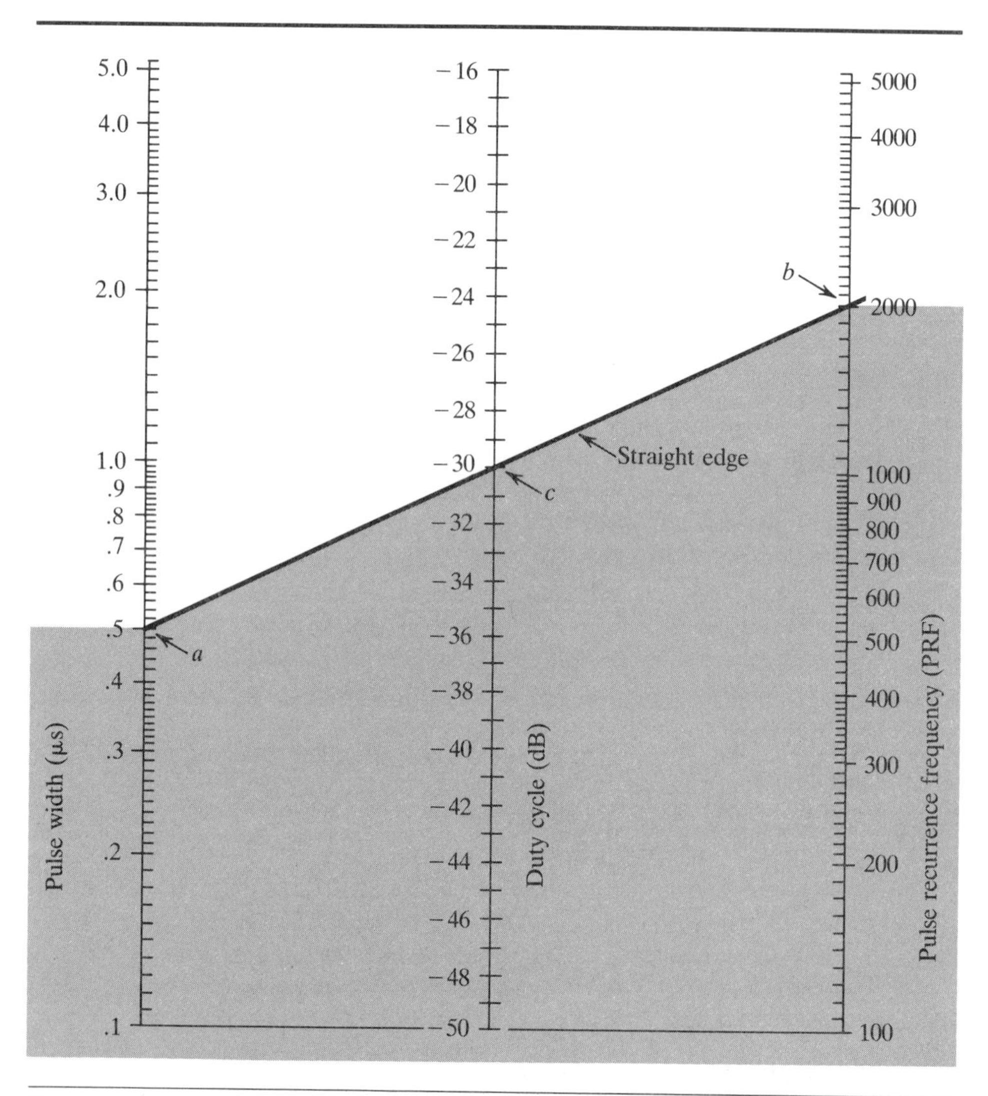

Figure 9-7-3 Nomograph for determining the duty factor (duty cycle) in dB from pulse width and pulse recurrence frequency.

9-8 RECIPROCITY THEOREM

The reciprocity theorem of antennas may be stated as follows:

> If a voltage is applied to the terminals of antenna No. 1 and the current is measured at the terminals of antenna No. 2, equal current in both amplitude and phase will be obtained at the terminals of antenna No. 1, if the same voltage is applied to the terminals of antenna No. 2.

We can prove this theorem for antennas if the antennas, antenna system, and free space (which can be replaced by a network) are bilateral, linear, and passive. From circuit theory, any four-terminal network can be reduced to an equivalent T or π section.

Figure 9-8-1(a) shows a two-antenna system; Fig. 9-8-1(b) shows an equivalent circuit network, where:

Z_a, Z_b, and Z_c represent the network between the two antennas;

Z_1, and Z_2 are the antenna impedances of antennas No. 1 and No. 2, respectively;

V_{t1} is the transmitted voltage of antenna No. 1;

V_{t2} is the transmitted voltage of antenna No. 2;

z_{t1} is the transmitted system impedance; and

z_{t2} is the receiver system impedance.

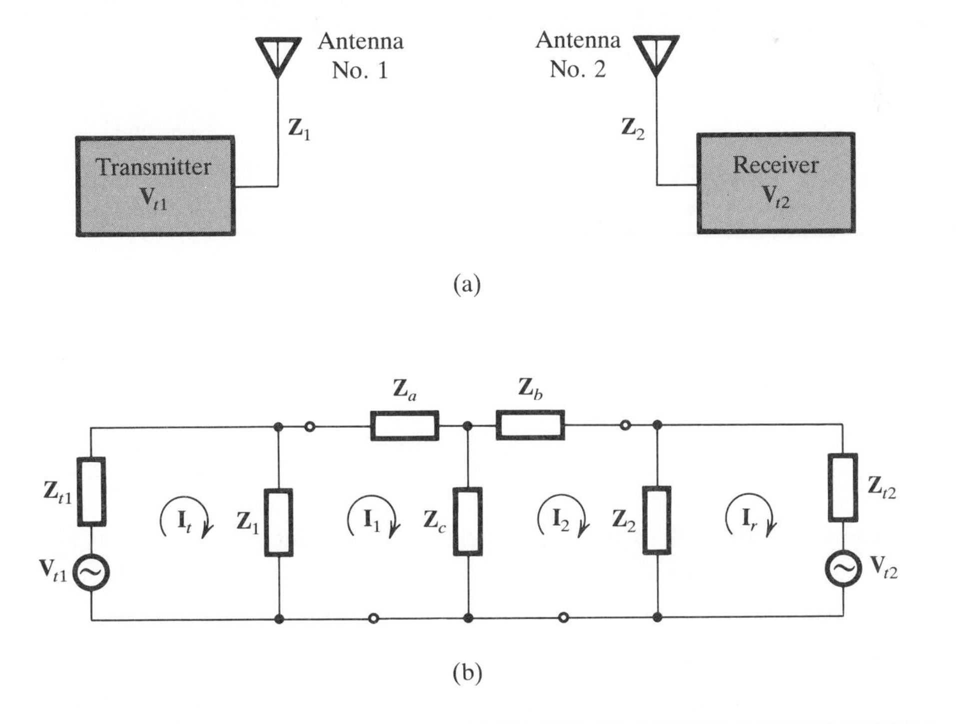

Figure 9-8-1 (a) A two-antenna system and (b) its equivalent network.

If antenna No. 1 acts as the transmitter and antenna No. 2 as the receiver, that is, $V_{t1} \neq 0$ and $V_{t2} = 0$, the system impedance determinant is

$$|Z| = \begin{vmatrix} Z_{11} & -Z_1 & 0 & 0 \\ -Z_1 & Z_{22} & -Z_c & 0 \\ 0 & -Z_c & Z_{33} & -Z_2 \\ 0 & 0 & -Z_2 & Z_{44} \end{vmatrix} \tag{9-8-1}$$

The transfer admittance of the current I_r in antenna No. 2 to the transmitter voltage V_{t1} in antenna No. 1 is

$$\frac{I_r}{V_{t1}} = \frac{\begin{vmatrix} -Z_1 & Z_{22} & Z_c \\ 0 & -Z_c & Z_{33} \\ 0 & 0 & -Z_2 \end{vmatrix}}{|Z|} = \frac{-Z_1 Z_2 Z_c}{|Z|} \tag{9-8-2}$$

If antenna No. 1 acts as the receiver and antenna No. 2 as the transmitter, that is, $V_{t1} = 0$ and $V_{t2} \neq 0$, the transfer admittance of the current I_t in antenna No. 1 to the transmitter voltage V_{t2} in antenna No. 2 is

$$\frac{I_t}{V_{t2}} = \frac{\begin{vmatrix} -Z_1 & 0 & 0 \\ Z_{22} & -Z_c & 0 \\ -Z_c & Z_{33} & -Z_2 \end{vmatrix}}{|Z|} = \frac{-Z_1 Z_2 Z_c}{|Z|} \tag{9-8-3}$$

From Eqs. (9-8-2) and (9-8-3), it is evident that if $V_{t1} = V_{t2}$, $I_r = I_t$. Thus we can conclude that the received current from a transmitter voltage is the same, regardless of which antenna is used for transmitting or receiving.

9-9 ANTENNA NOISE TEMPERATURE AND SIGNAL-TO-NOISE RATIO

9-9-1 Antenna Noise Temperature

The concept of noise temperature is derived from Nyquist's theory, which states that if a resistance R in a circuit is at absolute temperature T °K, a thermal-noise voltage will be generated in R, or

$$V_n = \sqrt{4KTBR} \tag{9-9-1}$$

where

$K = 1.38 \times 10^{-23}$ W · s/°K and is Boltzmann's constant;

T is absolute temperature, in °K;

B is bandwidth in H; and

R is resistance in Ω.

The available power (maximum power available) from any source is often used in noise-temperature and noise-figure measurements. It is the power available at an output terminal when the load impedance has matched the source impedance, as

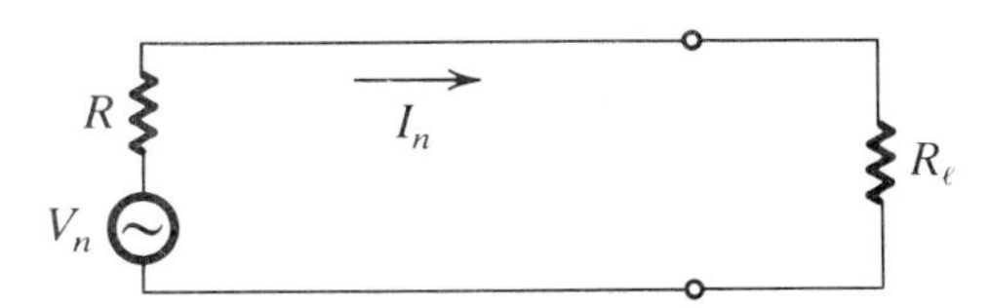

Figure 9-9-1 Matched network for maximum noise power.

shown in Fig. 9-9-1. Obviously, if the load matches the source impedance in the complex conjugate sense, the available power is

$$P_n = \frac{\mathbf{V}_n^2}{4R} \tag{9-9-2}$$

Substituting Eq. (9-9-2) into Eq. (9-9-1) gives the available power that is independent of the resistance R:

$$P_n = KTB \tag{9-9-3}$$

The noise power that is constant within the passband is often called the *white noise*. The usual noise that exists in an antenna and its receiving system is partly of thermal origin and partly from other noise-generating processes. However, these processes produce noise which, within the radio-frequency spectrum, has the same spectral and probabilistic nature. Thus we can lump them all together and regard them as thermal noise. The noise temperature of the antenna expresses the noise-power density available per unit bandwidth at the antenna terminals and is defined as

$$T_a = \frac{P_n}{KB_n} \tag{9-9-4}$$

Antenna temperature depends upon the noise temperatures of various radiating sources within the receiving-antenna pattern. However, antenna temperature is not basically independent of antenna gain and beamwidth in an overall sense. The noise temperature of a high-gain antenna averaged over all directions of free space is approximately the same if the sidelobe and backlobe levels are equivalent. In the microwave region the antenna temperature is also a function of the antenna-beam elevation angle. Figure 9-9-2 shows the curves of noise temperature for a lossless antenna without earth-directed sidelobes. The solid curves are for geometric-mean galactic temperature, sun noise 10 times quiet level, sun in unity-gain sidelobe, cool temperature-zone troposphere, 2.7 °K cosmic blackbody radiation, and 0 ground noise. The upper dashed curve is for maximum galactic noise, sun noise 100 times quiet level, and 0 elevation angle—the other factors being the same as for the solid curves. The lower dashed curve is for minimum galactic noise, 0 sun noise, and 90° elevation angle. The bump in the curves at about 500 MHz is caused by the sun-noise characteristic. The curves for low-elevation angles lie below those for high angles at frequencies below 400 MHz because of the reduction of galactic noise by atmospheric absorption. The maxima of 22.2 and 60 GHz are due to water-vapor and oxygen absorption resonance.

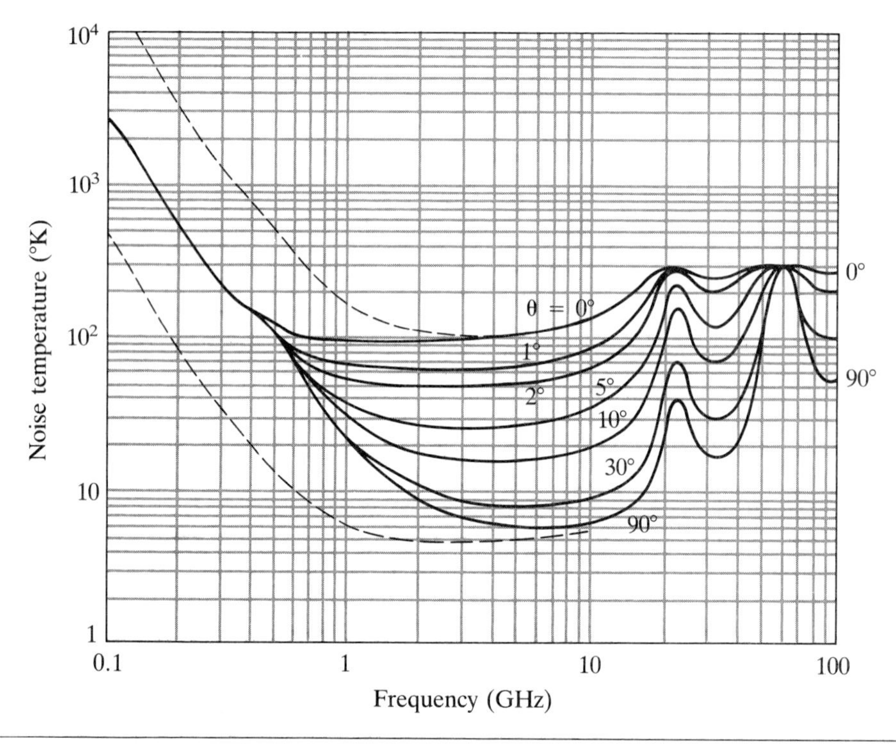

Figure 9-9-2 Antenna noise temperature versus frequency.

9-9-2 Signal-to-Noise Ratio

For receiving antennas, the most serious problem is that of the signal-to-noise ratio that an antenna can deliver. The noise delivered by a receiving antenna originates, in general, either within the antenna itself or from sources external to it. At present, state-of-the-art antenna structures are unable to reduce ambient noise without degrading the signal level.

In transmitting electromagnetic energy to free space, the main objective usually is to deliver a field intensity as large as possible to the point of reception. However, the receiving antenna receives only a very small fraction of energy radiated from the transmitting antenna, because the energy spreads through free space as the electromagnetic waves propagate. The efficiency of RF receiving antennas does not need to be high, since atmospheric, galactic, or man-made noise is usually much greater than receiver noise. An inefficient antenna attenuates the received noise as well as the signal, so that the signal-to-noise ratio remains the same as it would be with a 100 percent efficient antenna. The principal requirement for receiving-antenna efficiency is that sufficient signal and noise power be delivered to the receiver so that the predetection signal-to-noise ratio is not determined by the receiver noise figure. Mathematically, the signal-to-noise ratio is

$$\frac{S}{N} \equiv \frac{P_r}{P_n} \tag{9-9-5}$$

where P_r is the minimum detectable power, and P_n is noise power in the receiver system.

Alternatively, we can express the minimum detectable power as

$$P_r = \left(\frac{S}{N}\right) K T_s B \tag{9-9-6}$$

where T_s is the noise temperature of the system, which we discuss in detail in Section 12-1.

EXAMPLE 9-9-1 Antenna Noise Power and Noise Temperature

An antenna has a resistance of 500 Ω that generates a noise voltage of 1 mV at a bandwidth of 40 GHz.

Calculate:

a) the antenna noise power; and
b) the antenna temperature.

Solutions:

a) The antenna noise power is

$$P_n = \frac{V_n^2}{4R} = \frac{(10^{-3})^2}{4 \times 500} = 5 \times 10^{-10} \text{ W}$$

b) The antenna noise temperature is

$$T_{an} = \frac{P_n}{KB} = \frac{5 \times 10^{-10}}{1.38 \times 10^{-23} \times 40 \times 10^9}$$

$$= 906\ °\text{K} \quad \text{or} \quad 633\ °\text{C}$$

9-10 ANTENNA IMPEDANCE AND MATCHING TECHNIQUES

9-10-1 Antenna Impedance

When an antenna is connected to a transmission line, we customarily consider the antenna as a two-terminal load Z_a, as shown in Fig. 9-10-1. The impedance Z_a seen from the transmission line is called the *antenna terminal impedance* or, simply, the *antenna impedance.* If the antenna is lossless and remote from the ground or any other objects, the terminal impedance is the same as the self-impedance of the antenna. The real part of the self-impedance is called the self-resistance, which is also the radiation resistance;

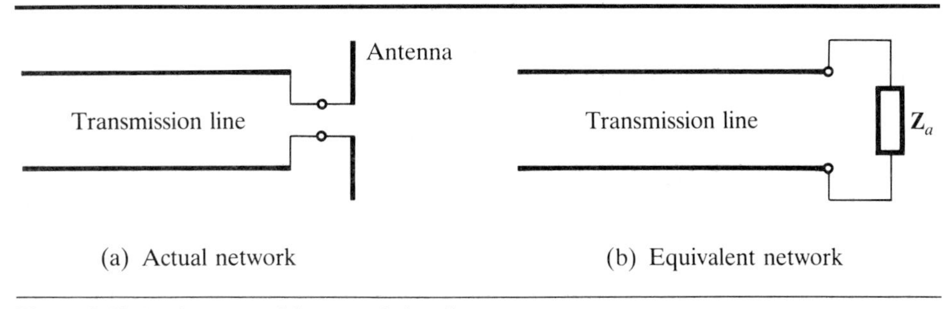

Figure 9-10-1 Antenna with transmission line.

the imaginary part is called the self-reactance. The self-impedance is the same for both reception and transmission.

In the case where there are nearby objects, or the antenna is close to the ground, the terminal impedance will normally consist of not only the radiation resistance of the antenna but also the mutual impedance between the antenna and the objects. Here, we discuss only the former part. The radiation resistance of an antenna can be determined either by the analytic method or by experimental measurements.

1. *Analytic Method*

The analytic approach involves first finding the field equations for an antenna and then determining the radiation resistance by equating the radiated power to I^2R_r, as shown in Eq. (9-3-5) with the assumption of zero loss resistance in the antenna. This assumption is true for most antennas, because the radiation is much larger than the loss resistance and antenna efficiency is almost 100 percent. Table 9-10-1 shows the radiation resistance for several antennas, as determined analytically.

Table 9-10-1 Radiation Resistance of Several Antennas

Antenna	Radiation Resistance (Ω)	Antenna	Radiation Resistance (Ω)
Short wire	$80\pi^2\left(\frac{L}{\lambda}\right)^2$	Short dipole	$20\pi^2\left(\frac{L}{\lambda}\right)^2$
Long wire	$30\pi L\omega\sqrt{\mu\varepsilon}$	Half-wave dipole	73
Single loop	$20\pi^2(\beta a)^4$	Monopole	$10\pi^2\left(\frac{L}{\lambda}\right)^2$

2. *Experimental Measurements*

Antenna impedance can be determined by transmission-line methods. The terminals of the balanced antenna are connected to the ends of unbalanced or coaxial transmission line. Measurements of the voltage or current standing-wave ratio along a slotted line by a probe and of the distance between two adjacent minima permit determination of the terminal impedance by well-known transmission-line equations. Furthermore, if a balanced antenna is connected to a balanced two-wire transmission line, the voltage standing-wave ratio can be measured by a small loop with a current indicator or a network analyzer.

9-10-2 Matching Techniques

For efficient operation of an antenna system, the voltage standing-wave ratio on the transmission line feeding the antenna has to be as close to one as possible. A high standing-wave ratio causes mismatching losses in the system and, in some cases, can prevent proper functioning of the equipment connected to the antenna. In general, the terminal impedance of the antenna is different from that of its associated transmission line, and some impedance-matching devices are necessary to reduce the voltage standing-wave ratio on the transmission line over the required band of frequencies.

Some antennas may have constant terminal resistances over their operating frequency bands. For this special case, we can transform the terminal resistance to any other value by means of suitable matching networks. However, the terminal impedance of most common antennas is not a constant pure resistance, but a reactance component that varies with the frequencies. This type of terminal impedance cannot be matched perfectly over a large frequency range. The matching techniques that can be used for antenna impedance are feeding methods, stubs, and baluns.

1. *Feeding methods*

A simple half-wave dipole antenna has a terminal resistance of about 73 Ω. An impedance transformer is required to match this antenna to an ordinary shielded two-wire line connected to a TV set with a characteristic impedance of 300 or 600 Ω. Commonly used feeding methods that would eliminate the mismatch are the folded-dipole and shunt-feed methods.

a) Folded-Dipole Method

The two-wire folded half-wave dipole antenna, as shown in Fig. 9-10-2(a), consists essentially of two half-wave dipoles placed very close to each other and connected at their outer ends.

If we let a source of voltage V applied to the antenna terminals be divided between the two dipole antennas, as shown in Fig. 9-10-2(b),

$$\frac{V}{2} = I_1 Z_{11} + I_2 Z_{12} \tag{9-10-1}$$

where

I_1 is the current at the terminals of dipole antenna 1;

I_2 is the current at the terminals of dipole antenna 2;

Z_{11} is the self-impedance of dipole antenna 1; and

Z_{12} is the mutual impedance of dipole antennas 1 and 2.

Since $I_1 = I_2$, $d \ll \lambda$, and $Z_{12} \simeq Z_{11}$,

$$V = 4 I_1 Z_{11} \tag{9-10-2}$$

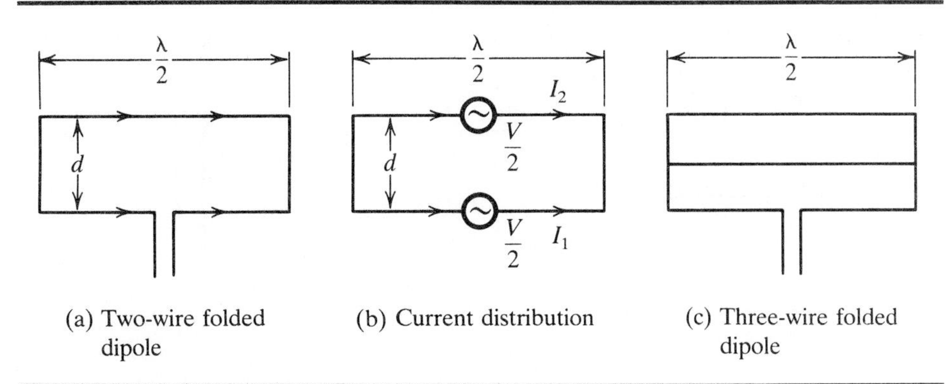

Figure 9-10-2 Folded-dipole antenna.

Therefore the terminal impedance Z of the two-wire folded dipole antenna is

$$Z = \frac{V}{I} = 4Z_{11} \tag{9-10-3}$$

The half-wave dipole antenna has a terminal resistance of about 73 Ω, so the terminal resistance of the two-wire folded dipole antenna becomes

$$Z = 290 \quad \Omega \tag{9-10-4}$$

and the two-wire folded dipole antenna can be connected to the open two-wire line without significant mismatching losses.

For a three-wire folded half-wave dipole antenna, as shown in Fig. 9-10-1(c), the terminal resistance is $73 \times 9 = 657\,\Omega$. In general, for a folded half-wave dipole antenna of N wires, the terminal resistance is

$$Z = N^2(73) \quad \Omega \tag{9-10-5}$$

b) Shunt-Feed Method

The shunt-feed method (or tapped-line matching or delta match) is shown in Fig. 9-10-3. In Fig. 9-10-3(a), a transmission line that is an odd number of quarter-wavelengths long, short-circuited at one end, and open-circuited at the other, is tapped on both sides at the same distance from the ends by a main transmission line with the same characteristic impedance. From transmission-line theory, we know that the reactance of the open-circuited line is

$$X_{oc} = -R_o \cot(\beta d) = -R_o \cot\left[\frac{2\pi}{\lambda}(\ell - x)\right] \tag{9-10-6}$$

and the reactance of the short-circuited line is

$$X_{sc} = R_o \tan(\beta d) = R_o \tan\left(\frac{2\pi}{\lambda}x\right) \tag{9-10-7}$$

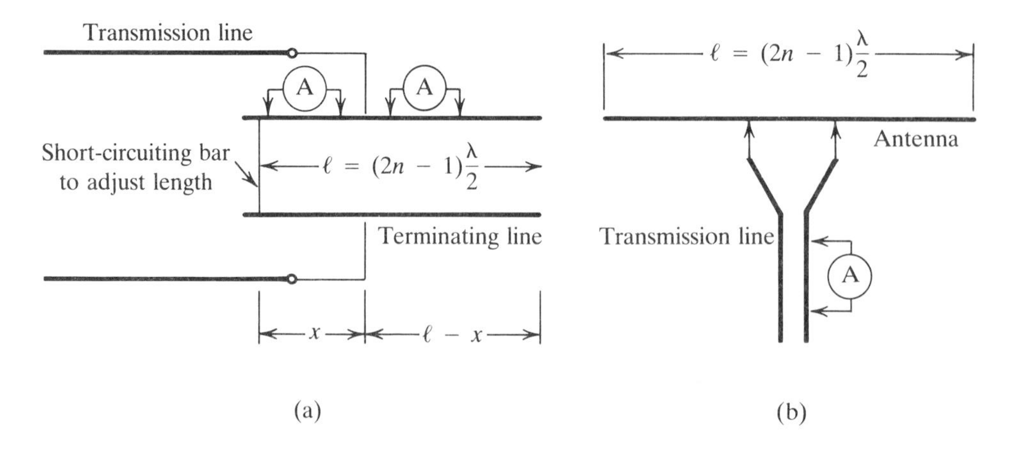

Figure 9-10-3 Shunt-feed matching.

If ℓ is an odd number of quarter-wavelengths long, the factor $2\pi\ell/\lambda$ would be an odd multiple of $\pi/2$:

$$\cot\left[\frac{2\pi(\ell - x)}{\lambda}\right] = \cot\left[\frac{(2n-1)\pi}{2} - \frac{2\pi x}{\lambda}\right] = \tan\left(\frac{2\pi x}{\lambda}\right) \tag{9-10-8}$$

The reactance of the short-circuited end would be equal and opposite to the reactance of the open-circuited end. This condition is independent of the tapped point. The proper length of the line can be determined experimentally by finding equal ammeter readings on both sides of the tapped point when the main line is energized Since the reactances of the two branches on either side of the taps are equal, the condition for antiresonance is set up, and a resistive load is provided at the tapped point for impedance match. If the line is lossy, the magnitude of the load can be adjusted by moving the positions of the taps as indicated in Figure 9-10-3(b). The taps are carefully adjusted until the ammeter, moved along the main line, has a constant reading. The constant reading indicates that the main transmission line has a VSWR of one and that impedance matching has been achieved.

2. *Stub Matching*

If the antenna input impedance is too low for direct connection to a transmission line, stub matching is needed, as shown in Fig. 9-10-4. Recall that we discussed impedance matching by means of a single stub or double stubs in Chapter 2.

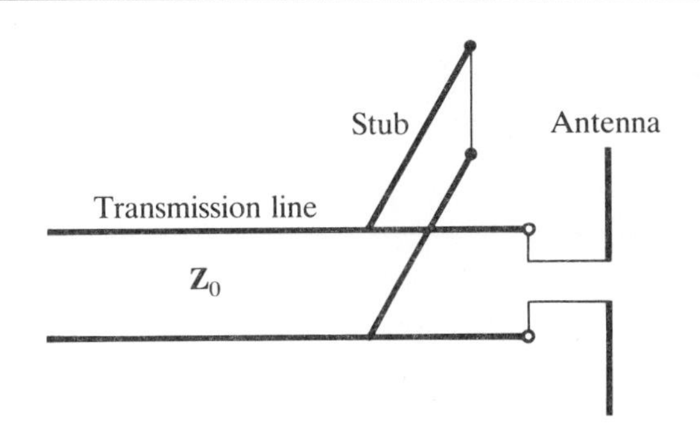

Figure 9-10-4 Stub matching.

3. *Baluns*

The problem of transmitting energy from a coaxial or unbalanced transmission line to a dual or balanced antenna, and vice versa, occurs frequently in broadband antenna systems. Balanced antennas, in general, are more suitable for broadband applications than unbalanced antennas are. Since balanced antennas are almost always fed by unbalanced voltage of the coaxial line to the balanced voltage required by the antenna, the device that accomplishes this *balance-to-unbalance* transformation is called the *balun*. Although many types of baluns are available for use in transformation, we discuss only three commonly used baluns here.

a) Bazooka balun

Figure 9-10-5(a) shows an unbalanced coaxial line connected to a balanced an-

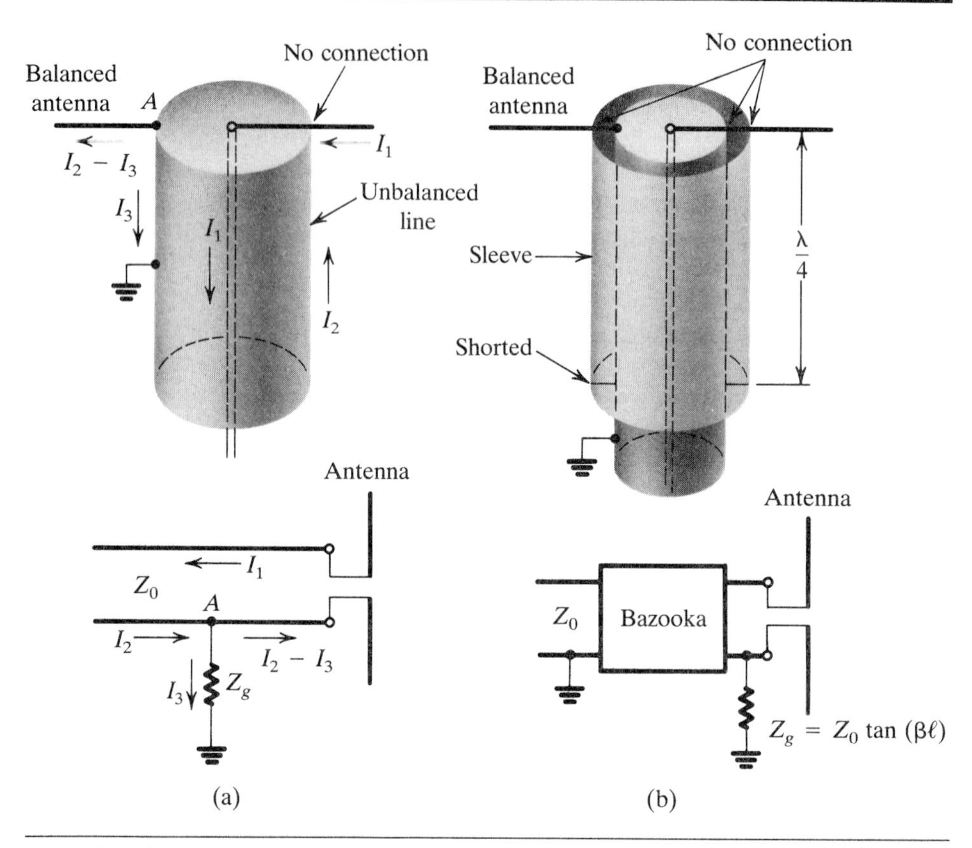

Figure 9-10-5 Bazooka balun.

tenna without a quarter-wave sleeve. Figure 9-10-5(b) shows the same configuration but with a sleeve added.

In Fig. 9-10-5(a), the balanced dipole antenna is connected directly to the end of an unbalanced coaxial line. The currents I_1 and I_2 must be equal and opposite. At junction A, current I_2 divides: into I_3, which flows down the outside of the conductor of the line, and $I_2 - I_3$, which flows on the second arm of the dipole. Current I_3 can be made very small by increasing the impedance Z_g with a quarter-wave sleeve around the outer conductor as shown in Fig. 9-10-5(b).

With the sleeve shorted to the outer conductor at the bottom, which is grounded, the impedance between the sleeve and the outer conductor of the line at the open end is

$$Z_g = jZ_0 \tan(\beta\ell) \tag{9-10-9}$$

where Z_0 is the characteristic impedance of the coaxial line that consists of the outer conductor of the main line and the sleeve.

When $\ell = \frac{1}{4}\lambda$, the impedance Z_g is very high, and there is no shunt path to ground from the outer conductor. Consequently, the current does not divide at the second arm of the antenna, and both conductors of the dual line carry equal currents. Since the impedance Z_g is a function of frequency, the bandwidth of this type of balun is quite limited; the balun is suitable for use only over a frequency band not more than 10 percent wide.

b) Colinear balun

A colinear balun and its equivalent circuit are shown in Fig. 9-10-6. We can increase the bandwidth of a bazooka by utilizing another section of transmission line having the same characteristics as the section formed by outer conductor B and sleeve C of the bazooka. As shown in Figure 9-10-6(a), the added section consists of B' and C', with B' connected to junction 1. Thus whatever impedance is between junction 2 and ground is due to B' and C', as illustrated in the equivalent circuit in Fig. 9-10-6(b). At the frequency at which the lengths of B and B' are exactly a quarter-wavelength, the impedance presented to the coaxial line is only the input impedance of the dual line, since the impedance of the sections of the balun is infinite. At any other frequency the balun shunts the dual line, with an impedance equal to

$$Z_{sh} = jZ_{0BC} \tan(\beta\ell) \tag{9-10-10}$$

where Z_{0BC} is the characteristic impedance of each leg of the balun. In general, the characteristic impedance Z_{0BC} of the dual line is greater than the impedance of the coaxial line.

The gap between junctions 1 and 2 should be as small as possible, which is consistent with consideration of voltage arc-over and lumped capacitance between them. Sizes used in practice range from $\frac{1}{8}$–$\frac{1}{4}$ in. If the gap is large compared to the shortest wavelength at which the balun is to be used, the section of inner conductor of the coaxial line is used to add inductance in series with one side, and some unbalance is created.

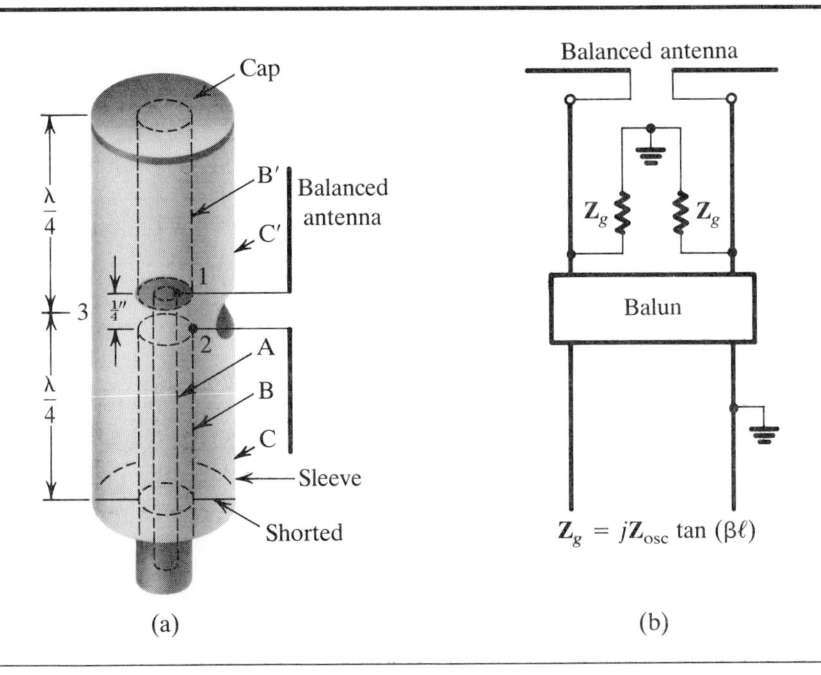

Figure 9-10-6 Colinear balun.

c) Folded balun

The folded balun operates in the same manner as the colinear balun, except that section B' is folded back beside section B to form a dual line within skirt C, as shown in Fig. 9-10-7.

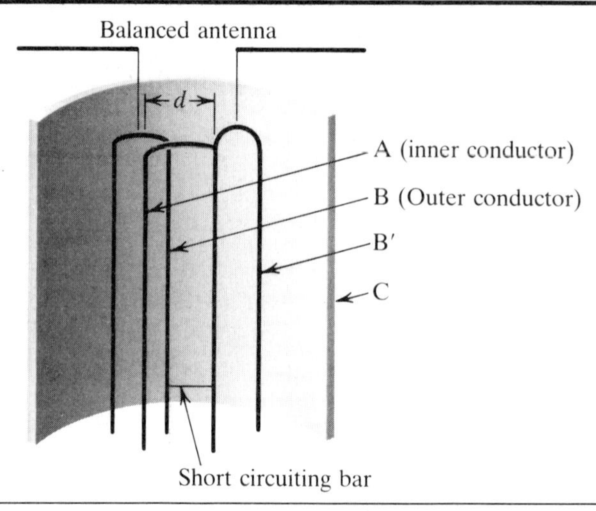

Figure 9-10-7 Folded balun.

As we described in the preceding section, the length d of the inner conductor of the coaxial line must be as short as possible to minimize unbalance. The principal advantage of the folded balun is that its length needs to be only one-half that of the colinear balun. Furthermore, a short-circuiting bar can be inserted as shown to permit adjustment of the length of the balun for operation at different frequencies.

Problems

9-1 What is the effective aperture of a microwave antenna that has a power gain of 15 dB at 8 GHz?

9-2 In an electromagnetic energy-transmission system, the transmitting antenna has a 25-dB gain and the receiving antenna has a 20-dB gain. The input power fed to the transmitting antenna at 1 GHz is 150 W. The distance between the two antennas over free space is 500 m. Determine the

a) power density at the receiving antenna aperture; and
b) power received by the receiving antenna.

9-3 A generator with an output of 0.0025 W supplies power to an amplifier, which in turn supplies power to a transmitting antenna. If the power level in the antenna is 16 dB above 0.001 W reference, what power gain in dB is required for the amplifier?

9-4 A quarter-wave monopole antenna has a driving-point impedance of 36.50 Ω. The transmission line connected to a TV set has a characteristic impedance of 300 Ω. The problem is to design a shorted stub with the same characteristic impedance in order to match the antenna to the line. The stub may be placed at a short distance from the antenna. The reception is on Channel 24 at a frequency of 531.25 MHz. Calculate the

a) susceptance contributed by the stub;
b) length of the stub; and
c) distance between the antenna and the point where the stub is placed.

9-5 A very short-wire antenna of $L = \lambda/10$ operates as a receiving antenna at a frequency of 30 MHz. The incoming electric field intensity is 20 mV/m. Determine the maximum possible power delivered by the antenna to a load.

9-6 A long wire antenna has a length of a half-wave and operates at 1 GHz. Determine the

a) antenna radiation resistance; and
b) effective antenna aperture.

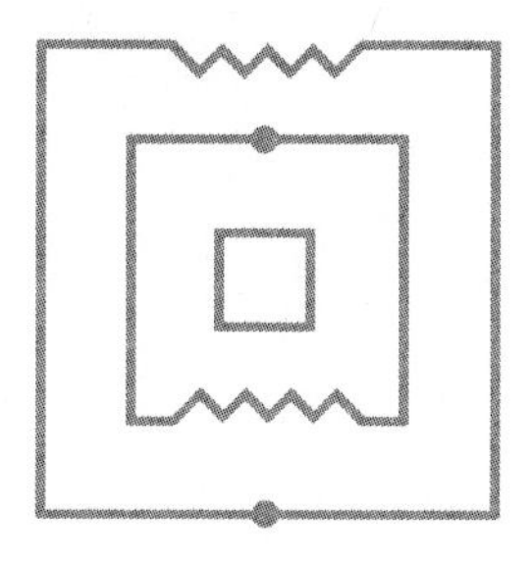

Chapter 10

Dipole Antennas and Slot Antennas

10-0 INTRODUCTION

The physical types of antennas are too numerous to be described here. In general, they are grouped according to their physical structures: dipole antennas, aperture antennas, slot antennas, array antennas, reflector antennas, traveling-wave antennas, and broadband antennas. In this chapter we discuss only the dipole antenna, monopole antenna, slot antenna, and cavity-backed slot antenna.

10-1 DIPOLE ANTENNAS

One of the most commonly used types of antenna is the dipole antenna, as shown in Fig. 10-1-1(a). The dipole antenna is energized at its center. From transmission-line theory, we know that the current on a wire is a harmonic function of βz; that is, the current on the dipole antenna must be zero at both ends of the wire, symmetrical in z, and have a standing-wave pattern on the wire with current nulls at the ends and at every half-wavelength from the ends, as shown in Fig. 10-1-1(b). However, for a very short dipole antenna ($L \ll \lambda$), the current distribution is approximately linear, not sinusoidal.

The magnetic vector potential for a dipole antenna is

$$A_z = \frac{1}{4\pi} \int_{-L/2}^{L/2} \frac{\mu I_0 e^{-j\beta|r-r'|}}{|r-r'|} dz \qquad (10\text{-}1\text{-}1)$$

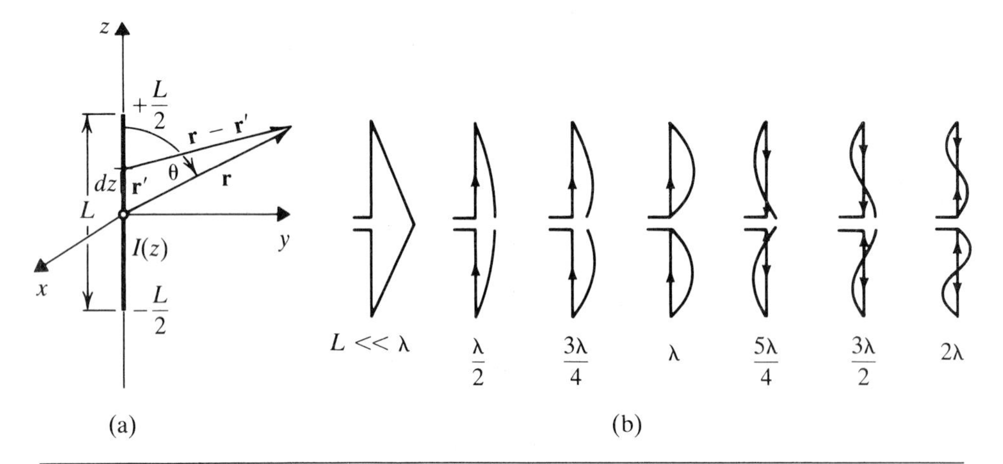

Figure 10-1-1 (a) The dipole antenna and (b) its current distributions.

where I_0 is the amplitude of the current, and

$$|r - r'| = (r^2 + z^2 - 2rz \cos \theta)^{1/2} \tag{10-1-2}$$

For $r \gg z$ and $r \gg L$,

$$|r - r'| \simeq r - z \cos \theta \tag{10-1-3}$$

and

$$A_z \doteq \frac{\mu e^{-j\beta r}}{4\pi r} \int_{-L/2}^{L/2} I_0 e^{j\beta z \cos \theta} \, dz \tag{10-1-4}$$

Note that we have to retain the second term of Eq. (10-1-3) in the phase term $e^{-j\beta|r-r'|}$, but do not need to retain it in the amplitude term $|r - r'|^{-1}$.

In order to evaluate the radiation field, we must know the current distribution on an antenna. An exact determination of the current requires solving a boundary-value problem. However, the radiation field is relatively insensitive to minor changes in current distribution, and we can obtain much useful information from an approximation of current distribution. For a dipole antenna, we can represent the current distribution on the antenna by

$$I(z) = I_0 \sin\left[\beta\left(\frac{L}{2} - z\right)\right] \qquad \text{for } z > 0 \tag{10-1-5a}$$

and

$$I(z) = I_0 \sin\left[\beta\left(\frac{L}{2} + z\right)\right] \qquad \text{for } z < 0 \tag{10-1-5b}$$

We can now evaluate the vector potential in the far field:

$$A_z = \frac{I_0 e^{-j\beta r}}{4\pi} \int_{-L/S}^{L/2} \sin\left[\beta\left(\frac{L}{2} - |z|\right)\right] e^{j\beta z \cos \theta} \, dz$$

$$= \frac{I_0 e^{-j\beta r}}{4\pi r} \, \frac{2\left[\cos\left(\beta \frac{L}{2} \cos \theta\right) - \cos\left(\beta \frac{L}{2}\right)\right]}{\beta \sin^2 \theta} \tag{10-1-6}$$

Then, the electric intensity at the far field is

$$E_\theta = \frac{j60I_0e^{-j\beta r}}{r}\left[\frac{\cos\left(\beta\frac{L}{2}\cos\theta\right)-\cos\left(\beta\frac{L}{2}\right)}{\sin\theta}\right] \tag{10-1-7}$$

and the magnetic field intensity at the radiation zone is

$$H_\phi = \frac{jI_0e^{-j\beta r}}{2\pi r}\left[\frac{\cos\left(\beta\frac{L}{2}\cos\theta\right)-\cos\left(\beta\frac{L}{2}\right)}{\sin\theta}\right] \tag{10-1-8}$$

EXAMPLE 10-1-1 Computer Plot for an Electric or Magnetic Field Pattern of a Dipole Antenna, Based on Eqs. (10-1-7) and (10-1-8), respectively:

$$E_\theta = \frac{j60I_0e^{-j\beta r}}{r}\left[\frac{\cos\left(\beta\frac{L}{2}\cos\theta\right)-\cos\left(\beta\frac{L}{2}\right)}{\sin\theta}\right]$$

$$H_\phi = \frac{jI_0e^{-j\beta r}}{2\pi r}\left[\frac{\cos\left(\beta\frac{L}{2}\cos\theta\right)-\cos\left(\beta\frac{L}{2}\right)}{\sin\theta}\right]$$

```
010C      PROGRAM TO PLOT RELATIVE ELECTRIC OR MAGNETIC
020C+     FIELD PATTERN OF DIPOLE ANTENNA
022C      THE VALUE OF L IS IN LAMBDA AND BETA IS TWO PI
024C+     OVER LAMBDA. SO THE PRODUCT OF BETAXL/2 IS
026C+     EQUAL TO NPI, WHERE N IS ANY REAL NUMBER
030C      DIPO2
040       DIMENSION LINE(41)
045       REAL L
047     1 ETHETA=0.0
048       PRINT*,"   INPUT L VALUE"
049       READ*,L
050       IF(L .LE. 0.0) STOP
055       DO 6, J=1,41
060     6 LINE(J)=1H.
070       PRINT 7, LINE
080     7 FORMAT(1H ,"ANGLE MAGNITUDE",41A1)
090       DO 8 J=1,41
100     8 LINE(J)=1H
110       LINE(20)=1H.
120       DTHETA=10.0*3.141593/180.0
125       THETA=DTHETA
128       ANGLE=10.0
130       PI=3.141593
140       B=2.0*PI
160       DO 10 J1=1,36
170       ETHETA=(COS(B*L/2.*COS(THETA))-COS(B*L/2.))/SIN(THETA)
180       N=IFIX(ETHETA*15.0+20.0)
190       LINE(N)=1HX
200       PRINT 9, ANGLE,ETHETA,LINE
210     9 FORMAT(1X,F5.1,1X,F8.5,41A1)
220       LINE(N)=1H
230       LINE(20)=1H.
```

(continued)

EXAMPLE 10-1-1 *(continued)*

```
240       THETA=THETA+DTHETA
250       ANGLE=ANGLE+10.0
260    10 CONTINUE
265       GO TO 1
270       STOP
280       END
READY.
RUN

PROGRAM    DIPO2

   INPUT L VALUE
? 0.5
 ANGLE MAGNITUDE...............................................
  10.0    .13741                         . X
  20.0    .27656                         .   X
  30.0    .41779                         .     X
  40.0    .55894                         .       X
  50.0    .69464                         .         X
  60.0    .81650                         .           X
  70.0    .91426                         .            X
  80.0    .97789                         .             X
  90.0   1.00000                         .              X
 100.0    .97789                         .             X
 110.0    .91426                         .            X
 120.0    .81650                         .           X
 130.0    .69464                         .         X
 140.0    .55894                         .       X
 150.0    .41779                         .     X
 160.0    .27656                         .   X
 170.0    .13741                         . X
 180.0   -.00000                        X.
 190.0   -.13741                      X  .
 200.0   -.27656                    X    .
 210.0   -.41779                  X      .
 220.0   -.55894                X        .
 230.0   -.69464              X          .
 240.0   -.81650            X            .
 250.0   -.91426           X             .
 260.0   -.97789          X              .
 270.0  -1.00000         X               .
 280.0   -.97789          X              .
 290.0   -.91426           X             .
 300.0   -.81650            X            .
 310.0   -.69464              X          .
 320.0   -.55894                X        .
 330.0   -.41779                  X      .
 340.0   -.27656                    X    .
 350.0   -.13741                      X  .
 360.0    .00000                         X
   INPUT L VALUE
? 0.0

SRU        0.924 UNTS.

RUN COMPLETE.
```

We derive the power density radiated in the r direction for a dipole antenna from Eqs. (10-1-7) and (10-1-8):

$$P_d = \frac{1}{2}\frac{|E|^2}{\eta} = \frac{15|I_0|^2}{\pi r^2}\left[\frac{\cos\left(\beta\frac{L}{2}\cos\theta\right) - \cos\left(\beta\frac{L}{2}\right)}{\sin\theta}\right]^2 \tag{10-1-9}$$

then obtain the total radiated power by integrating P_d over a large sphere:

$$P = \int_0^{2\pi}\int_0^{\pi} P_d r^2 \sin\theta \, d\theta \, d\phi$$

$$= 30|I_0|^2 \int_0^{\pi} \frac{\left[\cos\left(\frac{\beta L}{2}\cos\theta\right) - \cos\left(\frac{\beta L}{2}\right)\right]^2}{\sin\theta} d\theta \tag{10-1-10}$$

EXAMPLE 10-1-2 Computer Plot for Radiation Power Pattern of a Dipole Antenna Based on Eq. (10-1-9):

$$P_d = \frac{1}{2}\frac{|E|^2}{\eta} = \frac{15|I_0|^2}{\pi r^2}\left[\frac{\cos\left(\beta\frac{L}{2}\cos\theta\right) - \cos\left(\beta\frac{L}{2}\right)}{\sin\theta}\right]^2$$

```
PROGRAM    POWER

010C      PROGRAM TO PLOT RELATIVE RADIATION POWER
020C+     PATTERN OF DIPOLE ANTENNA
030C      THE VALUE OF L IS IN LAMBDA AND BETA IS TWO
040C+     PI OVER LAMBDA. SO THE PRODUCT OF BETAXL/2
050C+     IS EQUAL TO NPI, WHERE N IS ANY REAL NUMBER
060C      POWER
070       DIMENSION LINE(41)
080       REAL L
090       PRINT*,"  INPUT L VALUE"
100     1 READ*,L
110       IF(L .LE. 0.0) STOP
120       DO 6, J=1,41
130     6 LINE(J)=1H.
140       PRINT 7, LINE
150       DO 8 J=1,41
160     8 LINE(J)=1H
170       LINE(2)=1H.
180       DTHETA=10.0*3.141593/180.0
190       THETA=DTHETA
200       ANGLE=10.0
210       PI=3.141593
220       B=2.0*PI
230       DO 10 J1=1,36
240       PD=((COS(B*L/2.*COS(THETA))-COS(B*L/2.))/SIN(THETA))**2
250       N=IFIX(PD*15.+20.)
260       LINE(N)=1HX
270       PRINT 12,ANGLE,PD,LINE
280       LINE(N)=1H
290       LINE(20)=1H.
300       THETA=THETA+DTHETA
```

(*continued*)

EXAMPLE 10-1-2 *(continued)*

```
310        ANGLE=ANGLE+10.0
320    10  CONTINUE
330     7  FORMAT(1H ,"ANGLE MAGNITUDE", 41A1)
340    12  FORMAT(1X,F5.1,1X,F8.5,41A1)
350        GO TO 1
360        STOP
370        END
READY.
RUN

PROGRAM    POWER

   INPUT L VALUE
? 0.5
 ANGLE MAGNITUDE.........................................
  10.0    .01888 .                X
  20.0    .07649 .                .X
  30.0    .17455 .                . X
  40.0    .31242 .                .   X
  50.0    .48252 .                .      X
  60.0    .66667 .                .         X
  70.0    .83587 .                .           X
  80.0    .95626 .                .             X
  90.0   1.00000 .                .              X
 100.0    .95626 .                .             X
 110.0    .83587 .                .           X
 120.0    .66667 .                .       X
 130.0    .48252 .                .      X
 140.0    .31241 .                .   X
 150.0    .17455 .                . X
 160.0    .07649 .                .X
 170.0    .01888 .                X
 180.0    .00000 .                X
 190.0    .01888 .                X
 200.0    .07649 .                .X
 210.0    .17455 .                . X
 220.0    .31242 .                    X
 230.0    .48252 .                .      X
 240.0    .66667 .                .         X
 250.0    .83587 .                .           X
 260.0    .95626 .                .             X
 270.0   1.00000 .                .              X
 280.0    .95626 .                .             X
 290.0    .83587 .                .           X
 300.0    .66667 .                .        X
 310.0    .48252 .                .      X
 320.0    .31241 .                .   X
 330.0    .17455 .                . X
 340.0    .07648 .                .X
 350.0    .01888 .                X
 360.0    .00000 .                X
? 0.0

SRU       0.977 UNTS.

RUN COMPLETE.
```

Substituting the radiated power from Eq. (10-1-10) into

$$P = \frac{1}{2}|I_0|^2 R_r \tag{10-1-11a}$$

we obtain the radiation resistance R_r of a dipole antenna:

$$R_r = 60 \int_0^\pi \frac{\left[\cos\left(\frac{\beta L}{2}\cos\theta\right) - \cos\left(\frac{\beta L}{2}\right)\right]^2}{\sin\theta}\, d\theta \tag{10-1-11b}$$

The radiation resistance R_r is also referred to as the current maximum. Determining the radiation resistance R_r for dipole antennas with Eq. (10-1-11b) is tedious work. However, we can find the radiation resistance R_r from the curve in Fig. 10-1-2.

Because a small dipole antenna has the same radiation pattern as a short-wire antenna, a dipole antenna thus has the same directivity and effective aperture as a short-wire antenna. Therefore the effective aperture and the directivity of a dipole antenna, respectively, are

$$A_e = \frac{3\lambda^2}{8\pi} \tag{10-1-12a}$$

and

$$d = 1.5 \tag{10-1-12b}$$

The shape of a far-field pattern is expressed by the factor inside the brackets in Eqs. (10-1-7) and (10-1-8). The factors preceding the brackets represent the instantaneous magnitude of the fields as a function of the antenna current I_0 and the distance r. Because amplitude is independent of antenna length, the relative field pattern (as represented by the pattern factor) changes only as the length changes.

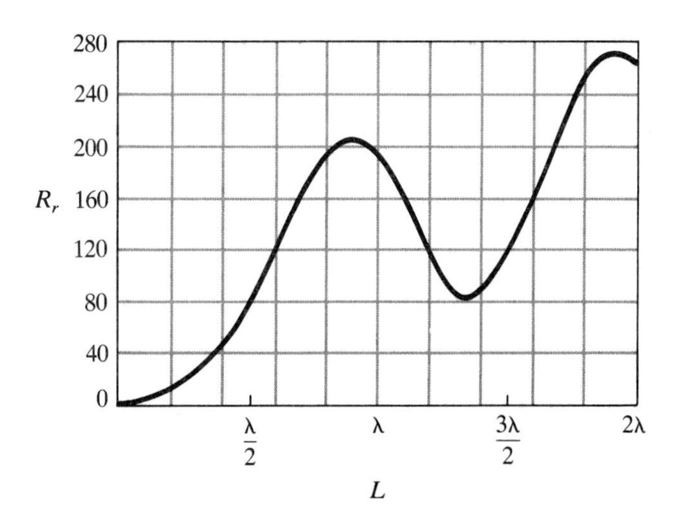

Figure 10-1-2 Radiation resistance of the dipole antenna.

1. *Half-wave antenna.* When the length of a dipole antenna is one half-wavelength long ($L = \lambda/2$), the field-pattern factor is

$$E_\theta = \frac{\cos\left(\dfrac{\pi}{2}\cos\theta\right)}{\sin\theta} \tag{10-1-13}$$

This pattern is illustrated in Fig. 10-1-3(a) on page 348. Only slightly more directional than the pattern of a short-wire or short-dipole antenna, it is represented by $\sin\theta$. The beamwidth is 78°, compared to 90° for a short-wire or short-dipole antenna.

2. *Full-wave antenna.* When the length of a dipole antenna is one wavelength long ($L = \lambda$), the field-pattern factor becomes

$$E_\theta = \frac{\cos(\pi\cos\theta) + 1}{\sin\theta} \tag{10-1-14}$$

This pattern is shown in Fig. 10-1-3(b) on page 348. More directional than the pattern of the half-wave dipole antenna, it has a beamwidth of about 47°.

3. *Three-halves–wave antenna.* When the length of a dipole antenna is three halves of a wavelength long ($L = \frac{3}{2}\lambda$), the field-pattern factor is

$$E_\theta = \frac{\cos\left(\dfrac{3}{2}\pi\cos\theta\right)}{\sin\theta} \tag{10-1-15}$$

These patterns are shown in Fig. 10-1-3(c) on page 348.

EXAMPLE 10-1-3 Computer Plot for the Electric Field Pattern of a Full-Wave Antenna, Based on Eq. (10-1-14):

$$E_\theta = \frac{\cos(\pi\cos\theta) + 1}{\sin\theta}$$

```
PROGRAM    FULWAV

010C       PROGRAM TO PLOT RELATIVE ELECTRIC FIELD
020C+      PATTERN OF FULL-WAVE ANTENNA
030C       FULWAV
040        DIMENSION LINE(41)
050        DO 4 J=1,41
060      4 LINE(J)=1H.
070        PRINT 6, LINE
080      6 FORMAT(/1H ,"DEGREES MAGNITUDE",41A1)
090        DO 8 J=1,41
100      8 LINE(J)=1H
110        LINE(20)=1H.
120        PI=3.141593
130        DO 12 J=10,360,10
140        THETA=FLOAT(J)*PI/180.0
150        IF(ABS(SIN(THETA)) .LT. 0.00001) GO TO 12
160        ETHETA=(COS(PI*COS(THETA))+1.0)/SIN(THETA)
170        N=IFIX(ETHETA*8.0+20.0)
180        LINE(N)=1HX
```

EXAMPLE 10-1-3 *(continued)*

```
190       PRINT 10, J,ETHETA,LINE
200    10 FORMAT(2X,I4,1X,F10.5,41A1)
210       LINE(N)=1H
220       LINE(20)=1H.
230    12 CONTINUE
240       STOP
250       END
READY.
RUN

PROGRAM    FULWAV

 DEGREES MAGNITUDE.........................................
    10     .00656                   X
    20     .05232                   X
    30     .17455                   .X
    40     .40163                   .  X
    50     .73927                   .    X
    60    1.15470                   .        X
    70    1.57092                   .           X
    80    1.88347                   .              X
    90    2.00000                   .              X
   100    1.88347                   .              X
   110    1.57092                   .           X
   120    1.15470                   .        X
   130     .73927                   .    X
   140     .40163                   .  X
   150     .17455                   .X
   160     .05232                   X
   170     .00656                   X
   190    -.00656                  X.
   200    -.05232                  X.
   210    -.17455                 X .
   220    -.40163               X   .
   230    -.73927             X     .
   240   -1.15470         X         .
   250   -1.57092      X            .
   260   -1.88347   X               .
   270   -2.00000   X               .
   280   -1.88346   X               .
   290   -1.57092      X            .
   300   -1.15470         X         .
   310    -.73927             X     .
   320    -.40163               X   .
   330    -.17455                 X .
   340    -.05232                  X.
   350    -.00656                  X.

SRU        0.758 UNTS.

RUN COMPLETE.
```

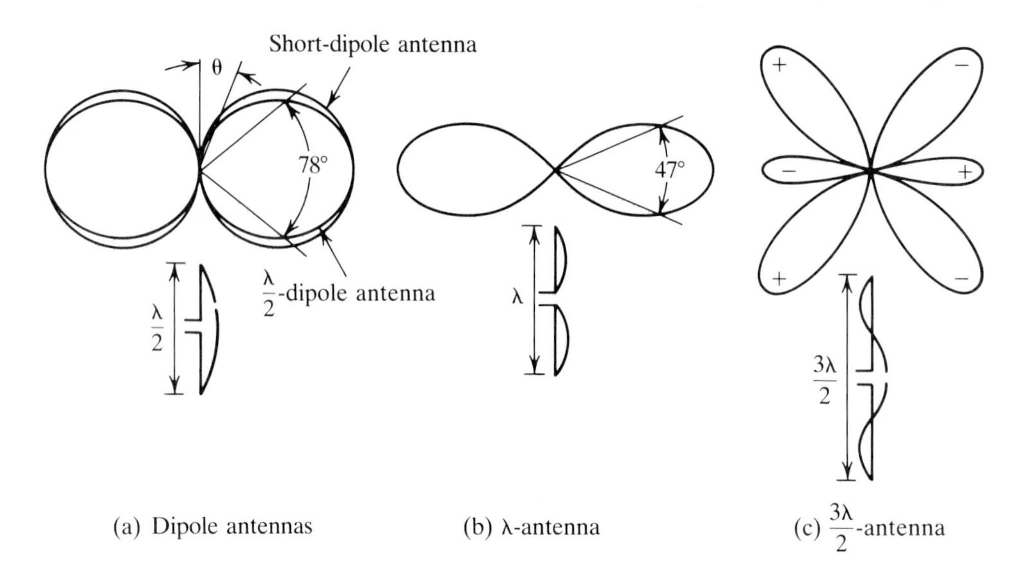

Figure 10-1-3 Far-field antennas of $\frac{1}{2}\lambda$, λ, and $\frac{3}{2}\lambda$. The antennas are center-fed, and the current distribution is assumed to be sinusoidal.

4. *Short-Dipole Antenna.* When the length of a dipole antenna is much shorter than the wavelength of the impressed signal, evaluating the radiation field pattern by means of Eqs. (10-1-7) and (10-1-8) is very difficult. Instead, we can find the field equations from a prescribed current distribution. For a short-dipole antenna, the current distribution is approximately triangular, as previously shown in Fig. 10-1-1(b). When the length of a dipole antenna is much shorter than one wavelength long ($L < \lambda/30$), we can represent its current distribution by

$$I(z) = \frac{2I_0}{L}\left(\frac{L}{2} - z\right) \qquad z > 0 \tag{10-1-16a}$$

and

$$I(z) = \frac{2I_0}{L}\left(\frac{L}{2} + z\right) \qquad z < 0 \tag{10-1-16b}$$

The magnetic vector potential for a short-dipole antenna is

$$\begin{aligned} A_z &= \mu \int_{-L/2}^{L/2} \frac{I_0 e^{j\omega\left(t - \frac{r}{c}\right)}}{4\pi r} dz \\ &= \frac{\mu I_0 e^{j(\omega t - \beta r)}}{2\pi L r}\left[\int_0^{L/2}\left(\frac{L}{2} - z\right)dz + \int_{-L/2}^{0}\left(\frac{L}{2} + z\right)dz\right] \\ &= \frac{1}{2}\frac{\mu L I_0 e^{-j\beta r}}{4\pi r} \end{aligned} \tag{10-1-17}$$

where the time-dependent term $e^{j\omega t}$ has been omitted.

You can see from Eq. (10-1-17) that the triangular current distribution gives a magnetic vector potential of magnitude one-half the magnitude of a magnetic potential that would be obtained if the current distribution were uniform over the same wire length. Thus the far field of a short-dipole antenna is one-half the field given in Eq. (9-2-6) for a short-wire antenna. Therefore the field components are

$$E_\theta = \frac{j\beta\eta_0 I_0 L e^{-j\beta r}}{8\pi r} \sin\theta \tag{10-1-18a}$$

and

$$H_\phi = \frac{j\beta I_0 L e^{-j\beta r}}{8\pi r} \sin\theta \tag{10-1-18b}$$

The power density, then, is

$$P_d = \frac{\eta_0}{8}\left(\frac{I_0 L \beta}{4\pi r}\right)^2 \sin^2\theta \quad \text{W/m}^2 \tag{10-1-19}$$

the total radiated power becomes

$$P = 10\pi^2 I_0^2 \left(\frac{L}{\lambda}\right)^2 \quad \text{W} \tag{10-1-20}$$

and the radiation resistance is

$$R_r = 20\pi^2 \left(\frac{L}{\lambda}\right)^2 \quad \Omega \tag{10-1-21}$$

EXAMPLE 10-1-4 Field Equations, Radiation Power, and Radiation Resistance of a Short-Dipole Antenna

A short-dipole antenna has the following parameters:

Antenna length	$L = 0.0333\ \lambda$
Current magnitude	$I_0 = 5$ A
Operating frequency	$f = 100$ MHz
Radiation range	$r = 100$ m

Calculate:

a) the electric field;
b) the magnetic field;
c) the power density at 100 m;
d) the total radiated power; and
e) the radiation resistance.

Solutions:

a) The electric field is

$$E_\theta = j\frac{2.0944 \times 377 \times 5 \times 0.1}{8\pi \times 100} e^{-j38.16^\circ} \sin\theta$$

$$= j0.157 e^{-j38.16^\circ} \sin\theta$$

b) The magnetic field is

$$H_\phi = j0.416 \times 10^{-3} e^{-j38.16^\circ} \sin\theta$$

c) The power density is

$$P_d = \frac{\eta_0}{8}\left(\frac{I_0 L\beta}{4\pi r}\right)^2 \sin^2\theta = \frac{377}{8}\left(\frac{5 \times 0.1 \times 2.0944}{4\pi \times 100}\right)^2 \sin^2\theta$$

$$= 32.72 \times 10^{-6} \sin^2\theta \quad \text{W/m}^2$$

d) The total radiated power is

$$P = 10\pi^2 I_0^2\left(\frac{L}{\lambda}\right)^2 = 10\pi^2 5^2\left(\frac{0.03\lambda}{\lambda}\right)^2$$

$$= 1.414 \text{ W}$$

e) The radiation resistance is

$$R_r = 20\pi^2\left(\frac{L}{\lambda}\right)^2 = 20\pi^2\left(\frac{0.03\lambda}{\lambda}\right)^2 = 0.113\ \Omega$$

EXAMPLE 10-1-5 Computer Plot for Radiation Resistance of a Short-Dipole Antenna, Based on Eq. (10-1-21):

$$R_r = 20\pi^2\left(\frac{L}{\lambda}\right)^2$$

```
PROGRAM    RESIS2

010C       PROGRAM TO PLOT RADIATION RESISTANCE OF
020C+      OF SHORT-DIPOLE ANTENNA
030C       DIRESI
040C       THE VALUE OF L IS IN LAMBDA
050        DIMENSION LINE(41)
060        REAL L, LAMBDA
070        DO 4 J=1,41
080      4 LINE(J)=1H.
090        PRINT 6, LINE
100      6 FORMAT(/1H ,"    L     RESISTANCE",41A1/
110+       1X,"(LAMBDA)  (OHMS)"/)
120        DO 8 J=1,41
130      8 LINE(J)=1H
140        LINE(2)=1H.
150        PI=3.141593
160        L=0.0
170        DELTAL=0.05
180        LAMBDA=1.0
190        DO 12 I=1,41
200        RR=20.0*(PI*L/LAMBDA)**2
210        N=IFIX(RR*0.05+3.0)
220        LINE(N)=1HX
230        PRINT 10, L,RR,LINE
240     10 FORMAT(2X,F5.2,1X,F9.3,41A1)
250        LINE(N)=1H
260        LINE(2)=1H.
270        L=L+DELTAL
280     12 CONTINUE
```

EXAMPLE 10-1-5 *(continued)*

```
290     STOP
300     END
READY.

PROGRAM   RESIS2

    L     RESISTANCE.......................................
 (LAMBDA)   (OHMS)

   0.00      0.000 .X
    .05       .493 .X
    .10      1.974 .X
    .15      4.441 .X
    .20      7.896 .X
    .25     12.337 .X
    .30     17.765 .X
    .35     24.181 . X
    .40     31.583 . X
    .45     39.972 . X
    .50     49.348 .  X
    .55     59.711 .  X
    .60     71.061 .   X
    .65     83.398 .    X
    .70     96.722 .    X
    .75    111.033 .     X
    .80    126.331 .      X
    .85    142.616 .       X
    .90    159.888 .       X
    .95    178.146 .        X
   1.00    197.392 .         X
   1.05    217.625 .          X
   1.10    238.844 .           X
   1.15    261.051 .             X
   1.20    284.245 .              X
   1.25    308.425 .               X
   1.30    333.593 .                X
   1.35    359.747 .                 X
   1.40    386.889 .                   X
   1.45    415.017 .                    X
   1.50    444.132 .                      X
   1.55    474.235 .                       X
   1.60    505.324 .                         X
   1.65    537.400 .                          X
   1.70    570.463 .                            X
   1.75    604.513 .                              X
   1.80    639.551 .                               X
   1.85    675.575 .                                 X
   1.90    712.586 .                                   X
   1.95    750.584 .                                     X
   2.00    789.569 .

SRU        0.702 UNTS.

RUN COMPLETE.
```

10-2 MONOPOLE ANTENNAS

The monopole antenna is a modification of the dipole antenna in which a plane conducting screen is placed at right angles to and at the bottom of the dipole antenna axis. Thc clcctric current flowing on the conducting screen simulates the missing half of the dipole antenna to create an image. A monopole antenna is commonly utilized instead of a dipole antenna for an omnidirectional, vertically polarized pattern. The monopole is shorter and cheaper to construct, the earth can be used as the conducting ground plane. Figure 10-2-1 shows a monopole antenna and its counterpart center-fed dipole antenna.

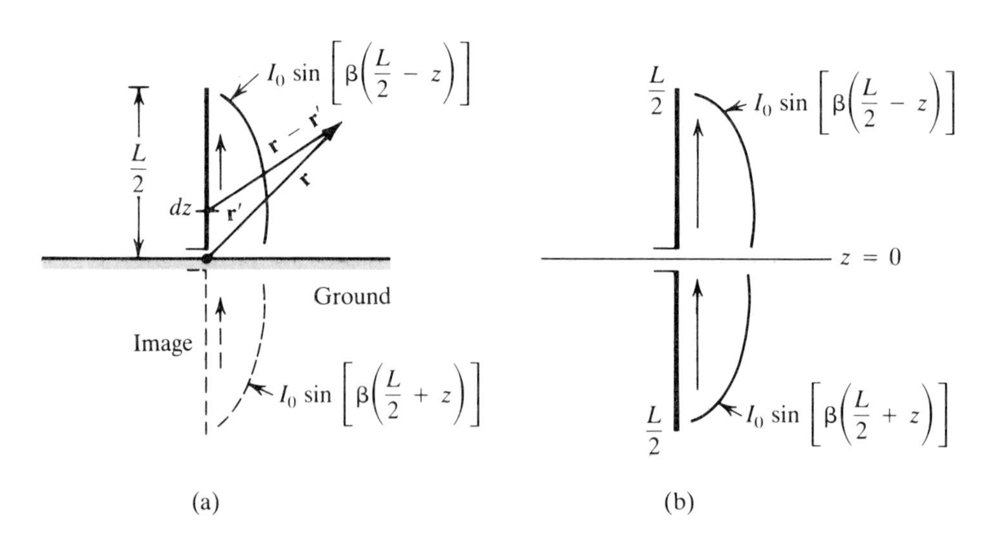

Figure 10-2-1 (a) Monopole and (b) dipole antennas.

Table 10-2-1 Characteristics of certain monopole antennas

Monopole Antennas	Radiation Resistance (Ω)	Radiated Power (W)	Directivity, Numeric (dB)	Effective Aperture (m^2)
Short monopole	$10\pi^2 \frac{L^2}{\lambda}$	$I^2 R_r$	3.0	$\frac{3}{4}\frac{\lambda^2}{\pi}$
$\frac{\lambda}{4}$ monopole	36.59	$36.5I^2$	3.3	$\frac{1.64\lambda^2}{2\pi}$
$\frac{\lambda}{2}$ monopole	100.00	$100I^2$		

Note: The current I is rms.

Experiments have shown that the pattern of a monopole antenna with a large ground plane is the pattern of an equivalent dipole antenna radiating into a single hemisphere. Because the total power radiated by a dipole antenna in an infinite ground is the integration of the power density Poynting vector over a hemisphere, the monopole antenna has only one-half the total power of a dipole antenna and one-half the radiation resistance of an equivalent dipole. Even though the maximum power density is the same, the directivity and the effective area of a monopole antenna are twice as large as those of a dipole antenna. Table 10-2-1 (on the preceding page) shows the characteristics of three types of monopole antennas.

10-3 SLOT ANTENNAS

The simple antenna that we analyze in terms of the field and aperture instead of the current and the wire is a slot antenna. The slot antennas discussed in the following sections are the short-slot antenna, half-wave slot antenna, and cavity-backed slot antenna.

10-3-1 Radiation Fields of Slot Antennas

Booker's principle states that the radiation pattern of a slot antenna is the same as that of a complementary dipole antenna but with the electric and magnetic fields interchanged. Figure 10-3-1 shows a half-wave dipole antenna and a half-wave slot antenna. The electric-field and magnetic-field equations for a half-wave slot antenna are derived from those of a half-wave dipole antenna; that is

$$E_\phi = \frac{jI_0 e^{-j\beta r}}{2\pi r}\left[\frac{\cos\left(\frac{\pi}{2}\cos\theta\right)}{\sin\theta}\right] \tag{10-3-1}$$

and

$$H_\theta = \frac{j60 I_0 e^{-j\beta r}}{r}\left[\frac{\cos\left(\frac{\pi}{2}\cos\theta\right)}{\sin\theta}\right] \tag{10-3-2}$$

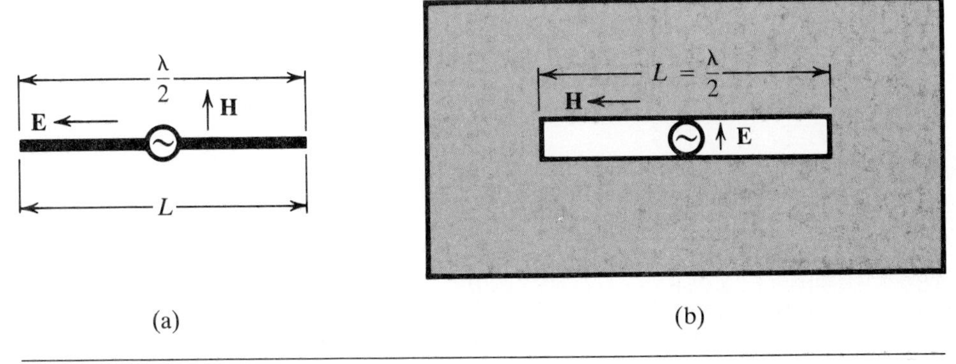

Figure 10-3-1 (a) Slot antenna and (b) complementary dipole antenna.

We can derive the field equations for a short-slot antenna from Eqs. (10-1-18a) and (10-1-18b) for a short-dipole antenna:

$$E_\phi = \frac{j\beta L I_0 e^{-j\beta r}}{8\pi r} \sin\theta \tag{10-3-3}$$

and

$$H_\theta = \frac{j\beta \eta_0 L I_0 e^{-j\beta r}}{8\pi r} \sin\theta \tag{10-3-4}$$

where

$L \ll \lambda/30$;

$r \gg 1/\beta$;

I_0 is the magnitude of the current; and

$\eta_0 = 120\pi$.

EXAMPLE 10-3-1 Computer Plot for the Electric-Field Pattern of a Half-Wave Slot Antenna, Based on Eqs. (10-3-1) and (10-3-2), respectively:

$$E_\phi = \frac{jI_0 e^{-j\beta r}}{2\pi r}\left[\frac{\cos\left(\frac{\pi}{2}\cos\theta\right)}{\sin\theta}\right]$$

$$H_\theta = \frac{j60 I_0 e^{-j\beta r}}{r}\left[\frac{\cos\left(\frac{\pi}{2}\cos\theta\right)}{\sin\theta}\right]$$

```
PROGRAM    SLOT

010C       PROGRAM TO PLOT RELATIVE ELECTRIC OR MAGNETIC
020C+      FIELD PATTERN OF A HALF-WAVE SLOT ANTENNA
030C       SLOT
040        DIMENSION LINE(41)
050        DO 4 J=1,41
060      4 LINE(J)=1H.
070        PRINT 6, LINE
080      6 FORMAT(/1H ,"DEGREES MAGNITUDE",41A1)
090        DO 8 J=1,41
100      8 LINE(J)=1H
110        LINE(20)=1H.
120        PI=3.141593
130        DO 12 J=10,360,10
150        THETA=FLOAT(J)*PI/180.0
160        IF(ABS(SIN(THETA)) .LT. 0.00001) GO TO 12
170        EPHI=COS(PI/2.*COS(THETA))/SIN(THETA)
180        N=IFIX(EPHI*15.0+20.0)
190        LINE(N)=1HX
200        PRINT 10, J, EPHI,LINE
210     10 FORMAT(2X,I4,1X,F10.5,41A1)
```

EXAMPLE 10-3-1 *(continued)*

```
220       LINE(N)=1H
230       LINE(20)=1H.
240   12  CONTINUE
250       STOP
260       END
READY.
RUN

PROGRAM   SLOT

 DEGREES MAGNITUDE.........................................
    10      .13741                 . X
    20      .27656                 .    X
    30      .41779                 .     X
    40      .55894                 .       X
    50      .69464                 .         X
    60      .81650                 .           X
    70      .91426                 .            X
    80      .97789                 .             X
    90     1.00000                 .             X
   100      .97789                 .             X
   110      .91426                 .            X
   120      .81650                 .           X
   130      .69464                 .         X
   140      .55894                 .       X
   150      .41779                 .     X
   160      .27656                 .   X
   170      .13741                 . X
   190     -.13741               X .
   200     -.27656             X   .
   210     -.41779           X     .
   220     -.55894         X       .
   230     -.69464        X        .
   240     -.81650      X          .
   250     -.91426     X           .
   260     -.97789    X            .
   270    -1.00000    X            .
   280     -.97789    X            .
   290     -.91426     X           .
   300     -.81650      X          .
   310     -.69464        X        .
   320     -.55894         X       .
   330     -.41779           X     .
   340     -.27656             X   .
   350     -.13741               X .

SRU       0.762 UNTS.

RUN COMPLETE.
```

10-3-2 Slot Antenna as an Efficient Radiator

The electric field **E** of a center-fed half-wave slot antenna is across the slot, and its magnetic field **H** is parallel to the slot. Conversely, the electric field **E** of a center-fed half-wave complementary dipole antenna is parallel to the dipole, and its magnetic field **H** is normal to the complementary dipole. These conditions are illustrated in Fig. 10-3-2.

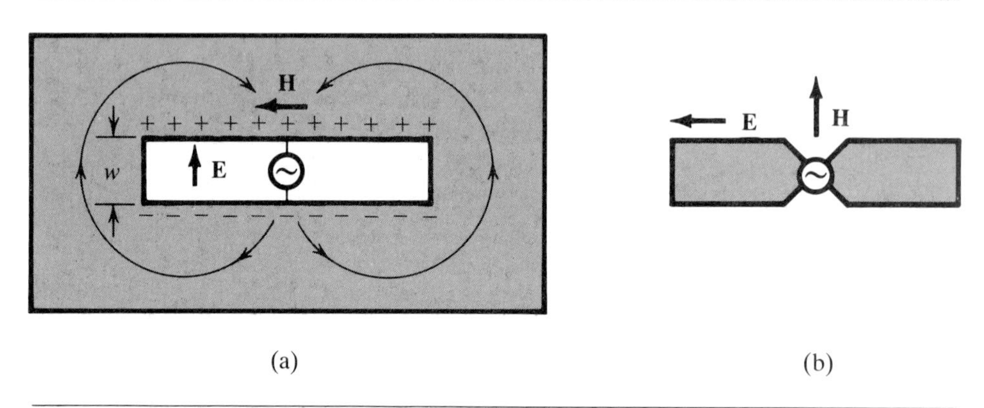

Figure 10-3-2 (a) Slot antenna and (b) complementary dipole antenna.

The slot antenna in an infinite, plane, thin, and perfect conducting sheet of metal is a very efficient radiator. The long edges of the slot are closely spaced ($w \ll \lambda$) and carry electric currents of the opposite phase, so their fields tend to cancel. The two-end edges carry currents in the same phase, but they are too short to radiate efficiently. However, in an infinite metal sheet, although the width of the slot is small, the electric currents are not confined to the edges of the slot but spread out over the sheet. If the sheet is finite, the impedance of a slot antenna is substantially the same as a dipole antenna, provided that the edge of the sheet is at least one wavelength from the slot.

10-3-3 Impedance, Directivity, and Effective Aperture of a Slot Antenna

The impedance of a slot antenna can be determined in terms of the impedance of its complementary dipole antenna. A slot antenna and its complementary dipole antenna are shown in Fig. 10-3-3.

We obtain the voltage across the gap of a complementary dipole antenna by integrating $\mathbf{E}_d \cdot d\ell$ along a line of the field **E** between points a and c, or

$$V_d = -\int_{abc} \mathbf{E}_d \cdot d\ell \tag{10-3-5}$$

The current into one arm of the dipole antenna is

$$I_d = \oint \mathbf{H}_d \cdot d\ell = 2\int_{efg} \mathbf{H}_d \cdot d\ell \tag{10-3-6}$$

where the factor 2 indicates that only one-half the closed line is integrated.

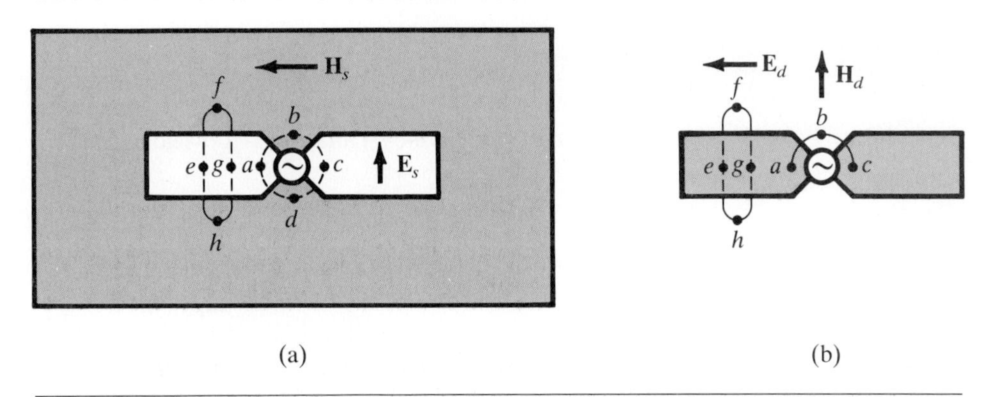

Figure 10-3-3 (a) Slot antenna and (b) complementary dipole antenna.

Next, we can express the impedance of the complementary dipole as

$$Z_d = \frac{V_d}{I_d} = \frac{-\int_{abc} \mathbf{E}_d \cdot d\ell}{2\int_{efg} \mathbf{H}_d \cdot d\ell} \tag{10-3-7}$$

Similarly, we can obtain the voltage across the slot or a slot antenna:

$$V_s = -\int_{gfe} \mathbf{E}_s \cdot d\ell = \int_{efg} \mathbf{E}_s \cdot d\ell \tag{10-3-8}$$

The current of the slot antenna is

$$I_s = \oint_{abcd} \mathbf{H}_s \cdot d\ell = 2\int_{abc} \mathbf{H}_s \cdot d\ell \tag{10-3-9}$$

Therefore we can write the impedance of a slot antenna as

$$Z_s = \frac{V_s}{I_s} = -\frac{\int_{efg} \mathbf{E}_s \cdot d\ell}{2\int_{abc} \mathbf{H}_s \cdot d\ell} \tag{10-3-10}$$

Because the electric and magnetic fields for a slot antenna and its complementary dipole antenna are interchangeable, the corresponding field intensities apparently are related by

$$\mathbf{E}_s = k_1 \mathbf{H}_d \tag{10-3-11}$$

and

$$\mathbf{H}_s = k_2 \mathbf{E}_d \tag{10-3-12}$$

where k_1 and k_2 are constants. At the far field in free space the electric and magnetic intensities are uniform plane waves. That is,

$$\frac{E_\phi}{H_\theta} = \frac{E_s}{H_s} = \eta_0 \tag{10-3-13}$$

and

$$\frac{E_\theta}{H_\phi} = \frac{E_d}{H_d} = -\eta_0 \tag{10-3-14}$$

Substituting Eqs. (10-3-11)–(10-3-14) into Eqs. (10-3-7) and (10-3-10) yields

$$\sqrt{Z_s Z_d} = \frac{\eta_0}{2} \tag{10-3-15}$$

Therefore the geometric mean of the driving-point impedances of a slot antenna and its complementary dipole antenna is equal to one-half the intrinsic impedance of free space.

Alternatively, we can express the impedance of a slot antenna in terms of the impedance of its complementary dipole, or

$$Z_s = \frac{\eta_0^2}{4Z_d} \tag{10-3-16}$$

Insofar as the power-density Poynting vector is concerned, it always remains the same for both the slot antenna and its complementary dipole antenna, because the electric and magnetic field intensities are interchangeable between them. Therefore the directivity and the effective aperture of a short-slot antenna and a half-wave slot antenna are the same as that of their complementary dipole antennas. Table 10-3-1 shows the characteristics of the types of commonly used slot antennas.

Table 10-3-1 Characteristics of certain slot antennas

Slot antenna	Directivity (Numeric)	Effective Aperture (m^2)	Radiation Resistance (Ω)
Short slot	1.5	$\frac{3\lambda^2}{8\pi}$	$\frac{\eta_0}{80\pi^2}\frac{\lambda^2}{L}$
Half-wave slot	1.65	$\frac{\lambda^2}{4\pi}(1.64)$	500
Long slot	1	$\frac{\lambda^2}{4\pi}$	$\frac{\eta_0^2}{4Z_d}$

EXAMPLE 10-3-2 Short-Slot Antenna

A short-slot antenna has the following parameters:

Slot antenna length	$L = 0.03\lambda$
Current magnitude	$I_0 = 4$ A
Operating frequency	$f = 300$ MHz
Radiation range	$r = 100$ m

Calculate:

a) the electric field;
b) the magnetic field;
c) the radiation resistance; and
d) the directivity in dB.

Solutions:

a) The electric field is

$$E_\theta = j\frac{\beta L I_0}{8\pi r}e^{-j\beta r}\sin\theta = j\frac{2\pi \times 0.03 \times 4}{8\pi(100)}e^{-j200\pi}\sin\theta$$

$$= j0.3 \times 10^{-3}e^{-j200\pi}\sin\theta$$

b) The magnetic field is

$$H_\phi = j\frac{\beta L I_0 \eta_0}{8\pi r}e^{-j\beta r}\sin\theta$$

$$= j0.3 \times 10^{-3} \times 377e^{-j200\pi}\sin\theta$$

$$= j0.113e^{-j200\pi}\sin\theta$$

c) The radiation resistance is

$$R_r = \frac{\eta_0}{80\pi^2}\frac{\lambda^2}{L} = \frac{377 \times 1^2}{80\pi^2 \times 0.03}$$

$$= 16\ \Omega$$

d) The directivity is

$$D = 1.6$$

$$= 2.04\ \text{dB}$$

10-3-4 Cavity-Backed Slot Antenna

We usually want a slot antenna to radiate on only one side of the sheet. We can do this by boxing the slot antenna in on the side from which we want no radiation. If done with a hemisphere of radius about a quarter-wavelength concentric with the center of a half-wave slot, no appreciable shunt susceptance appears across the center of the slot. To avoid throwing reactance across the slot, the reflecting sheet has to be about at least a quarter-wavelength behind the slotted sheet. Reduction of this separation creates inductive reactance across the slot, which could be balanced, however, by lengthening the slot beyond the usual half-wavelength.

Because the cavity-backed slot antenna radiates into one hemisphere only, the radiation conductance is one-half the radiation conductance of a sphere. Thus the terminal resistance of the center of a half-wave slot antenna with a cavity in a large sheet is about 1000 Ω. When a coaxial line is connected to the slot antenna, an off-center feed is required to provide a better impedance match. The connection should be shifted to the point of $\lambda/40$ distance from either end of the half-wave slot antenna. Moreover, because the cavity-backed slot antenna radiates only on one side, its directivity and effective aperture are twice those of a slot antenna that is not backed by a cavity.

We can best describe the resonant frequency of a backed cavity by reference to three cases:

1. A slot antenna backed by a rectangular cavity, as shown in Fig. 10-3-4. The resonant frequency for the TE and TM modes, from Eq. (5-7-4), is

$$f_r = \frac{1}{2\sqrt{\mu\varepsilon}}\sqrt{\left(\frac{m}{a}\right)^2 + \left(\frac{n}{b}\right)^2 + \left(\frac{p}{d}\right)^2} \tag{10-3-17}$$

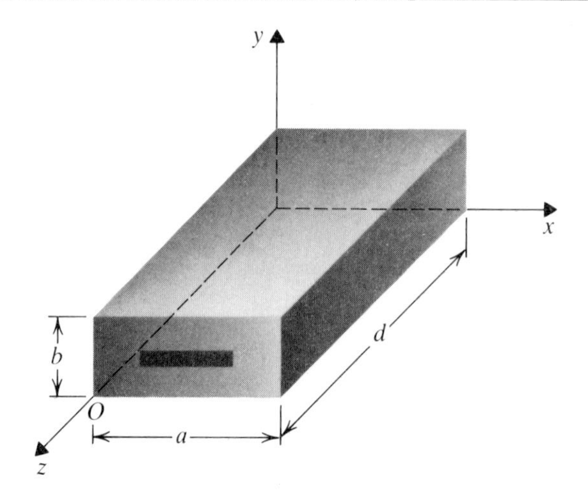

Figure 10-3-4 A slot antenna backed by rectangular cavity.

2. A slot antenna backed by a circular cavity, as shown in Fig. 10-3-5. The resonant frequencies from Eqs. (6-8-5) and (6-8-6), are

$$f_r = \frac{1}{2\pi\sqrt{\mu\varepsilon}}\sqrt{\left(\frac{X'np}{a}\right)^2 + \left(\frac{q\pi}{d}\right)^2} \quad \text{for TE modes} \qquad (10\text{-}3\text{-}18)$$

where

$n = 0, 1, 2, 3, \ldots;$
$p = 1, 2, 3, 4, \ldots;$ and
$q = 1, 2, 3, 4, \ldots$

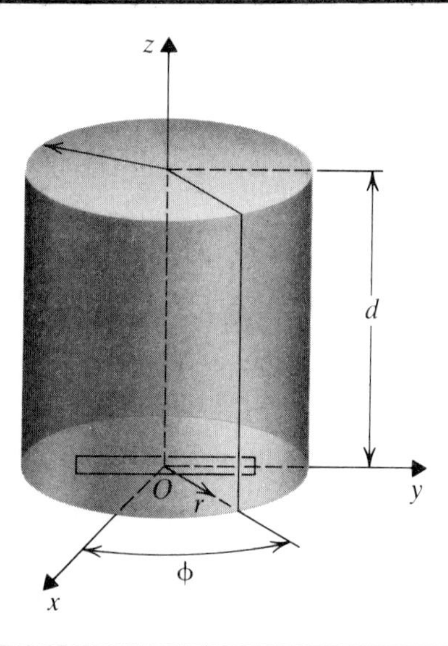

Figure 10-3-5 Slot antenna backed by circular cavity.

and

$$f_r = \frac{1}{2\pi\sqrt{\mu\varepsilon}}\sqrt{\left(\frac{X_{np}}{2}\right)^2 + \left(\frac{g}{d}\right)^2} \quad \text{for TM modes} \qquad (10\text{-}3\text{-}19)$$

where

$n = 0, 1, 2, 3, \ldots;$
$p = 1, 2, 3, 4, \ldots;$ and
$q = 0, 1, 2, 3, \ldots$

3. A slot antenna backed by a semicircular cavity, as shown in Fig. 10-3-6. The wave function of the TM modes in a semicircular cavity is

$$\psi_{\text{TM}} = J_n\left(\frac{X_{np}}{a}r\right)\sin(n\phi)\cos\left(\frac{q\pi}{d}z\right) \qquad (10\text{-}3\text{-}20)$$

where

$n = 1, 2, 3, 4, \ldots$, is the number of the periodicity in the ϕ direction;

$p = 1, 2, 3, 4, \ldots$, is the number of zeros of the field in the radial direction;

$q = 0, 1, 2, 3, \ldots$, is the number of the half-wave in the axial direction; and

J_n is the Bessel function of the first kind.

The resonant frequency for the TM modes in a semicircular cavity is

$$f_r = \frac{1}{2\pi\sqrt{\mu\varepsilon}}\sqrt{\left(\frac{X_{np}}{a}\right)^2 + \left(\frac{q\pi}{d}\right)^2} \qquad (10\text{-}3\text{-}21)$$

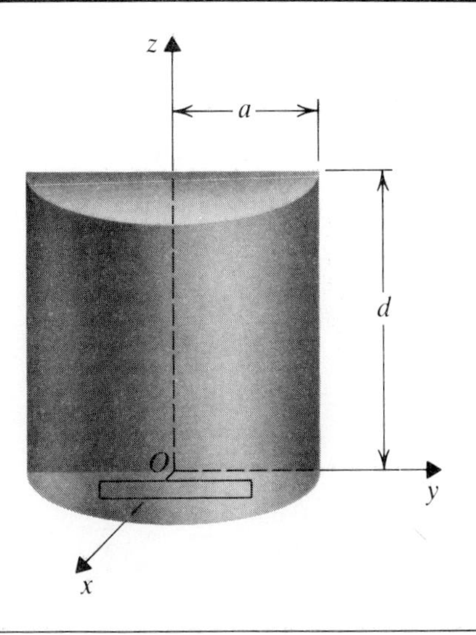

Figure 10-3-6 Slot antenna backed by semicircular cavity.

The wave function of the TE modes in a semicircular cavity is

$$\psi_{\text{TE}} = J_n\left(\frac{X'_{np}}{a} r\right) \cos(n\phi) \sin\left(\frac{q\pi}{d} z\right) \tag{10-3-22}$$

where

$n = 0, 1, 2, 3, \ldots;$
$p = 1, 2, 3, 4, \ldots;$ and
$q = 1, 2, 3, 4, \ldots$

The resonant frequency for the TE modes in a semicircular cavity is

$$f_r = \frac{1}{2\pi\sqrt{\mu\varepsilon}} \sqrt{\left(\frac{X'_{np}}{a}\right)^2 + \left(\frac{q\pi}{d}\right)^2} \tag{10-3-23}$$

Interestingly, for $d < a$, the TM_{110} mode is dominant, whereas for $d > a$, the TE_{111} mode is dominant.

EXAMPLE 10-3-3 Slot Antenna with a Cavity

A slot antenna backed by an air-filled semicircular cavity has the following parameters:

Cavity radius	$a = 10$ cm
Cavity height	$d = 5$ cm

Determine:

a) the dominant mode; and
b) the resonant frequency.

Solutions:

a) The dominant mode is TM_{110}, because $a > d$.
b) The resonant frequency is

$$\begin{aligned} f_r &= \frac{1}{2\pi\sqrt{\mu_0\varepsilon_0}} \sqrt{\left(\frac{X_{np}}{a}\right)^2 + \left(\frac{q\pi}{d}\right)^2} \\ &= \frac{3 \times 10^8}{2\pi} \sqrt{\left(\frac{3.832}{10 \times 10^{-2}}\right)^2 + \left(\frac{0 \times \pi}{d}\right)^2} \\ &= 1.83 \text{ GHz} \end{aligned}$$

Problems

10-1 A short-dipole antenna has the following parameters:

Antenna length	$L = 0.03\lambda$
Current magnitude	$I_0 = 4$ A
Operating frequency	$f = 1$ GHz
Radiation range	$r = 50$ m

Calculate:

a) the electric field;
b) the magnitude field;
c) the total radiated power; and
d) the radiation resistance.

10-2 A short-slot antenna has the following parameters:

Slot antenna length	$L = 0.03\lambda$
Current magnitude	$I_0 = 5$ A
Operating frequency	$f = 3$ GHz
Radiation range	$r = 100$ m

Determine

a) the electric field;
b) the magnetic field;
c) the radiation resistance; and
d) the directivity.

10-3 An airfilled rectangular cavity-backed slot antenna has dimensions of $a = 1.016$ cm, $b = 2.286$ cm, and $d = 6.858$ cm. Calculate

a) the resonant frequency for the TE dominant mode; and
b) the resonant frequency for the TM dominant mode.

10-4 An airfilled circular cavity-backed slot antenna has a radius of 2.024 cm and a length of 10.120 cm. Determine

a) which mode is the dominant mode; and
b) the resonant frequency of the dominant mode.

10-5 A semicircular cavity-backed slot antenna has a radius of 2.286 cm and a height of 4.572 cm.

a) Calculate the resonant frequency for the dominant mode TE_{111}, if the cavity is air-filled.
b) Repeat (a) for a dielectric-filled cavity having $\varepsilon_r = 2.25$.

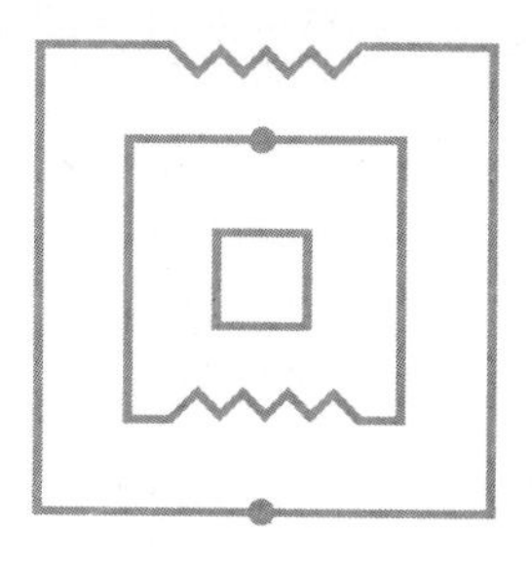

Chapter 11

Broadband and Array Antennas

11-0 INTRODUCTION

In Chapters 9 and 10 we discussed antenna parameters, dipole antennas, and slot antennas in detail. However, in order to have both high gain and wide bandwidth, antenna design must be different from the principles discussed in those two chapters. In electronic network theory the gain–bandwidth product is constant. That means realizing both high gain and broadband in a simple circuit is very difficult—and antennas are no exception. In this chapter, we describe the design principles for broadband and high-gain antennas [1].

11-1 LOG-PERIODIC ANTENNAS

11-1-1 Principles of Log-Periodic Antennas

If the electrical performance of a device is periodic because of a particular scaling of its dimensions (by some ratio τ of distances), it will have the same properties of performance at frequency f_1 and at all other frequencies

$$f_1\left(\frac{1}{\tau}\right), f_1\left(\frac{1}{\tau}\right)^2, \ldots, f_1\left(\frac{1}{\tau}\right)^n$$

Consequently, the impedance or any other characteristic is a periodic function with the period being the logarithm of the frequency, or log $(1/\tau)$. That is, the frequencies at which the performance will be periodically identical are related by the equation:

$$f_n = f_{n+1}\tau \tag{11-1-1}$$

or

$$\log(f_{n+1}) = \log f_n + \log\left(\frac{1}{\tau}\right) \tag{11-1-2}$$

Equation (11-1-2) shows that performance is a periodic function of the logarithm of the frequency. In other words, the frequencies at which the performance is identical are spaced equally when plotted on a log scale. The devices constructed according to this principle are called log-periodic antennas. [1]

11-1-2 Design and Construction of Log-Periodic Antennas

The structure of a log-periodic antenna and its equivalent circuit are shown in Fig. 11-1-1. You can see that the ratio τ is defined as the ratio of successive distances between the apex and the dipole elements. That is,

$$\tau = \frac{X_{n+1}}{X_n} \tag{11-1-3}$$

The ratio σ is defined as the ratio of the dipole element spacing to twice the length of the next-larger element. That is,

$$\sigma = \frac{d_n}{2L_n} \tag{11-1-4}$$

We can express the half-angle α subtended at the apex in terms of τ and σ as

$$\tan\alpha = \frac{L_n/2}{X_n} = \frac{d_n/2\sigma}{2X_n} = \frac{X_n - X_{n+1}}{4\sigma X_n} = \frac{X_n - \tau X_n}{4\sigma X_n} = \frac{1-\tau}{4\sigma}$$

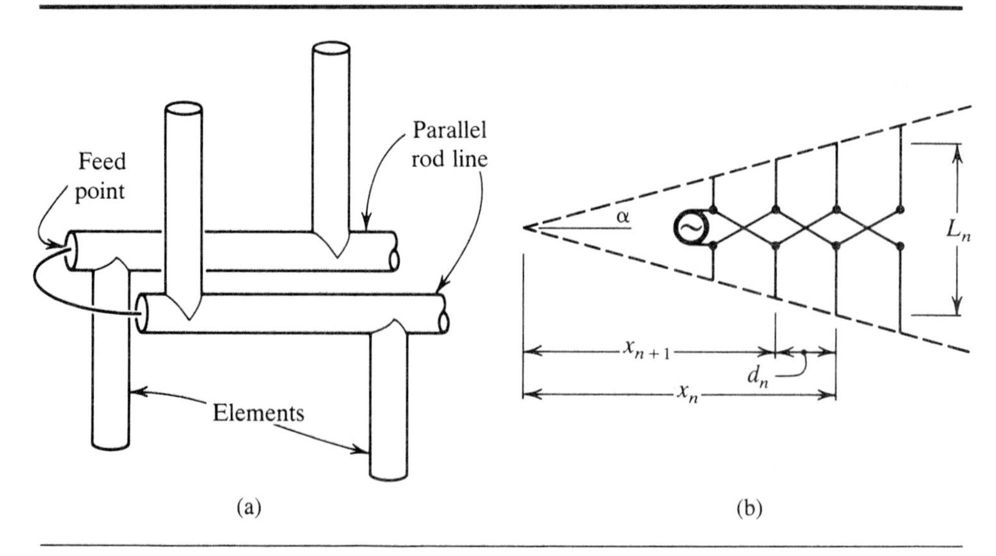

Figure 11-1-1 (a) Log-periodic dipole structure and (b) equivalent circuit.

SOURCE: COURTESY OF R. L. CARREL. [2]

Hence,

$$\alpha = \tan^{-1}\left(\frac{1-\tau}{4\sigma}\right) \tag{11-1-5}$$

Recall that the characteristic impedance of a parallel-rod line (often called the antenna boom) is

$$Z_0 = 120 \cosh^{-1}\left(\frac{S}{D}\right) \tag{11-1-6}$$

where S is the center-to-center spacing between the two booms, and D is the diameter of the rod. Alternatively, we can express the spacing S in terms of the diameter D, or

$$S = D \cosh \frac{Z_0}{120} \tag{11-1-7}$$

When designing a log-periodic antenna, the two decisive factors that we have to consider are the antenna gain and its input impedance. Figure 11-1-2 shows a plot of the contours of constant directivity versus τ and σ. Note that gains in the range of from 7.5 to 12 dB over isotropic antenna can be obtained if we choose the parameters properly.

The input impedance R_0 at the feed point for a given σ, τ, and element length-to-diameter ratio depends on the characteristic impedance Z_0 of the unloaded parallel-rod feeder line. The usual procedure is to work in terms of an average characteristic impedance for the elements. An approximate formula for the average characteristic impedance of the elements is

$$Z_{avg} = 120\left(\ln\frac{h}{a} - 2.25\right) \tag{11-1-8}$$

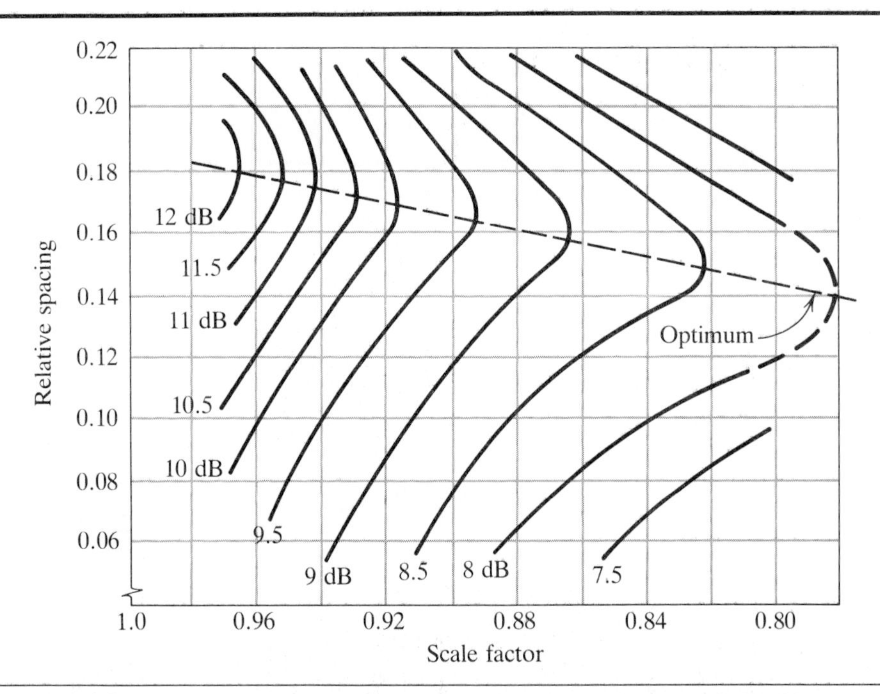

Figure 11-1-2 Contours of constant directivity versus τ and σ, for log-periodic dipole arrays.

SOURCE: COURTESY OF R. L. CARREL. [2]

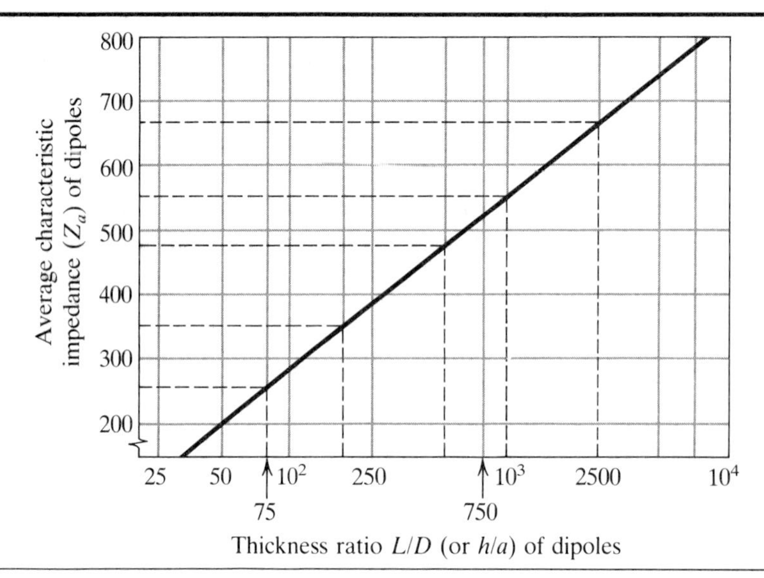

Figure 11-1-3 Average characteristic impedance, Z_{avg} of dipoles, as a function of length-to-diameter ratio.

SOURCE: COURTESY OF R. L. CARREL. [2]

where h/a is the half-length-to-radius ratio of the element antenna. When the half-length-to-radius ratio is given, we can find the loaded average characteristic impedance Z_{avg} from Fig. 11-1-3.

The loading produced by the element antennas depends on their spacing, as shown in Fig. 11-1-4. The mean relative spacing σ' is related to σ and τ by

$$\sigma' = \frac{\sigma}{\sqrt{\tau}} \tag{11-1-9}$$

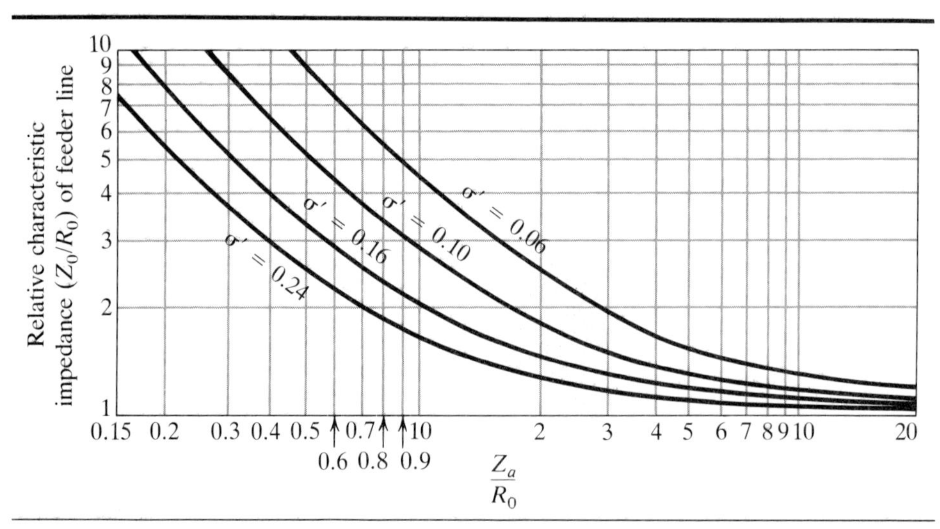

Figure 11-1-4 Relative characteristic impedance of feeder line, Z_0/R_0, as a function of relative impedance of element antennas.

SOURCE: COURTESY OF R. L. CARREL. [2]

When the relative characteristic impedance Z_{avg}/R_0 of the loaded element antennas and the mean relative spacing σ' are known, we can find the unloaded characteristic impedance Z_0 of the elements from Fig. 11-1-4.

Another factor that has to be considered is the bandwidth of the log-periodic antenna. You may be asked to design for a slightly larger bandwidth than actually required. This larger bandwidth B_s is related to the required bandwidth B by

$$B_s = BB_{ar} \tag{11-1-10}$$

where B_{ar} is the bandwidth at the active region. The factor B_{ar} is a function of σ and α, as shown in Fig. 11-1-5. Taking the longest element to be one half-wavelength at the low-frequency end of the band B_s, we can calculate the distance L (or d) between the shortest element and the longest element (see Fig. 11-1-1) by using the following

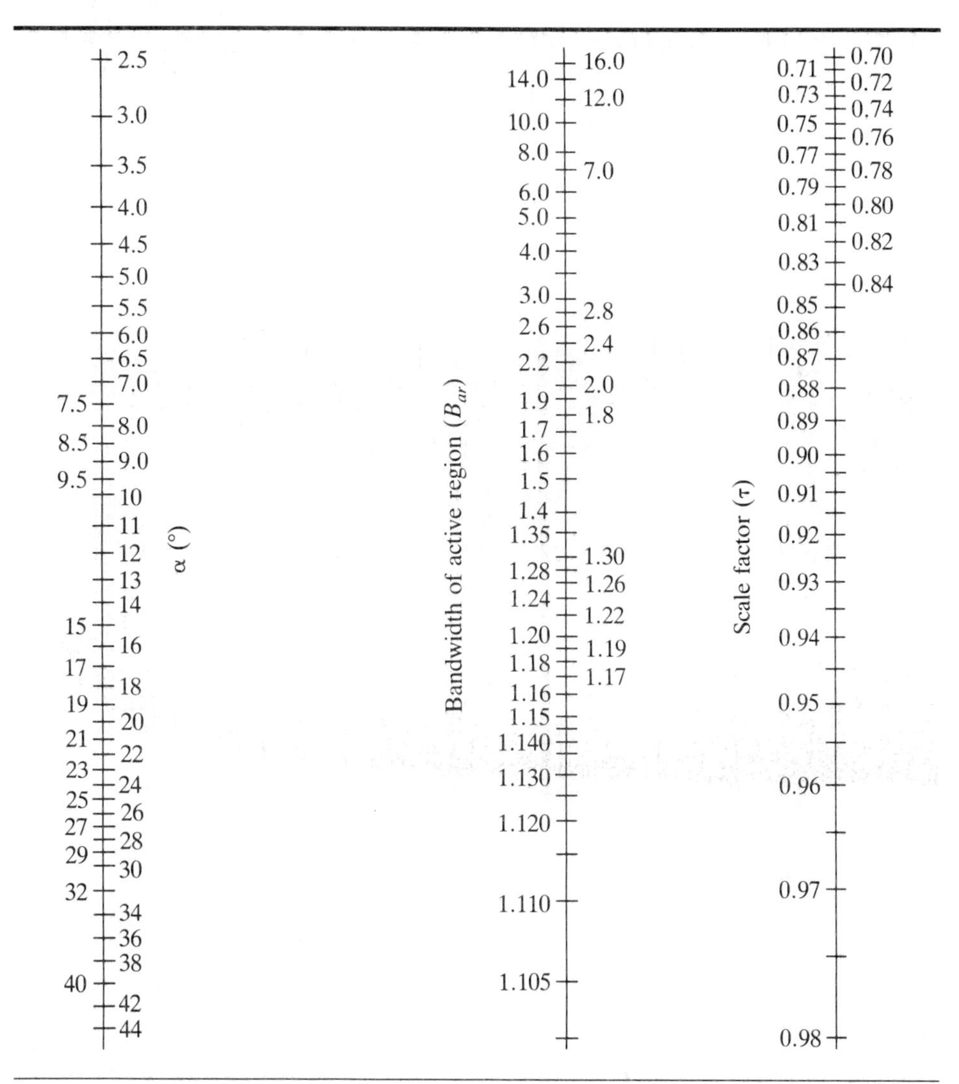

Figure 11-1-5 Nomograph for $B_{ar} = 1.1 + 7.7(1 - \tau)^2 \cot \alpha$.

SOURCE: COURTESY OF R. L. CARREL. [2]

equation:

$$L = \frac{\lambda_{\max}}{4}\left(1 - \frac{1}{B_s}\right)\cot\alpha \tag{11-1-11}$$

Because the ratio of successive element distances is

$$\frac{X_n}{X_1} = \tau^{N-1}$$

the number of element antennas becomes

$$N = 1 + \frac{\log B_s}{\log\left(\frac{1}{\tau}\right)} \tag{11-1-12}$$

11-1-3 Performance and Characteristics

Experiments have demonstrated that for a certain range of values for τ and α, a structure was indeed a broadband log-periodic antenna with a unidirectional pattern. Experiments have also shown that most of the radiation power was coming from those

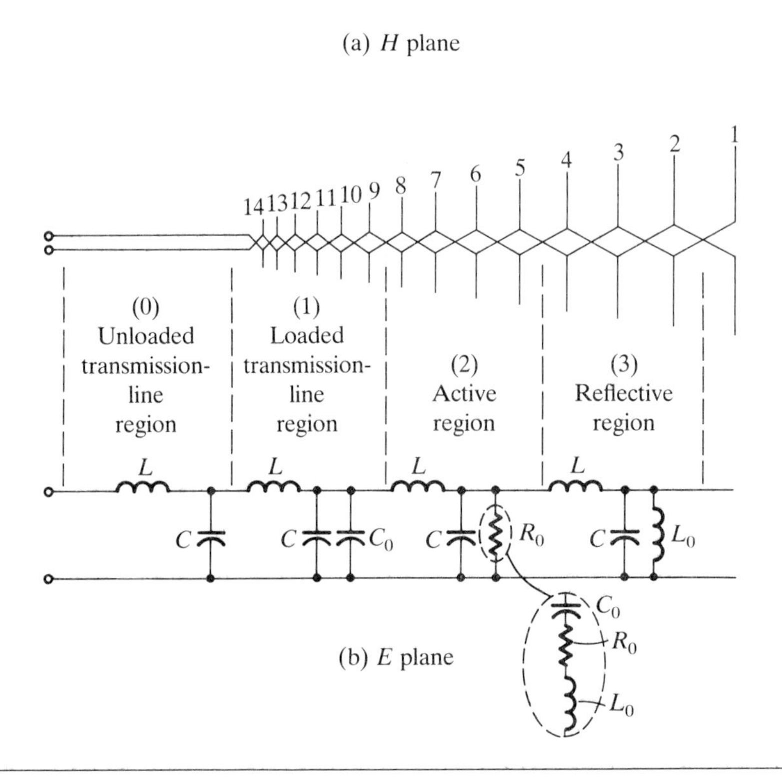

Figure 11-1-6 Log-periodic dipole array, showing main regions of operation.

dipole elements that were in the vicinity of one half-wavelength long and that the currents and voltages at the large end of the antenna were negligible within the operating band of frequencies. The clear conclusion was that the operating range of frequencies was limited on the high side by the frequencies corresponding to the size of the shortest elements and on the low side by the frequencies at which the largest dipole element is about one half-wavelength long.

Figure 11-1-6 shows the three main regions of performance for a 14-element log-periodic antenna:

> *Transmission-line region.* In the transmission-line region, the lengths of the element antennas are much shorter than one half-wavelength, so the elements present a relatively high capacitive impedance. In the unloaded-line region, region (0), the series inductance and shunt capacitance per unit length are shown as L and C, respectively. In the loaded-line region, region (1), the transmission line is loaded by a capacitance C_a per unit length. The amplitude of voltage along the line is almost constant, and the phase change between element positions increases gradually from about 20° to 30°, as shown in Figs. 11-1-7 and 11-1-8.

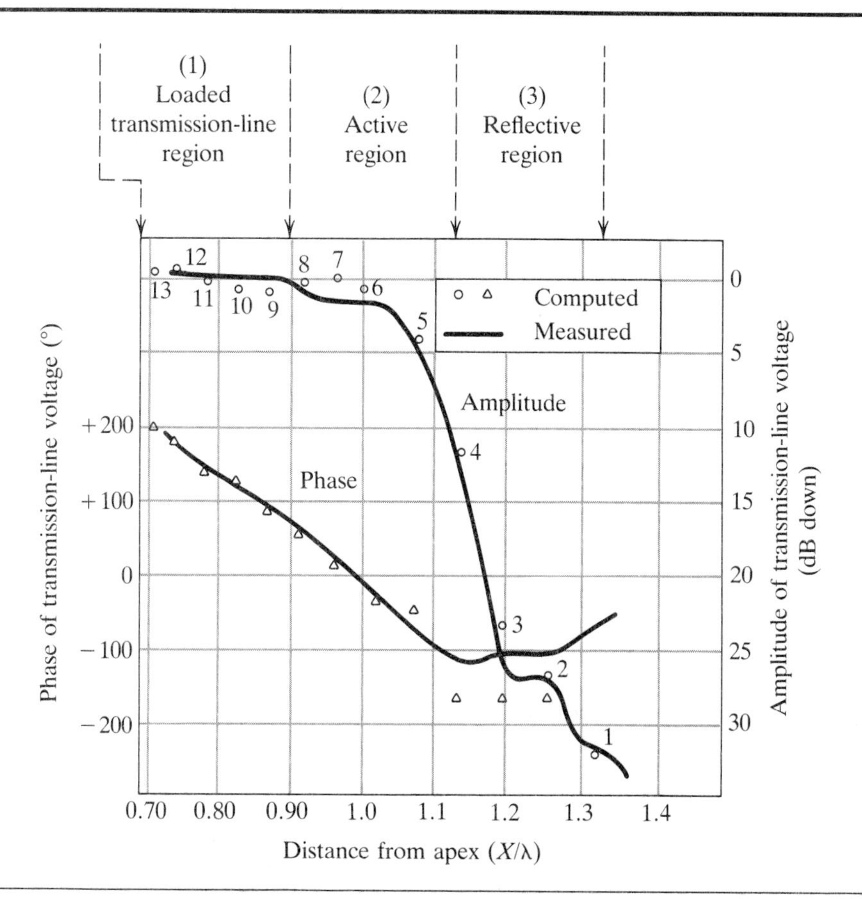

Figure 11-1-7 Amplitude and phase of transmission-line voltages along a particular log-periodic antenna in which element number 4 has length $L = \lambda/2$.

SOURCE: COURTESY OF R. L. CARREL. [2]

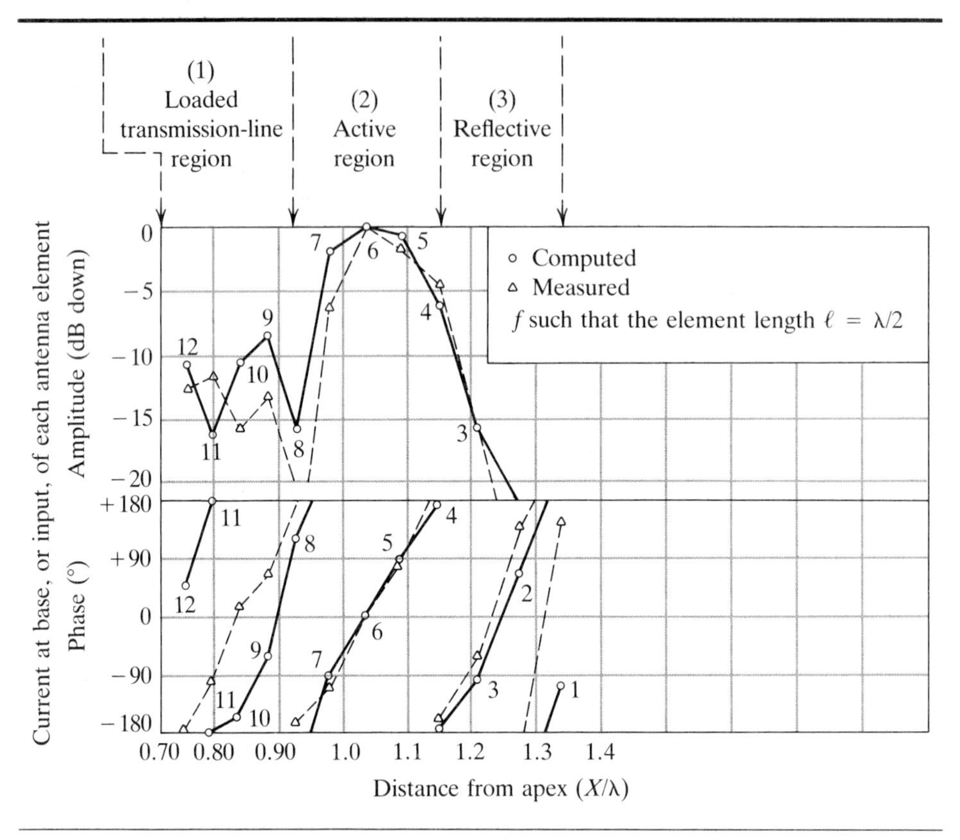

Figure 11-1-8 Element currents for the antenna shown in Fig. 11-1-7.

SOURCE: COURTESY OF R. L. CARREL. [2]

Active region. In the active region, the lengths of the element antennas are almost equal to one half-wavelength (resonant length), so the element impedance is almost pure resistance. The amplitude of transmission-line voltage drops rapidly because power is absorbed by the highly radiating elements; element current is large and has a phase shift of about 90° between adjacent elements.

Reflection region. In the reflection region, the lengths of the element antennas are greater than one half-wavelength, so the element impedance becomes inductive and the element current lags the base voltage. The amplitude of the transmission-line voltage drops to a very low level, as does the element current. The reflection region is also called the unexcited region.

The radiation pattern can be determined from the element currents and positions. The direction of the maximum in the radiation pattern is off the small end of the structure, which is opposite to the direction in which the electric and magnetic waves propagate down the parallel-rod feeder. Figure 11-1-9 shows the field radiation patterns for the log-periodic antenna represented in Fig. 11-1-7.

As the frequency varies within the operating frequency band, the input impedance at the feed point lies on and within a small circle centered on the real axis of the Smith Chart. The mean resistance is a function of σ, as shown in Fig. 11-1-10. The input resistance is also a function of the length-to-diameter ratio of the element antennas, as indicated in Fig. 11-1-11.

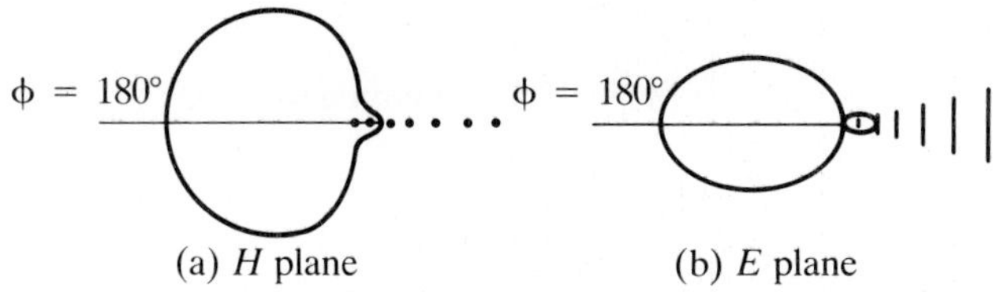

$\phi = 180°$ $\phi = 180°$

(a) H plane (b) E plane

Figure 11-1-9 Field patterns of a log-periodic antenna.

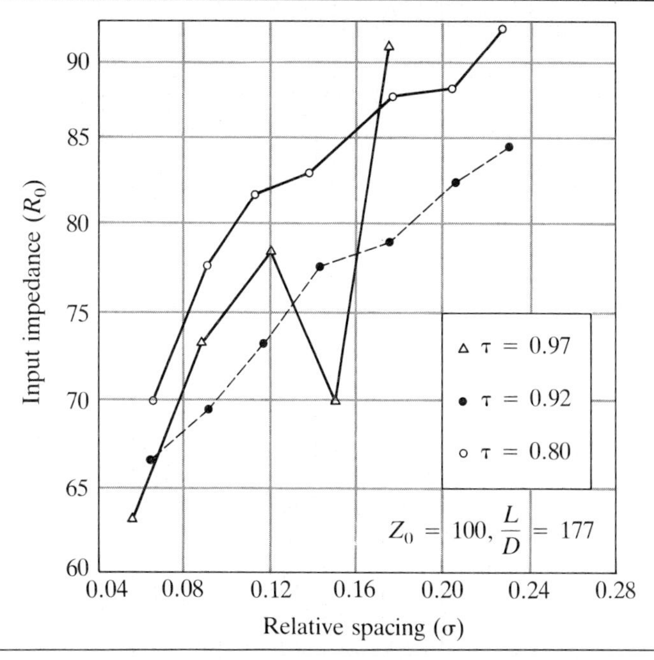

Figure 11-1-10 Input impedance, R_0, as a function of relative spacing σ, with τ as a parameter.

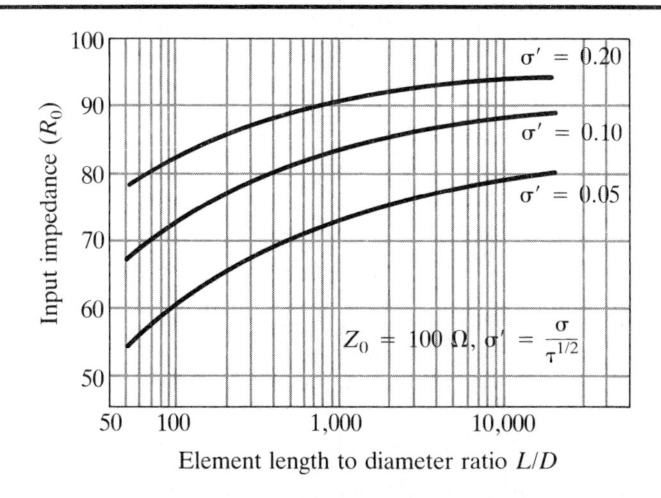

Figure 11-1-11 Input impedance, R_0, as a function of element length-to-diameter ratio, L/D, with Z_0 and σ' as parameters.

SOURCE: COURTESY OF R. L. CARREL. [2]

EXAMPLE 11-1-1 Design of a VHF Log-Periodic Antenna

The frequency range is 50–250 MHz. The antenna gain is to be 11 dB, and the input impedance is to be 75 Ω.

Solution:

First, determine τ and σ from Fig. 11-1-2 for the antenna gain of 11 dB:

$$\tau = 0.942 \quad \text{and} \quad \sigma = 0.174$$

Then calculate the angle α from Eq. (11-1-5):

$$\alpha = \tan^{-1}\left(\frac{1-\tau}{4\sigma}\right) \doteq 5^\circ$$

and determine B_{ar} from Fig. 11-1-5:

$$B_{ar} = 1.4$$

Next, calculate the larger bandwidth B_s from Eq. (11-1-10):

$$\begin{aligned} B_s &= BB_{ar} \\ &= (5)(1.4) \\ &= 7 \end{aligned}$$

and the number of element antennas from Eq. (11-1-12):

$$\begin{aligned} N &= 1 + \frac{\log B_s}{\log(1/\tau)} \\ &= 1 + \frac{0.846}{0.0254} \\ &= 35 \end{aligned}$$

Let's assume that 1-in. aluminum tubing is to be used. Thus the largest element would have a half-length of about 10 ft, and the half-length-to-radius ratio is

$$\begin{aligned} \frac{h}{a} &= \frac{10 \times 12}{0.50} \\ &= 240 \end{aligned}$$

The average characteristic impedance, from Fig. 11-1-3, is

$$Z_{avg} = 360\ \Omega$$

The mean relative spacing is

$$\begin{aligned} \sigma &= \frac{\sigma}{(\tau)^{1/2}} = \frac{0.174}{(0.942)^{1/2}} \\ &= 0.18 \end{aligned}$$

The characteristic impedance of the unloaded booms is found from Fig. 11-1-4:

$$\begin{aligned} Z_0 &= 1.1 \times 75 \\ &= 83\ \Omega \end{aligned}$$

The center-to-center spacing between the two booms is found from Fig. (11-1-7):

$$S = D \cosh\left(\frac{Z_0}{120}\right) = 2 \times 1.25$$

$$= 2.5 \text{ in.}$$

where D is taken to be 2 in.

11-2 PHASED-ARRAY ANTENNAS

Often, a single antenna cannot supply the high gain and high power demanded by a system. The alternative approach is to use several interconnected antennas arranged in space to produce a highly directional radiation pattern. This type of configuration of multiple radiating elements is referred to as an array antenna or, simply, as an array. Many small antennas can be used in an array to obtain a level of performance similar to that of a single large antenna. The trade-off involved is the set of mechanical problems associated with a single large antenna versus the set of electrical problems encountered in feeding several small antennas.

Advancements in solid-state technology have improved the quality and reduced the cost of the feed network required for array excitation. Arrays offer the unique capability of scanning the main beam electronically. The radiation pattern can be scanned through space by changing the phase of the exciting currents in each element antenna of the array. This array is then called a *phased array*. Phased arrays have many applications, particularly in radar and satellite communications.

Arrays can be constructed in many geometric configurations. The following are the five most common types:

1. Linear array, which is the most elementary type of array. In a linear array, the centers of the array elements lie along a straight line. The elements may be equally or unequally spaced.
2. Planar array, in which the centers of the array elements are located in a plane.
3. Circular array, in which the centers of the elements are located on a circle.
4. Conformal array, in which the locations of the elements must conform to some specific nonplanar surface, such as on an aircraft or a missle.
5. Adaptive array, which self-adapts to various incoming signal conditions, so as to maximize the signal from a particular source or to null out interfering signals. A self-phasing adaptive array, for example, brings all the signals received by various antenna elements from a particular source into phase. The basic principle involved is the mixing of all the incoming signals with separate local oscillators that have phases adjusted so that all the signals at the intermediate frequency are in phase and may be added together in a summing amplifier. All of this can be achieved by using phase-lock loop principles. Each local oscillator is a voltage-controlled oscillator (VCO) with an instantaneous phase that is controlled by an applied voltage. The phase of the mixed signal at the intermediate frequency is compared with that of a fixed-reference oscillator operating at the intermediate frequency.

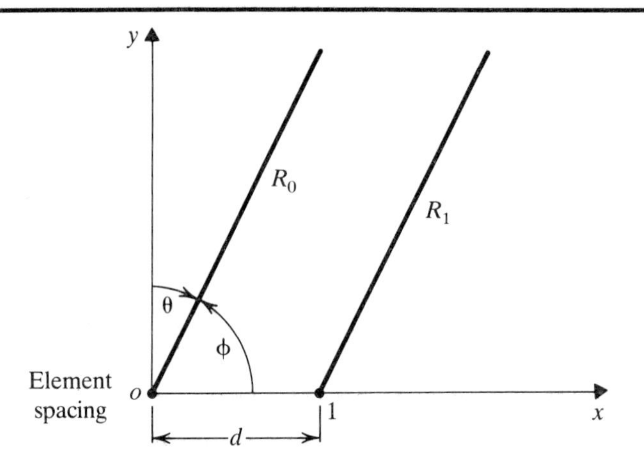

Figure 11-2-1 Two-element array antenna.

11-2-1 Two-Element Arrays

The simplest antenna array is a two-element array that consists of two identical element antennas, spaced a distance d apart, as shown in Fig. 11-2-1.

We assume that the two antennas are aligned along the x axis, a distance d apart, and the electric field of the radiation pattern for the antenna is in the θ direction. The two antennas are excited with a current of equal amplitude, but the phase in antenna 1 is leading that in antenna 0 by an angle δ. We can write the expressions for the electric fields of the radiation pattern, from Eq. (9-2-6), as

$$E_0 = E_m f(\theta, \phi)\frac{1}{R_0}e^{-j\beta R_0} \tag{11-2-1}$$

and

$$E_1 = E_m f(\theta, \phi)\frac{1}{R_1}e^{-j\beta R_1}e^{j\delta} \tag{11-2-2}$$

where

E_m is the amplitude of the electric field;
$f(\theta, \phi)$ is the element-antenna pattern, or factor; and
δ is the phase shift between element antennas.

The electric field of the radiation pattern in the far zone is the sum of the two individual electric fields, or

$$E = E_0 + E_1 = E_m f(\theta, \phi)\left[\frac{1}{R_0}e^{-j\beta R_0} + \frac{1}{R_1}e^{j(\delta - \beta R_1)}\right] \tag{11-2-3}$$

In the far zone, R_0 and R_1 are related by

$$R_1 \simeq R_0 - d \sin\theta \cos\phi \tag{11-2-4}$$

Substituting Eq. (11-2-4) into Eq. (11-2-3) yields the electric field in the radiation zone:

$$E = E_m f(\theta, \phi)\frac{1}{R_0}e^{-j\beta R_0}e^{j\psi/2}\left(2\cos\frac{\psi}{2}\right) \tag{11-2-5}$$

where

$\psi = \beta d \sin\theta \cos\phi + \delta$; and

$\beta = 2\pi/\lambda$

The magnitude of the electric field for a two-element array, then, is

$$|E| = \frac{2E_m}{R_0}|f(\theta, \phi)||F(\psi)| \tag{11-2-6}$$

where

$|f(\theta, \phi)|$ is the element-antenna pattern, or factor; and

$|F(\psi)| = \left|\cos\frac{\psi}{2}\right|$ and is the normalized array pattern, or factor; in the H plane.

The element factor is the magnitude of the element radiation factor, and the array factor is the function of the magnitudes and phases of the element excitations. In conclusion, the total field pattern of an array of identical elements is the product of the element pattern and the array pattern. We usually refer to this property as the *principle of pattern multiplication.*

EXAMPLE 11-2-1 Principle of Pattern Multiplication

Two short-dipole antennas are aligned along an x axis, are separated by distance d, and are pointed vertically in the z direction. Plot the electric-field patterns in the H plane ($\theta = 90°$) of the array for:

a) $d = \lambda/2$ and $\delta = 0$.
b) $d = \lambda/4$ and $\delta = -\pi/2$.

Solutions:

a) Using Eq. (10-1-18a), we can obtain the element electric-field pattern:

$$f(\theta) = E_\theta = E_0 \sin\theta$$

The array factor in the H plane ($\theta = 90°$) is

$$|F(\phi)| = \left|\cos\left(\frac{\pi}{2}\cos\phi\right)\right|$$
$$= \left|\cos\left(\frac{\psi}{2}\right)\right|$$

and the total electric pattern is

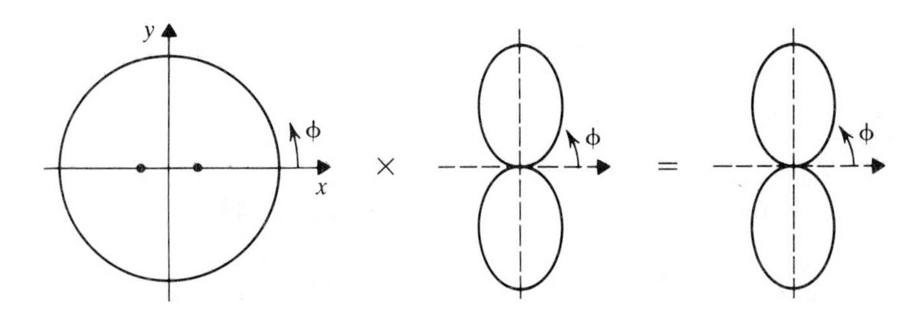

b) The element electric-field pattern, from Eq. (10-1-18a), is

$$f(\theta) = E_\theta = E_0 \sin\theta$$

The array factor in the H plane ($\theta = 90°$) is

$$|F(\phi)| = \left|\cos\left(\frac{\psi}{2}\right)\right| = \left|\cos\left(\frac{\pi}{4}\cos\phi - \frac{\pi}{4}\right)\right|$$

and the total electric pattern is

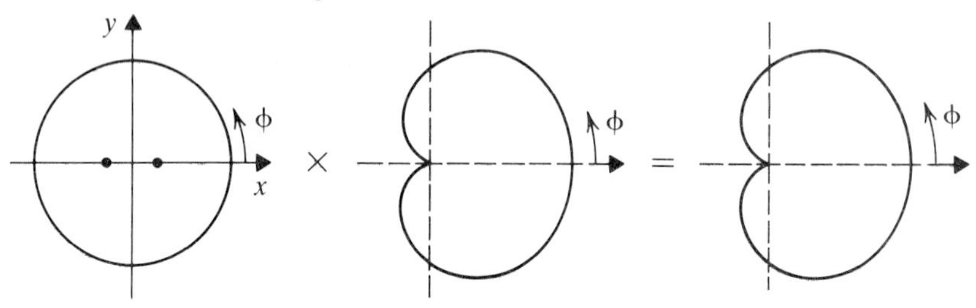

11-2-2 *N*-Element Arrays

An N-element linear array consists of several identical element antennas spaced equally along a straight line, as shown in Fig. 11-2-2. This type of array is usually called a *uniform linear array.*

We assume that the N elements are aligned along the x axis. Based upon the principle of pattern multiplication, the array pattern is the product of the element pattern and the array pattern. Thus the normalized array factor pattern in the H plane ($\theta = 90°$), or the xy plane, is

$$AF = F(\psi) = \frac{1}{N}\left|1 + e^{j\psi} + e^{j2\psi} + \cdots + e^{j(N-2)\psi} + e^{j(N-1)\psi}\right| \tag{11-2-7}$$

where

N is the number of element antennas;

$\psi = \beta d\cos\phi + \delta$ and is the progressive phase difference between two adjacent antennas;

δ is the phase shift between two neighboring element antennas; and

d is the distance between two element antennas.

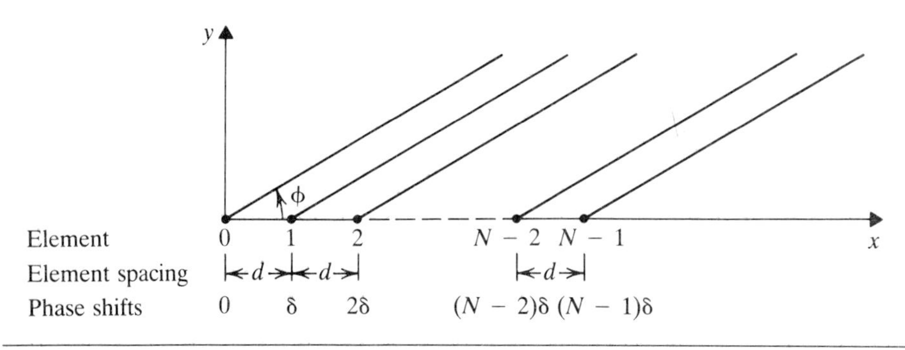

Figure 11-2-2 N-element linear array antenna.

Using polynomial theory, we can write the array factor shown in Eq. (11-2-7) as

$$AF = |F(\psi)| = \frac{1}{N}\left|\frac{1 - e^{jN\psi}}{1 - e^{j\psi}}\right| \tag{11-2-8}$$

or

$$AF = |F(\psi)| = \frac{1}{N}\left|\frac{\sin (N\psi/2)}{\sin (\psi/2)}\right| \tag{11-2-9}$$

Figure 11-2-3 shows a normalized array factor for a five-element uniform linear array.

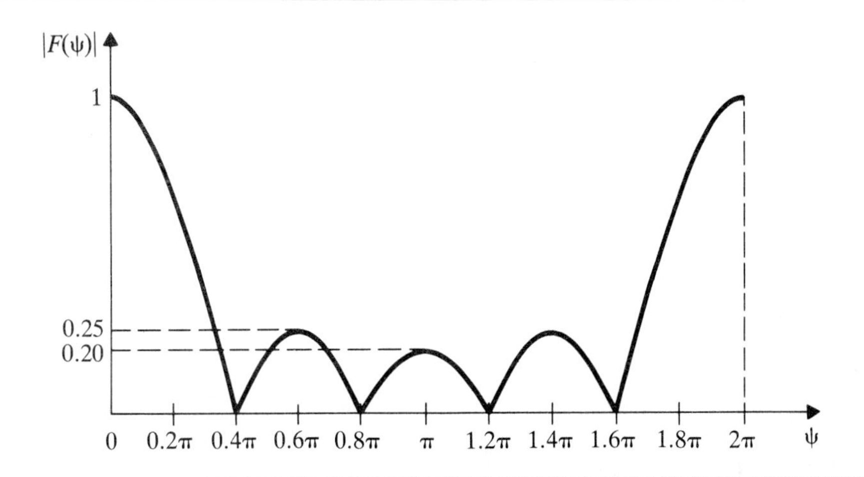

Figure 11-2-3 Normalized array factor of a five-element uniform linear array.

Several important properties can be derived from Eq. (11-2-9):

1. *Main-beam direction.* The maximum value of the main beam occurs at $\psi = 0$, which means that

$$\psi = \beta d \cos \phi_m + \delta = 0$$

Then

$$\cos \phi_m = -\frac{\delta}{\beta d} \tag{11-2-10}$$

This type of linear array has an endfire radiation pattern for $\delta = 0$.

2. *Null locations.* The array pattern has nulls at

$$\sin (N\psi/2) = 0$$

which means that

$$\frac{N\psi}{2} = \pm n\pi \qquad n = 1, 2, 3\ldots \tag{11-2-11}$$

3. *Sidelobe locations.* The sidelobes of a linear array occur at

$$\left|\sin\left(\frac{N\psi}{2}\right)\right| = 1$$

which means that

$$\frac{N\psi}{2} = \pm(2m + 1)\frac{\pi}{2} \qquad m = 1, 2, 3, \ldots \tag{11-2-12}$$

The first sidelobe appears at $m = 1$, which occurs at

$$\frac{N\psi}{2} = \pm\frac{3\pi}{2}$$

EXAMPLE 11-2-2 Uniform N-Element Linear Array in the H Plane (or xy Plane)

A uniform linear array of five identical elements is equally spaced at a distance of $d = \lambda/2$ and has an equal exciting amplitude.

Determine:

a) the normalized array factor for $\delta = 0$;
b) the normalized radiation pattern in the H plane; and
c) the location of the first null, in degrees.

Solutions:

a) The phase angle is

$$\psi = \beta d \cos\phi + \delta = \frac{2\pi}{\lambda}\frac{\lambda}{2}\cos\phi + 0 = \pi\cos\phi$$

Then the normalized array factor is derived from Eq. (11-2-9):

$$AF = |F(\psi)| = \frac{1}{5}\left|\frac{\sin\left(\frac{5\pi}{2}\cos\phi\right)}{\sin\left(\frac{\pi}{2}\cos\phi\right)}\right|$$

b) The normalized radiation pattern in the H plane is

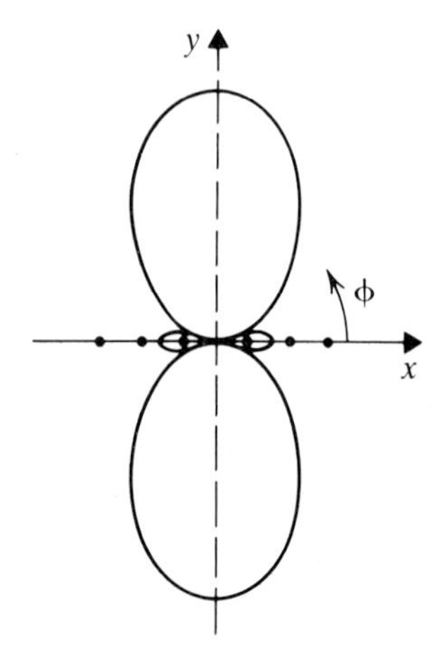

c) The first null, from Eq. (11-2-11), occurs at

$$\frac{N\psi}{2} = \frac{5}{2}\pi \cos \phi = \pi$$

$$\cos \phi = 0.4$$

$$\phi = 66.42^\circ$$

11-3 YAGI–UDA ANTENNAS

A Yagi–Uda antenna is a parasitic linear array consisting of several parallel dipoles and is commonly used for receiving television (TV) signals. A transmitting Yagi–Uda antenna has only one driver element; the rest are parasitic and are not directly connected to the source. In a receiving Yagi–Uda antenna, when the TV signal wave impinges on all elements of the array, only one active element will receive the signal. Figure 11-3-1 shows the structure of a six-element Yagi–Uda antenna.

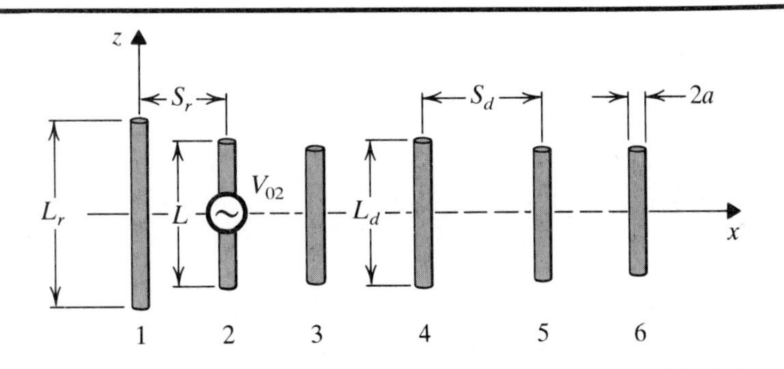

Figure 11-3-1 Physical structure of a six-element Yagi–Uda array antenna.

In Fig. 11-3-1, element 2 is the driven (or active) element, which is usually a resonant half-wave dipole. All the other elements are parasitic. Element 1 is a reflector, which is normally slightly longer than the driven element, whereas the other elements are directors, which are shorter than the driven element. When a parasitic element is spaced very close to the driven element, it is excited by the driven element with roughly equal amplitude. Thus the electric field incident on and tangent to the parasitic element is

$$E_{\text{inc}} = E_{\text{drv}} \tag{11-3-1}$$

A current is then induced or excited on the parasitic element and the resulting electric field, which is tangent to the wire, is equal in amplitude and opposite in phase to the incident wave. Because a good conductor cannot support a field,

$$E_{\text{inc}} + E_{\text{par}} = 0 \tag{11-3-2}$$

As a result, the electric field of a parasitic element is

$$E_{\text{par}} = -E_{\text{inc}} = -E_{\text{drv}} \tag{11-3-3}$$

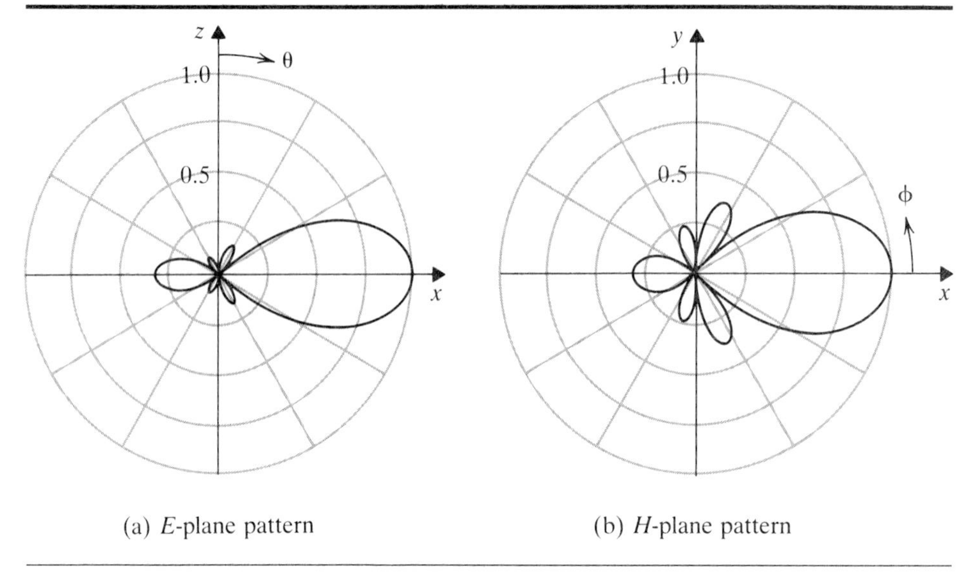

(a) E-plane pattern (b) H-plane pattern

Figure 11-3-2 Electric-field patterns of six-element Yagi–Uda array antenna for Channel 24.

Two closely spaced arrays with equal amplitude and opposite phase will produce an endfire pattern (from array theory). The electric-field patterns of a six-element Yagi–Uda antenna for Channel 24 (at the frequency of 531.25 MHz) are shown in Fig. 11-3-2 for both the E plane and the H plane. Physical data and characteristics of three types of equally spaced Yagi–Uda antennas are shown in Tables 11-3-1 and 11-3-2, respectively.

In practice, Yagi–Uda antennas are mounted on a tubular aluminum mast. It is not necessary to insulate the parasitic elements from the mast, because the excited

Table 11-3-1 Physical dimensions of several Yagi–Uda antennas

Number of Elements (N)	Spacing $S(\lambda)$	Reflector Length $L_r(\lambda)$	Driver Length $L(\lambda)$	Director Length $L_d(\lambda)$
4	0.30	0.475	0.453	0.446
5	0.30	0.482	0.459	0.451
6	0.30	0.472	0.449	0.437

Note: Conductor diameter $a = 0.005\lambda$.

Table 11-3-2 Characteristics of the several Yagi–Uda antennas in Table 11-3-1

Gain (dB)	Front-to-Back Ratio (dB)	Input Impedance (Ω)	E Plane HP_E (°)	E Plane SLL_E (dB)	H Plane HP_H (°)	H Plane SLL_H (dB)
10.7	5.2	$25.8 + j23.2$	56	−18.5	64	−7.3
9.3	2.9	$19.3 + j39.4$	40	− 9.5	42	−3.3
11.6	6.7	$61.2 + j7.7$	52	−14.8	56	−7.4

current on the mast is very small. The driven element must be insulated, however, to avoid shorting the terminals.

EXAMPLE 11-3-1 Design of a Six-Element Yagi–Uda Antenna for TV Channel 24

Television Channel 24 has a frequency range of 530–536 MHz. We want to design a Yagi–Uda antenna for reception of TV Channel 24, which involves

a) determining the midband frequency and its wavelength;
b) choosing the wire number to be used;
c) finding the lengths of the elements;
d) calculating the spacings; and
e) predicting the radiation pattern.

Solutions:

a) The midband frequency is 533 MHz, and its wavelength is 56.29 cm.
b) We choose number 9 wire, so

$$2a = 0.2906/56.29$$
$$= 0.0052\lambda$$

c) The lengths of the elements are

$$L_r = 0.47\lambda = 26.46 \text{ cm};$$
$$L = 0.45\lambda = 25.33 \text{ cm}; \quad \text{and}$$
$$L_d = 0.44\lambda = 24.77 \text{ cm}$$

d) The spacings are

$$S_r = 0.25\lambda = 14.07 \text{ cm}; \quad \text{and}$$
$$S_d = 0.30\lambda = 16.89 \text{ cm}$$

e) We predict that the radiation pattern will be as shown in Fig. 11-3-2.

11-4 ANTENNA MEASUREMENT TECHNIQUES

11-4-1 Overview of Techniques

The far-field radiation pattern of an antenna is an important characteristic. The three types of techniques for measuring the far-field pattern are: conventional far-field measurement, compact-range measurement, and near-field measurement [3]. We describe all three types generally and then consider in more detail the methods of near-field measurement because of their obvious advantages.

1. *Conventional Far-Field Measurement*

The distance between the probe antenna and the test antenna in conventional far-field measurement determines the extent to which the illuminating field approximates the desired plane-wave illumination. In other words, the distance must be larger than the value of $2D^2/\lambda$, as indicated in Eq. (9-2-5). Figure 11-4-1 is a schematic diagram of a conventional far-field measurement range.

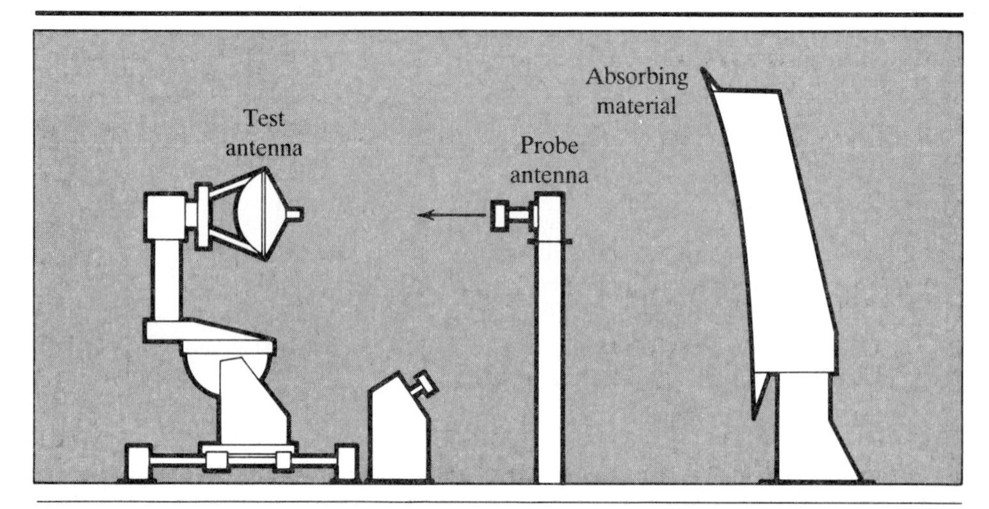

Figure 11-4-1 Schematic diagram of conventional far-field measurement.

Because the probe antenna and the test antenna must be separated by a considerable distance, an outdoor antenna range is often required. Consequently, control of the test environment is poor. The antenna test is subject to adverse weather conditions and distortion of test measurements by range reflections. If the distance requirement permits, conventional testing can be conducted in an indoor anechoic chamber. Indoor testing is more convenient, less expensive, and more efficient than outdoor testing. Control of environmental conditions and scattering reflections produces more-accurate test results.

2. *Compact-Range Measurement*

The compact-range measurement technique was invented in the late 1960s to overcome the high cost of conventional antenna measurement. The compact-range test utilizes what is essentially a continuous array of sources. A reflective parabolic surface reradiates power emitted by a feed source. Plane-wave illumination is thus created within a short antenna range, as shown in Fig. 11-4-2.

Compact-range testing offers distinct advantages over conventional testing. Specifically, testing is performed indoors, with all its advantages. Additionally, the chamber required is smaller than for conventional testing. Thus less absorbant material is needed, making the compact range more economical than a conventional indoor range.

The compact-range technique, however, does have limitations. The size of the test zone—in which the field approximates a plane wave—is limited by the size of the reflecting surface. Diffraction from the edges of the reflecting surface becomes a problem at low operating frequencies, whereas the tolerance of the reflecting surface becomes a problem at high operating frequencies. These conditions can cause depolarization of the test field, and as a result, degrade the test pattern.

3. *Near-Field Measurement*

Near-field scanning is an effective alternative to conventional antenna measurement. A small probe antenna is used to sample the near-field radiation pattern of the test antenna. The scanning operation is performed on a geometric surface that

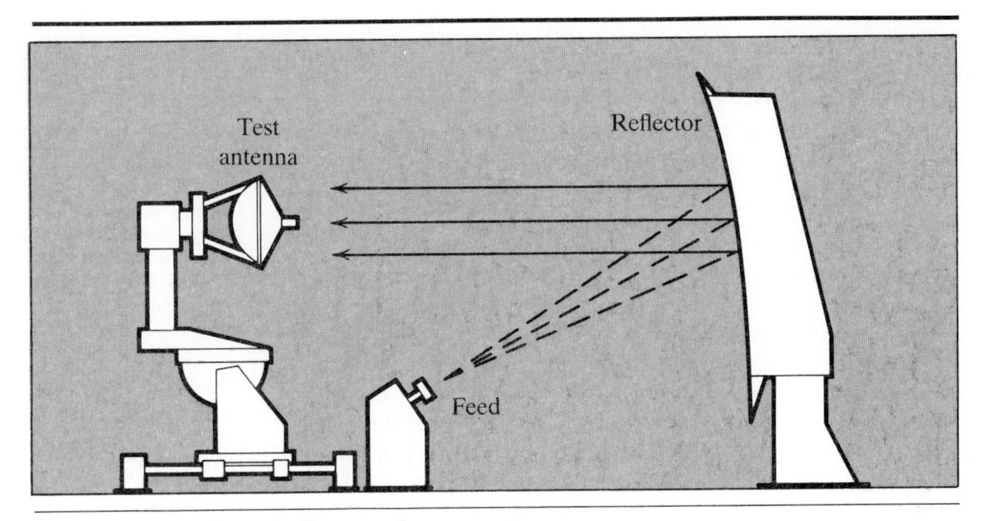

Figure 11-4-2 Schematic diagram of a compact range.

completely or partially encloses the volume containing the test antenna. When the near-field radiation pattern has been recorded, a computerized algorithm is used to calculate the far-field radiation pattern from the input data. Figure 11-4-3 shows a block diagram of the near-field measurement technique.

A primary advantage of near-field measurement is its ability to perform the measurement with the probe antenna very close to the test antenna. This method minimizes the space requirements for the anechoic chamber. Another primary advan-

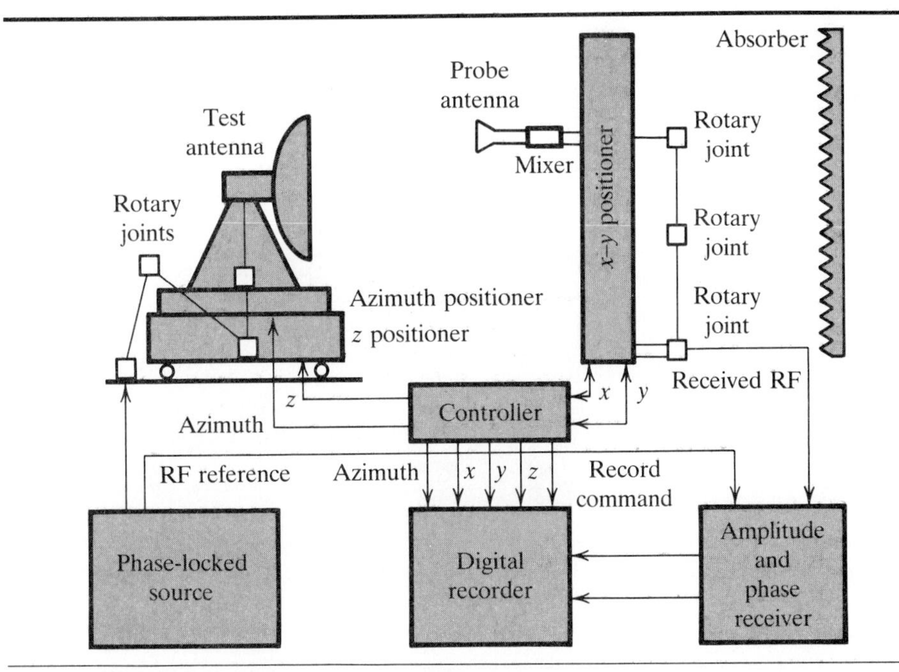

Figure 11-4-3 Block diagram of near-field measurement.

tage of near-field measurement is the high degree of accuracy that can be achieved. The disadvantage of near-field testing is the requirement for a computer and the time needed to execute the computation algorithm. Because of the finite speed and memory of computers, near-field computation algorithms are limited to test antennas with electrical sizes of 100–1000 wavelengths.

11-4-2 Near-Field Scanning Surfaces

Near-field scanning involves the sampling of a radiation distribution on a two-dimensional test-antenna surface. There are three surface geometries for near-field scanning: planar, cylindrical, and spherical.

1. *Planar Surface*

A planar surface is shown in Fig. 11-4-4. The sampling surface area must extend far enough to intersect the major portion of the radiation propagated by the test antenna.

The major advantage of a planar scanning surface is the simplicity of the transformation algorithm compared to that for cylindrical and spherical scanning surfaces. The relatively simple mathematics allows for speedy computation of the far-field magnitude and reduces production costs.

The planar scanning technique also has some disadvantages. The first is the probe positioning problem. Two perpendicular linear axes are normally used to position the probe in a plane. This configuration is difficult and quite expensive to build. A new method for solving this problem is to use a laser-beam alignment technique. The second disadvantage is that the sampling surface encloses only one hemisphere of the test antenna. From a practical standpoint, this might not be a major concern, because most antennas tested with near-field scanning measurement are highly directional; the

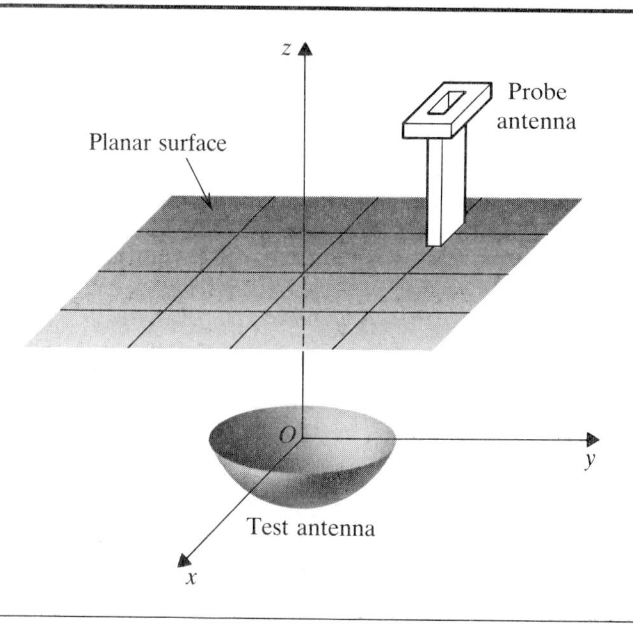

Figure 11-4-4 Planar scanning surface.

planar scanning surface can be arranged to intercept most of the radiated power. As a result, however, only a hemisphere of the far field of the test antenna can be plotted.

2. *Cylindrical Surface*

A cylindrical surface is shown in Fig. 11-4-5. The cylindrical coordinate system—with its origin at the test antenna—forms the basis for the sampling grid. The physical movement of the probe, limited to the z-axis and the θ-axis directions, is simulated by rotation of the test antenna. Thus cylindrical scanning requires only one linear axis and one rotary axis. This arrangement is simpler and less expensive than the dual linear axes most often used for planar scanning.

The disadvantage of cylindrical scanning is that it does not cover the two end poles of the cylindrical scanning surface. As the z axis is increased, the scanning surface will intercept all the energy radiated by the test antenna. Processing the data collected from cylindrical near-field measurement takes longer than the processing of the cartesian planar near-field–measurement data. As a result, the cost of cylindrical scanning measurement is higher than that for planar scanning measurement.

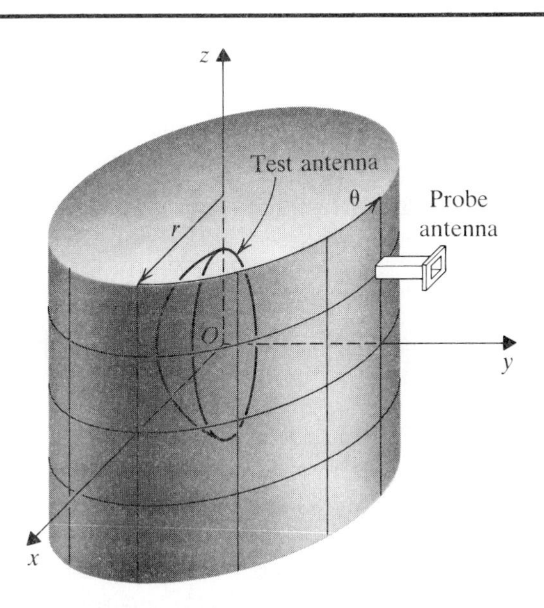

Figure 11-4-5 Cylindrical scanning surface.

3. *Spherical Surface*

A spherical surface is shown in Fig. 11-4-6. The probe antenna in spherical scanning is held stationary. Movements in the θ-coordinate and ϕ-coordinate directions are simulated by rotation of the test antenna. Spherical scanning does not require any linear motion and thus completely eliminates the scaling problem associated with planar and cylindrical scanning. The measurement covers the entire volume of the test antenna.

A disadvantage of spherical scanning is that the transformation algorithm is far more complicated. Consequently, the computation time is longer and the cost is higher than for planar and cylindrical scanning.

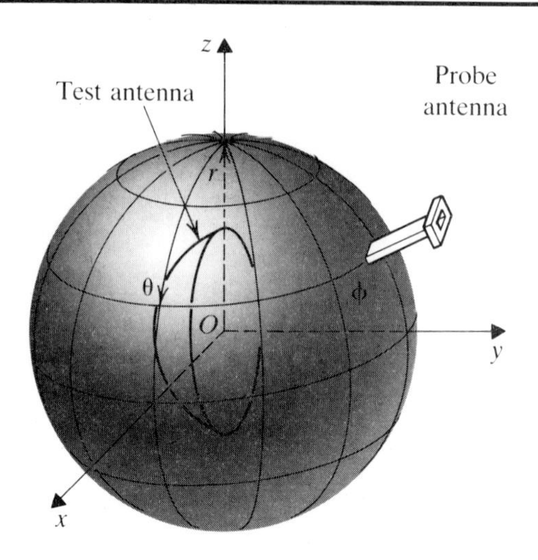

Figure 11-4-6 Spherical scanning surface.

11-4-3 Near-Field Measurement System

Antenna near-field measurement is a new technique used to obtain the antenna's far-field radiation pattern by near-field data transformation.

1. *Major features.* There are three major features:

   a) Measurements are carried out in the near field of the test antenna.
   b) Directional effects of the probe antenna on the primary field radiated by the test antenna are taken into account.
   c) Computed patterns of the test antenna span a solid angle and not just one or two principal plane cuts.

2. *Advantages.* The advantages of antenna near-field measurements over direct far-field pattern measurements are:

   a) Near-field measurements are time and cost effective, and the accuracy of the computed radiation pattern is comparable to or even better than that for the conventional far-field range.
   b) The near-field range provides a controlled environment and all-weather capability.
   c) The far-field range size limitations, transportation and mounting problems, and the requirement for large-scale positioners for large antenna systems are eliminated.

3. *Disadvantages.* The near-field measurement technique also has the following disadvantages:

   a) A more complicated and expensive measurement system is required.
   b) A more extensive procedure is required to calibrate the near-field measurement probes compared to that for the far-field probes.
   c) The test-antenna patterns are not obtained in the real-time spectrum.
   d) More complex computer software is required for computation of the pattern.

4. *Test procedures.* Near-field measurements can be carried out by using the following procedures:

   a) The probe antenna is characterized first. This process, from the system-theory viewpoint, is needed to establish the "transfer function" of one component of the system.
   b) The measurements of the tangential field are obtained at preselected intervals on a prescribed surface in the near-field of the test antenna by using two independent and orthogonal orientations of the probe antenna.
   c) Far-field pattern evaluation is normally performed last by using a fast Fourier transform (FFT) algorithm. When the pattern data over some prescribed section in the far field is desired, filters can be used in order to obtain data reduction and/or increase resolution.

1. *Near-Field Probe Design*

There are several different types of probe antennas, such as the dipole antenna, small-horn antenna, and open-end waveguide probe, that can be used for near-field measurements. The dipole antenna is usually used at frequencies below 1 GHz and requires a balun for impedance matching. The small-horn antenna can also be used as a probe antenna, but it must be carefully designed and constructed in order to obtain satisfactory data. The open-end waveguide probe is the simplest to construct and has demonstrated excellent results. In general, when an accurate measurement of the field-intensity distribution near a microwave antenna is required, the probe antenna must meet the following four requirements:

1. The probe antenna must have an aperture small enough to measure the field intensity essentially at a point.
2. The probe antenna must have the desired polarization to a high degree of accuracy. Because some linearly polarized component of the field is usually to be measured, the probe should be linearly polarized.
3. The probe antenna must deliver a signal voltage large enough to permit accurate measurement.
4. The probe antenna and its associated equipment must not distort seriously the measured field.

a) Open-End Waveguide Probe

Figure 11-4-7 (on the following page) shows schematically an open-end waveguide probe antenna. The standard dimensions of the probe are shown in Table 11-4-1. The length l of the open-end waveguide is about 5λ (6 in., or 15 cm, at 10 GHz).

Table 11-4-1 Physical dimensions of two open-end waveguide probes

Frequency	Type	Inside Height (h)	Inside Width (w)
X band	WR(90)	1.016 cm (0.400 in.)	2.286 cm (0.900 in.)
Ku band	WR(62)	0.790 cm (0.311 in.)	1.580 cm (0.622 in.)

b) Pyramidal Horn Probe

A pyramidal horn fed from a rectangular waveguide WR(90) is also often used as a probe antenna from 8 to 12 GHz at X band. Waveguide WR(90) has a width a of 0.90 in. (2.286 cm) and a height b of 0.40 in. (1.016 cm). The dimensions of the

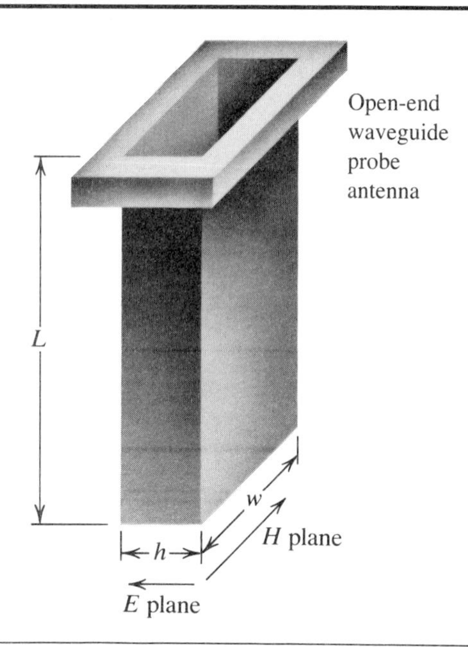

Figure 11-4-7 Schematic diagram of an open-end waveguide antenna.

pyramidal horn having a gain of 22 dB (162.18) at 9.3 GHz ($\lambda_0 = 3.226$ cm) are shown in Fig. 11-4-8 and are:

$A = 7.27$ in.	or	18.46 cm
$B = 5.73$ in.	or	14.55 cm
$\ell_E = 12.92$ in.	or	32.81 cm
$\ell_H = 13.86$ in.	or	35.21 cm
$R_E = R_H = 11.71$ in.	or	29.75 cm

The coaxial connector feed for TE_{10}-mode operation must be inserted at the wide side of the waveguide. Impedance matching can be achieved by adjusting the tuning stubs. The outside surfaces of the probe must be covered with microwave absorber to reduce the distortion of the fields.

2. *Antenna-Probe Characterization*

Accuracy of the computation of a far-field pattern from near-field measurements is critically dependent on the accuracy of the near-field probe characterization. Because the far-field amplitude and phase patterns of two orthogonal components of the near-field measuring probe determine its wave number spectrum for Fourier transfer computations, the far-field pattern of the probe antenna must be characterized first. Probe characterization procedures are as follows:

1. The near-field probe is mounted on a standard azimuth turntable to act as a transmitting antenna.
2. A receiving antenna is rotated in elevation by an inverted swing carriage to measure the far-field pattern of the probe antenna.
3. Two sets of amplitude and phase measurements are made by rotating the

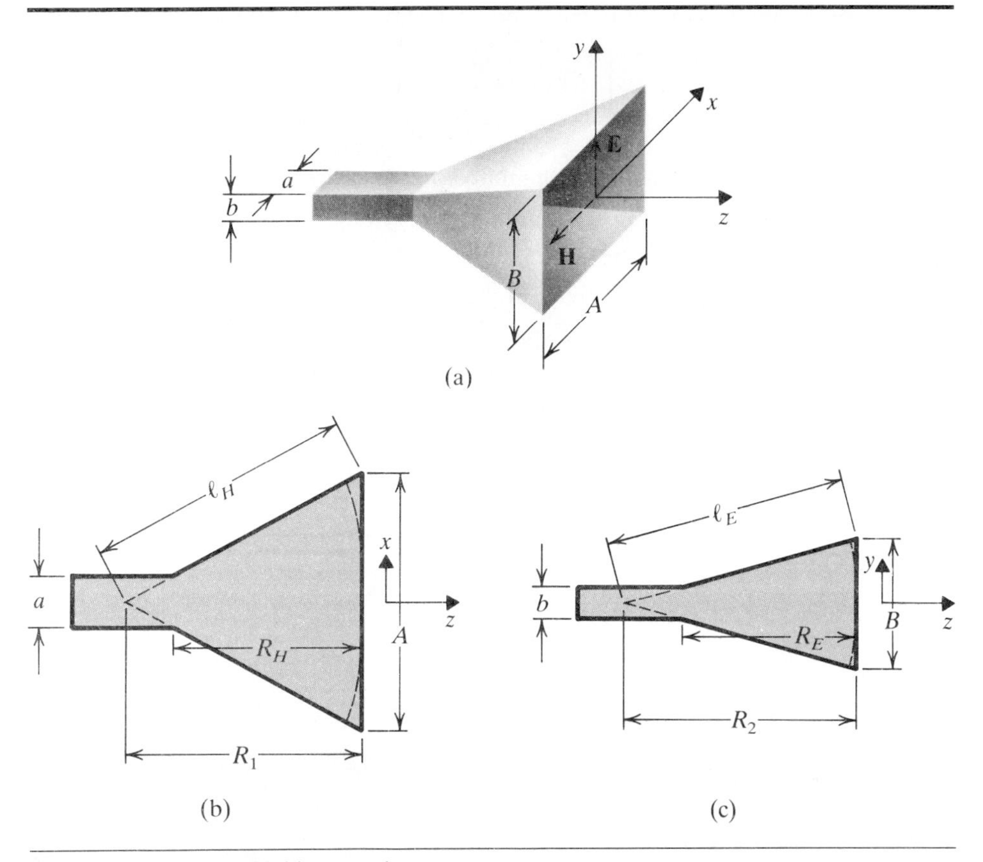

Figure 11-4-8 Pyramidal horn probe antenna.

receiving antenna 90° to obtain the far-field transverse components of the near-field probe.

4. The wave-number spectrum of the probe can be determined from the measured field components.

11-4-4 Far-Field Computation

Basically, the radiation fields of the test and probe antennas are expanded in terms of elementary plane modes or waves. Then the Lorentz theorem is used to calculate the probe output as a function of the expanded fields. The result is an algebraic equation that relates the known or measured field of the probe antenna to the unknown or desired field radiated by the test antenna. The unknown mode amplitudes are then determined from this equation and the far-field radiation pattern is finally computed from the measured amplitudes.

The basic theory of transforming the measured near-field data to the desired far-field radiation pattern is simple, but the mathematics of computating fast Fourier transforms (FFTs) is complicated. When the open-end waveguide probe detects the electric field on a planar surface located in the plane at $z = \Delta z$, the electric field can be expressed as

$$\mathbf{E}(x, y, \Delta z) = |\mathbf{E}| \underline{/\theta} \tag{11-4-1}$$

When the probe antenna moves in the x and y directions, the wave-number spectrum is

$$\mathbf{A}(k_x, k_y) = \frac{1}{2\pi} \int_{-\infty}^{\infty} \int_{-\infty}^{\infty} \mathbf{E}(x, y, \Delta z) \exp\left[-j(k_x x + k_y y)\right] dx\, dy \qquad (11\text{-}4\text{-}2)$$

where

$k_x = \pi/\Delta x = 2\pi/\lambda$ and is the wave number in the x direction;
$\Delta x = \frac{1}{2}\lambda$ and is the scanning space in the x direction;
$k_y = \pi/\Delta y = 2\pi/\lambda$ and is the wave number in the y direction;
$\Delta y = \frac{1}{2}\lambda$ and is the scanning space in the y direction; and
λ is the signal wavelength in free space.

When the wave-number spectrum is found, the far-field radiation pattern can be calculated by applying the FFT as

$$\mathbf{E}(x, y, z) = \frac{1}{2\pi} \int_{-\infty}^{\infty} \int_{-\infty}^{\infty} \mathbf{A}_x(k_x, k_y) \exp(j\mathbf{k} \cdot \mathbf{R})\, dk_x\, dk_y \qquad (11\text{-}4\text{-}3)$$

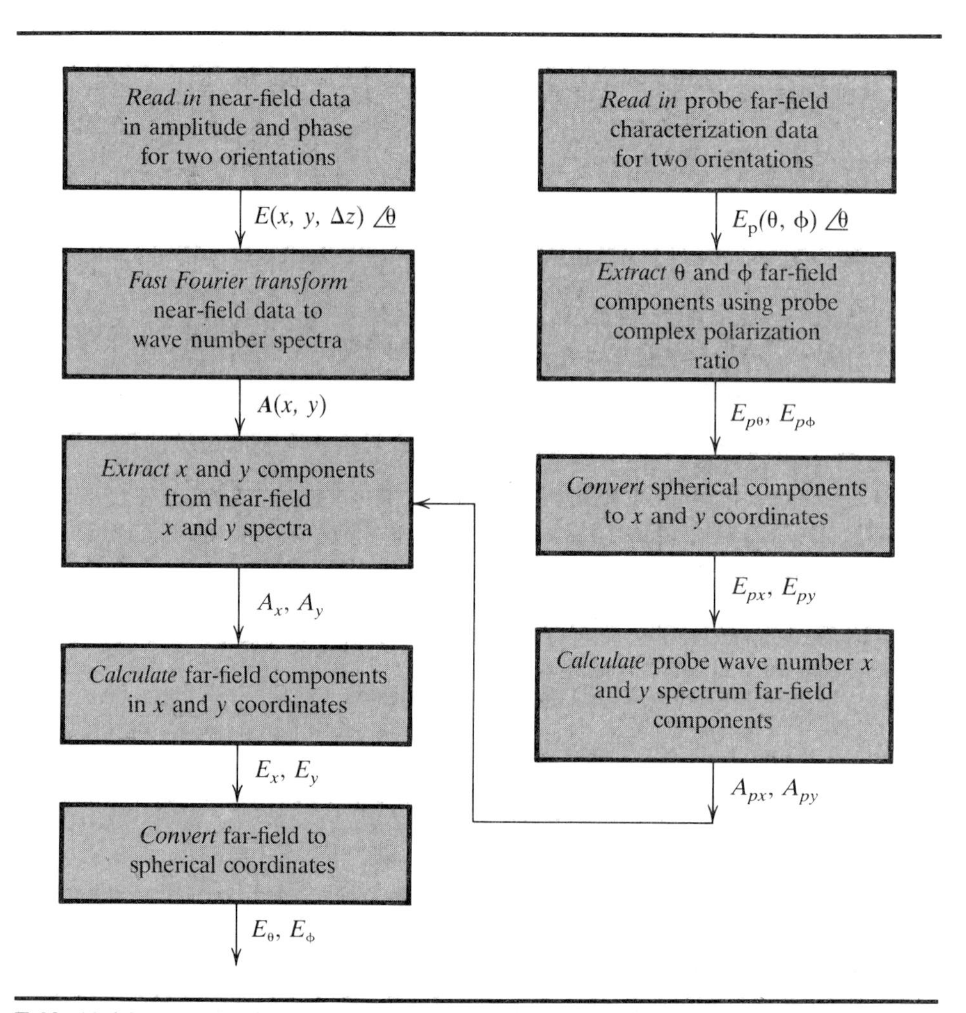

Table 11-4-2 Program flowchart in a planar scanning surface

where

$\mathbf{A}_x$ is the x-spectrum component in terms of the near and far fields;

$\mathbf{k} = k_x\mathbf{a}_x + k_y\mathbf{a}_y + (k_0^2 - k_x^2 - k_y^2)^{1/2}\mathbf{a}_z$ and is the wave-number vector;

$k_0 = 2\pi/\lambda$ and is the free-space wave number; and

$\mathbf{R} = (k_x^2 + k_y^2)^{1/2} \angle \tan^{-1}\left(\frac{k_x}{k_y}\right)$ and is the field-range vector.

The FFT computations can be carried out on a computer. The software may need to be modified to meet specific user requirements. Table 11-4-2 illustrates a program flow chart for determining the far field of a test antenna from the near-field measurements in a planar scanning surface.

REFERENCES

[1] Liao, Samuel Y., "Design and Analysis of a Polarization Diversity Antenna for High Frequency Reception." San Diego: Naval Electronics Laboratory Center, September 1969.

[2] Carrel, R. L., "Analysis and Design of the Log-Periodic Dipole Antenna." University of Illinois Antenna Lab Technical Report 52, 1962.

[3] Liao, Samuel Y., "Design and Analysis of Antenna Near-Field Measurement System." Report for Hughes Aircraft Company, El Segundo, Calif., August 1985.

Problems

11-1 Design a UHF log-periodic antenna. The frequency range is to be from 0.3 to 0.6 GHz. The antenna gain is to be 11 dB, and the input impedance is assumed to be 300 Ω.

11-2 Design a VHF log-periodic antenna. The frequency range is to be from 471 to 640 MHz. The antenna gain is to be 10 dB and the input impedance is assumed to be 300 Ω.

11-3 Two $3\lambda/2$-dipole antennas are aligned along the x axis, are separated by a distance of one half-wavelength, and are pointed vertically in the z direction. The phase in antenna 1 is leading that in antenna 0 by an angle of $\pi/2$.

a) Derive the element-factor equation.
b) Find the array-factor equation.
c) Plot the total electric-field pattern in the H plane ($\theta = 90°$) of the array antenna.

11-4 A uniform linear array of 10 identical elements is equally spaced at a distance of $d = \lambda/4$ and has an equal exciting amplitude. Calculate the

a) normalized array factor for $\sigma = 0$ and $\theta = 90°$ (x-y plane);
b) normalized radiation pattern in the H plane; and
c) location of the first null, in degrees.

11-5 Television channel 30 has a frequency range from 566 to 572 MHz. A six-element Yagi–Uda antenna is to be designed for reception of this channel. Determine the

a) midband frequency and its wavelength;
b) wire size number to be used;
c) element lengths;
d) element spacings; and
e) radiation pattern.

11-6 Television channel 47 has a frequency range from 668 to 674 MHz. A six-element Yagi–Uda antenna is to be designed for reception of this channel. Calculate the

a) midband frequency and its wavelength;
b) wire size number to be used;
c) element lengths;
d) element spacings; and
e) radiation pattern.

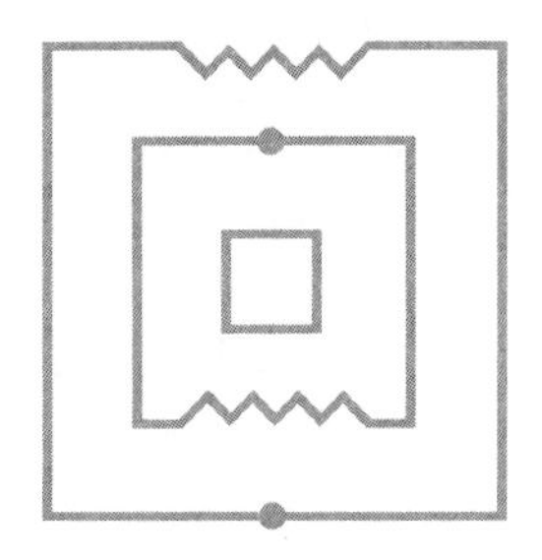

Chapter 12

Electromagnetic Energy Transmission System

12-0 INTRODUCTION

As we stated at the beginning of this book, transmission lines, waveguides, and antennas are the devices for transmitting and receiving electromagnetic energy between a transmitter and a receiver. Insofar as the entire system of energy transmission is concerned, care must be exercised to limit the noise figure and system temperature in the receiving system in order to obtain good reception. In this chapter we analyze several applications of the energy transmission system [1].

12-1 NOISE FIGURE AND SYSTEM NOISE TEMPERATURE

12-1-1 Noise Figure

We frequently specify the overall effect of many noise sources in a receiving system by the *noise figure* of the receiver. This quantity measures the noise generated within the receiving system, and antenna *noise temperature* measures the noise entering the system with the signal. We can define the noise figure F of any linear two-terminal–port network in terms of its performance by connecting a standard noise source to its input terminals. That is, we can define the noise figure in the following manner:

$$\text{Noise figure} \equiv \frac{\text{Available noise power at output}}{\text{Available noise power at input}}$$

or

$$F = \frac{N_0}{GKTB} \tag{12-1-1}$$

where KTB is the available noise power of the standard source in a bandwidth B, with $T = 290\ ^\circ\text{K}$ and G the available power gain of a network at the frequency of that band. The noise N_o at the output of the network within the same frequency band results from amplification of the input noise and from the noise N_n generated within the network, or

$$N_o = N_n + GKTB \tag{12-1-2}$$

Then the single-frequency noise figure becomes

$$F = 1 + \frac{N_n}{GKTB} \tag{12-1-3}$$

Another common form of the single-frequency noise figure is obtained if the source is supposed to supply both a noise N_i and a signal S_i, and the resulting output signal is S_o. The ratios of S_o to S_i and $GKTB$ to N_i within a narrow frequency band are equal, because each ratio is the gain of the network. Therefore we can write from Eq. (12-1-1),

$$F = \frac{N_o}{(S_o/S_i)N_i} = \frac{S_i/N_i}{S_o/N_o} \tag{12-1-4}$$

Equation (12-1-4) states that the noise figure is the ratio of the input signal-to-noise ratio to the output signal-to-noise ratio.

The noise figures of networks in cascade can be combined simply. Let G_1, G_2, $G_3, \ldots, G_n$ be the available-power gains of the first, second, third, ..., and nth stages of an n-stage network, respectively, and let $F_1, F_2, F_3, \ldots, F_n$ be the corresponding noise figures. The overall available-power gain is $G = G_1 G_2 G_3 \ldots G_n$. The overall noise figure in terms of Eq. (12-1-3) is

$$\begin{aligned} F &= 1 + \frac{\sum_1^n N_n}{GKTB} = 1 + \frac{N_1 G_2 G_3 + \cdots + N_2 G_3 G_4 + \cdots + N_3 G_4 + \cdots}{(G_1 G_2 G_3 \cdots G_n)KTB} \\ &= 1 + \frac{N_1}{G_1 KTB} + \frac{1}{G_1}\frac{N_2}{G_2 KTB} + \frac{1}{G_1 G_2}\frac{N_3}{G_3 KTB} + \cdots + \frac{1}{G_1 G_2 \cdots G_{n-1}}\frac{N_n}{G_n KTB} \end{aligned} \tag{12-1-5}$$

The component noise figure is

$$\frac{N_2}{G_2 KTB} = F_2 - 1, \frac{N_3}{G_3 KTB} = F_3 - 1, \ldots, \frac{N_n}{G_n KTB} = F_n - 1$$

so the total noise figure is

$$F = F_1 + \frac{F_2 - 1}{G_1} + \frac{F_3 - 1}{G_1 G_2} + \cdots + \frac{F_n - 1}{G_1 G_2 \cdots G_{n-1}} \tag{12-1-6}$$

Note that the effect of noise sources in the stages of a receiver other than the first is reduced because of division by the available-power gains of all preceding stages.

Noise figures are often expressed in decibels. Hence, if F represents a noise figure in terms of a power ratio, the noise figure in decibels, then, is

$$F(\text{dB}) = 10 \log (F) \quad \text{dB} \tag{12-1-7}$$

12-1-2 System Noise Temperature

In general, the noise temperature T_n of an electronic system can be expressed in terms of absolute temperature T_0, from Eq. (12-1-3), as

$$T_n = T_0(F - 1) \tag{12-1-8}$$

where $T_0 = 290\ ^\circ K$ is the standard noise temperature for noise measurements. When the noise temperature represents the available-noise power output of an entire receiving system, it is commonly called the *system noise temperature*, which is often used to calculate system noise power and the signal-to-noise ratio.

We can represent a receiving system as a cascade of an antenna, a transmission line, and a receiver, as shown in Fig. 12-1-1.

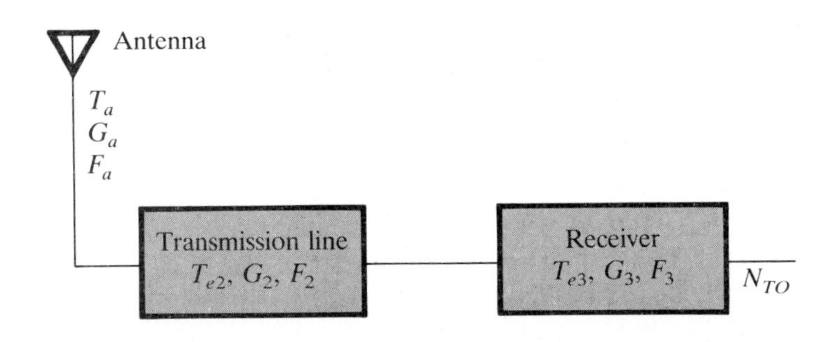

Figure 12-1-1 Three stages in a cascade.

We can then express the noise figure of the corresponding transmission system, from Eq. (12-1-4), as

$$F_{a3} = \frac{S_i/(KT_0B)}{S_o/N_{TO}} = \frac{S_iN_{TO}}{S_oKT_0B} = \frac{N_{TO}}{G_sKT_0B} \tag{12-1-9}$$

Therefore we can write the total output noise power as

$$N_{TO} = F_{a3}G_sKT_0B \tag{12-1-10}$$

where F_{a3} is the overall noise figure for the cascade network, and $G_s = G_aG_2G_3$ and is the overall system power gain. You can see from Fig. 12-1-1 that the total output noise power is

$$N_{TO} = G_sKT_sB \tag{12-1-11}$$

where T_s is the overall effective input noise temperature of the three stages in cascade.

We can also express the total output noise power in terms of its components in each stage:

$$N_{TO} = G_aG_2G_3KT_aB + G_2G_3KT_{e2}B + G_3KT_{e3}B \tag{12-1-12}$$

Solving Eqs. (12-1-11) and (12-1-12) simultaneously for the system noise temperature T_s results in

$$T_s = T_a + \frac{T_{e2}}{G_a} + \frac{T_{e3}}{G_aG_2} \tag{12-1-13}$$

The system noise temperature for n stages becomes

$$T_s = T_a + \sum_{i=2}^{n} \frac{T_{ei}}{G_{i-1}} \tag{12-1-14}$$

1. *Antenna Noise Temperature*

The noise temperature of an antenna expresses the noise density available at the antenna terminals as shown in Eq. (10-9-4). That is,

$$T_a = \frac{P_n}{KB_n} \tag{12-1-15}$$

Supposedly, the antenna is matched with a load resistor at temperature T and is in equilibrium with the environment represented by an antenna temperature, that is, $T_a = T$. The load resistor then receives noise power equal to

$$P_{en} = KTB \tag{12-1-16}$$

The environment supplies a noise figure, KTB, to the antenna and part of this noise, ηKTB, is delivered to the load (where η is the antenna efficiency). Thus the difference of $KTB - \eta KTB$ between the power supplied by the environment and the power actually reaching the load must be supplied by the ohmic resistance in the antenna. Then,

$$P_{ar} = KTB(1 - \eta) \tag{12-1-17}$$

Therefore the total output noise power for a lossy antenna is the sum of the noise power provided by the environmental antenna temperature and the noise power supplied by the ohmic resistance in the antenna. Thus

$$P_a = KT_a B\eta + KTB(1 - \eta) \tag{12-1-18}$$

The noise temperature of an antenna depends chiefly on the temperature of the medium that absorbs power in the main beam of the antenna. More specifically, the antenna temperature is determined in accordance with the following relationship:

$$\begin{aligned} T_a &= \alpha_1 T_1 + \alpha_2 T_2 + \alpha_3 T_3 + \cdots + \alpha_n \mathrm{T}_n \\ &= \sum_{i=1}^{n} \alpha_i \mathrm{T}_i \end{aligned} \tag{12-1-19}$$

For example, α_1 is associated with the ohmic losses in the antenna structure at its temperature T_1; α_2 is associated with the power absorbed in the atmosphere at its temperature T_2; α_3 is associated with the power absorbed in outer space at its temperaure T_3; α_4 is associated with the power absorbed in the sun at its temperature T_4; and so on.

2. *Transmission-Line Noise Temperature*

We assume that a lossy transmission line is matched at both ends with resistors at temperature T_t. The noise power available at the receiving end is KT_tB. The noise power generated at the sending end is also KTB and a part of this noise, $\eta_t KT_tB$, actually arrives at the load end (where η_t is the transmission-line efficiency). The difference of $KT_tB - \eta_t KT_tB$ between the noise power available at the load and that actually reaching the load must be the noise power supplied by the lossy line.

Thus

$$P_{nt} = KT_tB(1 - \eta_t) \tag{12-1-20}$$

The noise power provided by a lossy transmission line in terms of its effective temperature is

$$P_{nt} = KT_{eL}B\eta_t \tag{12-1-21}$$

Equating Eqs. (12-1-20) and (12-1-21), we have the transmission-line effective noise temperature:

$$T_{eL} = T_t\frac{(1 - \eta_t)}{\eta_t} \tag{12-1-22}$$

where T_t is the transmission-line thermal temperature.

3. *Receiver Noise Temperature*

The effective input noise temperature T_r and the noise figure F of a receiver are usually stated by the manufacturer. The relationship between them, as shown in Eq. (12-1-8) is

$$T_r = T_0(F - 1) \tag{12-1-23}$$

where $T_0 = 290\ ^\circ\text{K}$ is the reference room temperature.

12-2 ENERGY TRANSMISSION ANALYSIS

As we stated previously, the receiving antenna receives only a very small fraction of the energy delivered by the transmitting antenna, because the electric and magnetic plane waves spread all over free space as they progress from the source to the receiver site. In addition to the low signal-power level, there is a noise problem from the receiving antenna to the detector. The signal-to-noise ratio of the detector must meet the requirement for adequate reception. For instance, the required peak signal-to-noise ratio for telephony, double-sideband marginally commercial set by the Commercial Communication and Information Recommendation (CCIR) report 339 is

$$R_0 = \left(\frac{S}{N}\right)_0 = 27\ \text{dB/6-KHz band} \tag{12-2-1}$$

In other words, if the signal-to-noise ratio at the receiver is below the required level, the signal cannot be received. All the atmospheric noise information presented in CCIR 322 relates to a short vertical receiving antenna. Some allowance must be made for the effects of directivity on the signal-to-noise ratio, but experimental information on the effects of directivity is scarce. The atmospheric noise is assumed to be isotropically distributed, so the noise figure would be independent of directional properties. In practice, however, when the azimuthal direction of the antenna beam, for example, is in the direction of the thunderstorm, the noise figure is increased correspondingly, compared with the omnidirectional antenna. The directivity in the vertical plane differentiates in favor of or against the reception of noise from a strong source. Thus the movement of storms into and out of the antenna beam is expected to increase noise variability. The allowance for all such considerations is about 6 dB. Long-distance communication at high frequencies is normally achieved by the use of a highly directional antenna. Therefore an allowance of 6 dB must be made for the effects of directivity on the signal-to-noise ratio. It is sometimes necessary to change the signal-to-noise ratio from one band to another. The procedures are straightforward.

EXAMPLE 12-2-1 Determination of Signal-to-Noise Ratio

Determine the signal-to-noise ratio per 1 kHz band for the random (Rayleigh) fading condition.

Solution:

The required peak-to-peak signal-to-noise ratio under steady (nonfading) conditions in CCIR report 399 is

$$R_0 = 27 \text{ dB/6-KHz band}$$

We change the peak-to-peak ratio R_0 to R_0 (rms) at the half-power points, because the rms value of one peak is -3 dB and the rms value of the peak-to-peak ratio is -6 dB. Thus

$$R_0(\text{rms}) = 27 - 6 = 21 \text{ dB/6-kHz band}$$

Next, we change dB per 6 kHz to dB per 1 kHz:

$$\log\frac{1}{6} = -7.8 \doteq -8 \text{ dB} \quad \text{and} \quad R_0 = 21 - (-8) = 29 \text{ dB/1-kHz band}$$

The allowance for random (Rayleigh) fading is about 8 dB (CCIR report 266), so the required signal-to-noise ratio is

$$R_0 = 37 \text{ dB/1-kHz band}$$

The origins of the antenna noise temperature include: (1) the thermal noise from the reflector; (2) the thermal noise radiated from the ground; (3) the noise radiated from the galaxy; (4) the noise radiated from the atmosphere; and (5) the noise radiated from man-made objects. The relative importance of these types of noises depends upon their frequency and location:

1. At the frequency range of 1 MHz or less, the atmospheric noise originating from lightning is dominant. Precipitation noise from charged-particle bombardment is also important.
2. At the frequency range of 2 to 30 MHz, the noises from the galaxy and the sun are dominant.
3. At the frequency range of 10 to 200 MHz in heavily populated areas, man-made noise is usually dominant.
4. The combined noise temperature of galaxy noise plus the noise from oxygen and water vapor in the atmosphere is about 40 °K.

In general, we can describe an energy transmission circuit, as shown in Fig. 12-2-1, where

N_A is the atmospheric noise, in dBW above KTB;

S_i is the incoming signal, in dBW;

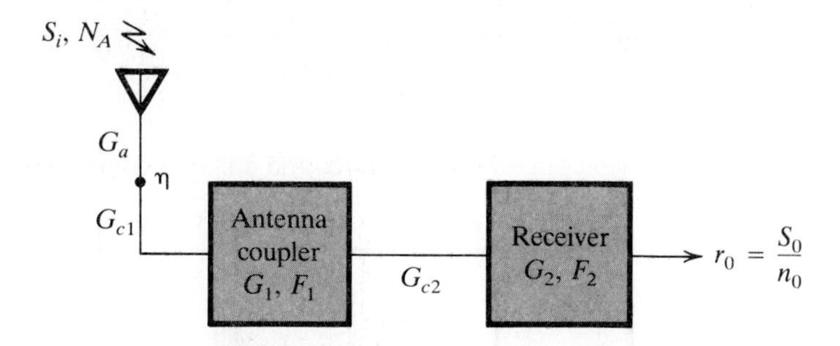

Figure 12-2-1 Energy transmission circuit.

G_a is the antenna gain, in dB;
η is the antenna efficiency (dimensionless);
$[\eta]$ is the antenna efficiency, in dB;
G_1 is the antenna coupler gain, in dB;
F_1 is the antenna coupler noise figure, in dB;
G_2 is the receiver gain, in dB;
F_2 is the receiver noise figure, in dB;
R_0 is the required signal-to-noise ratio at detector, in dB; and
G_{c1}, G_{c2} are the coaxial line losses, in dB.

The primary objective in analyzing an energy transmission circuit is to determine the minimum detectable incoming signal level for a required reception in terms of all system parameters. If we let lowercase letters represent the numerical amounts of the dB quantities just described,

f = antilog $(F/10)$;
g = antilog $(G/10)$;
r_0 = antilog $(R_0/10)$;
$[n_A]$ = antilog $(N_A/10)$;
n_A = atmospheric noise power = $(KTB)[n_A]$, in W; and
s_i = incoming signal power = antilog $(S_i/10)$, in W.

Then, if we let the total gain be

$$g = g_a g_{c1} g_1 g_{c2} g_2 \eta$$

and the system gain be

$$g_s = \eta g_{c1} g_1 g_{c2} g_2$$

the signal power at the detector is

$$s_0 = s_i g = s_i g_a \eta g_{c1} g_1 g_{c2} g_2 \quad \text{W} \tag{12-2-2}$$

The noise power at the detector is the sum of the three noise components. That is,

$$n_0 = n_{A0} + n_t + n_r \quad \text{W} \tag{12-2-3}$$

where

$n_{A0} = KTB[n_A]g_s$ and is the atmospheric noise power;

$n_t = KTB(1 - g_{c1})g_1 g_{c2} g_2 + KTB(1 - g_{c2})g_2$ and is the noise power generated by transmission lines; and

n_r is the noise power generated by the receiver and antenna coupler, as shown in Fig. 12-2-2.

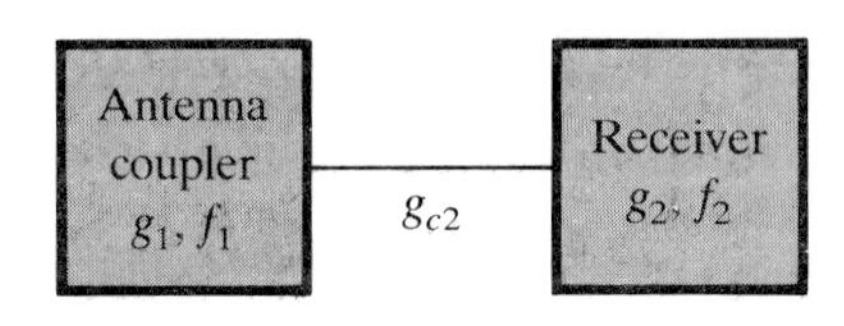

Figure 12-2-2 Equivalent circuit for antenna coupler and receiver.

The overall noise figure for the antenna coupler and receiver, from Eq. (12-1-6), is

$$f_{ar} = f_1 + \frac{f_2 - 1}{g_1 g_{c2}} \tag{12-2-4}$$

The noise figure n_r, then, is

$$n_r = f_{ar} g_1 g_{c2} g_2 KTB = (f_1 g_1 g_{c2} g_2 + f_2 g_2 - g_2)KTB \quad \text{W} \tag{12-2-5}$$

Substituting Eq. (12-2-5) into Eq. (12-2-3) gives us

$$\begin{aligned} n_0 = {} & KTB[n_A]g_s + KTB(1 - g_{c1})g_1 g_{c2} g_2 + KTB(1 - g_{c2})g_2 \\ & + KTB(f_1 g_1 g_{c2} g_2 + f_2 g_2 - g_2) \quad \text{W} \end{aligned} \tag{12-2-6}$$

Because the signal-to-noise ratio at the detector is

$$r_0 = \frac{s_0}{n_0} = \frac{s_i g}{n_{A0} + n_t + n_r} \tag{12-2-7}$$

the least detectable incoming signal power becomes

$$\begin{aligned} S_i &= \frac{n_{A0} + n_t + n_r}{g} r_0 \\ &= \frac{KTB([n_A]g_s + g_1 g_{c2} g_2 - g_1 g_{c1} g_{c2} g_2 - g_{c2} g_2 + f_1 g_1 g_{c2} g_2 + f_2 g_2) r_0}{g} \quad \text{W} \end{aligned} \tag{12-2-8}$$

The signal power can be expressed in dBW as

$$\begin{aligned} s_i = {} & 10 \log ([n_A]g_s + g_1 g_{c2} g_2 - g_1 g_{c1} g_{c2} g_2 - g_{c2} g_2 + f_1 g_1 g_{c2} g_2 + f_2 g_2) - 204 \\ & + 10 \log B + R_0 - G_a - G_{c1} - G_1 - G_{c2} - G_2 - [\eta] \quad \text{dBW} \end{aligned} \tag{12-2-9}$$

where 204 dBW per cycle bandwidth replaces 10 log (KT) at room temperature, 390 °K.

Equation (12-2-9) seems complicated. However, because the noise is a function of frequency, we can simplify and approximate Eq. (12-2-9) as follows:

Case 1. If $[n_A]g_s \gg (g_1 g_{c2} g_2 - g_1 g_{c1} g_{c2} g_2 - g_{c2} g_2 + f_1 g_1 g_{c2} g_2 + f_2 g_2)$, which is usually true for the frequency range of 4 to 15 MHz, and if $g_{c1} = g_{c2} < 1$, $g_1 \gg g_{c1}$, and $g_2 \gg g_{c2}$, Eq. (12-2-9) becomes

$$S_i = R_0 + N_A - G_a - 204 + 10 \log B \quad \text{dBW} \qquad (12\text{-}2\text{-}10)$$

Case 2. If $[n_A]g_s \ll (g_1 g_{c2} g_2 - g_1 g_{c1} g_{c2} g_2 - g_{c2} g_2 + f_1 g_1 g_{c2} g_2 + f_2 g_2)$, which is usually true for the frequency range of 16 to 30 MHz (but when the frequency is above 30 MHz, the galactic noise will be predominant), and if $N_r \gg n_t$, then for $f_1 g_1 g_{c2} g_2 \gg (f_2 - 1)g_2$, Eq. (12-2-9) becomes

$$S_i = R_0 + F - G_A - G_{c1} - [\eta] - 204 + 10 \log B \quad \text{dBW} \qquad (12\text{-}2\text{-}11)$$

For $(f_2 - 1)g_2 \gg f_1 g_1 g_{c2} g_2$, and Eq. (12-2-9) becomes

$$S_i = R_0 + F - G_a - G_{c1} - [\eta] - 204 + 10 \log B \quad \text{dBW} \qquad (12\text{-}2\text{-}12)$$

Then for $f_2 \gg 1$ and $f_2 = f_1 g_1 g_{c2}$, the expression for the signal is

$$S_i = R_0 + F_1 - G_a - G_{c1} - [\eta] - 201 + 10 \log B \quad \text{dBW} \qquad (12\text{-}2\text{-}13)$$

Case 3. If $[n_A]g_s \simeq (n_t + n_r)$, Eq. (12-2-9) becomes

$$S_i = R_0 + N_A - G_a - 201 + 10 \log B \ \text{dBW} \qquad (12\text{-}2\text{-}14)$$

12-3 ELECTRIC-FIELD MEASUREMENTS

As we stated in Section 0.4 regarding plane waves, the electric and magnetic intensities at the far-field in free space are always in phase and perpendicular to each other. The power density carried by these two waves at the observation point is expressed by the Poynting vector: [2]

$$P = \text{Re}(\mathbf{E} \times \mathbf{H}^*) \quad \text{W/unit area} \qquad (12\text{-}3\text{-}1)$$

where

E is the electric intensity, in V (rms) per unit length;

H is the magnetic intensity, in A (rms) per unit length;

* is the complex conjugate; and

Re is the real part.

We can determine the total power at the observation point by integrating the power-density function over an imaginary spherical surface through the observation point, with a radius of r from the source at the center of the sphere. That is,

$$\begin{aligned} P &= \iint_s \mathbf{p} \cdot d\mathbf{a} \\ &= \text{Re} \iint_s (\mathbf{E} \times \mathbf{H}^*) \cdot d\mathbf{a} \\ &= \frac{E^2}{\eta_0} \int_{\phi=0}^{2\pi} \int_{\theta=0}^{\pi} r^2 \sin\theta \, d\theta \, d\phi \\ &= \frac{E^2}{\eta_0}(4\pi r^2) \quad \text{W} \end{aligned} \qquad (12\text{-}3\text{-}2)$$

where

r is the distance, in m, between the observation point and source;
$\eta_0 = 377$ and is the intrinsic impedance of free space; and
E is the magnitude of the electric-field intensity, in V/m.

If the field-intensity meter reads volts (rms), as it often does, the time-average power density becomes

$$\langle p \rangle = (\mathbf{E} \times \mathbf{H}^*) = \frac{|E|^2}{\eta_0} \tag{12-3-3}$$

The energy stored in the electric field is

$$W_e = \varepsilon_0 \iiint |E|^2 \, dv \quad \text{J} \tag{12-3-4}$$

where $|E|$ is the magnitude of the electric-field intensity, and $\varepsilon_0 = 8.85 \times 10^{-12}$ F/m and is the permittivity, or capacitivity, of vacuum or air. The energy stored in the magnetic field is

$$W_m = \mu_0 \iiint |H|^2 \, dv \quad \text{J} \tag{12-3-5}$$

where $|H|$ is the magnitude of the magnetic-field intensity, and $\mu_0 = 4\pi \times 10^{-7}$ H/m is the permeability, or inductivity, of vacuum or air.

The units of measurement and the conversion of one unit to another are essential parts of making field-intensity measurements. Several commonly used units are:

1. dB—As we have described previously, the decibel (dB) is a dimensionless number expressing the ratio of two power levels. It is defined as

$$\text{dB} \equiv 10 \log_{10}\left(\frac{p_2}{p_1}\right) \tag{12-3-6}$$

   The two power levels are relative to each other. If power level p_2 is greater than power level p_1, dB is positive; if the opposite is true, dB is negative. Since $p = V^2/R$ when their voltages are measured across the same or equal resistors, the number of dB is

$$\text{dB} \equiv 20 \log_{10}\left(\frac{V_2}{V_1}\right) \tag{12-3-7}$$

   The voltage definition of dB has no meaning unless the two voltages under consideration appear across equal impedances. Above 10 GHz, the impedance of waveguides varies with frequency and the dB calibration is limited to power levels only.
2. dBW—The decibel above 1 watt (dBW) is another convenient measure to express power level p_2 with respect to a reference power level p_1 of 1 W. Similarly, if power level p_2 is lower than 1 W, the dBW is negative.
3. dBm—The decibel above 1 milliwat (dBm) is also a useful method of expressing power level p_2 with respect to a reference power level p_1 of 1 mW. Because the power level in the microwave region is quite low, the dBm unit is very useful in that frequency range.

4. dBμV—The decibel above 1 microvolt (dBμV) is a dimensionless voltage ratio in dB referred to a reference voltage of 1 μV. The field-intensity meters used for the measurements in the microwave region often are scaled in dBμV, because the power levels to be measured are usually quite low.
5. μV/m—Microvolts per meter (μV/m), or 10^{-6} volts per meter, express the electric-field intensity.
6. dBμV/m—The decibel above 1 microvolt per meter (dBμV/m) is a dimensionless field-intensity ratio in dB relative to 1 μV/m. This unit is often used in field-intensity measurements for the microwave region.
7. μV/m/MHz—Microvolts per meter per megahertz (μV/m/MHz), or 10^{-6} volts per meter per broadband electric-field–intensity distribution, represent a two-dimensional distribution in space and in frequency.
8. dBμV/m/MHz—The decibel above 1 microvolt per meter per megahertz (dBμV/m/MHz) is a dimensionless broadband electric-field–intensity distribution ratio relative to 1 μV/m/MHz.
9. μV/MHz—Microvolts per megacycle per second of bandwidth (μV/MHz) are units of 10^{-6} volt-seconds of broadband voltage distribution in the frequency domain. The use of this unit is based on the assumption that the voltage is evenly distributed over the bandwidth of interest.
10. dB/bandwidth—The decibel above a bandwidth is a dimensionless frequency-band distribution ratio relative to a certain frequency-range bandwidth.

Figure 12-3-1 shows the conversion of dB scales in terms of power and voltage. These scales are based on the following equations:

$$\text{dBW} = -30 + \text{dBm} = -60 + \text{dB}\mu\text{W} \tag{12-3-8a}$$

and

$$\text{dBV} = -60 + \text{dBmV} = -120 + \text{dB}\mu\text{V} \tag{12-3-8b}$$

Different types of measurements are often needed for transmission lines, waveguides, and antennas in actual use. Three of the most common types of measurements are:

Interference measurements. The noise and spurious signals conducted and/or

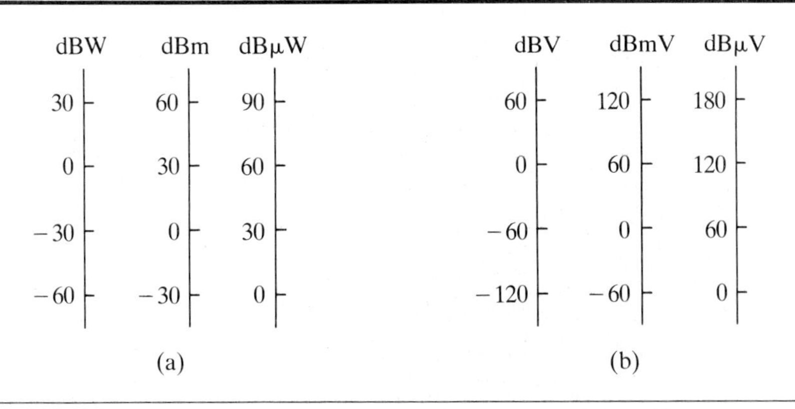

Figure 12-3-1 Conversion of dB scales in terms of (a) power and (b) voltage.

radiated from equipment provide data for designing or modifying equipment to a satisfactory level in order to reduce the interference signals generated.

Susceptibility measurements. The effects on electronic equipment of environmental interference provide data for designing or modifying equipment to make it less susceptible to interference. These measurements include tests for conducted and radiated interference and determine the ability of the subject equipment to operate in an environment that produces interference.

Electric field-intensity measurements. The field-intensity measurements can be made by using a receiver with a calibrated antenna. The receiver meter can be calibrated to read the absolute power level; the readings then indicate the field intensity at the site of the receiving antenna.

12-4 ELECTRIC-FIELD COMPUTATIONS

12-4-1 Conversion of Transmitting Power to Field Intensity

Figure 12-3-2 shows a setup for measuring the field intensity at distance R m from the transmitting antenna. The power density at R m from the transmitting antenna is

$$P_d = \frac{p_t g_a g_c}{4\pi R^2} \quad \text{W/m}^2 \tag{12-3-9}$$

where

p_t is the transmitting power (W);

R is the distance (m) from the power radiated by the transmitting antenna to the point of observation;

g_a is the antenna gain (numeric);

g_c is the cable loss (numeric); and

p_d is the power density in W/m^2 at the point of observation.

The field intensity at the observation point, from Eq. (12-3-3), is

$$E = 19.4\sqrt{p_d} \quad \text{V/m} \tag{12-3-10}$$

We usually find it helpful to express the power density and field intensity in dB terms.

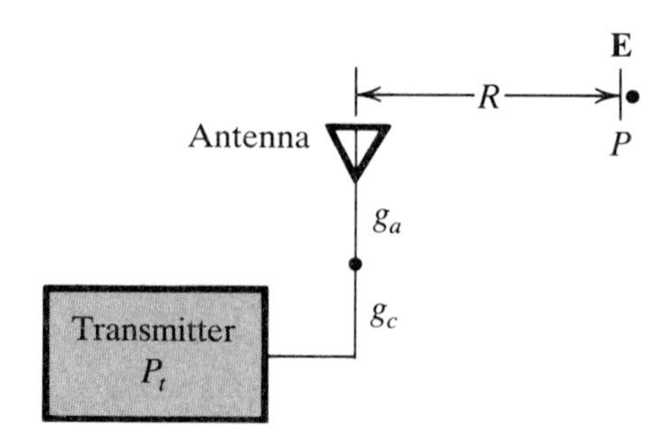

Figure 12-3-2 Field intensity in terms of transmitting power.

That is,

$$P_d = -11 \text{ dB} - 20 \log_{10}(R) + P_t(\text{dBW}) + G_a(\text{dB}) - G_c(\text{dB}) \quad \text{dBW/m}^2 \qquad (12\text{-}3\text{-}11)$$

and

$$E = 15 \text{ dB} - 20 \log_{10}(R) + P_t(\text{dBW}) + G_a(\text{dB}) - G_c(\text{dB}) \quad \text{dBV/m} \qquad (12\text{-}3\text{-}12)$$

Subtracting Eq. (12-3-11) from Eq. (12-3-12), we convert power density, in dBW/m^2, to field intensity, in dBV/m. That is,

$$E = 26 \text{ dB} + P_d(\text{dBW/m}^2) \quad \text{dBW/m} \qquad (12\text{-}3\text{-}12a)$$

Note that if values per square centimeter and per centimeter are desired, a term of -40 dB must be added to Eqs. (12-3-11), (12-3-12), and (12-3-12a). The uppercase letters P and G indicate dB values.

12-4-2 Conversion of Receiving Power to Field Intensity and Power Density

Figure 12-3-3 shows a setup for measuring field intensity from receiving power. The power density at the head of the antenna is

$$P_d = \frac{4\pi P_r}{\lambda^2 g_a g_c} \quad \text{W/m}^2 \qquad (12\text{-}3\text{-}13)$$

where

P_r is the receiving power (W);

g_a is the antenna gain (numeric);

g_c is the cable loss (numeric);

P_d is the power density at the antenna aperture (W/m^2);

λ is the wavelength of the source (m); and

$\lambda^2/4\pi$ is the isotropic antenna aperture.

The field intensity, then, is

$$E = 19.4\sqrt{P_d} \quad \text{V/m} \qquad (12\text{-}3\text{-}14)$$

We can express the power density and field intensity in dB as

$$P_d = 11 \text{ dB} - 20 \log_{10}(\lambda) + P_r(\text{dBw}) - G_a(\text{dB}) - G_c(\text{dB}) \quad \text{dBW/m}^2 \qquad (12\text{-}3\text{-}15)$$

and

$$E = 37 - 20 \log_{10}(\lambda) + P_r(\text{dBw}) - G_a(\text{dB}) - G_c(\text{dB}) \quad \text{dBV/m} \qquad (12\text{-}3\text{-}16)$$

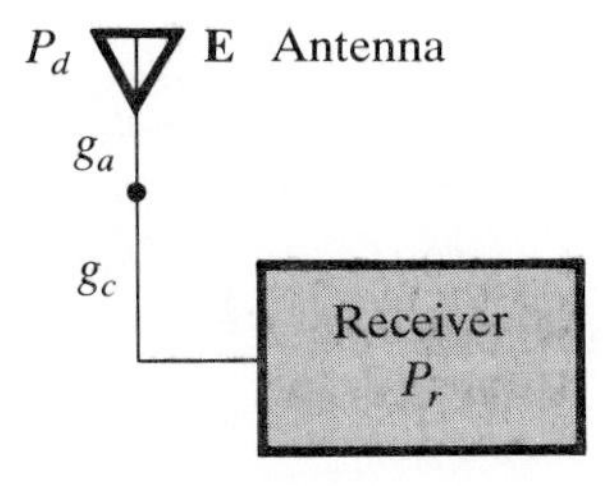

Figure 12-3-3 Power density measurement.

If we express the power density in mW/m^2, the field intensity is represented by

$$E = 0.613\sqrt{P_d} \quad V/m \tag{12-3-17}$$

12-4-3 Conversion of Receiving Voltage to Field Intensity and Power Density

The field-intensity meter at the receiving position is usually scaled in either μV or $dB\mu V$, so conversion of the receiving voltage to field intensity is required. Figure 12-3-4 shows how the receiving voltage is used to measure field intensity.

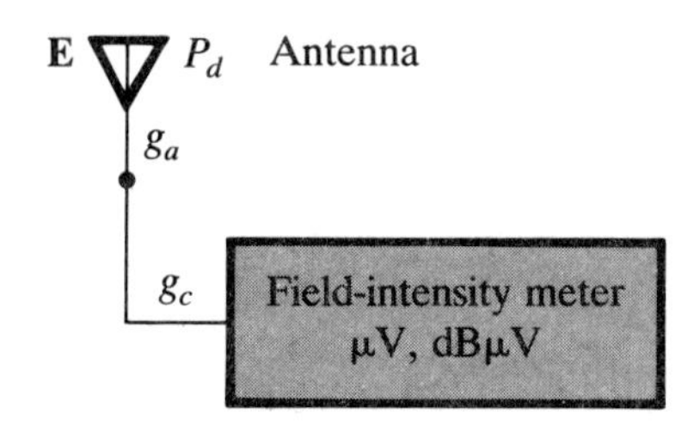

Figure 12-3-4 Field intensity in terms of receiving voltage.

The input impedance of the field-intensity receiving meter is normally specified by the manufacturer to be 50 Ω, because the coaxial line connected to the meter usually has a characteristic impedance of 50 Ω. Hence, the input impedance of the meter will match the impedance coaxial line perfectly. If a different input impedance is specified, a coaxial line with the same characteristic impedance must be chosen for impedance matching. Figure 12-3-5 shows the equivalent network for input impedance matching.

The input power to the intensity meter is

$$P_r = \frac{V^2}{Z_\ell} \quad W \tag{12-3-18}$$

the power density at the antenna aperture is

$$P_d = \frac{0.251V^2}{\lambda^2 g_a g_c} \quad W/m^2 \tag{12-3-19}$$

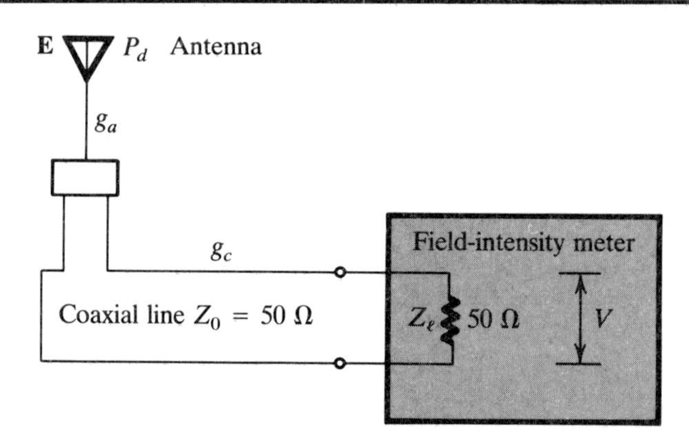

Figure 12-3-5 Input impedance matching to coaxial line.

and the field intensity at the antenna aperture becomes

$$E = \frac{4\pi V}{\lambda} \frac{30}{g_a g_c Z_\ell} \quad \text{V/m} \tag{12-3-20}$$

where

E is the electric-field intensity, in V/m (rms);
V is the intensity-meter reading, in V (rms);
λ is the wavelength of the source, in m;
g_a is the antenna gain (numeric);
g_c is the coaxial line loss (numeric); and
Z_ℓ is the input impedance, in Ω, of the meter—as specified by the manufacturer.

Similarly, we can express the power density and field intensity in dB:

$$P_d = -6 \text{ dB} - 20 \log_{10}(\lambda) + V(\text{dBV}) - G_a(\text{dB}) - G_c(\text{dB}) \quad \text{dBW/m}^2 \tag{12-3-21}$$

where

λ is the wavelength of the source expressed in meters;
V is the intensity-meter reading in volts (rms); and
$Z_\ell = 50$ Ω, which is the input impedance of the meter;

and

$$E = 19.8 \text{ dB} - 20 \log_{10}(\lambda) + V(\text{dBV}) - G_a(\text{dB}) - G_c(\text{dB}) \quad \text{dBV/m} \tag{12-3-22}$$

where λ is the wavelength in meters.

The algebraic sum of the first three terms in Eq. (12-3-22) is called the *antenna factor* (AF), or

$$AF = 19.8 \text{ dB} - 20 \log_{10}(\lambda) - G_a(\text{dB}) \quad \text{dB} \tag{12-3-23}$$

Note that the antenna factor is a function of frequency. The antenna manufacturer normally furnishes a graph of the AF for that antenna. Figure 12-3-6 shows the antenna

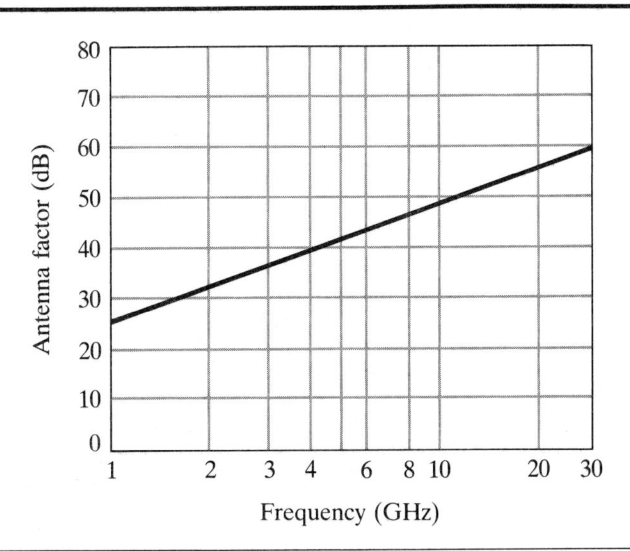

Figure 12-3-6 Antenna-factor graph for log-spiral antenna.

factor for a log-spiral antenna. The field intensity is the algebraic sum of the intensity-meter reading in dBμV and the antenna factor, from Eq. (12-3-22), if the coaxial line is lossless. That is,

$$E = V(\text{dBV}) + AF(\text{dB}) - G_c(\text{dB}) \quad \text{dBV/m} \tag{12-3-24}$$

EXAMPLE 12-3-1 Field-Intensity Measurement

The receiving antenna has a gain of 10 dB, and the coaxial line connecting the antenna to the receiving intensity meter is assumed to be lossless. The characteristic impedance of the coaxial line and the input impedance of the meter are both 50 Ω. The signal frequency is 3 GHz, and the meter reading is 100 μV.

Determine:

a) the power density;
b) the field intensity at the receiving antenna aperture; and
c) the antenna factor.

Solutions:

a) Power density.

1. The intensity-meter reading is 40 dBμV, or −80 dBV.
2. The power input to the receiver with an input impedance of 50 Ω, from Eq. (12-3-18), is

$$p_r = \frac{V^2}{R} = \frac{(100 \times 10^{-6})^2}{50}$$
$$= 200 \times 10^{-12} \text{ W}$$

3. The power density at the antenna aperture is determined from Eq. (12-3-19):

$$p_d = \frac{0.251 V^2}{\lambda^2 g_{a2} g_{c2}} = \frac{(0.251)(100 \times 10^{-6})^2}{(0.1)^2(10)}$$
$$= 251 \times 10^{-10} \text{ W/m}^2$$

4. The power density can be expressed in dB of Eq. (12-3-21) or

$$p_d = -6 \text{ dB} - 20 \log_{10}(0.1) - 80 \text{ (dBV)} - 10 \text{ (dB)} = -76 \text{ dBW/m}^2$$
$$= -6 \text{ dB} - 20 \log_{10}(10) - 20 \text{ (dB}\mu\text{V)} - 10 \text{ (dB)} = -56 \text{ dB}\mu\text{W/cm}^2$$
$$= -6 \text{ dB} - 20 \log_{10}(0.1) - 20 \text{ (dB}\mu\text{V)} - 10 \text{ (dB)} = -16 \text{ dB}\mu\text{W/m}^2$$

b) Field intensity.

1. The field intensity is obtained from Eq. (12-3-10):

$$E = \sqrt{120\pi p_d}$$
$$= \sqrt{(120\pi)(251 \times 10^{-10})}$$
$$= 3076.7 \times 10^{-6} \text{ V/m}$$

2. The field intensity can be expressed in dB by use of Eq. (12-3-22):

$$E = 19.8 \text{ dB} - 20 \log_{10}(0.1) - 80 \text{ (dBV)} - 10 \text{ (dB)} = -50.2 \text{ dBV/m}$$
$$= 19.8 \text{ dB} - 20 \log_{10}(10) + 40 \text{ (dB}\mu\text{V)} - 10 \text{ (dB)} = 29.8 \text{ dB}\mu\text{V/cm}$$
$$= 19.8 \text{ dB} - 20 \log_{10}(0.1) + 40 \text{ (dB}\mu\text{V)} - 10 \text{ (dB)} = 69.8 \text{ dB}\mu\text{V/m}$$

c) Antenna factor.

1. The antenna factor, from Eq. (12-3-23), is

$$AF = 19.8 \text{ dB} - 20 \log_{10}(10) - 10 \text{ (dB)}$$
$$= -10.2 \text{ dB for } \lambda, \text{ in cm}$$

or

$$AF = 19.8 \text{ dB} - 20 \log_{10}(0.1) - 10 \text{ (dB)}$$
$$= 29.8 \text{ dB for } \lambda, \text{ in m}$$

REFERENCES

[1] Liao, Samuel Y., "Design and Analysis of a polarization Diversity Antenna for High Frequency Reception." San Diego: Naval Electronics Laboratory Center, September 1969.

[2] Liao, Samuel Y., "Measurements and Computations of Electric Field Intensity and Power Density," *IEEE Trans. on Instrumentation and Measurement*, vol. IM-26, no. 1, March 1977.

Problems

12-1 A receiving antenna has an efficiency of 90%, a directive gain of 20 dB, and an effective antenna temperature $T_a = 200$ °K. The antenna is connected to a preamplifier with a power gain of 20 dB and a noise figure of 6 dB. The preamplifier is followed by a 30-m transmission line having a loss of 3 dB. The ambient temperature is 300 °K. The transmission line leads to another amplifier, which has a gain of 23 dB and a noise figure of 10 dB. Find the signal-to-noise ratio at the output of this system, assuming an incident field intensity of 200 μV/m at 2 GHz with a bandwidth of 10 MHz.

12-2 A transmitting antenna has a gain of 6 dB, and the coaxial line connecting the antenna to a transmitter is assumed to be lossless. The characteristic impedance of the coaxial line and the output impedance of the transmitter are both 50 Ω. The signal frequency is 200 MHz, and the transmitter power is 2 kW. Determine the power density in dBW/m^2 and the field intensity in dBV/m at the receiving antenna aperture for a distance of 10 m.

12-3 A generator having an output of 4.5 mW supplies power to an amplifier, which in turn supplies power to a transmitting antenna. If the input power level to the antenna is 16 dB above an mW reference, what power gain in dB is required for the amplifier?

12-4 The output of a transistor amplifier is 5.2 W.

a) Find the output power in dBm.

b) How many dB will be added to the output level if the power is made 4 times greater?

c) A transformer of 70% efficiency is inserted between the output of the amplifier and the load. What is the loss of the transformer in dB?

12-5 A receiving antenna has a gain of 15 dB and the coaxial line connecting the antenna to the receiving field-intensity meter has a power loss of 3 dB. The characteristic impedance of the coaxial line and the input impedance of the meter are both 50 Ω. The signal frequency is 5 GHz and the meter reading is 500 μV.

a) Convert the intensity-meter reading of 500 μV to dBμV and dBV.

b) Calculate the power input to the receiver in W.

c) Determine the power density at the antenna aperture in W/m^2.

d) Convert the power density from W/m^2 to dBW/m^2, $dB\mu W/m^2$, and $dB\mu W/cm^2$.

e) Calculate the electric-field intensity at the antenna aperture in V/m.

f) Convert the field intensity from V/m to dBV/m and dBμV/cm.

g) Find the antenna factor.

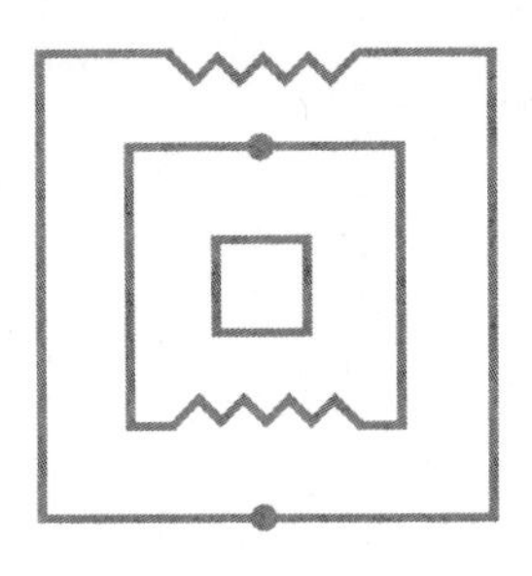

Appendixes

APPENDIX A Equations for Transmission Lines, Waveguides, and Antennas

APPENDIX B Hyperbolic Functions

APPENDIX C Constants of Materials

APPENDIX D Characteristics of Transmission Lines

APPENDIX E Characteristic Impedances of Common Transmission Lines

APPENDIX F First-Order Bessel Function Values

APPENDIX G Even-Mode and Odd-Mode Characteristic Impedances for Coupled Microstrip

APPENDIX H Values of Complete Elliptic Integrals of the First Kind

APPENDIX I Television (TV) Channel Frequencies

APPENDIX J Wire Data

APPENDIX K Hankel Functions

APPENDIX L Commercial Lasers and LED Sources

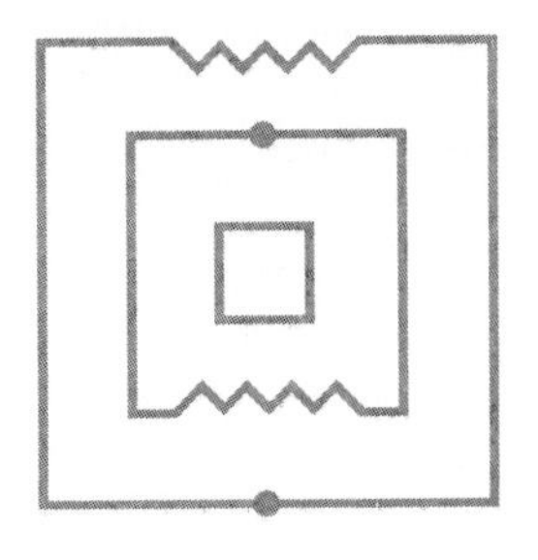

Appendix A

Equations for Transmission Lines, Waveguides, and Antennas

Device	Transmission Lines	Waveguides and Antennas
Theory	Distributed Circuit Theory	Electromagnetic Field Theory
Laws	Ohm's law: $R = \frac{V}{I}$ Kirchhoff's laws: $\sum_j v_j = 0$ and $\sum_j i_j = 0$	Maxwell's equations $\nabla \times \mathbf{E} = -\frac{\partial \mathbf{B}}{\partial t} = -j\omega\mu\mathbf{H}$ $\nabla \times \mathbf{H} = \sigma\mathbf{E} + \frac{\partial \mathbf{D}}{\partial t} = (\sigma + j\omega\varepsilon)\mathbf{E}$ $\nabla \cdot \mathbf{B} = 0$ $\nabla \cdot \mathbf{D} = \rho_v$
Voltage	$\frac{\partial v}{\partial z} = Ri - L\frac{\partial i}{\partial t}$	$\nabla \times \mathbf{E} = -\mu\frac{\partial \mathbf{H}}{\partial t}$
Current	$\frac{\partial i}{\partial z} = Gv - C\frac{\partial v}{\partial t}$	$\nabla \times \mathbf{H} = \sigma\mathbf{E} + \varepsilon\frac{\partial \mathbf{E}}{\partial t}$

(*continued*)

Device	Transmission Lines	Waveguides and Antennas
Theory	Distributed Circuit Theory	Electromagnetic Field Theory
Wave Equations	$\frac{\partial^2 v}{\partial z^2} = RGv + (RC + LG)\frac{\partial v}{\partial t} + LC\frac{\partial^2 v}{\partial t^2}$	$\nabla^2 \mathbf{E} = \mu\sigma \frac{\partial \mathbf{E}}{\partial t} + \mu\varepsilon \frac{\partial^2 \mathbf{E}}{\partial t^2}$
	$\frac{\partial^2 i}{\partial z^2} = RGi + (RC + LG)\frac{\partial i}{\partial t} + LC\frac{\partial^2 i}{\partial t^2}$	$\nabla^2 \mathbf{H} = \mu\sigma \frac{\partial \mathbf{H}}{\partial t} + \mu\varepsilon \frac{\partial^2 \mathbf{H}}{\partial t^2}$
	$\frac{\partial^2 v}{\partial z^2} = (R + j\omega L)(G + j\omega C)V$	$\nabla^2 \mathbf{E} = j\omega\mu(\sigma + j\omega\varepsilon)\mathbf{E}$
	$\frac{\partial^2 i}{\partial z^2} = (R + j\omega L)(G + j\omega C)I$	$\nabla^2 \mathbf{H} = j\omega\mu(\sigma + j\omega\varepsilon)\mathbf{H}$
	$\frac{\partial^2 v}{\partial z^2} = \gamma^2 v$	$\nabla^2 \mathbf{E} = \gamma^2 \mathbf{E}$
	$\frac{\partial^2 i}{\partial z^2} = \gamma^2 i$	$\nabla^2 \mathbf{H} = \gamma^2 \mathbf{H}$
Propagation Constant	$\gamma = \sqrt{(R + j\omega L)(G + j\omega C)}$	$\gamma = \sqrt{j\omega\mu(\sigma + j\omega\varepsilon)}$
Characteristic Impedance	$\mathbf{Z}_0 = \sqrt{\frac{Z}{Y}} = \sqrt{\frac{R + j\omega L}{G + j\omega C}}$	$\eta = \sqrt{\frac{j\omega\mu}{\sigma + j\omega\varepsilon}}$
	$\mathbf{Z}_0 = \sqrt{\frac{L}{C}}$ (lossless or low-loss)	$\eta = \sqrt{\frac{\mu}{\varepsilon}}$ (lossless)
Attenuation and Phase Constants	$(R \ll \omega L$ and $G \ll \omega C)$	$\alpha = \frac{1}{2}\left(\sigma\sqrt{\frac{\mu}{\varepsilon}}\right)$
	$\alpha = \frac{1}{2}\left(R\sqrt{\frac{C}{L}} + G\sqrt{\frac{L}{C}}\right)$	$= \frac{1}{2}\sigma\eta$
	$= \frac{1}{2}\left(\frac{R}{\mathbf{Z}_0} + G\mathbf{Z}_0\right)$	$\beta = \omega\sqrt{\mu\varepsilon}$
	$\beta = \omega\sqrt{LC}$	
Phase Velocity	$v_p = \frac{\omega}{\beta} = \frac{1}{\sqrt{LC}}$	$v_p = \frac{1}{\sqrt{\mu\varepsilon}}$
Wavelength	$\lambda = \frac{2\pi}{\beta} = \frac{1}{f\sqrt{LC}}$	$\lambda = \frac{2\pi}{\beta} = \frac{1}{f\sqrt{\mu\varepsilon}}$
Power	$p = \frac{1}{2}\lvert I\rvert\lvert V\rvert$ $= \frac{1}{2}\lvert I\rvert^2 R$ $= \frac{1}{2}VI\cos(\theta_i - \theta_v)$	$\mathbf{p} = \frac{1}{2}(\mathbf{E} \times \mathbf{H}^*)$ $= \frac{1}{2}\eta H^2$ $P_{avg} = \frac{1}{2}\mathrm{Re}(\mathbf{E} \times \mathbf{H}^*)$

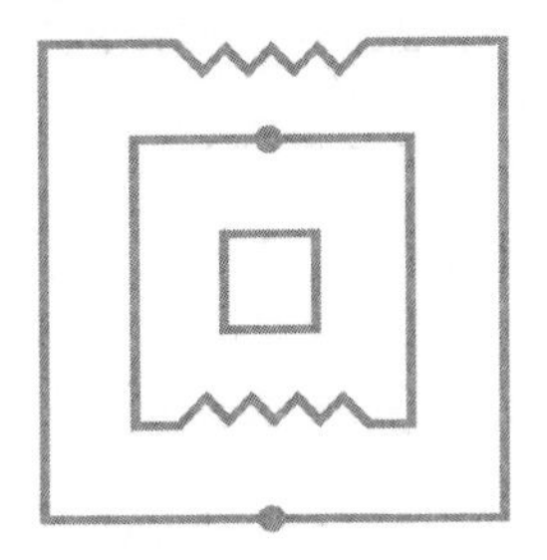

Appendix B

Hyperbolic Functions

The general solutions of the transmission-line equation, the components of traveling waves, $V_+(z)$ and $V_-(z)$, and many of the equations derived from them contain terms involving $e^{-\gamma z}$ and $e^{\gamma z}$. It is advantageous to replace the exponential factors by their corresponding hyperbolic functions. Euler's formula for exponential functions with an imaginary exponent is

$$e^{\pm j\gamma} = \cos\gamma \pm j\sin\gamma \tag{B-1}$$

Adding, subtracting, and rearranging Eq. (B-1) yield the sine and cosine functions in terms of the exponential functions, or

$$\sin\gamma = \frac{e^{j\gamma} - e^{-j\gamma}}{2j} \tag{B-2}$$

and

$$\cos\gamma = \frac{e^{j\gamma} + e^{-j\gamma}}{2} \tag{B-3}$$

Then

$$\tan\gamma = \frac{\sin\gamma}{\cos\gamma} = \frac{e^{j\gamma} - e^{-j\gamma}}{j(e^{j\gamma} + e^{-j\gamma})} \tag{B-4}$$

$$\cot\gamma = \frac{\cos\gamma}{\sin\gamma} = \frac{j(e^{j\gamma} + e^{-j\gamma})}{e^{j\gamma} - e^{-j\gamma}} \tag{B-5}$$

$$\sec \gamma = \frac{1}{\cos \gamma} = \frac{2}{e^{j\gamma} + e^{-j\gamma}} \tag{B-6}$$

and

$$\csc \gamma = \frac{1}{\sin \gamma} = \frac{2j}{e^{j\gamma} - e^{-j\gamma}} \tag{B-7}$$

The sine and cosine functions are sometimes called *circular functions*, because they can be defined for real values of γ by means of a circle. When $\gamma = \alpha + j\beta$, the sine and cosine functions have an imaginary term for their arguments. It might simplify the computation to change the circular functions to the hyperbolic functions. The hyperbolic functions in Euler's formula are

$$e^{\pm\gamma} = \cosh \gamma \pm \sinh \gamma \tag{B-8}$$

where

$$\sinh \gamma = \frac{e^{\gamma} - e^{-\gamma}}{2} \tag{B-9}$$

$$\cosh \gamma = \frac{e^{\gamma} + e^{-\gamma}}{2} \tag{B-10}$$

$$\tanh \gamma = \frac{\sinh \gamma}{\cosh \gamma} = \frac{e^{\gamma} - e^{-\gamma}}{e^{\gamma} + e^{-\gamma}} \tag{B-11}$$

$$\coth \gamma = \frac{\cosh \gamma}{\sinh \gamma} = \frac{e^{\gamma} + e^{-\gamma}}{e^{\gamma} - e^{-\gamma}} \tag{B-12}$$

$$\operatorname{sech} \gamma = \frac{1}{\cosh \gamma} = \frac{2}{e^{\gamma} + e^{-\gamma}} \tag{B-13}$$

and

$$\operatorname{csch} \gamma = \frac{1}{\sinh \gamma} = \frac{2}{e^{\gamma} - e^{-\gamma}} \tag{B-14}$$

The limiting values of some hyperbolic functions with real arguments are

$$\sinh (0) = 0 \tag{B-15}$$

$$\cosh (0) = 0 \tag{B-16}$$

$$\tanh (0) = 0 \tag{B-17}$$

$$\sinh (\infty) \to \infty \tag{B-18}$$

and

$$\tanh (\infty) \to 1 \tag{B-19}$$

Moreover, the hyperbolic identities are

$$\cosh^2 \gamma - \sinh^2 \gamma = 1 \tag{B-20}$$

$$\sinh (\alpha \pm j\beta) = \sinh \alpha \cos \beta \pm j \cosh \alpha \sin \beta \tag{B-21}$$

and

$$\cosh(\alpha \pm j\beta) = \cosh\alpha \cos\beta \pm j \sinh\alpha \sin\beta \tag{B-22}$$

in which the following relationships between the circular and hyperbolic functions are used:

$$\sinh(j\beta) = j \sin\beta \tag{B-23}$$

$$\sin(j\beta) = j \sinh\beta \tag{B-24}$$

$$\cosh(j\beta) = \cos\beta \tag{B-25}$$

and

$$\cos(j\beta) = \cosh\beta \tag{B-26}$$

Finally, some differentiation formulas that are needed occasionally are

$$\frac{d}{d\gamma}\sinh\gamma = \cosh\gamma \tag{B-27}$$

$$\frac{d}{d\gamma}\cosh\gamma = \sinh\gamma \tag{B-28}$$

and

$$\frac{d}{d\gamma}\tanh\gamma = \operatorname{sech}^2\gamma \tag{B-29}$$

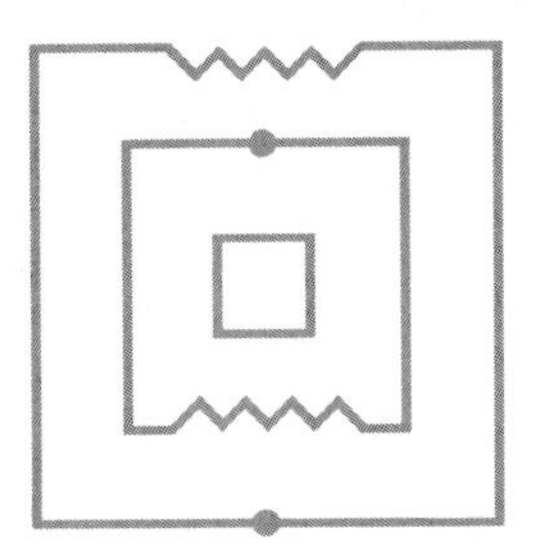

Appendix C

Constants of Materials

Table C-1 Conductivity σ (℧/m)

Conductor	σ	Insulator	σ
Silver	6.17×10^7	Quartz	10^{-17}
Copper	5.80×10^7	Polystyrene	10^{-16}
Gold	4.10×10^7	Rubber (hard)	10^{-15}
Aluminum	3.82×10^7	Mica	10^{-14}
Tungsten	1.82×10^7	Porcelain	10^{-13}
Zinc	1.67×10^7	Diamond	10^{-13}
Brass	1.50×10^7	Glass	10^{-12}
Nickel	1.45×10^7	Bakelite	10^{-9}
Iron	1.03×10^7	Marble	10^{-8}
Bronze	1.00×10^7	Soil (sandy)	10^{-5}
Solder	0.70×10^7	Sand (dry)	2×10^{-4}
Steel (stainless)	0.11×10^7	Clay	10^{-4}
Nichrome	0.10×10^7	Ground (dry)	10^{-4}–10^{-5}
Graphite	7.00×10^4	Ground (wet)	10^{-2}–10^{-3}
Silicon	1.20×10^3	Water (distilled)	2×10^{-4}
Water (sea)	3–5	Water (fresh)	10^{-3}
		Ferrite (typical)	10^{-2}

Table C-2 Dielectric constant (relative permittivity) ε_r

Material	ε_r	Material	ε_r
Air	1	Sand (dry)	4
Alcohol (ethyl)	25	Silica (fused)	3.8
Bakelite	4.8	Snow	3.3
Glass	4–7	Sodium chloride	5.9
Ice	4.2	Soil (dry)	2.8
Mica (ruby)	5.4	Styrofoam	1.03
Nylon	4	Teflon	2.1
Paper	2–4	Water (distilled)	80
Plexiglass	2.6–3.5	Water (sea)	20
Polyethylene	2.25	Water (dehydrated)	1
Polystyrene	2.55	Wood (dry)	1.5–4
Porcelain (dry process)	6	Ground (wet)	5–30
Quartz (fused)	3.80	Ground (dry)	2–5
Rubber	2.5–4	Water (fresh)	80

Table C-3 Relative permeability (magnetic constant) μ_r

Diamagnetic Material	μ_r	Ferromagnetic Material	μ_r
Bismuth	0.99999860	Nickel	50
Paraffin	0.99999942	Cast iron	60
Wood	0.99999950	Cobalt	60
Silver	0.99999981	Machine steel	300
		Ferrite (typical)	1,000
Paramagnetic Material	μ_r	Transformer iron	3,000
Aluminum	1.00000065	Silicon iron	4,000
Beryllium	1.00000079	Iron (pure)	4,000
Nickel chloride	1.00004	Mumetal	20,000
Manganese sulphate	1.0001	Superalloy	100,000

Table C-4 Properties of free space

Velocity of light in vacuum c	2.997925×10^{8} m/s
Permittivity ε_0	8.854×10^{-12} F/m
Permeability μ_0	$4\pi \times 10^{-7}$ H/m
Intrinsic impedance η_0	377 or 120π Ω

Table C-5 Physical constants

Charge of electron e	1.60×10^{-19} C
Mass of electron m	9.1×10^{-31} kg
Charge to mass ratio of electron $\frac{e}{m}$	1.76×10^{11} C/kg

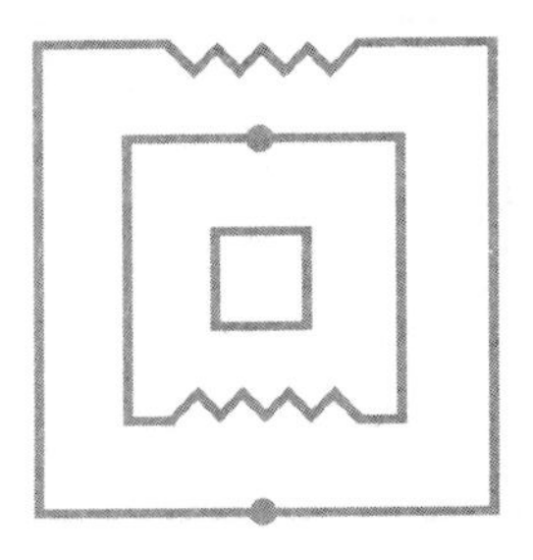

Appendix D

Characteristics of Transmission Lines

Table D-1 Transmission lines at radio frequencies

Class of Cables	Type	Inner Conductor	Nominal Overall Diameter (in.)	Nominal Capacitance (pF/ft)	Maximum Operating Voltage (rms)
50 Ω	RG-8A/U	7/0.0296	0.405	29.5	4000
50 Ω	RG-55A/U	0.035	0.216	28.5	1900
50 Ω	RG-58C/U	19/0.0071	0.195	28.5	1900
50 Ω	RG-122/U	27/0.005	0.160	29.3	1900
50 Ω	RG-174/U	7/0.0063	0.100	30.0	1000
50 Ω	RG-196A/U	7/0.0040	0.080	29.0	1000
High attenuation	RG-126/U	7/0.0203	0.275	29.0	3000
High attenuation	RG-21A/U	0.053	0.332	29.0	2700
High delay	RG-65A/U	0.008	0.405	44.0	1000
Twin conductor	RG-86/U	7/0.0285	0.30 × 0.65	7.8	—
Twin conductor	RG-130/U	7/0.0285	0.625	17.0	8000

Table D-2 Power transmission lines at 60 Hz

Circuit Voltage (kV, $L-L$)	Conductor Size (Thousands of circular mils, or AWG)	Equivalent Spacing (ft)	R (at 50 °C—Ω/ phase/mi)	X (Ω/ phase/mi)	Shunt C (megohms/ phase/mi)	Surge Imped. (Ω, L–N)
69	2/0	8	0.481	0.784	0.182	378
69	336.4	19	0.306	0.808	0.191	393
115	336.4	13	0.306	0.762	0.180	370
115	336.4	22	0.306	0.826	0.196	402
138	397.5	15	0.259	0.764	0.181	371
138	397.5	24	0.259	0.821	0.195	399
161	397.5	17	0.259	0.779	0.185	379
161	397.5	25	0.259	0.826	0.196	402

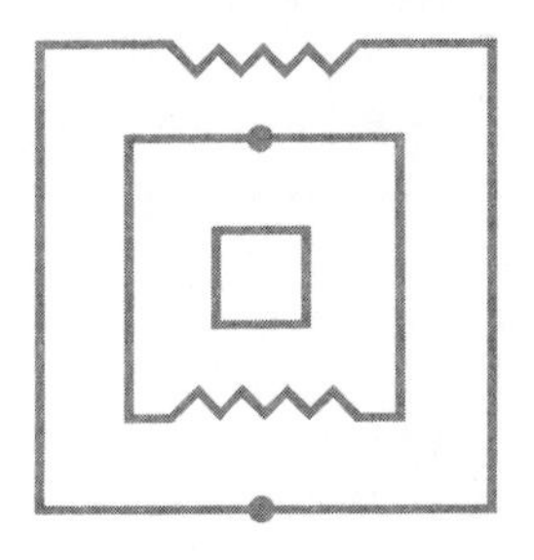

Appendix E

Characteristic Impedances of Common Transmission Lines

Line	Geometry	Characteristic Impedance
Two wire	d, D	$Z_0 \approx \frac{\eta}{\pi} \ln\left(\frac{2D}{d}\right) \qquad D \gg d$
Coaxial	E_r, a, b	$Z_0 = \frac{\eta}{2\pi} \ln\left(\frac{b}{a}\right) = \frac{60}{\sqrt{\varepsilon_r}} \ln\left(\frac{b}{a}\right)$
Confocal elliptic	$2a$, $2c$, $2b$	$Z_0 = \frac{\eta}{2\pi} \ln\left(\frac{b + \sqrt{b^2 - c^2}}{a + \sqrt{a^2 - c^2}}\right)$
Parallel plate	w, b	$Z_0 \approx \eta \frac{b}{\omega} \qquad \omega \gg b$
Collinear plate	D, w	$Z_0 \approx \frac{\eta}{\pi} \ln\left(\frac{4D}{\omega}\right) \qquad D \gg \omega$

(continued)

Line	Geometry	Characteristic Impedance
Wire above ground plane	h, d	$Z_0 \approx \frac{\eta}{2\pi} \ln\left(\frac{4h}{d}\right) \qquad h \gg d$
Shielded pair	D, d, s	$Z_0 \approx \frac{\eta}{\pi} \ln\left(\frac{2s}{d} \frac{D^2 - s^2}{D^2 + s^2}\right) \qquad D \gg d, \; s \gg d$
Wire in trough	w, d, h	$Z_0 \approx \frac{\eta}{2\pi} \ln\left(\frac{4\omega}{\pi d} \tanh\left(\frac{\pi h}{\omega}\right)\right) \qquad h \gg d, \; \omega \gg d$

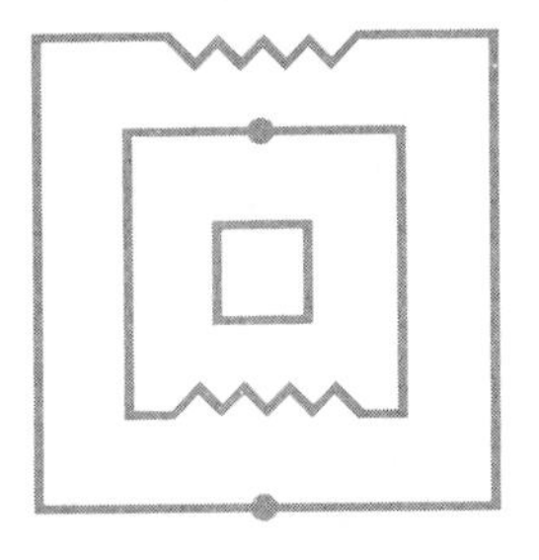

Appendix F

First-Order Bessel-Function Values

x	$J_1(x)$	x	$J_1(x)$	x	$J_1(x)$	x	$J_1(x)$	x	$J_1(x)$
.00	0.000	.92	0.413	1.86	0.582	2.86	0.389	3.84	−0.003
.02	+.010	.94	.420	1.88	.581	2.88	.382	3.86	.011
.04	.020	.96	.427	1.90	.581	2.90	.375	3.88	.019
.06	.030	.98	.433	1.92	.580	2.92	.368	3.90	.027
.08	.040	1.00	.440	1.94	.580	2.94	.361	3.92	.035
.10	.050	1.02	.446	1.96	.579	2.96	.354	3.94	.043
.12	.060	1.04	.453	1.98	.578	2.98	.346	3.96	.051
.14	.070	1.06	.459	2.00	.577	3.00	.339	3.98	.058
.16	.080	1.08	.465	2.02	.575	3.02	.331	4.00	.066
.18	.090	1.10	.471	2.04	.574	3.04	.324	4.10	.103
.20	.099	1.12	.477	2.06	.572	3.06	.316	4.20	.139
.22	.109	1.14	.482	2.08	.570	3.08	.309	4.30	.172
.24	.119	1.16	.488	2.10	.568	3.10	.301	4.40	.203
.26	.129	1.18	.493	2.12	.566	3.12	.293	4.50	.231
.28	.139	1.20	.498	2.14	.564	3.14	.285	4.60	.256
.30	.148	1.22	.503	2.16	.561	3.16	.277	4.70	.279
.32	.158	1.24	.508	2.18	.559	3.18	.269	4.80	.298
.34	.167	1.26	.513	2.20	.556	3.20	.261	4.90	.315
.36	.177	1.28	.517	2.22	.553	3.22	.253	5.00	.327
.38	.187	1.30	.522	2.24	.550	3.24	.245	5.05	.334
.40	.196	1.32	.526	2.26	.547	3.26	.237	5.10	.337
.42	.205	1.34	.530	2.28	.543	3.28	.229	5.16	.341

(continued)

x	$J_1(x)$	x	$J_1(x)$	x	$J_1(x)$	x	$J_1(x)$	x	$J_1(x)$
.44	.215	1.36	.534	2.30	.540	3.30	.221	5.20	.343
.46	.224	1.38	.538	2.32	.536	3.32	.212	5.26	.345
.48	.233	1.40	.542	2.34	.532	3.34	.204	5.30	.346
.50	.242	1.42	.545	2.36	.528	3.36	.196	5.32	.346
.52	.251	1.44	.549	2.38	.524	3.38	.186	5.34	.346
.54	.260	1.46	.552	2.40	.520	3.40	.179	5.36	.346
.56	.269	1.48	.555	2.42	.516	3.42	.171	5.38	.346
.58	.278	1.50	.558	2.44	.511	3.44	.162	5.40	.345
.60	.287	1.52	.561	2.46	.507	3.46	.154	5.47	.343
.62	.295	1.54	.563	2.48	.502	3.48	.146	5.50	.341
.64	.304	1.56	.566	2.50	.497	3.50	.137	5.56	.337
.66	.312	1.58	.568	2.52	.492	3.52	.129	5.60	.334
.68	.321	1.60	.570	2.54	.487	3.54	.121	5.66	.323
.70	.329	1.62	.572	2.56	.482	3.56	.112	5.70	.324
.72	.337	1.64	.573	2.58	.476	3.58	.104	5.80	.311
.74	.345	1.66	.575	2.60	.471	3.60	.095	5.90	.295
.76	.353	1.68	.576	2.62	.465	3.62	.087	6.00	.277
.78	.361	1.70	.578	2.64	.459	3.64	.079	6.10	.256
.80	.369	1.72	.579	2.66	.454	3.66	.070	6.20	.233
.82	.3765	1.74	.580	2.68	.448	3.68	.062	6.30	.208
.84	.384	1.76	.580	2.70	.442	3.70	.054	6.40	.182
.86	.3915	1.78	.581	2.72	.435	3.72	.045	6.60	.125
.88	.399	1.80	.581	2.74	.429	3.74	.037	6.70	.095
.90	.406	1.82	.582	2.76	.423	3.76	.029	6.80	.065
		1.84	.582	2.78	.416	3.78	.021	6.90	.035
				2.80	.410	3.80	.013	7.00	.005
				2.82	.403	3.82	.005	7.01	.000
				2.84	.396	3.83	.000		

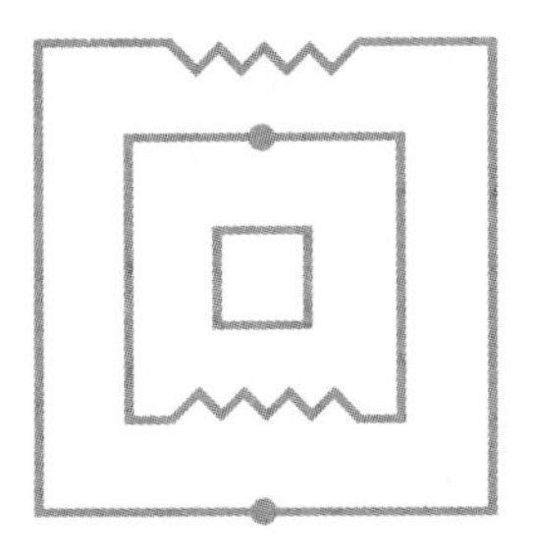

Appendix G

Even-Mode and Odd-Mode Characteristic Impedances for Coupled Microstrip

The characteristic impedances of coupled microstrip lines for even- and odd-modes are functions of the ratio w/h, and they can be determined using the curves on the following pages. The curves are plotted for $\varepsilon_r = 1.0, 6.0, 9.0$, and 12.0.

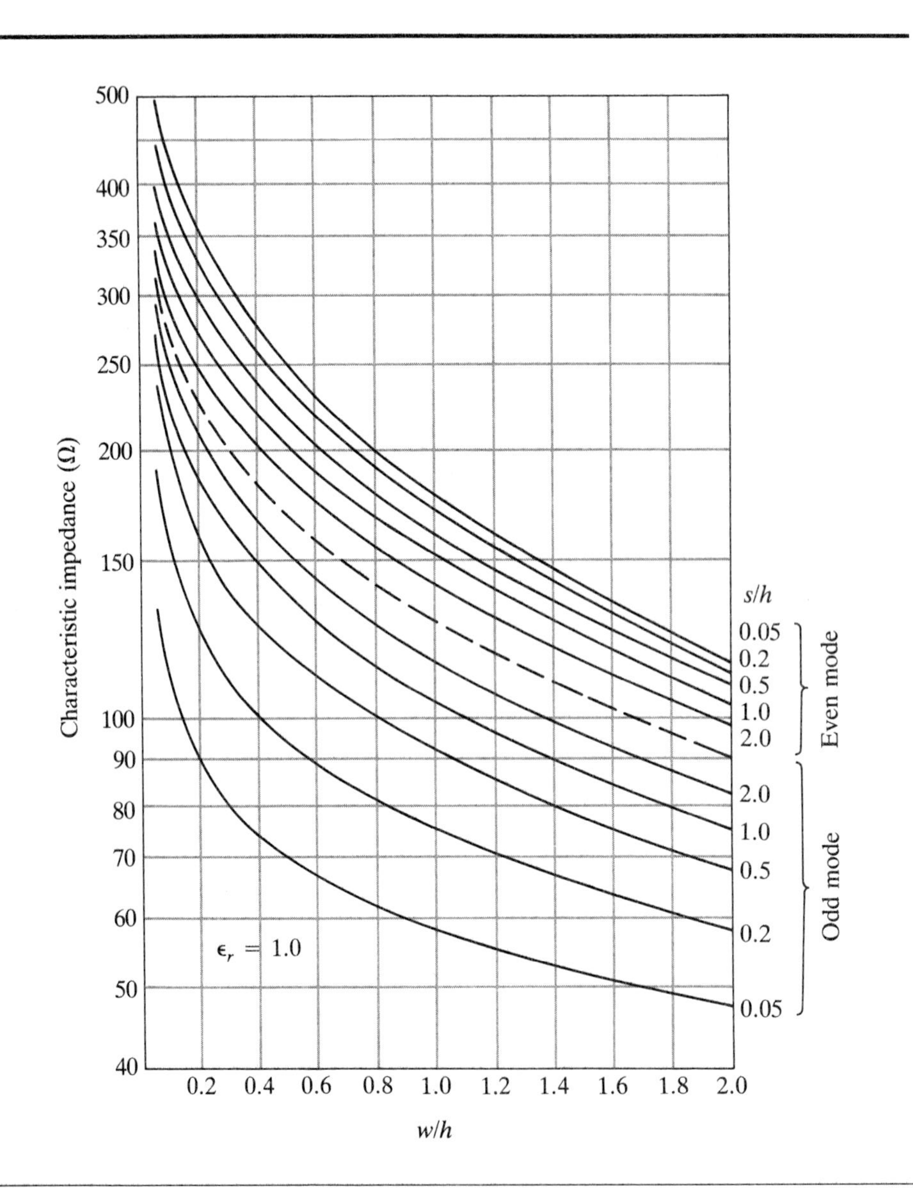

Figure G-1 Characteristic impedances for $\varepsilon_r = 1.0$

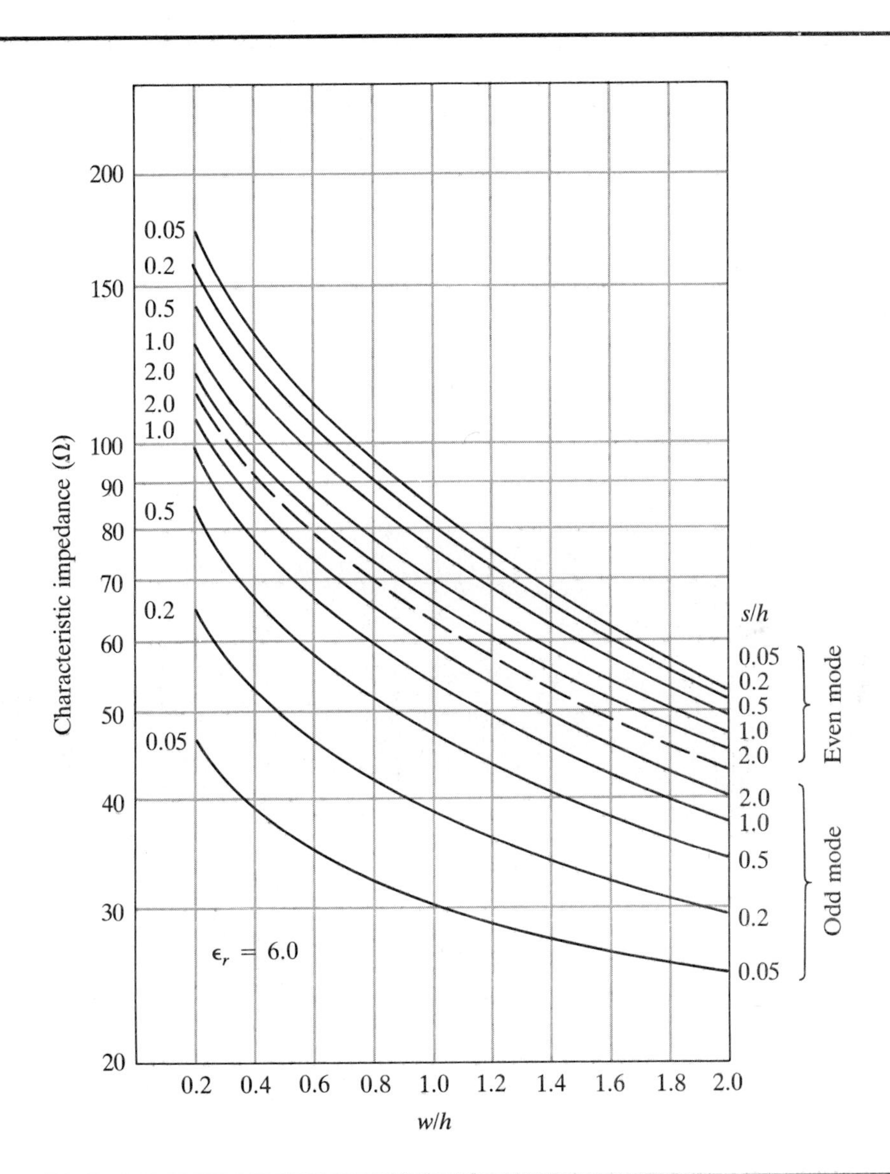

Figure G-2 Characteristic impedances for $\varepsilon_r = 6.0$

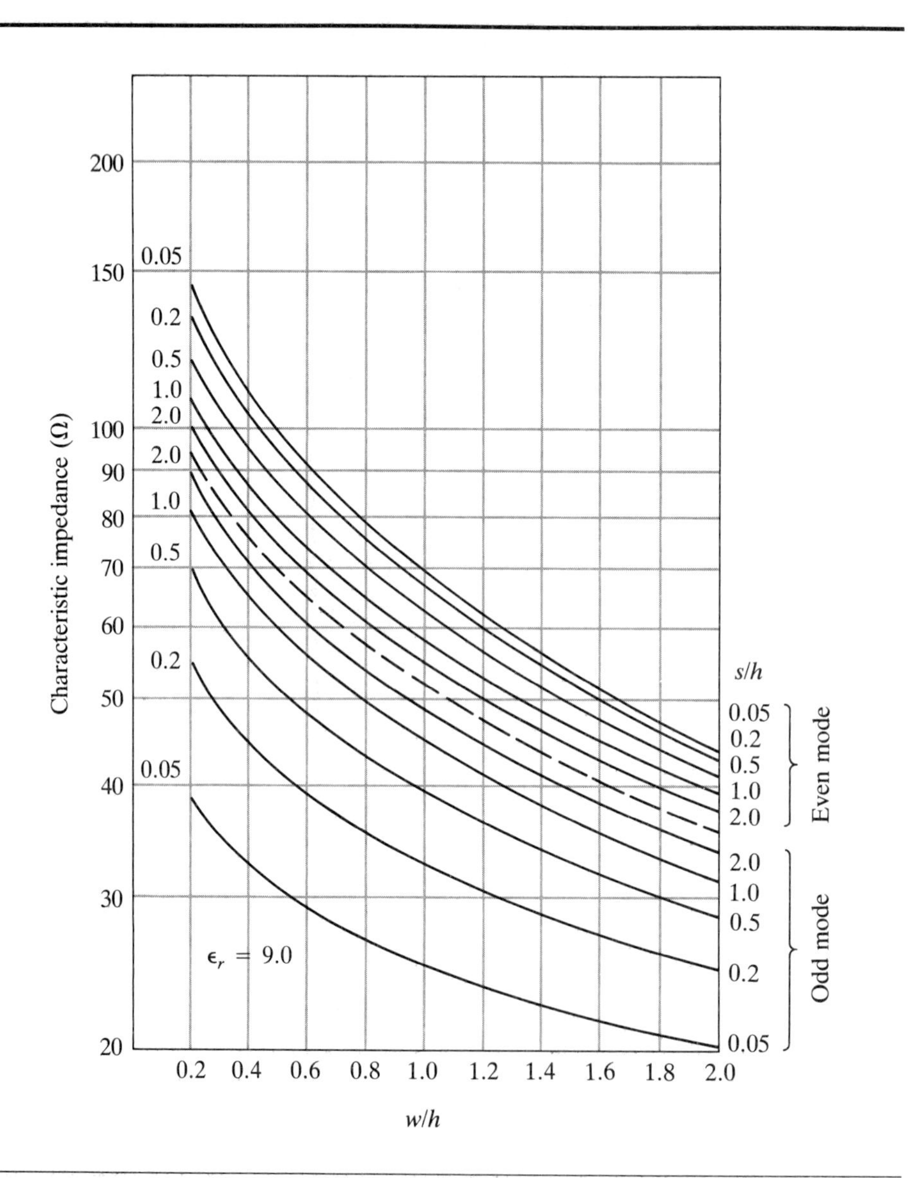

Figure G-3 Characteristic impedances for $\varepsilon_r = 9.0$

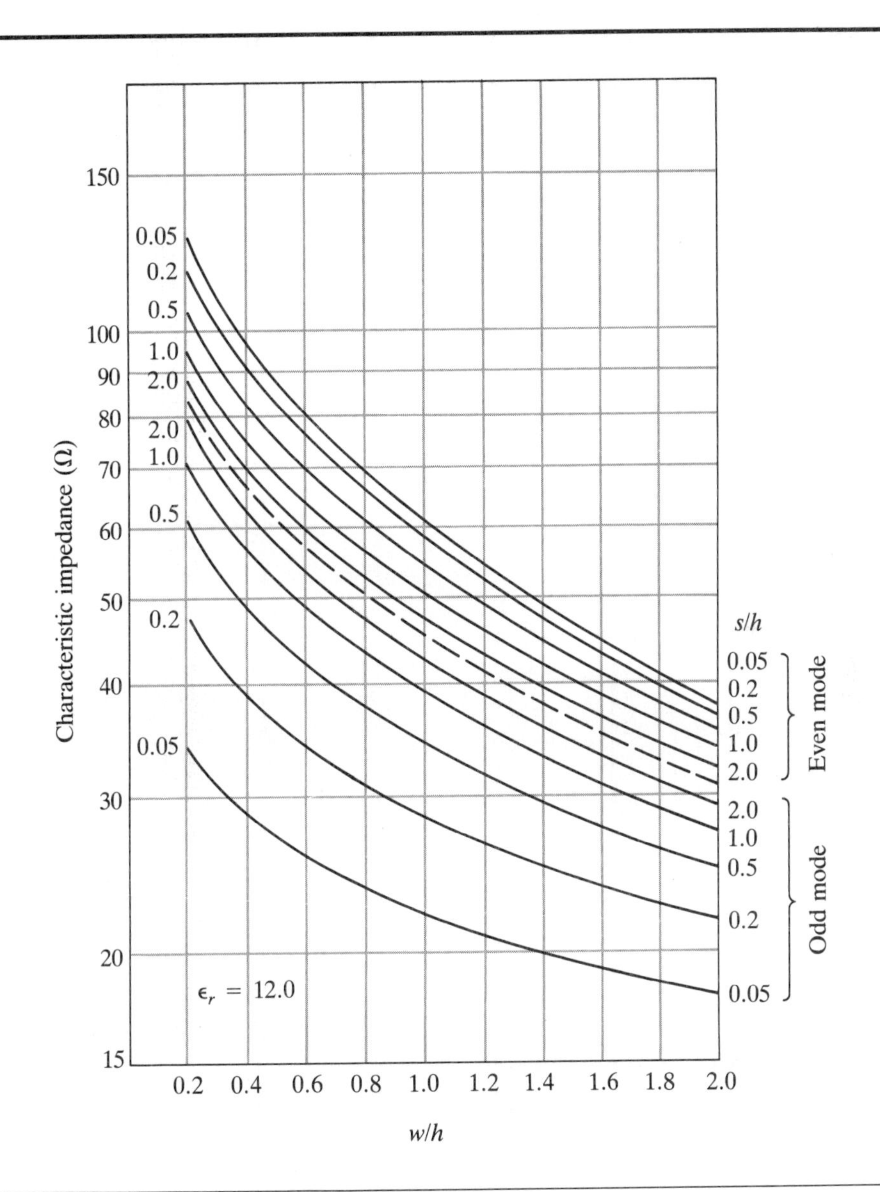

Figure G-4 Characteristic impedances for $\varepsilon_r = 12.0$

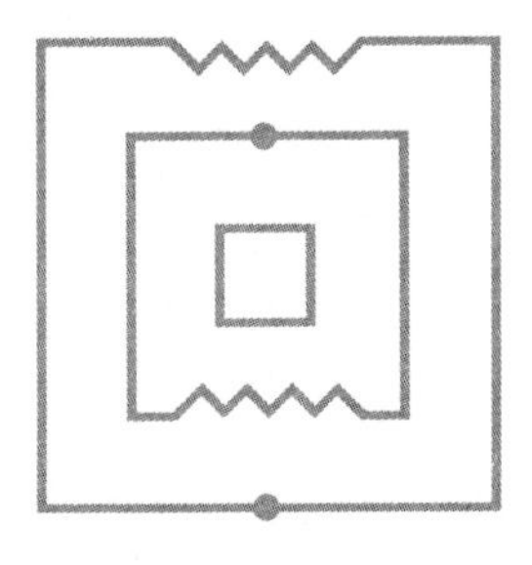

Appendix H

Values of Complete Elliptic Integrals of the First Kind

		K											
k^2		*0*	*1*	*2*	*3*	*4*	*5*	*6*	*7*	*8*	*9*	*d*	k'^2
0,0	1,5	708	747	787	828	869	910	952	994	*037	*080	41	0,9
0,1	1,6	124	169	214	260	306	353	400	448	497	546	47	0,8
0,2		596	647	698	751	804	857	912	967	*024	*081	53	0,7
0,3	1,7	139	198	258	319	381	444	508	573	639	706	63	0,6
0,4		775	845	916	989	*063	*139	*216	*295	*375	*457	76	0,5
0,5	1,8	541	626	714	804	895	989	*085	*184	*285	*398	94	0,4
0,6	1,9	496	605	718	834	953	*0076	*0203	*0334	*0469	*0609	123	0,3
0,7	2,	0754	0904	1059	1221	1390	1565	1748	1940	2140	2351	175	0,2
0,8		2572	2805	3052	3314	3593	3890	4209	4553	4926	5333	397	0,1
0,9		5781	6278	6836	7471	8208	9083	*0161	*1559	*3541	*6956	875	0,0
k'^2		*10*	*9*	*8*	*7*	*6*	*5*	*4*	*3*	*2*	*1*	*d*	k^2
		K′											

SOURCE: FROM EUGENE JAHNKE AND FRITZ EMDE, *TABLES OF FUNCTIONS WITH FORMULAE AND CURVES*, 4TH ED. NEW YORK: DOVER PUBLICATIONS, 1945, P. 78.

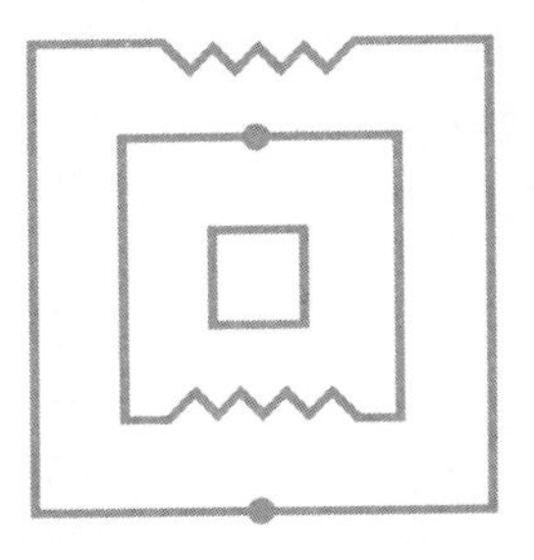

Appendix I

Television (TV) Channel Frequencies

Channel Number	Frequency Range (MHz)	Video Carrier (MHz)	Audio Carrier (MHz)
VHF			
2	54–60	55.25	59.75
3	60–66	61.25	65.75
4	66–72	67.25	71.75
5	76–82	77.25	81.75
6	82–88	83.25	87.75
7	147–180	175.25	179.75
8	180–186	181.25	185.75
9	186–192	187.25	191.75
10	192–198	193.25	197.75
11	198–204	199.25	203.75
12	204–210	205.25	209.75
13	210–216	211.25	215.75
UHF			
14	470–476	471.25	475.75
15	476–482	477.25	481.75
16	482–488	483.25	487.75
17	488–494	489.25	493.75
18	494–500	495.25	499.75
19	500–506	501.25	505.75
20	506–512	507.25	511.75

(continued)

Channel Number	Frequency Range (MHz)	Video Carrier (MHz)	Audio Carrier (MHz)
21	512–518	513.25	517.75
22	518–524	519.25	523.75
23	524–530	525.25	529.75
24	530–536	531.25	535.75
25	536–542	537.25	541.75
26	542–548	543.25	547.75
27	548–554	549.25	553.75
28	554–560	555.25	559.75
29	560–566	561.25	565.75
30	566–572	567.25	571.75
31	572–578	573.25	577.75
32	578–584	579.25	583.75
33	584–590	585.25	589.75
34	590–596	591.25	595.75
35	596–602	597.25	601.75
36	602–608	603.25	607.75
37	608–614	609.25	613.75
38	614–620	615.25	619.75
39	620–626	621.25	625.75
40	626–632	627.25	631.75
41	632–638	633.25	637.75
42	638–644	639.25	643.75
43	644–650	645.25	649.75
44	650–656	651.25	655.75
45	656–662	657.25	661.75
46	662–668	663.25	667.75
47	668–674	669.25	673.75
48	674–680	675.25	679.75
49	680–686	681.25	685.75
50	686–692	687.25	691.75
51	692–698	693.25	697.75
52	698–704	699.25	703.75
53	704–710	705.25	709.75
54	710–716	711.25	715.75
55	716–722	717.25	721.75
56	722–728	723.25	727.75
57	728–734	729.25	733.75
58	734–740	735.25	739.75
59	740–746	741.25	745.75
60	746–752	747.25	751.75
61	752–758	753.25	757.75
62	758–764	759.25	763.75
63	764–770	765.25	769.75
64	770–776	771.25	775.75
65	776–782	777.25	781.75
66	782–788	783.25	787.75
67	788–794	789.25	793.75
68	794–800	795.25	799.75
69	800–806	801.25	805.75
70	806–812	807.25	811.75

Channel Number	Frequency Range (MHz)	Video Carrier (MHz)	Audio Carrier (MHz)
71	812–818	813.25	817.75
72	818–824	819.25	823.75
73	824–830	825.25	829.75
74	830–836	831.25	835.75
75	836–842	837.25	841.75
76	842–848	843.25	847.75
77	848–854	849.25	853.75
78	854–860	855.25	859.75
79	860–866	861.25	865.75
80	866–872	867.25	871.75
81	872–878	873.25	877.75
82	878–884	879.25	883.75
83	884–890	885.25	889.75

Note: Each channel has a 6-MHz bandwidth. The video carrier has the lower frequency plus 1.25 MHz, and the Audio carrier has the upper frequency minus 0.25 MHz.

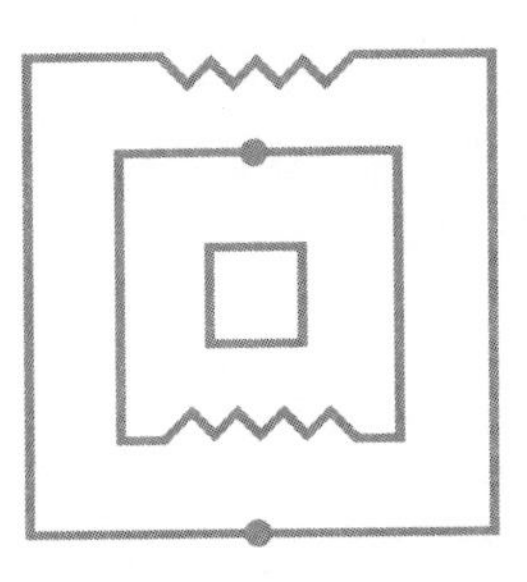

Appendix J

Wire Data

Wire Size (AWG)	Diameter (mm)	Diameter (in.)	Copper Wire d-c Resistance (Ω/100 m)	Current Capacity of Copper Wire (A)
8	3.264	0.1285	0.1952	73
9	2.906	0.1144	0.2462	—
10	2.588	0.1019	0.3103	55
11	2.305	0.0907	0.3914	—
12	2.053	0.0808	0.4935	41
13	1.828	0.0720	0.6224	—
14	1.628	0.0641	0.7849	32
16	1.291	0.0508	1.2480	22
18	1.024	0.0403	1.9840	16
20	0.812	0.0320	3.1550	11
22	0.644	0.0253	5.0170	—

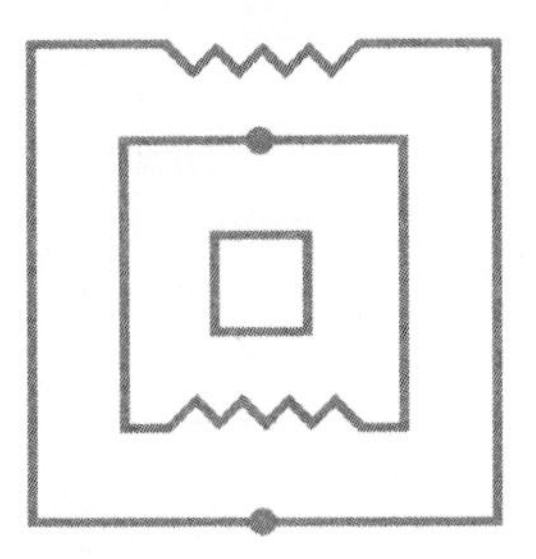

Appendix K

Hankel Functions

The Hankel functions of the first and second kinds are combinations of the Bessel functions:

$$H_v^{(1)}(x) = J_v(x) + jN_v(x) \qquad \text{incoming wave} \tag{K-1}$$

and

$$H_v^{(2)}(x) = J_v(x) - jN_v(x) \qquad \text{outgoing wave} \tag{K-2}$$

For large argument, they become

$$H_v^{(1)}(x) \xrightarrow[x\to\infty]{} \sqrt{\frac{2}{j\pi x}}\, j^{-v} e^{jx} \tag{K-3}$$

and

$$H_v^{(2)}(x) \xrightarrow[x\to\infty]{} \sqrt{\frac{2j}{\pi x}}\, j^{v} e^{-jx} \tag{K-4}$$

When $x = ju$ is imaginary, the modified Bessel functions are

$$I_v(u) = j^v J_v(-ju) \tag{K-5}$$

and

$$K_v(u) = \frac{\pi}{2}(-j)^{v+1} H_v^{(2)}(-ju) \tag{K-6}$$

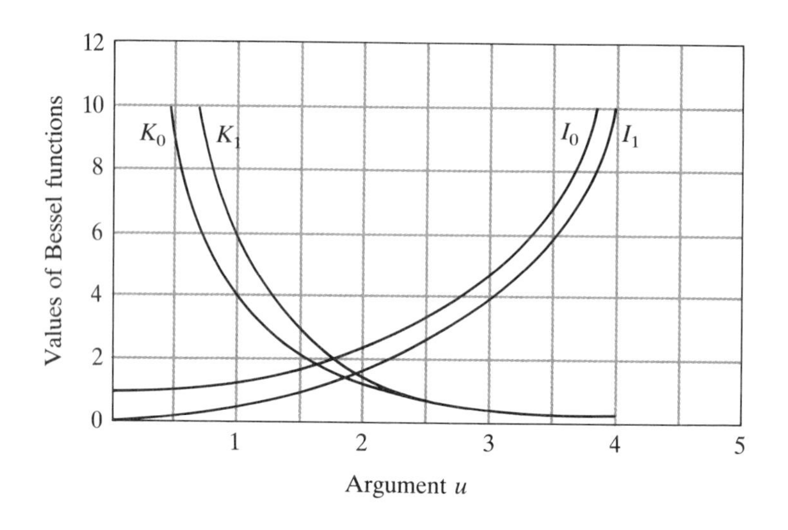

Figure K-1 Modified Bessel functions.

SOURCE: REPRINTED FROM R. F. HARRINGTON, *TIME-HARMONIC ELECTROMAGNETIC FIELDS*, 1964, BY PERMISSION OF MCGRAW-HILL BOOK COMPANY.

For large argument, I_v and K_v become

$$I_v(u) \xrightarrow[u \to \infty]{} \frac{e^u}{\sqrt{2\pi u}} \tag{K-7}$$

and

$$K_v(u) \xrightarrow[u \to \infty]{} \sqrt{\frac{\pi}{2u}}\, e^{-u} \tag{K-8}$$

Figure K-1 shows several types of Hankel functions.

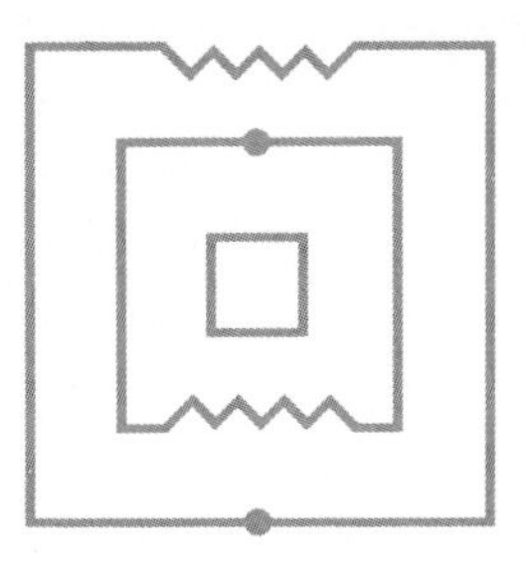

Appendix L

Commercial Lasers and LED Sources

Maker	Device	Drive Current (A)	Emission Wavelength λ_0 (μm)	Spectral Width (nm)	Power (mW)
		Lasers			
AEG	CQX-20	0.40	0.82	2.5	5.0
ITT	T901-L	0.35	0.84	2.0	7.5
LASER D. LABS	LCW10	0.20	0.85	2.0	14.0
RCA	C30130	0.25	0.82	2.0	6.0
		LEDs			
ASEA	IA83	0.10	0.94	40.0	10.0
ITT	T851-S	0.20	0.84	38.0	1.5
LASER D. LABS	IRE150	0.10	0.82	35.0	1.5
MERET	TL-36C	0.30	0.91	38.0	12.0
MONSANTO	ME60	0.05	0.90	40.0	0.4
PHILIPS	CQY58	0.05	0.88	45.0	0.5
RCA	SG1009	0.10	0.94	45.0	3.5
TI	TIXL472	0.05	0.91	23.0	1.0

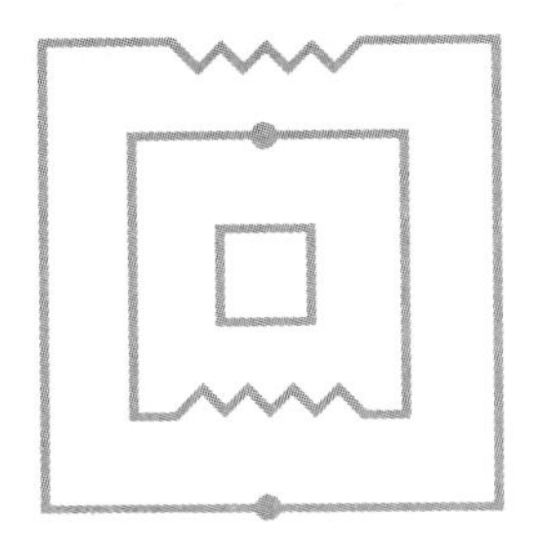

BIBLIOGRAPHY

PART ONE: TRANSMISSION LINES

1. Adler, Richard B., et al., *Electromagnetic Energy Transmission and Radiation*. New York: John Wiley & Sons, 1960.
2. Brown, Robert G., et al., *Lines, Waves, and Antennas*, 2nd ed. New York: Ronald, 1973.
3. Johnson, Walter C., *Transmission Lines and Networks*. New York: McGraw-Hill, 1950.
4. King, R. W. P., Mimno, H. R., and Wing, A. H., *Transmission Lines, Antennas, and Wave Guides*. New York: Dover, 1965.
5. Magnusson, Phillip C., *Transmission Lines and Wave Propagation*, 2nd ed. Boston: Allyn and Bacon, 1970.
6. Matick, Richard E., *Transmission Lines for Digital and Communication Networks*. New York: McGraw-Hill, 1969.
7. Moore, Richard K., *Traveling-Wave Engineering*. New York: McGraw-Hill, 1960.
8. Seshadri, S. R., *Fundamentals of Transmission Lines and Electromagnetic Fields*. Reading, Mass.: Addison-Wesley, 1971.
9. Sinnema, William, *Electronic Transmission Technology*. Englewood Cliffs, N.J.: Prentice-Hall, 1979.
10. Smith, Philip H., "Transmission Line Calculator," *Electronics*, vol. 12, pp. 29–31, 1939.

———, "An Improved Transmission Line Calculator," *Electronics*, vol. 17, pp. 130–133, 318–325, 1944.

———, "Smith Charts—Their Development and Use." A series published at intervals by the Kay Electric Co. No. 1 is dated March 19, 1962, and No. 9 is dated December 1966.

PART TWO: WAVEGUIDES

1. Adler, Richard B., et al., *Electromagnetic Energy Transmission and Radiation*. New York: John Wiley & Sons, 1960.
2. Brown, Robert G., et al., *Lines, Waves, and Antennas*, 2nd ed. New York: Ronald, 1973.
3. Galagan, Steve, "Understanding Microwave Absorbing Materials and Anechoic Chambers," *Microwaves*: Part 1, December 1969; Part 2, January 1970; Part 3, April 1970; and Part 4, May 1970.
4. Kao, Charles K, *Optical Fiber Systems*. New York: McGraw-Hill, 1982.
5. Liao, Samuel Y., *Microwave Devices and Circuits*, 2nd ed. Englewood Cliffs, N.J.: Prentice-Hall, 1985.
6. Magnusson, Philip C., *Transmission Lines and Wave Propagation*, 2nd ed. Boston: Allyn and Bacon, 1970.
7. Midwinter, John E., *Optical Fiber for Transmission*. New York: John Wiley & Sons, 1979.
8. Owyang, Gilbert H., *Foundations of Optical Waveguides*. New York: Elsevier, 1981.
9. Schelkunoff, S. A., *Electromagnetic Waves*. New York: D. Van Nostrand, 1944.
10. Schulz, Richard B., et al., "Shielding Theory and Practice." *Proc. 9th Tri-Service Conference on Electromagnetic Compatibility*, IIT Research Institute, October 1963, pp. 597–636.
11. Southworth, G. C., *Principles and Applications of Waveguide Transmission*. New York: D. Van Nostrand, 1950.

PART THREE: ANTENNAS

1. Booker, H. G., "Slot Aerials and Their Relation to Complementary Aerials (Babinet's Principle)," *IEEE Journal*, vol. IIIA, pp. 620–626 (1946).
2. Carrel, R. L., "Analysis and Design of the Log-Periodic Dipole Antenna." University of Illinois Antenna Lab. Tech. Rept. 52, 1962.
3. Elliott, Robert S., *Antenna Theory and Design*. Englewood Cliffs, N.J.: Prentice-Hall, 1981.
4. King, R. W. P., *Theory of Linear Antennas*. Cambridge, Mass: Harvard University Press, 1956.
5. ———, and Harrison, C. W., *Antenna and Waves*. Cambridge, Mass: M.I.T. Press, 1969.
6. ———, Mimno, H. R., and Wing, A. H., *Transmission Lines, Antennas, and Wave Guides*. New York: Dover, 1965.
7. Kraus, J. D., *Antennas*. New York: McGraw-Hill, 1950.
8. Jasik, Henry, *Antenna Engineering Handbook*. New York: McGraw-Hill, 1961.

9. Jordan, E. C., and Balmain, K. G., *Electromagnetic Waves and Radiation Systems*, 2nd ed. Englewood Cliffs, N.J.: Prentice-Hall, 1968.
10. Schelkunoff, S. A., *Advanced Antenna Theory*. New York: John Wiley & Sons, 1952.
11. ———, and Friis, H. T., *Antennas: Theory and Practice*. New York: John Wiley & Sons, 1952.
12. Silver, S., *Microwave Antenna Theory and Design*. New York: McGraw-Hill, 1949.
13. Stutzman, Warren L., and Thiele, Gary A., *Antenna Theory and Design*. New York: John Wiley & Sons, 1981.
14. Walter, C. H., *Traveling Wave Antennas*. New York: McGraw-Hill, 1965.
15. Weeks, W. L., *Antenna Engineering*. New York: McGraw-Hill, 1968.
16. Williams, H. P., *Antenna Theory and Design*. London: Pitman & Sons, 1950.
17. Wolff, E. A., *Antenna Analysis*. New York: John Wiley & Sons, 1966.

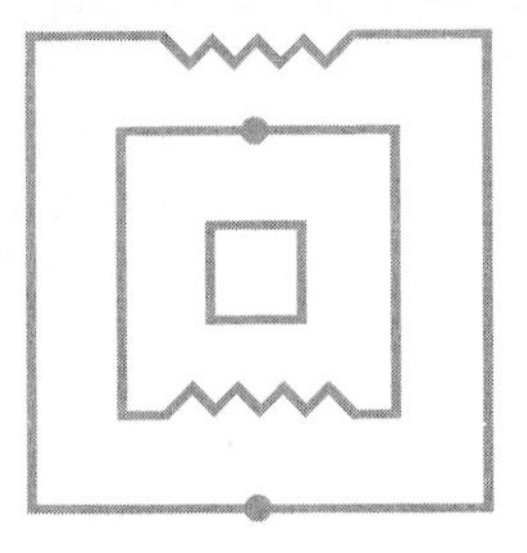

Index

A

A-band, 4
Admittance, 47, 52
Antenna, 2
  aperture or area, 316–318
  array, 375–383
  array factor, 377
  bandwidth, 312
  beamwidth, 312
  broadband, 365
  directivity, 314–315
  effective aperture, 317
  efficiency, 315–316
  factor, 409
  gain, 313–314
  lobes, 302–304
Antenna factor, 409
Antenna measurements, 383–392
Antenna noise temperature, 398
Antenna regions
  far field, 298
  induction field, 299–300
  near field, 316–318
  radiation field, 300
  static field, 299–300
Antenna types
  broadband, 365
  dipole, 339–352
  log–periodic, 365–375
  monopole, 352–353
  phased-array, 375–377
  short-wire, 295–312
  slot, 353–362
  Yagi–Uda, 381

B

B-band, 4
Balanced line, 122

Balun, 122–123
Bessel functions, 213–214, 427–428
Boundary conditions, 16–17
Broadband antenna, 365

C

C-band, 4
Characteristic impedance, 45–47, 53–54
  coaxial line, 4, 425
  colinear plate, 425
  coupled microstrip lines, 429–433
  parallel striplines, 4, 425
  shielded line, 426
  two open-wire line, 4, 425
Chart, *see* Smith chart
Computer programs
  antenna: field pattern
    for dipole, 341–342
    for full-wave dipole, 346–347
    for half-wave slot, 345–355
    for short-wire, 309–310
  antenna: power pattern
    for dipole, 343–344
    for short-wire, 304–305
  antenna: radiation resistance
    for short-dipole, 350–351
    for short-wire, 306–307
  transmission line
    characteristic impedance, 55
    inductive loading, 165
    line impedance, 63–65
    line voltage and current, 171–172
    microstrip line, 140–141
    propagation constant, 55
    reflection coefficient, 69
    standing-wave ratio, 81
  waveguide: circular
    TE mode, 219–220
    wave propagation, 223
  waveguide: rectangular
    power transmission, 198–199
    TE mode, 193–194
Connectors, coaxial, 83–84
Constants
  material, 421–422
  physical, 422
Coordinate systems
  cylindrical, 9
  rectangular, 9
  spherical, 9
Coplanar waveguide, 281–285
Cutoff frequency
  in circular waveguide
    TE mode, 217
    TM mode, 222
  in rectangular waveguide
    TE mode, 191
    TM mode, 187

D

D-band, 4
Decibel (dB), 48
Degenerate modes in waveguide, 192
Dielectric constants, 422
Dielectric planar waveguides, 268–289
  basic equations, 269–271
  coplanar, 281–285
  dielectric-slab, 272–281
  thin film-on-conductor, 285–287
  thin film-on-dielectric, 288–289
Digital transmission lines, 159–160
Dipole antenna, 339–353
  elemental, 339–341
  full-wave, 346
  half-wave, 346
  short-wave, 348
Directivity, antenna, 314–315
Distortion effect, 162–163
Distortionless line, 163–164
Dominant modes
  circular waveguide, 222
  rectangular waveguide, 192
  resonant cavity, 205
Duty cycle, 322

E

E-band, 4
Effective aperture, 317
Efficiency, antenna, 315–316
ELV frequency, 3

F

F-band, 4
Frequency-independent antenna, *see* Log-periodic antenna
Fresnel zone, 300

G

G-band, 4
Gain, antenna, 312–313
Ground wave, 23
Group velocity, 160–161

H

H-band, 4
Hankel functions, 443
Hyperbolic functions, 417–419

I

I-band, 4
IEEE frequency bands, 4
Impedance, characteristic, *see* Characteristic impedance
Impedance matching
  double-stub, 112–117
  quarter-wavelength transformer, 123–124
  single-stub, 107–112
Incident angle, 20–21

J

J-band, 4

K

K-band, 4
Ka-band, 4
Ku-band, 4

L

L-band, 3
Laser sources, 445
LED sources, 445
Lobes, antenna, 302–304
Log-periodic antenna, 365–375
  characteristic impedance, 367
  design, 366–370
Loss tangent, 31

M

M-band, 4
Matching, impedance
  balun, 122–123
  dielectric bead, 120–122
  double-stub, 112–117
  n-junction, 124–126
  quarter-wave transformer, 123–124
  series-stub, 118–120
  single-stub, 107–112
Maxwell's equations, 6
Measurement units, 404–405
Microstrip line, 135–148
Modes
  TE, *see* Waveguides
  TEM, *see* Waveguides
  TM, *see* Waveguides
Monopole antenna, 352–353

N

Neper, 48
Noise figure, 395–396
Noise temperature, 397–399

O

Optical-fiber waveguides
  characteristics, 240
  cutoff wavelength, 245–246
  graded-index, 254–256
  light-gathering power, 250–251
  multimode step-index, 252–254
  numerical aperture, 246–249
  optical fibers, 236–240
  step-index, 251–252
  wave equations, 242–243
  wave modes, 244
  wave patterns, 256
Oblique incidence of plane wave, 20–23
Optical-fiber transmission line, 176

P

P-band, 4
Permeability, 422
Permittivity, 422
Phase-array antenna, 375–377
  array factor, 377
  n-element array, 378–380
  two-element array, 376–378
Phase velocity, 160
Phasor, 41
Physical constants, 5
Planar waveguides, *see* Dielectric planar waveguides
Plane waves
  in good conductor, 27–28
  in ionosphere, 24
  in lossless dielectric, 26
  in lossy dielectric, 30
  in lossy media, 27
  normal incidence, 17–19
  oblique incidence, 20–22
  in poor conductor, 30
  reflection, 17–19
  uniform, 14–16
Potential
  retarded, 296
  scalar electric, 296
  vector magnetic, 296
Poynting theorem, 10–13
Propagation constants, 10, 416
  attenuation constant, 41, 46–47
  in free space, 25
  in lossless medium, 44–45
  in lossy medium, 46–52
  phase constant, 41, 46–47
Pulsed transmission line, 167–171

Q

Q-band, 3–4
Q-factor
  microstrip line, 148–149
  rectangular cavity, 206–207

R

Radiation from antenna, 300
Radiation resistance, 304
Reciprocity law for antenna, 325–326
Reflection coefficients
  current, 66–68
  impedance, 18, 22
  resistance, 18, 22
  voltage, 66–68
Resonant frequency
  circular resonator, 232
  rectangular resonator, 205
Retarded potential, 296

S

S-band, 3–4
Scattering parameters, 126
Shielded line, 426
Short-wire antenna, 295–312
Skin depth, 166
Skin effect, 166
Slot antenna, 353–362
  cavity-backed, 359–360
  semicircular-cavity, 360–361
Slotted line, 80
Smith chart
  admittance chart, 93–94
  impedance chart, 93–94
  impedance matching, 104–122
Snell's law, 21
Standing wave, 72–76
Standing-wave ratio, 76–80
Step-function response, 167–171
Strip lines, 135–153
  coplanar, 151–152
  microstrip, 136–141
  parallel, 149–151
  shielded, 152–154
Stub matching
  double stub, 112–117
  single stub, 107–112
Superconducting transmission line, 172–175

T

TE mode, *see* Waveguides
Television channel frequencies, 437–439
TM mode, *see* Waveguides

Transformers
  balun, 122
  quarter-wave, 123
Transmission chart, *see* Smith chart
Transmission coefficient, 22, 71–72
Transmission lines, 2, 35–37
  characteristic impedance, 45–47, 53–54, 416, 423, 425–426
  coaxial line, 36, 425
  digital line, 159–160
  half-wave line, 61
  line equations and solutions, 39–54
  lossless line, 45–46
  lossy line, 46–48
  optical-fiber line, 176
  shielded line, 426
  slotted line, 80
  stripline, 36
  superconducting line, 172–175
Transmission-line equations, 39–54
  in frequency domain, 43–44
  for lossless line, 45
  for lossy line, 46
  in time domain, 40–42

U

UHF, 3

V

V-band, 3–4
Velocity
  group, 160–161
  phase, 160
VF, 3
VHF, 3
Voltage on transmission line, 47
Voltage standing-wave ratio, *see* Standing-wave ratio

W

W-band, 4
Wave equations, 6–9
  in cylindrical coordinates, 8–10
  in rectangular coordinates, 8–10
  in spherical coordinates, 8–10
  in transmission line, 41–44
Wave propagation
  in good conductor, 27–28
  in free space, 17–19
  in ionosphere, 24–25
  in lossless dielectric, 26
  in lossy media, 27
Waveguides
  circular
    basic equations, 211–213
    cutoff frequency, 218–222
    mode excitation, 227–229
    power transmission, 226–227
    propagation constant, 219–222
    standard size, 230
    TE modes, 182, 214–218
    TEM mode, 224–226
    TM modes, 182, 220–223
  rectangular
    basic equations, 183–184
    cavity, 203–204
    cutoff frequency, 184
    mode excitation, 199–201
    power transmission, 194–196
    propagation constant, 184
    standard size, 202–203
    TE modes, 182, 189–191
    TM modes, 182, 185–188
    wave impedance, 188

X

X-band, 3–4

Y

Yagi–Uda antenna, 381